Ecosystem Services in the Planning and Sustainable Development of Urban Green Spaces

Ecosystem Services in the Planning and Sustainable Development of Urban Green Spaces

Guest Editors

Wei Hou
Chunxiao Zhang
Cheng Li

MDPI

Basel • Beijing • Wuhan • Barcelona • Belgrade • Novi Sad • Cluj • Manchester

Guest Editors
Wei Hou
Research Center of Natural Resources Survey and Monitoring
Chinese Academy of Surveying and Mapping
Beijng
China

Chunxiao Zhang
Department of Geographic Information Science
China University of Geosciences
Beijing
China

Cheng Li
Department of Architecture and Urban Planning
China University of Mining and Technology
Xuzhou
China

Editorial Office
MDPI AG
Grosspeteranlage 5
4052 Basel, Switzerland

This is a reprint of the Special Issue, published open access by the journal *Sustainability* (ISSN 2071-1050), freely accessible at: https://www.mdpi.com/journal/sustainability/special_issues/3W4171G383.

For citation purposes, cite each article independently as indicated on the article page online and as indicated below:

Lastname, A.A.; Lastname, B.B. Article Title. *Journal Name* **Year**, *Volume Number*, Page Range.

ISBN 978-3-7258-8451-3 (Hbk)
ISBN 978-3-7258-8452-0 (PDF)
https://doi.org/10.3390/books978-3-7258-8452-0

Contents

About the Editors

Wei Hou

Wei Hou is a Professor and deputy director of the Natural Resource Survey and Monitoring Research Center at the Chinese Academy of Surveying and Mapping. He was selected as a young scientific talent under the 'High-Level Scientific and Technological Innovation Talent Program' of the Ministry of Natural Resources and is a young academic leader at the Chinese Academy of Surveying and Mapping. He is dedicated to natural resources surveying and monitoring, with a particular focus on multi-scale simulation and the analysis of ecological security patterns. His primary interests also encompass the accounting of carbon budgets in territorial spaces. He has expertise in publishing academic research and applying for patents, having published over 40 research papers and obtained 13 authorized patents. His outstanding achievements include three awards, and notably the First Prize of the Surveying and Mapping Science and Technology Progress Award.

Chunxiao Zhang

Chunxiao Zhang is an Associate Professor, Ph.D. Supervisor, and Director of the Department of Geographic Information Science at China University of Geosciences (Beijing). Selected for a national high-level young talent program, she is dedicated to geographic intelligent simulation and urban thermal environments. Her primary interests encompass virtual geographic environments, physics-informed AI modeling, the impact of urbanization on climate, and geospatial intelligence. She has served as principal investigator for four projects funded by the National and Beijing Natural Science Foundations. She has published over 50 SCI/SSCI papers as a first or corresponding author, holds six authorized invention patents, and has edited three textbooks and one monograph. Her outstanding achievements include the Second Prize of the Surveying and Mapping Science and Technology Award, the "New GIS Talent" Award, the "Zhu Xun" Young Teacher Award, and multiple teaching achievement awards.

Cheng Li

Cheng Li is an Associate Professor and the Director of the Department of Urban and Rural Planning in the School of Architecture and Design at China University of Mining and Technology, where he joined in 2015 after receiving his doctoral degree from TU Dortmund University in 2014. He is dedicated to regional development and planning, with a particular focus on human–environment interactions and urbanization and ecological effects. His primary interests also encompass remote sensing, geographic information systems, and ecological restoration in mining areas. He has expertise in leading national and provincial research projects, having published two English academic monographs and over 30 academic papers. His outstanding achievements include multiple awards for both scientific research and teaching.

Preface

The motivation for compiling this reprint stems from the urgent need to update our understanding of urban sustainability. Rapid urbanization poses significant challenges to urban environments, necessitating a shift from static assessments to dynamic, data-driven, and policy-relevant research. We observed that while the "supply" side of ecosystem services is well-documented, the "demand" side—how people interact with, perceive, and benefit from these spaces—requires deeper exploration. By integrating advanced analytical methods with social science perspectives, this work strives to provide a holistic view of how UGSs can be governed to maximize social equity and environmental resilience.

Wei Hou, Chunxiao Zhang, and Cheng Li
Guest Editors

sustainability

MDPI

stematic Review

ssessing Cultural Ecosystem Services in Sponge City nfrastructure: A Systematic Review and Framework Proposal

uan Han, Roziya Binti Ibrahim * and Mohd Sallehuddin Bin Mat Noor

Faculty of Design and Architecture, Universiti Putra Malaysia, Serdang 43400, Selangor, Malaysia; gs61866@student.upm.edu.my (N.H.); mohdsallehuddin@upm.edu.my (M.S.B.M.N.)
* Correspondence: roziya@upm.edu.my

Abstract: Urbanization has significantly transformed ecological landscapes and created challenges in sustaining both environmental functionality and cultural values. In response, China's Sponge City Infrastructure (SPI) aims to enhance urban water resilience by integrating green and blue infrastructure. While the ecological benefits of SPI have been widely studied, the cultural ecosystem services (CES) it provides remain underexplored. This study systematically reviews 61 empirical articles to evaluate how CES has been addressed in SPI-related research. Bibliometric analysis was conducted to identify CES research trends and to systematically categorize CES types, assessment methods, and evaluation indicators in SPI-related studies. The findings reveal a dominant use of non-monetary assessment methods, led by questionnaire surveys (47.5%), while monetary approaches were rarely applied. However, several limitations were identified, including the geographic concentration of studies in a few major cities, the scarcity of research on abstract CES categories (e.g., inspiration and sense of place), and the lack of measurable indicators in nearly half of the reviewed studies. To address these issues, this study proposes a context-specific CES assessment framework aligned with China's socio-cultural conditions and planning priorities in sponge city development. The framework, based on the reviewed literature, provides a preliminary tool for evaluating CES in sponge city contexts. This work contributes to the integration of cultural ecosystem services into urban ecological planning and offers insights for sustainable development in rapidly urbanizing regions.

Keywords: sponge cities infrastructure; cultural ecosystem services; urban sustainability; green infrastructure; CES assessment; urban planning

1. Introduction

Intensifying climate change has triggered a global surge in extreme rainfall events, significantly exacerbating the risk of urban flooding—a major threat to ecological sustainability and human well-being worldwide [1–3]. In response, the United Nations' 2030 Agenda for Sustainable Development—particularly SDG 11 (Sustainable Cities and Communities) and SDG 6 (Clean Water and Sanitation)—highlights the importance of inclusive, resilient, and water-sensitive urban development. National-level strategies have emerged accordingly, including the European Union's Green Infrastructure Strategy, the United States' Low Impact Development (LID), Australia's Water Sensitive Urban Design (WSUD), and the Netherlands' Room for the River program, all of which integrate hydrological functions with urban spatial planning [4,5].

China, facing distinct socio-ecological challenges due to its rapid urbanization and population density, introduced the SPI in 2013 as a novel approach to urban water management [6,7]. By mimicking natural hydrological processes, SPI incorporates green and blue infrastructure into urban planning, aiming to mitigate urban flooding, improve water quality, and enhance urban resilience [8,9]. It has since been embedded into the national "Ecological Civilization" agenda, which emphasizes the synergistic advancement of ecological and cultural values in urban development [10,11]. This policy alignment is reflected in key documents such as the Sponge City Construction Technical Guide [12] and the 14th Five-Year Ecological and Environmental Protection Plan [13].

Despite SPI's multifunctional design and its potential to provide CES—such as recreation, aesthetics, identity, and spiritual well-being—current research and evaluation frameworks primarily focus on ecological outcomes, including runoff reduction, habitat restoration, and pollution control [14,15]. CES, defined as the non-material benefits people obtain from nature, remains underexplored in the context of SPI, especially within China [16,17].

Although SPI like rain gardens, wetlands, and permeable pavements inherently offer cultural value, empirical studies addressing CES in this context remain limited both domestically and internationally [18,19]. Existing CES evaluation frameworks, largely developed in Western contexts, often lack cultural relevance for China's socio-ecological realities [20,21]. Furthermore, abstract CES categories—such as cultural diversity, traditional knowledge systems, or inspiration—are rarely operationalized, due to conceptual ambiguity and a lack of context-sensitive indicators [21–23]. Figure 1 illustrates how SPI components—such as wetlands, rain gardens, and permeable pavements—can potentially support various CES. A single type of infrastructure may provide multiple CES at the same time, and the specific services it supports can vary depending on the design and surrounding environment.

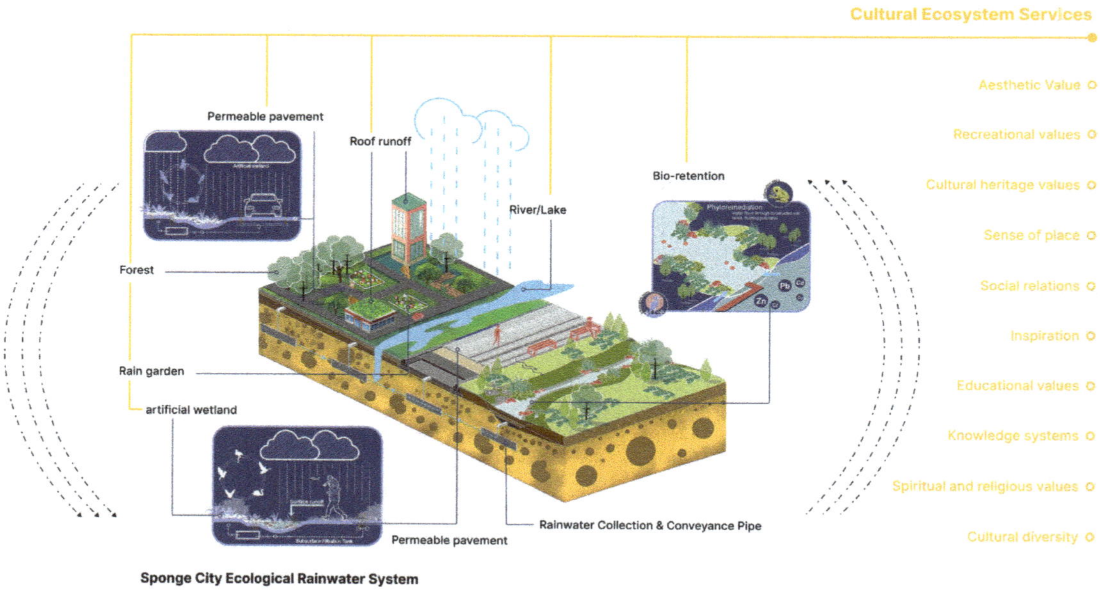

Figure 1. The relationship between sponge cities and CES.

To address these gaps, this study conducts a systematic review of empirical research on CES within the context of sponge city development. Specifically, it analyzes the geographic distribution, research themes, assessment methods, CES categories, and indicator usage in the selected studies. Drawing on these findings, a CES evaluation framework is developed for application in SPI planning. This framework is specifically developed for the assessment of cultural ecosystem services within sponge city infrastructure. It serves as a practical instrument to support governments and urban planners in effectively incorporating cultural services into ecological planning and promoting public well-being.

2. Methods

2.1. Data Collection

This study performed a systematic literature search with reference to PRISMA 2020 guidelines. Protocol registration: https://osf.io/fge5r. The search focused on identifying empirical studies related to CES in the context of SPI within China.

Searches were conducted in the Web of Science Core Collection and Scopus. The search strategy combined multiple related terms across four conceptual dimensions: (1) core CES concepts (e.g., "cultural ecosystem services", "cultural services", "social values"); (2) urban infrastructure terms (e.g., "sponge cities", "green infrastructure", "stormwater management", "urban green space"); (3) specific CES types (e.g., "recreation", "aesthetic value", "sense of place", "cultural heritage"); and (4) landscape features (e.g., "urban park", "blue-green infrastructure", "wetland", "urban nature"). Boolean operators (AND/OR) were used to structure the queries, and search terms were applied to the titles, abstracts, and keywords of the records. Although the primary search strategy was based on English-language sources, two high-relevance Chinese-language studies retrieved from CNKI were also included due to their empirical relevance and methodological contributions to CES assessment in sponge city contexts.

The search was restricted to English-language, peer-reviewed journal articles published between 2013 and April 2025. This time frame was selected to capture the literature surrounding the development and implementation of China's sponge city policy, which was formally initiated in 2015. Grey literature, conference papers, and non-peer-reviewed materials were excluded.

All search results were imported into Zotero, and duplicates were removed manually using its built-in deduplication function. Subsequent title and abstract screening, full-text review, and inclusion decisions were made based on predefined eligibility criteria (see Section 2.2). This systematic review was conducted in accordance with the PRISMA 2020 guidelines. Although the protocol was not registered in a public database, all study selection and reporting procedures followed PRISMA recommendations.

2.2. Study Selection Criteria

A set of inclusion and exclusion criteria was developed to ensure the relevance, quality, and comparability of the studies included in this systematic review.

The selection criteria were established based on the research objectives, focusing on empirical assessments of CES in the context of SPI within China.

Inclusion Criteria:

Studies were included if they met the following requirements:

- Published between 2013 and April 2025 to capture the emergence and development of CES research in relation to China's sponge city initiative, formally launched in 2015;
- Written in English;
- Published as peer-reviewed journal articles reporting original empirical research;
- Focused on the evaluation of CES;
- Conducted in China and specifically focused on green or blue infrastructure associated with sponge city initiatives (e.g., urban parks, wetlands, blue–green corridors);
- Provided empirical data through case studies, surveys, participatory mapping, or other forms of primary data collection.

Exclusion Criteria:

Studies were excluded if they:

- Were conducted in regions other than China;
- Were conceptual or theoretical papers, editorials, or policy commentaries lacking empirical analysis;
- Focused exclusively on provisioning services (e.g., food, water supply) or regulating services (e.g., flood control) without incorporating any CES components;
- Were limited to hydrological or engineering aspects without discussing cultural values;
- Were non-peer-reviewed materials, including conference abstracts, dissertations, technical reports, and other non-peer-reviewed publications.

The study selection was conducted in two stages: initial screening based on titles and abstracts, followed by a full-text review for eligible studies. Any ambiguity regarding eligibility was resolved through detailed reading and manual cross-validation by the authors. Zotero was used for reference management and deduplication, while Microsoft Excel was employed to track inclusion decisions and document key screening criteria.

2.3. *Identification of Paper Categorization*

The following attributes were used to provide a comprehensive description of each publication: (1) the geographic distribution and research type; (2) CES categories examined; (3) the assessment methods applied; and (4) the CES indicator usage.

2.3.1. Geographic Location and Type of the Study

This study first classified the geographic distribution and thematic focus of the 61 reviewed publications to uncover spatial patterns and research emphasis. Specifically, for each study, the city or region where the data were collected was recorded. For studies involving multiple locations, each study site was recorded separately to enable disaggregated geographic analysis. The reviewed publications reflect diverse disciplinary orientations and research objectives, which influence their specific positioning within the sponge city research agenda.

For instance:

(1) Blue and Green Areas: Dou et al. [24], focused on blue areas (e.g., water and wetlands) and green areas (e.g., forests, street trees, cropland, and grasslands), which are among the measures employed by sponge cities in addressing stormwater management.
(2) Urban Green Spaces: Mao et al. [25] studied roadside parks, residential green areas, and natural woodlands, highlighting their dual role in both stormwater regulation and cultural ecosystem services, particularly in providing recreational opportunities.
(3) Typologies of Green Spaces: Panduro and Veie [26], and Bell et al. [27], proposed detailed classifications of urban green spaces—such as parks, gardens, and green corridors—which offer a structured framework for understanding the spatial characteristics of CES in sponge city contexts.

In this study, geographic distribution was categorized to include all relevant spatial levels discussed above, such as individual parks, neighborhoods, and entire cities, to ensure comprehensive representation of green and blue spaces related to CES in sponge cities.

2.3.2. CES Categories

The Millennium Ecosystem Assessment (MEA) framework has been increasingly recognized as a robust and comprehensive structure for categorizing ecosystem services. To ensure terminological consistency and analytical comparability, this study adopted the MEA classification system as the primary theoretical basis for identifying and interpreting CES categories (see Table 1).

Table 1. Categories of CES. Classification and definitions quoted from the Millennium Ecosystem Assessment (MEA, 2005).

Category	Concept
Cultural diversity	Cultural diversity reflects how the variety of ecosystems contributes to the development of diverse cultures.
Spiritual and religious values	Many religions attach spiritual and religious values to ecosystems or their components.
Knowledge systems	Ecosystems influence the types of knowledge systems developed by different cultures.
Educational values	Ecosystems and their components and processes provide the basis for both formal and informal education in many societies.
Inspiration	Ecosystems provide a rich source of inspiration for art, folklore, national symbols, architecture, and advertising.
Aesthetic values	Many people find beauty or aesthetic value in various aspects of ecosystems, as reflected in the support for parks, 'scenic drives,' and the selection of housing locations.
Social relations	Ecosystems influence the types of social relations that are established in cultures. Fishing societies, for example, differ in many respects in their social relations from nomadic herding or agricultural societies.
Sense of place	Many people value the 'sense of place' that is associated with recognized features of their environment, including aspects of the ecosystem.
Cultural heritage values	Many societies place high value on the maintenance of either historically important landscapes ('cultural landscapes') or culturally significant species.
Recreational values	People often choose where to spend their leisure time based in part on the characteristics of the natural or cultivated landscapes in a particular area.

A systematic review of the selected literature was conducted to identify the CES categories employed across empirical studies. A set of ten categories was identified for analysis, drawing from the MEA, 2005 and its subsequent extensions as commonly adopted in the reviewed literature. These categories include Cultural Diversity; Recreational Values; Aesthetic Values; Spiritual and Religious Values; Knowledge Systems; Educational Values; Inspiration; Social Relationships; Sense of Place; and Cultural Heritage Values. Categories that lacked conceptual clarity, consistent definitions, or measurable indicators were excluded from the comparative analysis. This approach ensured greater consistency, clarity, and cross-study comparability in the evaluation of CES representation in sponge city research.

2.3.3. CES Indicators

To support robust evaluation of CES, this study systematically identified 186 distinct indicators from the reviewed empirical literature. These indicators were assigned to ten CES categories based on their primary measurement intent and thematic alignment. For instance, Recreation and Ecotourism were commonly assessed using visitor frequency, activity participation rates, and satisfaction surveys, whereas Aesthetic Values were measured through landscape quality ratings, visual preference surveys, and the frequency of landscape photography. To reduce redundancy and enhance categorical clarity, each indicator was reviewed for thematic coherence and grouped under the CES category that best reflected its primary evaluative purpose. Indicators that lacked clear definitions, measurable dimensions, or contextual relevance were excluded to maintain consistency and analytical rigor. A detailed list of all indicators, definitions, and classification references is provided in Table S2. In this study, information such as authors, publication year, study region, CES categories, evaluation methods, and evaluation indicators was systematically extracted for analysis. As all data were obtained from published sources, no imputation or supplementary data handling procedures were necessary.

2.3.4. Classification of CES Assessment Methods

To classify the empirical methods used for evaluating CES, this study adapted the typology proposed by Scholte et al., which provides a structured overview of data collection techniques [28]. Based on this framework and refined according to the methods observed in the reviewed literature—CES evaluation methods were classified into the following categories:

(1) Market-based (e.g., market prices);
(2) Observation-based (e.g., direct observation, document analysis, social media data);
(3) Stated preferences (e.g., interviews and questionnaires);
(4) Expert-based methods (e.g., Delphi technique); and
(5) Participatory and technology-assisted approaches (e.g., PGIS, PPGIS, VEP, and Virtual Reality).

A summary of these methods and their descriptions is provided in Table 2.

Table 2. Classification of CES evaluation methods and descriptions.

Category	Method	Method Description
Method Description	Market Prices	Estimates the economic value of CES based on the prices of marketable goods. For example, the entrance fee to a park can be used to approximate the value of recreation and ecotourism.
	Observation	Directly observes human behaviors and actions to infer the social value of CES. For instance, counting visits to a park can help deduce the importance of recreation in that area.
	Document Analysis	Examines texts, images, or other forms of documentation to glean information about human preferences for CES. For example, analyzing the number of photographs taken by the public or featured in advertisements can indicate aesthetic value.
	Social Media-Based	Utilizes social media data from various platforms to assess CES. For example, the number of wildlife photographs posted on photo-sharing sites (e.g., Flickr) can serve as a proxy for recreation and ecotourism values.
Stated Preferences	Interviews	Gathers in-depth insights into how and why people value CES through direct, often face-to-face, discussions. Respondents freely express their feelings and thoughts, facilitating a deeper understanding of services such as a sense of place or inspiration.
	Questionnaires	Uses a series of questions to collect information on CES from respondents. For example, researchers may employ Likert scales, asking participants to select from a given set of CES to reveal their perceived benefits.
	Expert-Based	Leverages the specialized knowledge and extensive experience of experts who are familiar with technical terminology and concepts. Experts identify key factors they deem relevant to CES issues.
	Participatory GIS (PGIS)	Integrates participatory mapping methods with Geographic Information Systems (GIS), involving stakeholders in spatial data collection and analysis.
	Public Participation GIS (PPGIS)	Emphasizes local-level engagement, encouraging knowledge production by communities and non-governmental groups.
	Visitor-Employed Photography (VEP)	Involves participants using location-enabled applications to freely photograph specific sites. Participants then evaluate various CES value types depicted in the photographs, enabling the quantification of perceived values [27].
	Virtual Reality (VR)	Employs panoramic VR or immersive virtual environments to create realistic, interactive scenes, allowing participants or subjects to virtually "enter" and experience specific natural or urban settings.

To reflect evolving practices in CES research, the framework was further expanded by incorporating emerging approaches such as participatory mapping, participatory geographic information systems (PGIS), public participation GIS (PPGIS), social media data analysis, building on the work of Márquez et al. and Yang and Cao [21,29]. These additions highlight the increasing adoption of digital and participatory tools in urban ecosystem service assessments.

Following the typology proposed by Hirons et al. and Christie et al. [30,31], a distinction was also made between monetary and non-monetary assessment approaches. For monetary methods, revealed preference methods assess value by observing market behavior related to CES, while stated preference methods construct a hypothetical market and ask respondents about their willingness to pay for, accept, or obtain services. For non-monetary methods, revealed preference methods typically involve indirect measurements of human preferences through behavioral observations or document analysis, whereas stated preference methods directly inquire about individuals' values related to CES.

This classification was iteratively refined during the literature review to accommodate methodological heterogeneity across the 61 studies. All evaluation methods used in the selected studies were recorded. For studies employing multiple methods, each method was documented and categorized separately to ensure an accurate representation of methodological diversity.

3. Results

3.1. Study Selection and Publication Trends

The study selection process is illustrated in the PRISMA 2020 flow diagram (Figure 2). A total of 341 records were retrieved from Web of Science and 745 from Scopus, yielding 1086 total records. After removing 77 duplicates using Zotero, 1009 unique records remained. An initial screening based on titles and abstracts excluded 814 records, resulting in 195 studies eligible for full-text screening.

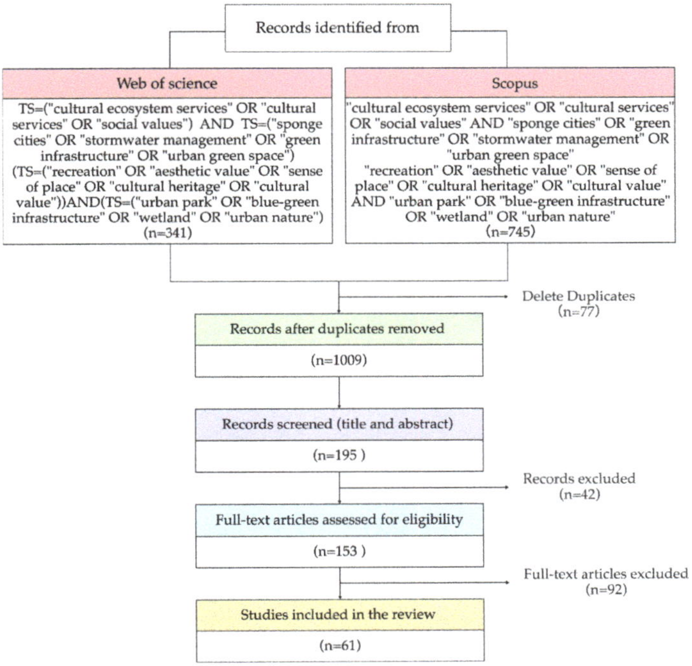

Figure 2. PRISMA 2020 flow diagram for study selection.

Following full-text assessment, 42 articles were excluded due to a lack of empirical data, irrelevance to CES, or not being conducted within the Chinese SPI context. Consequently, 153 full-text articles were assessed for eligibility, and 61 studies met the inclusion criteria and were included in the final review.

From 2013 to 2025, a total of 61 empirical studies on CES in the context of SPI were identified. As shown in Figure 3, the number of publications remained relatively low between 2013 and 2019, with fewer than five studies published per year. A noticeable increase began in 2020, reaching a peak in 2024 with 15 studies. As of April 2025, only three relevant studies have been identified. Therefore, the analysis of publication trends in this study is limited to the period up to 2024. From 2019 to 2024, the number of studies shows a consistent upward trajectory.

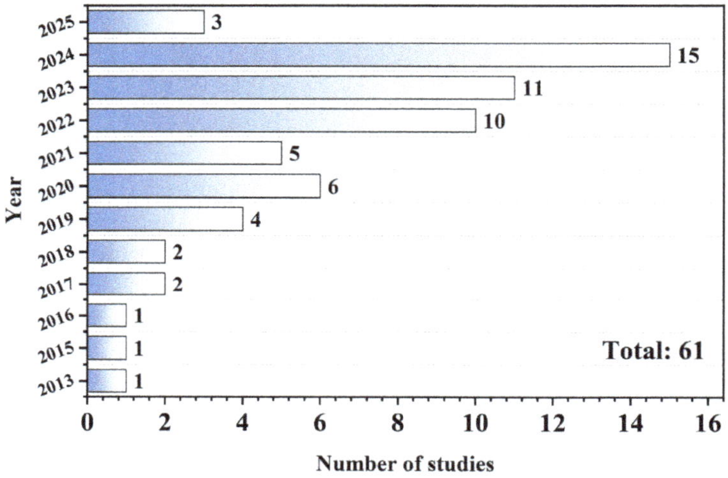

Figure 3. Number of published studies on CES between 2013 to April 2025.

3.2. Geographic Distribution and Thematic Focus of CES Studies

The included studies were analyzed based on key characteristics such as study location, thematic focus, CES categories addressed, and assessment methodologies. A structured summary of these attributes—covering authorship, publication year, study location, CES classification, and evaluation methods—is presented in Table S1 and served as the basis for subsequent analysis.

The spatial distribution of the reviewed literature reveals a clear concentration of CES studies in economically developed and research-intensive Chinese cities. As illustrated in Figure 4, Beijing has the highest number of studies (10), followed by Shanghai (8), Chengdu (6), and both Guangzhou and Hangzhou (3 each). Several other cities—such as Wuhan, Nanjing, and Shenzhen—were represented by only one or two studies. Only five papers included multi-city comparative designs, indicating limited geographic diversification in current CES research that explicitly addresses Cultural Ecosystem Services in the context of Sponge City development.

This distribution indicates that researchers tend to favor tangible, accessible, and clearly defined green public spaces, especially parks [32]. In contrast, few CES studies adopt a city-scale or infrastructure-oriented perspective on sponge city development, or investigate underrepresented spatial types such as rooftops, drainage corridors, and permeable surfaces [29,33]. This imbalance may constrain the ability to assess CES within the full complexity of SPI spatial configurations.

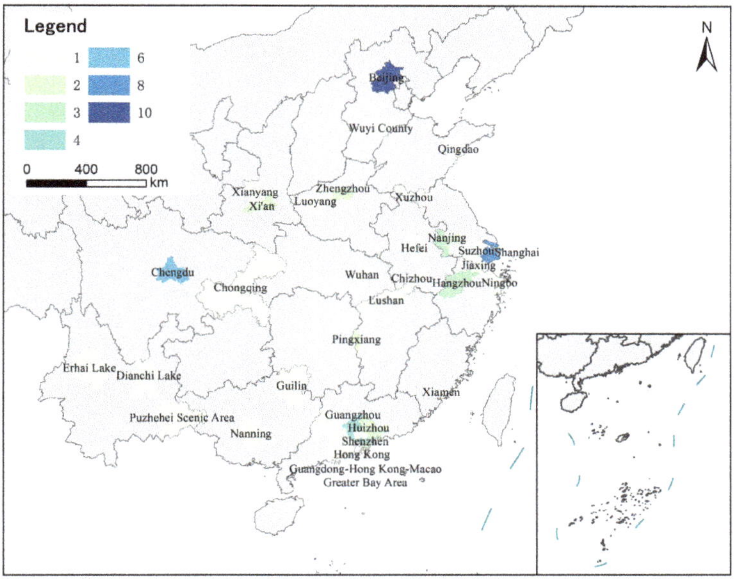

Figure 4. Geographic distribution of the reviewed studies ($N = 61$).

Figure 5 illustrates the thematic focus of the 61 CES-related empirical studies based on their primary research area. Parks are the most frequently studied spaces, accounting for nearly half of the studies (48.5%), followed by urban/town-level areas (27.3%). Studies explicitly focused on sponge cities, green infrastructure, and urban green spaces are much less common (each ≤8%). Blue areas such as rivers or wetlands are rarely addressed (only 3%). This indicates that CES studies under the SPI context are still predominantly park-centered, with broader urban water-related infrastructure largely underexplored.

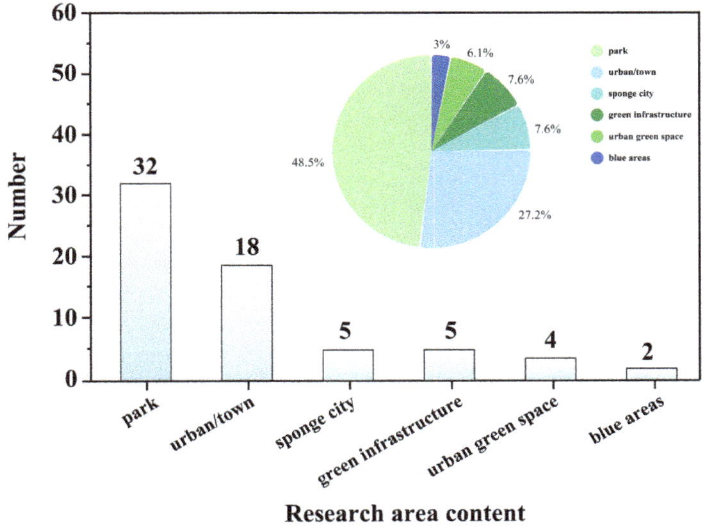

Figure 5. Thematic focus of CES studies.

3.3. CES Category

As illustrated in Figure 6, the reviewed literature shows a significant imbalance in the frequency of CES categories assessed. Recreational services were by far the most evaluated, with 58 times, followed by aesthetic value (48), educational services (33), and cultural heritage (29). These categories are often more tangible, directly observable, and relatively easy to quantify in urban environments.

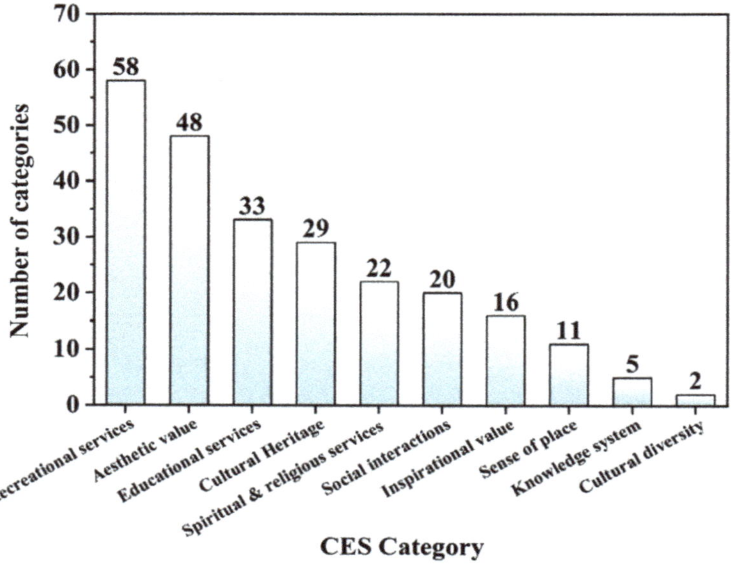

Figure 6. CES category.

In contrast, less emphasis was placed on more abstract CES dimensions. Categories such as spiritual and religious services (22 studies), social interactions (20), inspirational value (16), and sense of place (11) appeared to be underrepresented, especially when compared with frequently assessed categories such as recreation (45) and aesthetic value (41). This indicates a research bias toward more tangible and directly perceived cultural benefits. The least addressed categories were knowledge systems (5) and cultural diversity (2), which may be due to the conceptual complexity and cultural specificity of these values, making them difficult to operationalize through standard indicators.

These results demonstrate that tangible CES categories, such as recreation (45 studies) and aesthetic value (41), dominate current research in SPI contexts. In contrast, more abstract or culturally sensitive categories—such as knowledge systems (5) and cultural diversity (2)—remain significantly underrepresented. Full definitions of all CES categories are provided in Table S2.

3.4. CES Assessment Methodology

As illustrated in Figure 7, non-monetary methods represent the predominant approach for assessing CES in sponge city studies. Questionnaire-based surveys are the most frequently employed tool, used in 36 out of 61 studies, reflecting their ease of implementation and suitability for capturing public perceptions at scale. Social media-based analysis ranks second, featuring in 15 studies, with platforms such as Weibo, Ctrip, and Mafengwo providing real-time, user-generated data for sentiment and behavior analysis.

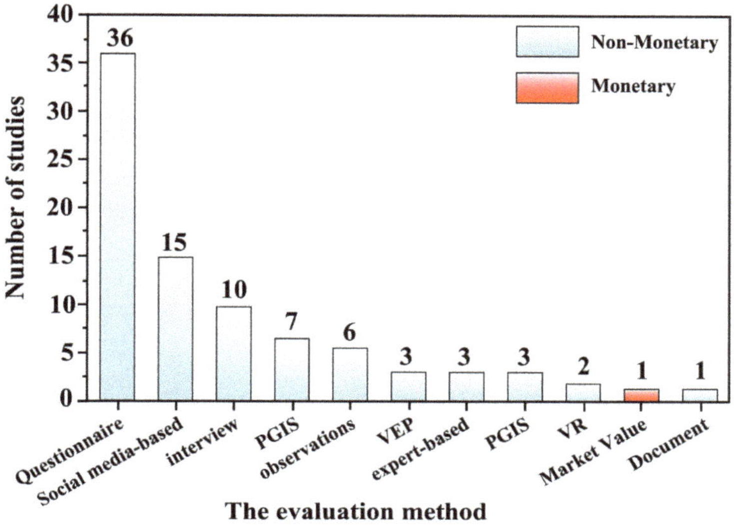

Figure 7. Assessment methods applied in CES.

Other commonly used non-monetary approaches include in-depth interviews (10 studies), participatory GIS (PGIS) (7 studies), and field observations (6 studies). Less commonly applied methods—such as visitor-employed photography (VEP), expert-based evaluation, and participatory public GIS (PPGIS)—were each used in three studies. Two studies employed VR-based participatory mapping to visualize user preferences in spatial planning contexts.

By contrast, monetary valuation methods remain extremely limited. Only one study employed economic techniques to quantify CES, highlighting the challenges of monetizing intangible cultural benefits in urban green infrastructure. One additional study relied on document-based analysis.

These findings highlight the prevalence of qualitative and spatially explicit non-monetary methods—such as surveys, interviews, and participatory GIS—in CES assessments under the SPI framework. However, they also reveal a methodological gap in integrating economic valuation with cultural perception, suggesting the need for mixed-method approaches that combine qualitative depth with quantitative rigor.

3.5. CES Assessment Indicators

A total of 187 CES indicators were extracted from the reviewed literature and categorized based on the Millennium Ecosystem Assessment (MA) framework as presented in Table S2. However, as shown in Figure 8, only 54.1% of the studies provided clearly defined and measurable indicators for the respective CES categories, highlighting the persistent inconsistency in indicator reporting.

As shown in Figure 8, the studies were classified based on whether they explicitly reported indicators that were both clearly defined and quantitatively or qualitatively measurable for the CES categories under investigation. "Yes" refers to studies that provided at least one such indicator per CES category, while "No" includes studies that discussed CES conceptually or descriptively without specifying measurable indicators. The nearly even split between the two (54.1% vs. 45.9%) indicates a lack of standardization in how CES indicators are defined and reported across the literature.

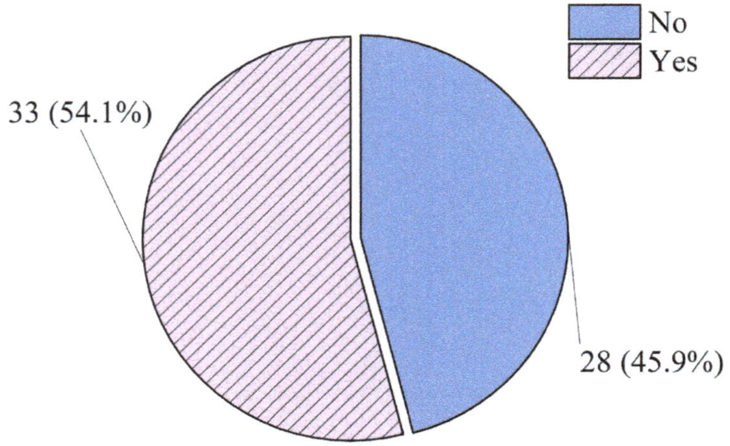

Figure 8. The proportion of indicators in the reviewed research.

As shown in Figure 9, recreational services were the most frequently represented category, with 46 indicators, followed by aesthetic value (32 indicators), cultural heritage (28), educational services (25), and spiritual and religious services (21). In contrast, more intangible or abstract categories, such as social interactions (14), inspiration (10), and sense of place (9), were much less commonly assessed. Notably, knowledge systems were represented by only two indicators, and cultural diversity was entirely absent from the reviewed studies.

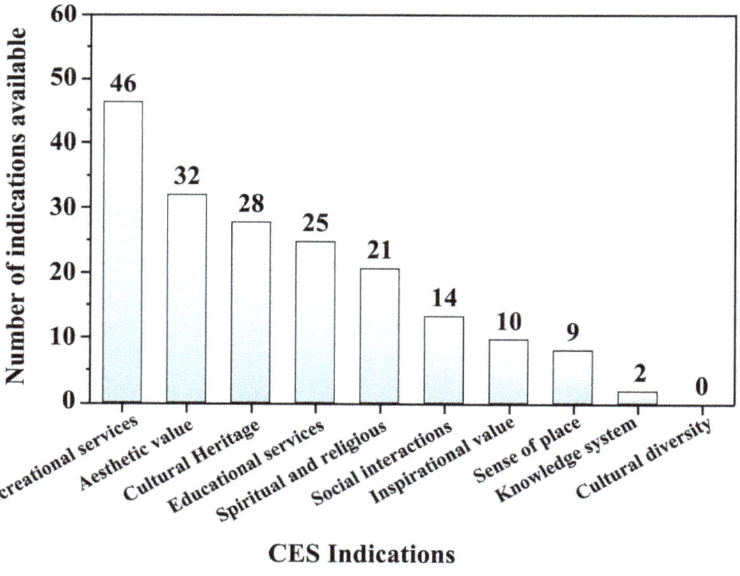

Figure 9. The number of indicators used for each CES category in the reviewed study.

This highly uneven distribution suggests a strong emphasis on tangible, easily quantifiable services, while symbolic and culturally embedded values remain underrepresented. For a complete list of indicator definitions and applications, please refer to Table S2.

4. Discussion

This discussion section synthesizes the key findings from the systematic review and critically analyzes the current state of CES research under the SPI framework. It focuses on five main dimensions: publication trends, geographic and thematic distributions, CES category coverage, assessment methods, and indicator use. Based on the identified gaps and contextual challenges, the section concludes by proposing a localized CES evaluation framework tailored to China's SPI context.

4.1. The Number of Articles Related to This Topic

The steady growth in CES-related publications within sponge city research since 2013 reflects a broader shift in academic and policy interest. Early stages of SPI implementation were dominated by engineering-led approaches, focusing on hydrological control and ecological restoration [5,6]. However, after the gradual completion and initial evaluation of pilot projects around 2017, scholars began to pay attention to the broader socio-cultural aspects of urban ecological infrastructure.

This shift is particularly evident in the notable rise in publications between 2020 and 2024. During this period, China's ecological civilization policies began to emphasize cultural values alongside ecological performance [34–37]. The resulting regulatory and funding support helped foster empirical research beyond traditional ecological indicators [37]. In addition, the growing recognition of CES—particularly recreational and aesthetic values—as integral to public acceptance and urban livability has encouraged interdisciplinary collaboration among planners, designers, and environmental scientists [19,38].

Despite the growing interest in CES, its incorporation into SPI research frameworks has been relatively recent, with most studies emerging after 2017 when pilot projects began to be evaluated [39]. This highlights the importance of integrating cultural ecosystem services more systematically during the early stages of sponge city planning and evaluation [40,41]. Moving forward, CES is likely to become a central component of sponge city evaluation as the discourse around urban sustainability continues to evolve.

4.2. Geographic Location and Research Focus

The reviewed studies demonstrate a strong geographic concentration in Tier 1 and New Tier 1 cities in China, particularly in metropolitan areas such as Beijing, Shanghai, Chengdu, and Guangzhou (see Figure 4). Tier 1 cities refer to the most developed and internationally recognized metropolitan areas (e.g., Beijing, Shanghai), while New Tier 1 cities are emerging urban centers with rapid economic growth, such as Chengdu and Hangzhou. Due to their advanced infrastructure and the concentration of academic and research institutions [42], these cities have become leading centers for CES research in the context of SPI. In contrast, small and medium-sized cities—many of which serve as designated sponge city pilot zones [9,43]—are significantly underrepresented in CES-related empirical studies, with only a limited number of publications focusing on these urban contexts. Moreover, the lack of cross-departmental accountability mechanisms makes it difficult to implement cultural functions across less conventional spatial types [25,44] This creates a spatial mismatch between where SPI is most comprehensively implemented and where academic research is primarily conducted.

A similar pattern of thematic bias is evident in the spatial types studied (Figure 5). Nearly half of the reviewed studies (48.5%) concentrate on park, whereas other forms of green and blue infrastructure—such as green corridors, community wetlands, and blue spaces—are comparatively underrepresented [45,46]. This narrow focus may be attributed to both practical and institutional constraints: urban parks are readily accessible, widely recognized, and conducive to short-term evaluations [47]. However, this preference may obscure the multifunctional cultural benefits provided by less visible or integrated elements of sponge city infrastructure, such as green corridors, community wetlands, and blue spaces [19,48,49].

To address these limitations, future research and policy implementation should emphasize adaptive governance approaches, integrate CES metrics into early-stage planning, and establish clearer inter-agency coordination pathways. Future research could broaden its spatial scope to include small and medium-sized cities and consider a more integrated, system-level perspective on SPI to better reflect the cultural ecosystem service functions embedded in its infrastructure. Such an approach would help reveal the layered and spatially distributed cultural values within sponge city landscapes.

4.3. Categories Assessment of CES

The analysis of CES categories within the context of sponge city infrastructure reveals a pronounced imbalance in the current research focus. As shown in Figure 6, recreational services and aesthetic value dominate the literature, with a substantial number of studies concentrating on these easily observable and quantifiable categories. In contrast, more abstract cultural categories—such as spiritual values, sense of place, cultural diversity, and knowledge systems—remain significantly underrepresented.

In the context of China's SPI research, abstract categories such as "cultural diversity", "knowledge systems", "inspiration", and "sense of place" are often underrepresented. This imbalance may be attributed to several interrelated factors: first, the difficulty of incorporating these categories into performance-oriented evaluation systems; second, the dominance of natural and engineering sciences in the research paradigm; and third, the diminishing visibility of local cultures amid rapid urbanization [50–52]. Additionally, topics related to religion, ethnicity, and intangible cultural heritage are often considered institutionally sensitive, which further constrains exploration of these categories [53]. Even when the MA framework is formally referenced, some Chinese studies restructure or consolidate complex categories for practical purposes—for instance, merging "inspiration" with aesthetics, or treating "knowledge systems" as part of environmental education or cultural heritage [54–56]. Such adjustments result in surface-level alignment with international classification systems while differing substantially in practical interpretation and application [57].

Notably, the conceptual treatment of CES also diverges between Chinese and Western scholarship. In contrast, Chinese CES research has generally emphasized planning-related or aesthetic–cultural perspectives, focusing more on landscape use, user satisfaction, and heritage preservation [58,59]. This tendency may reflect the influence of national planning priorities and the practical orientation of environmental governance, which differs from the interpretive and theory-driven approaches more common in Western contexts [60].

On the other hand, emerging categories such as "infrastructure appreciation" and "future services" have begun to appear in the SPI-related literature, reflecting a growing interest in the diverse values of CES. However, these categories have yet to be supported by consistent theoretical or operational frameworks [61]. Overall, the imbalances in CES category assessment are evident in both Chinese SPI studies and international CES research, though the contexts and emphases vary.

In summary, CES research under the SPI framework in China exhibits common biases toward tangible and measurable categories [62]. However, their underlying analytical traditions and policy environments differ significantly. Moving forward, there is a pressing need to develop CES classification systems that are both globally comparable and locally adaptive. Such frameworks should integrate abstract cultural values through participatory, context-sensitive methods, and align with both international standards (e.g., MA) and the socio-political realities of Chinese urban governance [63,64].

4.4. Assessment Methods

In the context of SPI, CES is predominantly assessed through non-monetary approaches, reflecting the intangible and value-laden nature of cultural benefits in urban ecological planning. Among the 10 non-monetary methods analyzed in this study, questionnaires are most frequently employed, appearing in 36 instances—a finding that underscores their simplicity and practicality [65]. Recent global reviews have shown that questionnaire-based assessments remain the primary tool in CES research, reflecting a broader trend toward structured, quantifiable methods.

However, despite their popularity, questionnaires face well-documented limitations. These include restricted ability to capture context-specific meanings, challenges in expressing symbolic or emotional dimensions, and limited sensitivity to intergroup differences [64,66]. As a result, scholars increasingly emphasize the need to integrate multisensory perspectives, long-term behavioral observation, and socio-cultural variables into future CES assessment frameworks.

In recent years, technology-driven non-monetary methods, such as social media data analysis and participatory GIS, have garnered increasing attention [67]. These methods enable researchers to access real-time behavioral data and spatial preferences from diverse user groups [68]. However, their application requires significant technical capacity and often lacks standardized analytical protocols, which may limit cross-context comparability and local adaptability [61].

Other common non-monetary methods—such as field observation, expert scoring, and semi-structured interviews—can offer richer cultural insights, particularly in uncovering subjective perceptions and community narratives. However, their use is often constrained by small sample sizes and difficulties in integration with spatial planning instruments [69,70].

Although monetary valuation methods—such as contingent valuation and hedonic pricing—are widely used in broader ecosystem service research [71,72], their application in CES remains limited, particularly within the context of SPI. This is largely due to the intangible and symbolic nature of cultural values, ethical concerns, and the methodological difficulty of translating non-material benefits into economic terms [73].

To address the above challenges, CES assessment in the context of SPI may benefit from adopting integrated and mixed-method approaches to balance quantitative scope and qualitative depth [28,74]. Such as participatory GIS, virtual reality simulations, qualitative interviews, or observational approaches. These methods are better suited to capturing intangible cultural values that are difficult to quantify—such as sense of place, spiritual experience, and cultural identity—and can complement Likert-scale surveys to enhance the depth and legitimacy of CES assessments in the context of sponge city planning.

4.5. CES Evaluation Indicators Challenges in the Application

Despite the increasing attention to CES in sponge city research, the development and application of robust and standardized indicators remain limited [27,31]. As highlighted in the results (Section 3.5), although 187 CES indicators were identified across 61 studies, only 54.1% of the publications provided specific and measurable indicators, while the rest referred to general CES categories without defining how they were operationalized [75].

Even among those studies that did utilize indicators, inconsistencies in definitions and classification were common [76,77]. For example, the same activity—such as urban cycling—was categorized as both recreation and social interaction in different studies, resulting in ambiguity and potential data overlap [67]. This reflects the broader challenge of aligning indicator use with clearly defined CES typologies.

Assessing CES in sponge city contexts remains difficult due to the lack of clear, consistent, and practical indicators [28,78]. While CES categories such as recreation, aesthetics, and education are widely used, many studies fail to define how these categories are measured in practice [15,75]. In nearly half of the reviewed papers, only general categories were listed without any corresponding indicators, making results hard to compare or apply (See Figure 8).

Abstract CES categories, such as inspiration, sense of place, and cultural identity, remain particularly underrepresented [79]. These categories are difficult to quantify and often lack clear measurement tools or baseline data [23,78]. Even more marginal are knowledge systems and cultural diversity, which were represented in only two indicators—or were entirely absent—in the reviewed literature (see Figure 9). Such gaps are not exclusive to Chinese cases but are also observed in international research, where symbolic and non-material CES dimensions remain poorly captured [80].

In the context of SPI, which seeks to integrate ecological and cultural functions, the lack of context-specific CES indicators may contribute to the underrepresentation of certain cultural values in planning and evaluation processes [81]. This may result in limiting the potential for infrastructure design to reflect and support local cultural needs and identities.

To address these challenges, future studies could focus on developing CES indicators that are simple, context-sensitive, and culturally grounded. International frameworks such as MA and IPBES provide valuable reference points [82,83], but local adaptation and participatory validation are essential to ensure relevance. Involving community stakeholders, urban planners, and interdisciplinary experts in indicator design can help improve both the scientific robustness and the cultural legitimacy of CES assessments in sponge cities [84]. Thus, the current body of evidence remains fragmented and uneven, limiting the generalizability of CES evaluation findings across diverse SPI contexts.

4.6. Toward a Localized CES Evaluation Framework for SPI: Structure and Implications

This study identifies several key challenges in existing research on CES within the context of sponge cities, as summarized in Sections 4.1–4.5. These challenges include the spatial and thematic concentration of research areas, imbalanced use of CES categories, limited methodological diversity, and inconsistencies in indicator systems—as demonstrated in the empirical results presented in Sections 3.2–3.5. Collectively, these issues underscore the limitations of existing CES evaluation frameworks in addressing cultural services in sponge city development. Therefore, it is necessary to propose a localized and practice oriented CES classification and assessment framework to reflect the characteristics of cultural services accurately in the context of SPI.

The proposed framework adopts a structured and standardized approach to address the widespread inconsistency in CES indicator usage identified in previous studies. Each CES category is clearly defined and linked to a set of representatives, perception-based indicators, enhancing conceptual clarity, comparability, and replicability. To support practical implementation, all indicators are designed to be evaluated through 5-point Likert-scale surveys, enabling consistent data collection across different sponge city contexts.

The design of this framework is informed by three main sources of input. First, it is informed by a systematic review of 61 empirical studies, which helped identify the definitions, frequency of use, and indicator types associated with various CES categories. Second, it references key national planning and environmental policy documents such as the Ecological Civilization Strategy, the Technical Guide for Sponge City Construction [12] and the National Territorial Spatial Planning Outline (2020–2035), all of which emphasize the integration of ecological and cultural values in urban development. Although this study does not include a detailed policy analysis section, these documents serve as guiding references that inform the framework's alignment with national planning priorities and practical applicability. Third, the framework incorporates insights from pilot project practices, especially observed cultural and educational design interventions such as wetland education zones, landscape art installations, and interactive community spaces. These sources collectively support the framework's structural logic and categorization strategy.

The proposed framework organizes CES into eight context-relevant categories, specifically tailored for application within the SPI context. These eight services include aesthetic value, recreational value, educational value, cultural heritage, spiritual and religious value, social relations, sense of place, and inspiration. Each category was selected based on its empirical prevalence, contextual relevance, and alignment with the objectives of SPI planning.

Compared with the Millennium Ecosystem Assessment (MA) framework, this localized version deliberately excludes the categories of "cultural diversity" and "knowledge systems", due to their limited operational feasibility, data scarcity, and heightened cultural sensitivity within China's urban governance system. In both the international and domestic CES literature, these two categories lack mature, context-specific indicators that can be reliably used in quantitative evaluation—particularly in urban infrastructure settings like sponge city parks [69,77]. Moreover, the intangible and often overlapping nature of these dimensions—especially in relation to "cultural heritage" and "educational value"—makes them difficult to delineate clearly within the perception-based questionnaire format adopted in this study [78]. In the Chinese context, elements such as local dialects, traditional crafts, and indigenous knowledge are typically incorporated under heritage protection or community education programs, further reinforcing their conceptual integration with existing categories [85,86].

While maintaining conceptual clarity and international comparability, the framework places greater emphasis on cultural adaptability and practical applicability. By ensuring that each included CES category contains at least three measurable sub-indicators, the structure of the evaluation tool is optimized for statistical reliability and ease of public understanding during field surveys. Over-fragmentation at this stage may reduce respondent clarity and compromise data quality [77,86]. However, future iterations of the framework may consider disaggregating these two dimensions—once reliable indicators emerge from empirical observation or policy development—to enhance global alignment [69,77].

Each category is accompanied by SPI-specific definitions and recommended example indicators, aiming to provide both theoretical grounding and actionable tools for CES evaluation in real-world sponge city contexts. The structure and contents of this framework are summarized in Table 3, which presents each CES category along with its SPI-relevant definition and selected example indicators.

Table 3. Proposed cultural ecosystem services evaluation framework for sponge city planning in China. Each CES category is defined in the context of SPI, with illustrative indicators suggested for practical assessment.

Category	Definition	Example Indicators
Educational values	Sponge cities provide essential material for formal and informal education.	Science education facilities: the presence of educational infrastructure (labs, field classrooms, informational signage); on-site educational activities: organized ecological tours, environmental education programs, and outdoor classrooms; informal learning: casual nature observation by individuals or families.
Leisure and entertainment	Sponge cities support opportunities for leisure and recreation.	Sports activities: walking, jogging, playing sports (e.g., football, basketball), swimming, kite flying, square dancing; passive leisure: dog walking, sitting/relaxing, reading, meditation; water-based recreation: crab fishing, boating, angling, feeding fish.

Table 3. *Cont.*

Category	Definition	Example Indicators
Social relations	SPI promotes public social interaction and cohesion.	Social spaces: availability of gathering spots for family and friends; community participation: events or design features encouraging community involvement and cohesion; community activities: frequency of outdoor events (movies, concerts, group camping, etc.).
Cultural heritage	Sponge cities help preserve and showcase historical and cultural heritage.	Conservation of relics: protection of historical monuments, buildings, and cultural sites; maintenance of heritage sites: regular upkeep (inspections, restoration) of cultural landmarks; intangible heritage: support for traditional crafts, oral histories, and performing arts.
Sense of place	Emotional attachment and identity derived from the uniqueness of SPI sites.	Attachment: residents' emotional attachment to local SPI features; belonging: the sense of belonging or pride in the sponge city neighborhood; memorability: the degree to which the SPI landscape offers unforgettable experiences or lasting impressions.
Aesthetic value	Enjoyment of natural beauty and pleasant surroundings in sponge cities.	Natural scenery: visual appeal of landscapes (water bodies, wetlands, greenery); natural sounds: presence of pleasing sounds (birdsong, flowing water, insect calls); natural scents: occurrence of pleasant natural aromas (flowers, wood, fresh grass).
Spiritual and religious	Spiritual fulfillment and religious expression supported by SPI environments.	Physical relaxation: sponge city spaces facilitating stress relief and physical well-being; psychological restoration: opportunities for mental relaxation and recovery; spiritual activities: spaces or elements enabling religious practices or personal contemplation.
Inspiration	Stimulus for artistic creation and intellectual inspiration from SPI.	Artistic inspiration: settings inspiring art, photography, music, or design; cultural narratives: inspiration for folklore or local cultural stories; intellectual stimulation: new ideas or learning prompted by the sponge city environment (innovative thinking in daily life).

To operationalize this framework, all indicators have been translated into perception-based survey items for practical use in public satisfaction assessments. A sample questionnaire is provided in Table S3, demonstrating how the framework supports planning feedback and policy decision-making in sponge city design.

This framework aims to bridge the gap between global CES classifications and the operational needs of SPI, offering a practical approach tailored to China's socio-cultural and planning context. By incorporating China's unique socio-cultural characteristics, the framework enhances the local adaptability of CES classification systems and helps to address the disconnect between internationally adopted CES frameworks and local planning practices. Moreover, it provides methodological guidance for aligning global CES evaluation systems—such as MA and IPBES—with the practical needs of sponge city development in China, while offering relevant experience for other rapidly urbanizing countries. Through SPI-specific examples and operational definitions, the framework also facilitates the translation of abstract concepts—such as "sense of place" and "inspiration"—into planning-oriented tools via perception-based indicators, thereby improving the integration of cultural services into infrastructure decision-making. Additionally, the localized approach demonstrated in this framework offers a practical pathway for grounding global CES frameworks within specific cultural contexts and exhibits strong potential for scalability and international applicability.

Despite its advantages, several limitations should be acknowledged regarding the practical application of this CES framework. First, the framework is developed based on empirical studies that are predominantly concentrated in China's Tier 1 and New Tier 1 cities, such as Beijing, Shanghai, and Chengdu (see Section 3.4). This geographic concentration may limit the generalizability of the findings, as cultural perceptions, institutional structures, and public values can vary significantly across smaller or less-developed urban areas. Additionally, it is inherently shaped by China's policy-driven planning system and centralized governance structure. These institutional factors often create practical implementation barriers such as fragmented responsibilities, limited cross-sectoral coordination, and bureaucratic inertia, potentially hindering effective integration of intangible CES dimensions into operational planning processes [6,69].

Nevertheless, the underlying structure of the proposed framework remains transferable and aligns conceptually with internationally recognized CES frameworks such as the Millennium Ecosystem Assessment (MA) and IPBES. Like these global frameworks, the proposed model emphasizes the multidimensional and context-sensitive nature of CES, advocating for place-based, culturally specific adaptations rather than universal categories [77,87]. Through localized modifications—such as adjusting indicator definitions, incorporating traditional knowledge, or customizing participatory evaluation approaches—this framework may effectively be adapted to diverse socio-cultural and governance contexts.

To support integration into China's planning systems, the framework is suitable for incorporation into post-construction evaluations of sponge city parks, community satisfaction surveys, and pilot assessments conducted jointly with planning and environmental agencies.

Future research should explicitly explore the framework's applicability through comparative studies involving diverse international contexts, particularly within rapidly urbanizing regions such as Southeast Asia, Latin America, and Africa. Such studies should focus on identifying and validating culturally relevant indicators, engaging local communities through participatory methodologies, and addressing governance variations. Additionally, empirical studies are encouraged to investigate practical localization strategies, aiming to develop operational protocols for adapting global CES typologies into specific cultural and institutional settings, thus enhancing the framework's international scalability and comparative analytical value.

This review was limited by the exclusive use of Web of Science and Scopus databases, English-language restrictions, and the absence of protocol registration, which may have led to the omission of relevant studies published in other languages or in the grey literature.

5. Conclusions

This review has synthesized the current state of research on CES in the context of SPI and identified key challenges across spatial scope, category representation, methodological diversity, and indicator standardization. To address these gaps, the study proposes a localized CES evaluation framework that integrates China's socio-cultural specificities while maintaining conceptual alignment with global standards such as MA and IPBES.

By offering SPI-oriented definitions and perception-based indicators, the framework enables the practical integration of abstract CES dimensions—such as sense of place and inspiration—into planning processes. It provides not only a theoretical reference but also a pragmatic tool for policymakers and urban designers seeking to embed cultural values in ecological infrastructure.

The proposed framework exemplifies how global CES typologies can be meaningfully localized in rapidly urbanizing contexts, serving as a transferable model for other countries facing similar sustainability transitions. Future research should build on this foundation through empirical validation, multi-city comparisons, and participatory approaches that deepen cultural relevance and policy impact.

Future studies should explore the applicability of the proposed framework across different cultural and administrative settings to enhance its scalability and transferability.

Supplementary Materials: The following supporting information can be downloaded at: https://www.mdpi.com/article/10.3390/su17115130/s1, Table S1: Study Metadata of Empirical Research on Cultural Ecosystem Services; Table S2: CES Indicators Categorized by Service Type in the Reviewed Studies; Table S3: Operationalization of the CES Evaluation Framework via Questionnaire Items.

Author Contributions: Conceptualization, N.H.; methodology, N.H.; formal analysis, N.H.; writing—original draft preparation, N.H.; writing—review and editing, M.S.B.M.N. and R.B.I.; supervision, R.B.I.; project administration, R.B.I. and M.S.B.M.N. All authors contributed to the development of the manuscript. The initial draft was prepared by N.H., with both co-authors providing substantial comments and revisions throughout the manuscript. All authors have read and agreed to the published version of the manuscript.

Funding: This research received no external funding.

Acknowledgments: The author would like to express heartfelt gratitude to Roziya Binti Ibrahim and Mohd Sallehuddin Bin Mat Noor for their invaluable academic guidance, critical insights, and continuous support throughout this research.

Conflicts of Interest: The authors declare no conflicts of interest.

References

1. Xiong, L.; Yan, L.; Du, T.; Yan, P.; Li, L.; Xu, W. Impacts of Climate Change on Urban Extreme Rainfall and Drainage Infrastructur Performance: A Case Study in Wuhan City, China. *Irrig. Drain.* **2019**, *68*, 152–164. [CrossRef]
2. Trenberth, K.E. Changes in Precipitation with Climate Change. *Clim. Res.* **2011**, *47*, 123–138. [CrossRef]
3. Shastri, H.; Ghosh, S.; Paul, S.; Shafizadeh-Moghadam, H.; Helbich, M.; Karmakar, S. Future Urban Rainfall Projection Considering the Impacts of Climate Change and Urbanization with Statistical–Dynamical Integrated Approach. *Clim. Dyn.* **201** *52*, 6033–6051. [CrossRef]
4. Khodadad, M.; Aguilar-Barajas, I.; Khan, A.Z. Green Infrastructure for Urban Flood Resilience: A Review of Recent Literature o Bibliometrics, Methodologies, and Typologies. *Water* **2023**, *15*, 523. [CrossRef]
5. Qi, Y.; Chan, F.K.S.; Thorne, C.; O'Donnell, E.; Quagliolo, C.; Comino, E.; Pezzoli, A.; Li, L.; Griffiths, J.; Sang, Y.; et al. Addressin Challenges of Urban Water Management in Chinese Sponge Cities via Nature-Based Solutions. *Water* **2020**, *12*, 2788. [CrossRe
6. Xia, J.; Zhang, Y.; Xiong, L.; He, S.; Wang, L.; Yu, Z. Opportunities and Challenges of the Sponge City Construction Related t Urban Water Issues in China. *Sci. China Earth Sci.* **2017**, *60*, 652–658. [CrossRef]
7. Liu, D. China's Sponge Cities to Soak up Rainwater. *Nature* **2016**, *537*, 307. [CrossRef]
8. Chan, F.K.S.; Griffiths, J.A.; Higgitt, D.; Xu, S.; Zhu, F.; Tang, Y.-T.; Xu, Y.; Thorne, C.R. "Sponge City" in China—A Breakthroug of Planning and Flood Risk Management in the Urban Context. *Land Use Policy* **2018**, *76*, 772–778. [CrossRef]
9. Guo, Z.; Zhang, X.; Winston, R.; Smith, J.; Yang, Y.; Tao, S.; Liu, H. A Holistic Analysis of Chinese Sponge City Cases by Regio Using PLS-SEM Models to Understand Key Factors Impacting LID Performance. *J. Hydrol.* **2024**, *637*, 131405. [CrossRef]
10. Tang, Y.T.; Chan, F.K.S.; O'Donnell, E.C.; Griffiths, J.; Lau, L.; Higgitt, D.L.; Thorne, C.R. Aligning Ancient and Modern Approache to Sustainable Urban Water Management in China: Ningbo as a "Blue-Green City" in the "Sponge City" Campaign. *J. Flood Ris Manag.* **2018**, *11*, e12451. [CrossRef]
11. Wang, H.; Mei, C.; Liu, J.; Shao, W. A New Strategy for Integrated Urban Water Management in China: Sponge City. *Sci. Chin Technol. Sci.* **2018**, *61*, 317–329. [CrossRef]
12. Ministry of Ecology and Environment of the People's Republic of China. *14th Five Year Plan for Environmental Health Wor* Ministry of Ecology and Environment of the People's Republic of China: Beijing, China, 2022.
13. Ministry of Ecology and Environment of the People's Republic of China. 14th Five-Year Plan and 2035 Long-Term Objective 2021. Available online: https://climate-laws.org/document/14th-five-year-plan_0496 (accessed on 24 May 2025).
14. Satz, D.; Gould, R.K.; Chan, K.M.A.; Guerry, A.; Norton, B.; Satterfield, T.; Halpern, B.S.; Levine, J.; Woodside, U.; Hannahs, N et al. The Challenges of Incorporating Cultural Ecosystem Services into Environmental Assessment. *Ambio* **2013**, *42*, 675–68 [CrossRef] [PubMed]

Dang, H.; Li, J. Supply-Demand Relationship and Spatial Flow of Urban Cultural Ecosystem Services: The Case of Shenzhen, China. *J. Clean. Prod.* **2023**, *423*, 138765. [CrossRef]

Quevedo, J.M.D.; Kohsaka, R. A Systematic Review of Cultural Ecosystem Services of Blue Carbon Ecosystems: Trends, Gaps, and Challenges in Asia and Beyond. *Mar. Policy* **2024**, *159*, 105898. [CrossRef]

Taguchi, V.; Weiss, P.; Gulliver, J.; Klein, M.; Hozalski, R.; Baker, L.; Finlay, J.; Keeler, B.; Nieber, J. It Is Not Easy Being Green: Recognizing Unintended Consequences of Green Stormwater Infrastructure. *Water* **2020**, *12*, 522. [CrossRef]

Andersson, E.; Tengö, M.; McPhearson, T.; Kremer, P. Cultural Ecosystem Services as a Gateway for Improving Urban Sustainability. *Ecosyst. Serv.* **2015**, *12*, 165–168. [CrossRef]

Wen, L.; Li, B.; Peng, Y.; Zhou, Y.; Weng, A.; Jin, Y.; Cai, G.; Lin, Y.; Chen, B. Exploring the Evolution of Coupled Natural-Cultural Ecosystem Services and Their Geographically Scaled Driven Modeling in a Coastal City of Southeast China. *J. Environ. Manag.* **2024**, *361*, 121265. [CrossRef]

Jiang, W. Ecosystem Services Research in China: A Critical Review. *Ecosyst. Serv.* **2017**, *26*, 10–16. [CrossRef]

Yang, L.; Cao, K. Cultural Ecosystem Services Research Progress and Future Prospects: A Review. *Sustainability* **2022**, *14*, 11845. [CrossRef]

Pascua, P.; McMillen, H.; Ticktin, T.; Vaughan, M.; Winter, K.B. Beyond Services: A Process and Framework to Incorporate Cultural, Genealogical, Place-Based, and Indigenous Relationships in Ecosystem Service Assessments. *Ecosyst. Serv.* **2017**, *26*, 465–475. [CrossRef]

Stokowski, P.A. Languages of Place and Discourses of Power: Constructing New Senses of Place. *J. Leis. Res.* **2002**, *34*, 368–382. [CrossRef]

Dou, Y.; Zhen, L.; De Groot, R.; Du, B.; Yu, X. Assessing the Importance of Cultural Ecosystem Services in Urban Areas of Beijing Municipality. *Ecosyst. Serv.* **2017**, *24*, 79–90. [CrossRef]

Mao, Q.; Wang, L.; Guo, Q.; Li, Y.; Liu, M.; Xu, G. Evaluating Cultural Ecosystem Services of Urban Residential Green Spaces from the Perspective of Residents' Satisfaction with Green Space. *Front. Public Health* **2020**, *8*, 226. [CrossRef]

Panduro, T.E.; Veie, K.L. Classification and Valuation of Urban Green Spaces—A Hedonic House Price Valuation. *Landsc. Urban Plan.* **2013**, *120*, 119–128. [CrossRef]

Bell, S.; Montarzino, A.; Travlou, P. Mapping Research Priorities for Green and Public Urban Space in the UK. *Urban For. Urban Green.* **2007**, *6*, 103–115. [CrossRef]

Scholte, S.S.; Van Teeffelen, A.J.; Verburg, P.H. Integrating Socio-Cultural Perspectives into Ecosystem Service Valuation: A Review of Concepts and Methods. *Ecol. Econ.* **2015**, *114*, 67–78. [CrossRef]

Marquez, L.A.M.; Rezende, E.C.N.; Machado, K.B.; do Nascimento, E.L.M.; Castro, J.D.B.; Nabout, J.C. Trends in Valuation Approaches for Cultural Ecosystem Services: A Systematic Literature Review. *Ecosyst. Serv.* **2023**, *64*, 101572. [CrossRef]

Christie, M.; Fazey, I.; Cooper, R.; Hyde, T.; Kenter, J.O. An Evaluation of Monetary and Non-Monetary Techniques for Assessing the Importance of Biodiversity and Ecosystem Services to People in Countries with Developing Economies. *Ecol. Econ.* **2012**, *83*, 67–78. [CrossRef]

Hirons, M.; Comberti, C.; Dunford, R. Valuing Cultural Ecosystem Services. *Annu. Rev. Environ. Resour.* **2016**, *41*, 545–574. [CrossRef]

Jim, C.Y. Green-Space Preservation and Allocation for Sustainable Greening of Compact Cities. *Cities* **2004**, *21*, 311–320. [CrossRef]

Haase, D.; Larondelle, N.; Andersson, E.; Artmann, M.; Borgström, S.; Breuste, J.; Gomez-Baggethun, E.; Gren, Å.; Hamstead, Z.; Hansen, R.; et al. A Quantitative Review of Urban Ecosystem Service Assessments: Concepts, Models, and Implementation. *Ambio* **2014**, *43*, 413–433. [CrossRef]

Ministry of Housing and Urban-Rural Development of China. *Technical Guide for Sponge City Construction (Updated Pilot Version)*; Government of China: Beijing, China, 2022.

State Council of the People's Republic of China. *The 14th Five-Year Plan for National Economic and Social Development of the People's Republic of China (2021–2025)*; Government of China: Beijing, China, 2021.

Ministry of Ecology and Environment of China. *The 14th Five-Year Plan for Ecological and Environmental Protection (2021)*; Government of China: Beijing, China, 2021.

Ministry of Natural Resources of the People's Republic of China. *National Territorial Spatial Planning Outline of China (2020–2035)*; Government of China: Beijing, China, 2020.

Neidig, J.; Anguelovski, I.; Lliso, B.; Pascual, U. Pluralizing Environmental Values for Urban Planning: How to Uncover the Diversity of Imaginaries about Socio-natures from Vitoria-Gasteiz (Basque Country, Spain). *People Nat.* **2023**, *5*, 1262–1283. [CrossRef]

Ma, Y.; Jiang, Y. Mainstreaming the Framework of Ecosystem Services to Enhance China's Policy Implementation for Sponge City Development. *Sustain. Dev.* **2023**, *31*, 2291–2306. [CrossRef]

40. Raymond, C.; Breil, M.; Nita, M.; Kabisch, N.; de Bel, M.; Enzi, V.; Frantzeskaki, N.; Geneletti, G.; Lovinger, L.; Cardinaletti, M. et al. *An Impact Evaluation Framework to Support Planning and Evaluation of Nature-Based Solutions Projects*; Centre for Ecology an Hydrology: Wallingford, UK, 2017.
41. Andersson, E.; Barthel, S.; Borgström, S.; Colding, J.; Elmqvist, T.; Folke, C.; Gren, Å. Reconnecting Cities to the Biospher Stewardship of Green Infrastructure and Urban Ecosystem Services. *Ambio* **2014**, *43*, 445–453. [CrossRef] [PubMed]
42. Xu, J.; Huang, G.; Ye, Y.; Liu, Z. Spatial-Temporally and Industrially Heterogeneous Effects of New Infrastructure Constructio on Fostering Emerging Industries in Chinese Cities. *Heliyon* **2024**, *10*, e23774. [CrossRef]
43. Yin, D.; Chen, Y.; Jia, H.; Wang, Q.; Chen, Z.; Xu, C.; Li, Q.; Wang, W.; Yang, Y.; Fu, G.; et al. Sponge City Practice in China: Review of Construction, Assessment, Operational and Maintenance. *J. Clean. Prod.* **2021**, *280*, 124963. [CrossRef]
44. Tate, C.; Wang, R.; Akaraci, S.; Burns, C.; Garcia, L.; Clarke, M.; Hunter, R. The Contribution of Urban Green and Blue Spaces t the United Nation's Sustainable Development Goals: An Evidence Gap Map. *Cities* **2024**, *145*, 104706. [CrossRef]
45. Valente De Macedo, L.S.; Barda Picavet, M.E.; Puppim De Oliveira, J.A.; Shih, W.-Y. Urban Green and Blue Infrastructure: Critical Analysis of Research on Developing Countries. *J. Clean. Prod.* **2021**, *313*, 127898. [CrossRef]
46. Dong, Q.; Zeng, P.; Long, X.; Peng, M.; Tian, T.; Che, Y. An Improved Accessibility Index to Effectively Assess Urban Par Allocation: Based on Working and Residential Situations. *Cities* **2024**, *145*, 104736. [CrossRef]
47. Veerkamp, C.J.; Schipper, A.M.; Hedlund, K.; Lazarova, T.; Nordin, A.; Hanson, H.I. A Review of Studies Assessing Ecosyster Services Provided by Urban Green and Blue Infrastructure. *Ecosyst. Serv.* **2021**, *52*, 101367. [CrossRef]
48. Wang, S.; Palazzo, E. Sponge City and Social Equity: Impact Assessment of Urban Stormwater Management in Baicheng Cit China. *Urban Clim.* **2021**, *37*, 100829. [CrossRef]
49. Zhao, J. Research Hotspots and Trends in Chinese Academic Abstracts (2000–2022)—A Visualization Analysis Based on CiteSpac *J. Educ. Educ. Res.* **2024**, *10*, 232–239. [CrossRef]
50. Zhang, L.; Leydesdorff, L. The Scientometric Measurement of Interdisciplinarity and Diversity in Research Portfolios of Chines Universities. *J. Data Inf. Sci.* **2021**, *6*, 13–35. [CrossRef]
51. Wang, T. Cultural Diversity and Educational Policy in Chinese Society. *City Divers.* **2020**, *1*, 1926. [CrossRef]
52. Yuan, X. Citespace-Based Knowledge Mapping Analysis of China's Cultural Heritage Institutions and Regulations. *Front. Bu Econ. Manag.* **2023**, *10*, 352–358. [CrossRef]
53. Lin, H.; Hong, X.-C.; Wen, C.; Hu, F. The Historical Sensing of Urban Forest Based on the Indicators of CES and Landscap Categories: A Case of Kushan Scenic Area in CHINA. *Ecol. Indic.* **2024**, *166*, 112440. [CrossRef]
54. Pan, Y.; Qu, Y. Cultural Ecosystem Services in Land Use/Land Cover Change: A Literature Review and Prospects for Futur Research. *Land* **2024**, *13*, 2027. [CrossRef]
55. Cheng, X.; Van Damme, S.V.; Uyttenhove, P. Assessing the Impact of Park Renovations on Cultural Ecosystem Services. *Lan* **2022**, *11*, 610. [CrossRef]
56. Liu, J.; Han, H.; Chen, Z. Study on Practice Framework and Evaluation Indicator System of Landsense Design for the Enhancemer of Cultural Ecosystem Services in Historical Blocks. *Landsc. Archit. Front.* **2024**, *12*, 62. [CrossRef]
57. Jones, R.; Cox, S.; Jones, T. Place Meaning and Place Attachment. In *International Encyclopedia of Geography*; Richardson, D Castree, N., Goodchild, M.F., Kobayashi, A., Liu, W., Marston, R.A., Eds.; Wiley: Hoboken, NJ, USA, 2022; pp. 1–6. ISBN 978-0-470-65963-2.
58. Liao, X. Spiritually Cultural Differences between China and the West Manifested by Material Cultures—Human-Oriente Philosophy versus Harmonious Coexistence of Man and Nature. *Open J. Soc. Sci.* **2023**, *11*, 305–313. [CrossRef]
59. Cushing, D.F.; Miller, E. Place Attachment Theory. In *Creating Great Places*; Routledge: New York, NY, USA, 2019; pp. 67–79. ISBN 978-0-429-28963-7.
60. Blicharska, M.; Smithers, R.J.; Hedblom, M.; Hedenås, H.; Mikusiński, G.; Pedersen, E.; Sandström, P.; Svensson, J. Shades of Gre Challenge Practical Application of the Cultural Ecosystem Services Concept. *Ecosyst. Serv.* **2017**, *23*, 55–70. [CrossRef]
61. Jiang, W.; Marggraf, R. Making Intangibles Tangible: Identifying Manifestations of Cultural Ecosystem Services in a Cultura Landscape. *Land* **2021**, *11*, 26. [CrossRef]
62. Nie, X.; Huang, C.; Wang, H. A New Method to Classify Cultural Ecosystem Services and Visualize Their Economic Value: Case Study of Guilin, a Famous Tourist Destination in China. *Ecosyst. Serv.* **2025**, *72*, 101708. [CrossRef]
63. Xu, H.; Zhao, G.; Fagerholm, N.; Primdahl, J.; Plieninger, T. Participatory Mapping of Cultural Ecosystem Services for Landscap Corridor Planning: A Case Study of the Silk Roads Corridor in Zhangye, China. *J. Environ. Manag.* **2020**, *264*, 110458. [CrossRe
64. Raymond, C.M.; Brown, G.; Weber, D. The Measurement of Place Attachment: Personal, Community, and Environmenta Connections. *J. Environ. Psychol.* **2010**, *30*, 422–434. [CrossRef]
65. Henshaw, V. *Urban Smellscapes: Understanding and Designing City Smell Environments*; Routledge: New York, NY, USA, 2013.
66. Dai, P.; Zhang, S.; Chen, Z.; Gong, Y.; Hou, H. Perceptions of Cultural Ecosystem Services in Urban Parks Based on Social Networ Data. *Sustainability* **2019**, *11*, 5386. [CrossRef]

Poorthuis, A.; Chen, Q.; Zook, M. A Nationwide Dataset of De-Identified Activity Spaces Derived from Geotagged Social Media Data. *Environ. Plan. B Urban Anal. City Sci.* **2024**, *51*, 2264–2275. [CrossRef]

Klain, S.C.; Chan, K.M.A. Navigating Coastal Values: Participatory Mapping of Ecosystem Services for Spatial Planning. *Ecol. Econ.* **2012**, *82*, 104–113. [CrossRef]

Chivulescu, S.; Hapa, M.; Pitar, D.; Lorenţ, A.; Marmureanu, L.; Leca, S.; Radu, R.; Cazacu, R.; Dobre, A.C.; Pascu, I.S.; et al. Integrating Monetary and Non-Monetary Valuation for Ecosystem Services in Piatra Craiului National Park, Southern Carpathians: A Comprehensive Approach to Sustainability and Conservation. *Front. For. Glob. Change* **2024**, *7*, 1280793. [CrossRef]

Costanza, R.; d'Arge, R.; De Groot, R.; Farber, S.; Grasso, M.; Hannon, B.; Limburg, K.; Naeem, S.; O'Neill, R.V.; Paruelo, J.; et al. The Value of the World's Ecosystem Services and Natural Capital. *Nature* **1997**, *387*, 253–260. [CrossRef]

De Groot, R.; Brander, L.; Van Der Ploeg, S.; Costanza, R.; Bernard, F.; Braat, L.; Christie, M.; Crossman, N.; Ghermandi, A.; Hein, L. Global Estimates of the Value of Ecosystems and Their Services in Monetary Units. *Ecosyst. Serv.* **2012**, *1*, 50–61. [CrossRef]

Wright, W.C.C.; Eppink, F.V.; Greenhalgh, S. Are Ecosystem Service Studies Presenting the Right Information for Decision Making? *Ecosyst. Serv.* **2017**, *25*, 128–139. [CrossRef]

Tian, T.; Sun, L.; Peng, S.; Sun, F.; Che, Y. Understanding the Process from Perception to Cultural Ecosystem Services Assessment by Comparing Valuation Methods. *Urban For. Urban Green.* **2021**, *57*, 126945. [CrossRef]

Song, L.; Wu, M.; Wu, Y.; Xu, X.; Xie, C. Research on the Evaluation of Cultural Ecosystem Services in Zhengzhou Urban Parks Based on Public Perceptions. *Sustainability* **2023**, *15*, 11964. [CrossRef]

Tandarić, N.; Ives, C.; Watkins, C. Can We Plan for Urban Cultural Ecosystem Services? *J. Urban Ecol.* **2020**, *6*, juaa016. [CrossRef]

Chan, K.M.; Gould, R.K.; Pascual, U. Editorial Overview: Relational Values: What Are They, and What's the Fuss about? *Curr. Opin. Environ. Sustain.* **2018**, *35*, A1–A7. [CrossRef]

Fish, R.; Church, A.; Winter, M. Conceptualising Cultural Ecosystem Services: A Novel Framework for Research and Critical Engagement. *Ecosyst. Serv.* **2016**, *21*, 208–217. [CrossRef]

Daniel, T.C.; Muhar, A.; Arnberger, A.; Aznar, O.; Boyd, J.W.; Chan, K.M.A.; Costanza, R.; Elmqvist, T.; Flint, C.G.; Gobster, P.H.; et al. Contributions of Cultural Services to the Ecosystem Services Agenda. *Proc. Natl. Acad. Sci. USA* **2012**, *109*, 8812–8819. [CrossRef]

Milcu, A.I.; Hanspach, J.; Abson, D.; Fischer, J. Cultural Ecosystem Services: A Literature Review and Prospects for Future Research. *Ecol. Soc.* **2013**, *18*, art44. [CrossRef]

Chan, K.M.A.; Guerry, A.D.; Balvanera, P.; Klain, S.; Satterfield, T.; Basurto, X.; Bostrom, A.; Chuenpagdee, R.; Gould, R.; Halpern, B.S.; et al. Where Are Cultural and Social in Ecosystem Services? A Framework for Constructive Engagement. *BioScience* **2012**, *62*, 744–756. [CrossRef]

Raymond, C.M.; Singh, G.G.; Benessaiah, K.; Bernhardt, J.R.; Levine, J.; Nelson, H.; Turner, N.J.; Norton, B.; Tam, J.; Chan, K.M.A. Ecosystem Services and Beyond: Using Multiple Metaphors to Understand Human–Environment Relationships. *BioScience* **2013**, *63*, 536–546. [CrossRef]

IPBES. *Methodological Assessment Regarding the Diverse Conceptualization of Multiple Values of Nature and Its Benefits*; Intergovernmental Science-Policy Platform on Biodiversity and Ecosystem Services: Bonn, Germany, 2022.

Millennium Ecosystem Assessment (MEA). *Ecosystems and Human Well-Being*; Island Press: Washington, DC, USA, 2005; Volume 5.

Raymond, C.M.; Frantzeskaki, N.; Kabisch, N.; Berry, P.; Breil, M.; Nita, M.R.; Geneletti, D.; Calfapietra, C. A Framework for Assessing and Implementing the Co-Benefits of Nature-Based Solutions in Urban Areas. *Environ. Sci. Policy* **2017**, *77*, 15–24. [CrossRef]

Plieninger, T.; Dijks, S.; Oteros-Rozas, E.; Bieling, C. Assessing, Mapping, and Quantifying Cultural Ecosystem Services at Community Level. *Land Use Policy* **2013**, *33*, 118–129. [CrossRef]

Tengberg, A.; Fredholm, S.; Eliasson, I.; Knez, I.; Saltzman, K.; Wetterberg, O. Cultural Ecosystem Services Provided by Landscapes: Assessment of Heritage Values and Identity. *Ecosyst. Serv.* **2012**, *2*, 14–26. [CrossRef]

Hernández-Morcillo, M.; Plieninger, T.; Bieling, C. An Empirical Review of Cultural Ecosystem Service Indicators. *Ecol. Indic.* **2013**, *29*, 434–444. [CrossRef]

sustainability

MDPI

Article

Coordination Analysis Between Urban Livability and Population Distribution in China's Major Urban Agglomerations

Yingfeng Ran [1], Wei Hou [1,*], Jingli Sun [2], Liang Zhai [1,*], Chuan Du [1] and Jingyang Li [3]

1 Chinese Academy of Surveying and Mapping, Beijing 100036, China; wyyxcky@126.com (Y.R.); duchuan@casm.ac.cn (C.D.)
2 Department of Natural Resources of Shanxi Province, Taiyuan 030024, China; zrzytsjl@163.com
3 Department of Housing and Urban-Rural Development of Shanxi Province, Taiyuan 030013, China; jingyangli@yeah.net
* Correspondence: houwei@casm.ac.cn (W.H.); zhailiang@casm.ac.cn (L.Z.); Tel.: +86-01063880622 (W.H.)

Abstract: The mismatch between urban livability and population distribution can result in overcrowding and excessive pressure on ecosystem services if population growth surpasses urban capacity. Conversely, if urban expansion outpaces population needs, it can lead to underutilized infrastructure and inefficient land use. This study aims to assess the coordination between urban livability and population distribution in five major urban agglomerations in China: Beijing–Tianjin–Hebei (BTH), Yangtze River Delta (YRD), Pearl River Delta (PRD), Mid-Yangtze River (MYR), and Chengdu–Chongqing (CC). A comprehensive index for urban livability is established, from the aspects of social–economic development and ecosystem service. Additionally, a Coordination Distance Index (*CDI*) is developed to measure the relationship between urban livability and population distribution. Data from 2010, 2015, and 2020 are analyzed to evaluate the coordination levels and trends across the five urban agglomerations. The results show that from 2010 to 2020, most cities within these urban agglomerations experience improvements in their coordination levels, with the most notable advancements in the PRD and YRD regions. By 2020, the PRD and YRD were classified as having "high coordination", while BTH, MYR, and CC were categorized as having "moderate coordination". However, certain cities, such as Chengde in BTH, Shanghai in YRD, Ya'an in CC, and Zhuhai in PRD, still exhibited "low coordination", highlighting areas requiring spatial planning adjustments. This study introduces a method for quantitatively assessing the coordination between urban livability and population distribution, providing essential insights for policymakers and urban planners to refine urbanization development strategies and population regulation policies in China's major urban agglomerations.

Keywords: urban livability; urban green spaces; population distribution; coordination distance index; urban agglomeration

1. Introduction

Global urbanization is currently experiencing an unprecedented surge, particularly in developing countries, where rapid economic growth and demographic shifts are driving massive migrations to cities [1,2]. The expansion of metropolitan areas is not only altering landscapes but also transforming social and economic dynamics on a global scale [3]. As cities grow, they face various challenges in ecological management, including the sustainable provision of ecosystem services, the mitigation of environmental degradation, and the maintenance of biodiversity [1,4]. These issues underscore the need for integrated urban planning and policy frameworks that can balance the demands of urban development with the preservation of ecological integrity, ensuring that cities remain livable and resilient for future generations [5,6].

According to World Urbanization Prospects (WUP) data released by the United Nations, 55% of the world's population lives in urban areas, and this proportion is expected to increase to 68% by 2050 [7]. The successful management of the relationship between urbanization and population dynamics not only ensures equitable access to ecosystem services but also mitigates potential negative impacts of urbanization, such as increased pollution, urban heat islands, decreased water resources, and heightened vulnerability to climate change impacts [8–10]. To achieve the UN's Sustainable Development Goals (SDGs) by 2030 [11], particularly SDG 11 (making cities and human settlements inclusive, safe, resilient, and sustainable) and SDG 15 (protecting, restoring, and sustainably managing habitats), it is important to fully understand and quantitatively measure urban livability and its correlation with population distribution. Thus, fostering a balanced relationship between urban livability and demographic changes is not only crucial for the well-being of current urban residents but also for the long-term sustainability of urban development.

Taking into account the variations in geographical location, climate, and socioeconomic context across different cities, researchers have focused on studying city-specific urban livability challenges to align with the goals of sustainable development [12,13]. In India, rapid urbanization and population growth pose serious challenges to urban livability. Paul and Sen [14] point out that improving urban livability in Indian cities requires not only infrastructure enhancements but also attention to sociocultural factors, such as strengthening community safety and enhancing public services, to improve residents' overall life satisfaction. Another study finds that due to the high population density, limited resources, and weak public infrastructure, the quality of life is generally low, further weakening the livability, and demonstrates the discoordination of development in Kolkata city of India [15]. The research by Parker and Simpson [16] in Australia shows that urban livability benefits from prioritizing green spaces. By planning for ample parks, recreational areas, and public green spaces, Australia not only provides residents with a high-quality living environment but also effectively alleviates environmental pressure caused by population growth. Through sustainable green infrastructure planning, Australia has managed to maintain urban livability while accommodating population growth. In Brazil [12], the rise in population density in metropolitan areas has led to insufficient infrastructure and public services, particularly in informal communities. Future urban planning should place greater emphasis on community involvement and sustainable development to improve residents' living conditions and enhance urban livability. In the Metro Vancouver region of Canada, Martino et al. [17] propose that urban livability can be understood through four systemic dimensions: accessibility, social diversity, affordability, and economic vitality, and they use machine learning methods to analyze the relationship between urban livability and population dynamics; the result shows that in areas with different population densities, the main factors affecting livability vary. This indicates the need to emphasize different factors based on population density to achieve more coordinated development.

China has gone through an unprecedented urbanizing period and the number of cities increased from 193 to 691 from 1978 to 2022 [18,19]. Currently, there are 19 major urban agglomerations formed by closely located cities in China. The generation and development of the regional urban agglomeration brings many positive impacts, such as economic agglomeration, technological innovation, and cultural integration [20–23]. However, the rapid concentration of large populations within urban agglomerations can lead to urban livability issues such as resource depletion and degradation of life quality due to the increasing pressure on urban social security, healthcare services, and other critical infrastructures [24–27]. The Coupling Coordination Degree Model (CCDM) is often adopted by Chinese researchers to explore the interrelationships between urban livability and population dynamics [28,29]. It measures the level of coordination within complex systems by evaluating the coupling and coordination degrees among two or more subsystems [30–33]. Studies by Xu and Yin [34] and Zeng et al. [35] have applied CCDM to examine the relationship between ecological aspects of urban livability and population changes in the Yangtze River Delta (YRD) urban agglomeration. Their findings indicate that, despite fluctuations, the region generally demonstrates a moderate to high degree of coupling coordination. Some scholars use the indicator of elasticity coefficient to explore the relationship between urban livability and population growth [36]. This indicator measures the ratio of the rate of change between two systems and is used to assess the relative change relationship between variables [37]. Zhang et al. [29] conducted a study in Chongqing using the elasticity coefficient, revealing that during urban development, as the population grows, urban land expansion must consider the necessary requirements for maintaining livability. Cao et al. [38] found that even under favorable economic conditions, urban populations might still face out-migration. This implies that urban development strategies should not solely concentrate on economic growth but also integrate measures to improve various aspects of urban livability to ensure more harmonious and sustainable development.

In China's 'Fourteenth Five-Year Plan' outlined in 2020, the emphasis was on promoting long-term balanced population development [39]. This highlights the urgent need for aligning demographic trends with urban expansion. The challenge of addressing coordinated development within urban agglomerations has become increasingly significant [40]. However, existing studies mostly emphasized the role of economic factors in shaping urban livability, often considering urban livability as an isolated phenomenon [28]. There is a scarcity of quantitative analyses examining the dynamic interrelationships between urban livability and population distribution in China's urban agglomerations. This study aims to address this gap by developing a set of indices to understand the coordination and development trends within China's major urban agglomerations. Our objective is to provide valuable insights for urbanization development and population management.

2. Materials and Methods

We conduct the coordination analysis at two spatial scales. First, for each city within the urban agglomeration, we evaluate the coordination level between its livability and population. To do this, an evaluation framework for urban livability is established from the aspects of social–economic development and ecosystem services. We then combine this with population distribution data to propose an index called the Coordination Distance Index (*CDI*), which assesses the coordination degree between urban livability and population distribution. Finally, we develop a comprehensive index to describe the coordination level of the entire urban agglomeration, enabling comparisons among different urban agglomerations.

2.1. Study Area

Five major national urban agglomerations in China are selected as study areas. The Yangtze River Delta (YRD) urban agglomeration is one of the most economically developed and highly urbanized regions in China. The Mid-Yangtze River (MYR) urban agglomeration is the fastest-growing urban area and the most economically advanced region in Central China. The Chengdu–Chongqing (CC) urban agglomeration is a key platform for development and the largest urban agglomeration in the Western region. The Beijing–Tianjin–Hebei urban agglomeration (BTH) is a core regional component of the Chinese economy and a significant contributor to national competitiveness. The Pearl River Delta (PRD) urban agglomeration is recognized as one of the world's most rapidly growing urban areas. These urban agglomerations serve as exemplars of China's urban development, situated across the eastern, central, western, northern, and southern regions of the country (Figure 1). They encompass 95 cities, including four municipalities directly under the central government and portions of nine provinces. With a combined population of approximately 560 million inhabitants, these regions occupy roughly 10% of China's total land area. By the end of 2020, they accommodated 39% of China's population and contributed to 57% of the nation's Gross Domestic Product (GDP), thereby playing a pivotal role in terms of economic development and population concentration.

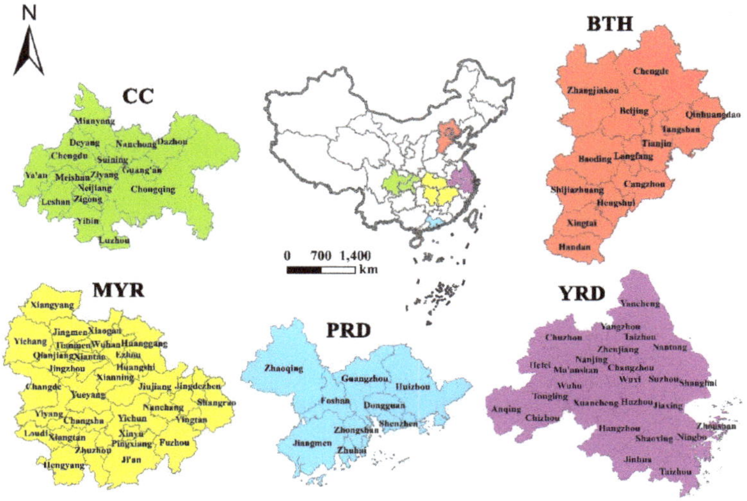

Figure 1. Locations of the five national urban agglomerations.

This paper employs data from the "Statistical Yearbook", the "China Energy Statistical Yearbook", and the "National Economic and Social Development Statistical Bulletin" in the years 2010, 2015, and 2020. Specifically, the datasets include GDP, per capita GDP, urban per capita disposable income, urban resident population, total resident population, number of hospital beds, and number of registered students for the relevant provinces and cities. Additionally, spatial datasets are also applied in this study, such as net primary productivity (NPP) products for the years 2010, 2015, and 2020 (sourced from http://modis.gsfc.nasa.gov, accessed on 24 November 2024), as well as land cover maps at a 30 m resolution adapted from the research [41], including nine categories: arable land, forest, shrub, grassland, water body, ice/snow, wasteland, impervious surface, and wetland.

2.2. Indices for Urban Livability

In consideration of the urban livability goals outlined by the government and informed by existing research [42,43], this study acknowledges the varying degrees of urbanization development among different urban agglomerations in China. Based on the principles of representativeness, data availability, and comparability in selecting indices, we have chosen a set of indices that reflect urban livability across two indicators: social–economic development (SD) and ecosystem service (ES). The selected indicators and descriptions can be found in Table 1. NPP denotes the net quantity of photosynthetic organic matter assimilated by vegetation per unit area over a given period, serving as an indicator of the region's vegetation carbon sequestration capability [44]. Indicators for carbon emission reflect the potential pressure to achieve carbon neutrality. The consumption of fossil fuels constitutes 80% to 90% of total carbon emissions [45]. For the purposes of this study, carbon emissions for each city have been calculated based on the consumption of eight primary energy sources, including coal, coke, crude oil, gasoline, kerosene, diesel, fuel oil, and natural gas. The areas of construction land, forest, grassland, and water bodies are derived through the statistical analysis of land cover datasets. The sub-indicator per capita urban green/blue space reflects the relationship between the supply and demand of urban ecosystem services.

Table 1. Indicators for urban livability.

Indicators	Sub Indicators	Description	Type
Social-economic Development (SD)	GDP per capita	Statistics of per capita GDP in each city	Positive
	Disposable income of urban residents	Statistics of per capita disposable income in each city	Positive
	Urban population ratio	Urban population/total resident population	Positive
	Per capita urban construction area	Construction land area/total resident population	Positive
	Number of hospital beds per 10,000 people	Number of beds/total resident population (per 10,000 people)	Positive
	Number of enrolled students per 10,000 people	Number of enrolled students (nursery school, primary school, middle school, university)/total resident population (per 10,000 people)	Positive
Ecosystem Service (ES)	Per capita NPP	Total NPP/total resident population	Positive
	Per capita urban green/bule space	(Forest + grassland + water area)/total resident population	Positive
	Per capita carbon emissions	Total carbon emissions/total resident population	Negative
	Energy efficiency	GDP/total carbon emissions	Positive

Due to the variances in scale across diverse indices and the inclusion of both positive and negative index types, a linear normalization method is used for preprocessing the indices [46]. This approach ensures that all indicators are adjusted to same range, facilitating a comprehensive analysis.

Normalization of positive sub indicators:

$$X_k' = \frac{X_k - min(X_k)}{max(X_k) - min(X_k)} \quad (1)$$

Normalization of negative sub indicators:

$$X_k' = \frac{max(X_k) - X_k}{max(X_k) - min(X_k)} \quad (2)$$

where X_k' represents the normalized value of the sub indicator for each city; X_k represents the original value of the index; and the functions $min()$, $max()$ are used to calculate the minimum and maximum indicator values across all cities.

In this study, the indicators are considered equally important to the livability of cities. Therefore, an equal-weight method is adopted to calculate the comprehensive Urban Livability Index (*ULI*). The *ULI* for each city is calculated by summing the values of the ten indices of SD and ES, each weighted equally.

$$ULI = \sum_{k=1}^{n} \omega \times X_k' \quad (3)$$

where ULI stands for Urban Livability Index; X_k' is the normalized value of each sub indicator; n is the number of indices; and ω is the indicator weight, which is 0.1 in this study.

2.3. Index for the Coordination Between Urban Livability and Population Distribution

The Z-score method transforms data of varying scales into a new dataset standardized to have a mean of 0 and a standard deviation of 1, thus mitigating the impact of differing scales on comparative analyses. This method is robust, ensuring that extreme values do not disproportionately influence the entire dataset [47,48]. Consequently, the indicators of ULI and population size for each city within the urban agglomerations are standardized using the Z-score method. The formula for this transformation is as follows:

$$ULI_j' = \frac{ULI_j - mean(ULI)}{std(ULI)} \quad (4)$$

$$POP_j' = \frac{POP_j - mean(POP)}{std(POP)} \quad (5)$$

where POP_j is the population of city j, and POP_j' is the standardized value of the POP_j; ULI_j' is the standardized value of the ULI for city j. The functions $mean()$ and $std()$ denote the calculations of the mean and standard deviation, respectively.

If a city's ULI' is equal to its POP', it indicates that the urban livability of the city is in balance with its population size within the urban agglomeration. The distance between a city point (POP', ULI') to the coordination line ($ULI' = POP'$) reflects the degree of coordination between urban livability and population distribution. It is quantified in this paper as the Coordination Distance Index (CDI).

$$CDI_j = \frac{ULI_j' - POP_j'}{\sqrt{2}} \tag{6}$$

If $ULI_j' < POP_j'$, it indicates an overcrowding of the population in the city compared to its urban livability, and vice versa. In addition, we calculated the Urban Agglomeration Coordination Distance Index ($UACDI$), reflecting the overall coordination between urban livability and population distribution across cities within the urban agglomeration.

$$UACDI = \frac{\sum_{j=1}^{J} (a_j \times abs(CDI_j))}{A} \tag{7}$$

where CDI_j and a_j represent the Coordination Distance Index and area for a city j. J is the city number within the urban agglomeration and A is the total area of the urban agglomeration. The function $abs()$ denotes the absolute value. A higher value of $UACDI$ indicates a greater imbalance and lower coordination degree for the urban agglomeration. The CDI range is categorized into distinct levels for assessing the degree of coordination for an individual city and urban agglomeration (see Table 2).

Table 2. Classification of coordination level between urban livability and population distribution.

CDI Range	Coordination Level	Implication
$CDI < -1$	Low	The current population is excessively high compared to the level of urban livability.
$-1 \leq CDI < -0.5$	Moderate	The current population is relatively high compared to the level of urban livability.
$-0.5 \leq CDI < 0.5$	High	The current population is well-coordinated with the level of urban livability.
$0.5 \leq CDI < 1$	Moderate	The current population is relatively low compared to the level of urban livability.
$1 \leq CDI$	Low	The current population is excessively low compared to the level of urban livability

3. Results

3.1. Coordination Evaluation for Cities Within Urban Agglomerations

Figure 2 shows the distribution of ULI and POP for each city in the five urban agglomerations, with values standardized using the Z-Score method based on Formulas (4) and (5). The coordination line (denoted as purple line in Figure 2) indicates a state of absolute coordination between urban livability and population. The dashed lines represent the distance to the coordination line on both sides with a distance interval of 0.5. The closer a city's point is to the coordination line, the higher the degree of coordination. If a city's point is positioned above the coordination line (greenish dots in Figure 2), it means that the city has a relatively smaller population in comparison to its level of urban livability within the urban agglomeration. Conversely, if a city's point is positioned below the coordination line (reddish dots in Figure 2), it indicates a relatively larger population distribution in comparison to its level of urban livability.

During the study period, most cities were close to the coordination line, with only a minority showing "low coordination". Specifically, in the YRD, there were four cities in a low coordination state in 2010, three in 2015, and three in 2020. Among them, Shanghai had a significantly larger population relative to its level of urban livability. Nanjing's level of urban livability exceeded its population and increased over time, resulting in a low coordination state in 2020. On the other hand, Suzhou's coordination improved from 2010 to 2015 and remained at a high coordination level from 2015 to 2020. In the MYR, many cities were in a low coordination state during the study period. Although overall coordination improved, there were still 12, 11, and 7 cities in a low coordination state in 2010, 2015, and 2020, respectively. The coordination levels of the three core cities, i.e., Nanchang, Changsha, and Wuhan, showed varying trends. Nanchang and Changsha progressed from low coordination to moderate and high coordination levels, respectively, while Wuhan experienced a decline from high coordination to moderate coordination. In the CC, most cities were close to the coordination line during the study period. In 2010, 2015, and 2020, there were only two, two, and three cities in a low coordination state, respectively. Among these, Chongqing and Ya'an exhibited low coordination, with Chongqing's population significantly exceeding its urban livability level, while Ya'an's urban livability level was higher compared to its population. Chengdu, another core city, maintained a high coordination state between 2010 and 2015 but deteriorated to a low coordination state by 2020. In the BTH, most cities were in a high or moderate coordination state during the study period. In 2010, 2015, and 2020, there were two, two, and one city in low coordination state, respectively. Among these, Chengde consistently remained in a low coordination state, with its urban livability level higher than its population. Baoding improved from low coordination to high coordination. On the other hand, Shijiazhuang showed little change in coordination, and Beijing and Tianjin consistently maintained a high coordination state. In the PRD, only one city, Zhuhai, was in a low coordination state as its urban livability level exceeded its population during the study period. The other cities were in high or moderate coordination states. Specifically, in 2020, seven cities were in a high coordination state, while Guangzhou and Shenzhen consistently maintained a high coordination state throughout the period.

In general, the number of cities with low coordination decreased in the YRD, MYR, and BTH. However, the CC region saw an increase of one city with low coordination, while the Pearl River Delta (PRD) remained unchanged. Among the five major urban agglomerations, Ya'an in the CC has the lowest coordination level, primarily due to its relatively good livability but small population. The second lowest is Chongqing, which, despite having a relatively high livability, suffers from an excessively large population as a mega-city. In the MYR, the proportion of cities with low coordination reached 39% in 2010, making it the agglomeration with the highest proportion of low coordination cities during the study period. In contrast, in the PRD, 78% of cities exhibited high coordination levels by 2020, making it the agglomeration with the largest proportion of highly coordinated cities.

The *CDI* for each city was calculated (using Formula (6)) based on the ULI' and POP' of the city, and the evaluation results (according to Table 2) are shown in Figure 3. In the YRD, cities in a high coordination state were concentrated in the northern, central, and southern regions, while cities in the eastern and western regions were mostly in a moderate coordination or low coordination state. In the MYR, most cities were in a moderate coordination or low coordination state, with high coordination cities concentrated in the northern and central areas. The northern, central, and southern parts of CC exhibited high coordination. The megacity Chongqing in the eastern region of CC, located in the mountainous regions of China, was densely populated with highly concentrated settlements and industrial activities. During the process of rapid economic development, conflicts often arose between policies aimed at protecting the environmental quality of mountainous areas and the demands of economic growth, resulting in relatively low overall development coordination [49]. In the BTH, cities in a high coordination state were mainly concentrated in the central and southern areas, with Beijing being a prime example of high coordination in the region. As the capital, Beijing experienced rapid economic growth, with indicators such as GDP per capita and per capita disposable income ranking the highest in the BTH region. As its natural environment improved gradually, Beijing maintained a good livability and achieved a high level of coordinated development. This is partly related to its strict population control policy adapted in the Beijing Master Plan (2016–2035) [50]. In the PRD, most cities were in a high coordination state, primarily Guangzhou and Shenzhen located in the north. Zhuhai, Jiangmen and Zhongshan were in a low coordination or moderate state in the south, while Zhongshan improved to a high coordination state in 2020.

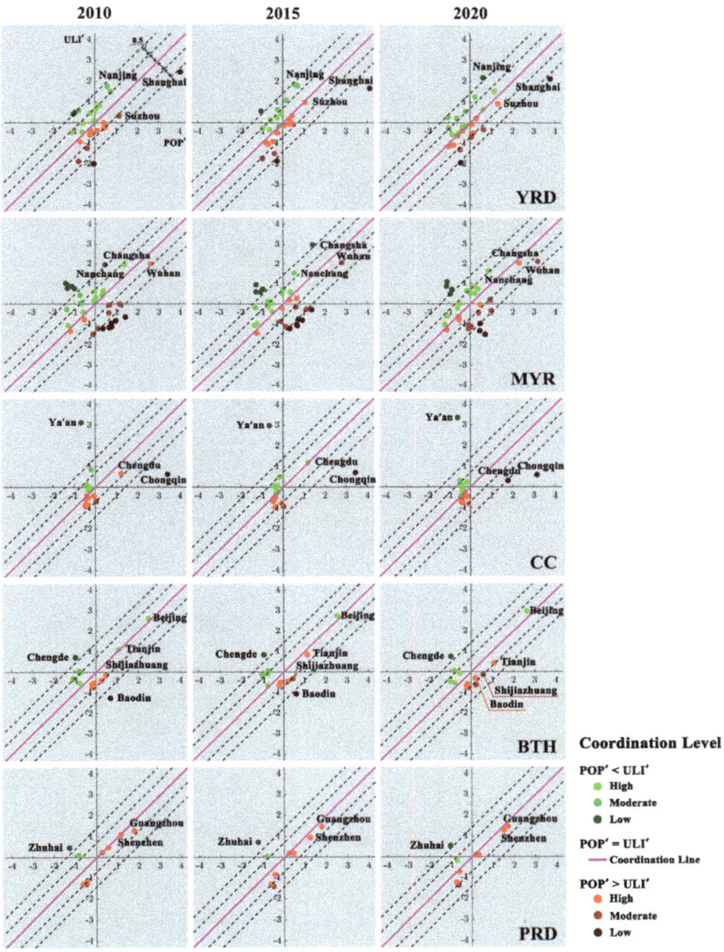

Figure 2. Scatter plots based on the standardized value of *ULI* and *POP* for each urban agglomeration in 2010, 2015, and 2020. (The greenish dots indicate that their populations are high compared to the level of urban livability; the reddish dots indicate their populations are low compared to the level of urban livability; and the purple line indicates absolute coordination between population and urban livability).

There were variations in the spatial distribution of coordination among cities within the urban agglomerations. In the YRD and CC, the central regions had better coordination, while the eastern and western regions had relatively lower coordination. The coordination distribution in MYR was more fragmented, with some cities in the northeast, central, and southern regions showing higher coordination. In BTH, the southern region had higher coordination, while the northern region had lower coordination. In contrast, the northern region in PRD had better coordination, while the southern region had lower coordination.

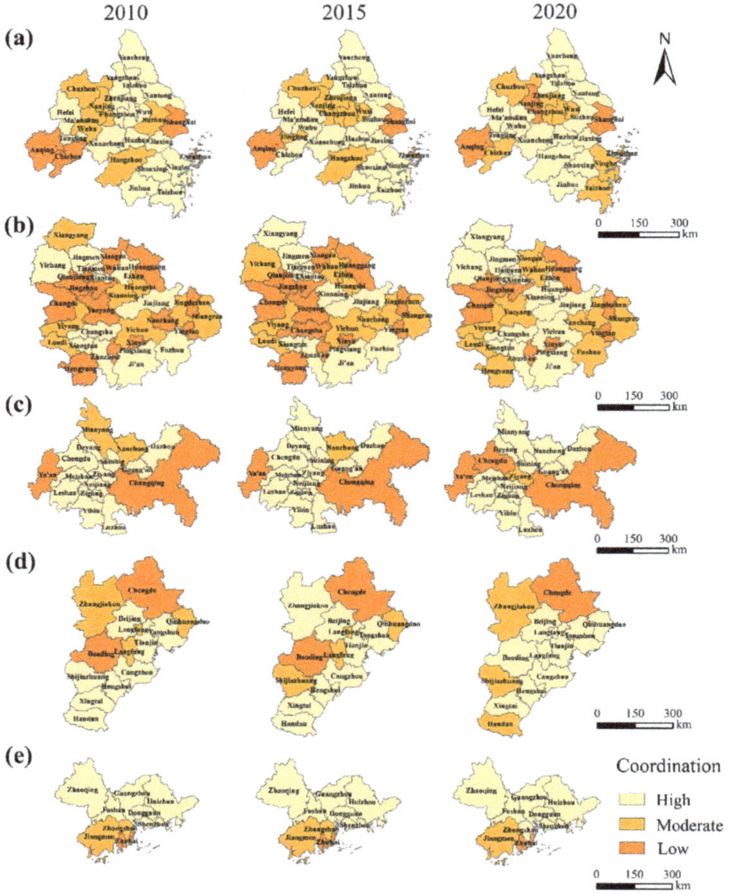

Figure 3. The coordination level of cities in the five urban agglomerations in 2010, 2015, and 2020 ((**a**) YRD, (**b**) MYR, (**c**) CC, (**d**) BTH, (**e**) PRD).

3.2. Coordination Levels of Urban Agglomerations

At the urban agglomeration scale, the *UACDI* was calculated based on the *CDI* and area of each city (using Formula (7)), and the results are shown in Figure 4. The index *UACDI* represents the accumulated gaps of all cities within the urban agglomeration to the absolute coordination. Therefore, the larger *UACDI* value, the lower the coordination level of the urban agglomeration. In addition, the area of the cities is also considered in the calculation of *UACDI*, as larger cities have a greater impact on the coordination of the entire urban agglomeration. Among the five urban agglomerations, the PRD exhibited the highest coordination level (indicated by the lowest *UACDI* value), whereas the CC urban agglomeration demonstrated the lowest level of coordination (reflected by the highest *UACDI* value) across all years. The YRD, BTH, and MYR regions ranked second, third, and fourth, respectively, in terms of coordination level. The *UACDI* value for the CC region was

1.039 in 2010, declining to 0.991 in 2015 and increasing to 0.996 in 2020. Despite improving from a low to a moderate coordination level, the CC's *UACDI* value remained relatively high compared to the other urban agglomerations. This is mainly because its key city Chongqing maintained relatively high *CDI* values, with its area constituting 34% of the total CC region. This significant contribution leads to CC's overall higher *UACDI* values from 2010 to 2020. The MYR region's *UACDI* value decreased to 0.609 in 2020, indicating moderate coordination but still above the critical threshold of 0.5, because many cities, including Nanchang and Wuhan, failed to achieve high coordination. The BTH region experienced a slight improvement in coordination level from 2010 to 2020. Cities, such as Beijing and Tianjin, had achieved a high level of coordination in 2020. However, the BTH region's *UACDI* value remained relatively high, suggesting that substantial efforts are still required to enhance the coordination level of individual cities, e.g., Shijiazhuang and Chengde. Both the YRD and PRD regions maintained relatively lower *UACDI* values compared to the other urban agglomerations and showed significant improvement. By 2020, although cities like Shanghai in the YRD had lower coordination, the overall coordination level was high because most cities, such as Hangzhou and Suzhou, had achieved high coordination. In contrast, most cities in the PRD, such as Guangzhou and Shenzhen, had reached a high level of coordination, resulting in a higher overall coordination level for the PRD.

Overall, between 2010 and 2020, the coordination levels in the PRD, YRD, BTH, and MYR regions showed continuous improvement. In contrast, the CC region's coordination level improved between 2010 and 2015 but then deteriorated slightly in 2020. The PRD experienced the most significant improvement in coordination development, while the CC saw the least improvement.

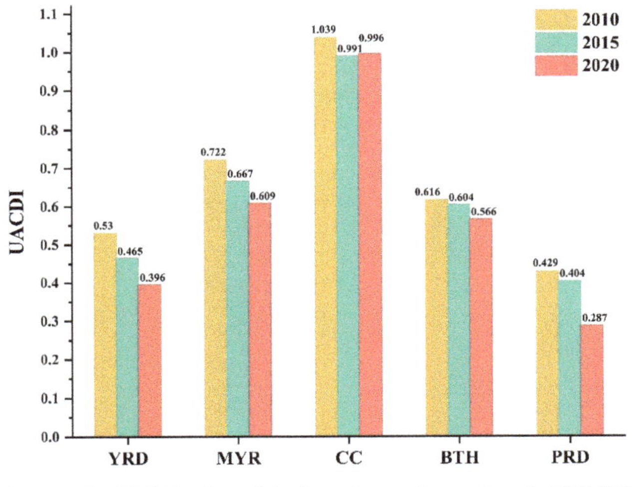

Figure 4. The *UACDI* values of the five urban agglomerations in 2010, 2015, and 2020.

4. Discussion

4.1. Indicators on Urban Livability Evaluation

In this study, the indicators used to evaluate urban livability are from two dimensions: social–economic development (SD) and ecosystem service (ES). We calculated the normalized indicators for each dimension across the urban agglomerations in the years 2010, 2015, and 2020 (see Figure 5). Specifically, the urban agglomeration YRD exhibited high levels of social–economic development, ranking first among the five urban agglomerations in 2020. This was the main driver of the rapid growth in its urban livability. However, regarding ES, significant efforts are needed for further improvement, such as strengthening environmental protection and nature conservation. The urban agglomeration MYR demonstrated relatively small improvements in the aspects of ES and it ranked second among the five urban agglomerations in 2020, but its urban livability remained low compared to the others. It needs to promote economic development and enhance livability to better meet the needs of the population [51]. From 2010 to 2020, the urban agglomeration CC made progress across all two dimensions. However, these improvements were relatively modest and could not accommodate the large population, leading to limited growth of *ULI* and low coordination. In 2020, due to the large proportion of urban green/blue space (UGS), CC's ES ranked first. For the BTH region, the indicator of SD was the key factor in maintaining a high level of *ULI*. The overall *ULI* showed an upward trend, meeting the current demands of the population. From 2010 to 2020, its SD indicator increased significantly, while ES saw less growth, showing similar development patterns to the other urban agglomerations. In the PRD region, the levels of SD and ES were relatively high compared to the other urban agglomerations, which produced the highest *ULI* in 2020. This allowed the PRD region to support a large population.

Many existing studies on urban livability suggest that the evaluation of urban livability encompasses multiple dimensions, with the natural environment making a significant contribution to livability and public green spaces serving as a key factor [16,28,52]. Additionally, the cultural environment, living standards [52,53], and residents' socioeconomic attributes also have a major impact on perceived livability [28]. In assessing socioeconomic development, this study incorporates not only GDP measures but also factors such as urban construction land, the number of hospital beds per 10,000 people, and the number of enrolled students per 10,000 people. Regarding the ecological environment, the evaluation considers not only green spaces and water resources but also includes carbon sequestration and carbon emissions to provide a comprehensive assessment of livability.

The economic development in the coastal regions (i.e., BTH, YRD and PRD) is relatively faster than that located in the central and western regions. Our results show that there is considerable potential for enhancing urban livability by focusing on the development of urban green spaces and carbon sinks in the future development of coastal regions. The amount of urban green spaces is an important factor influencing the ES, as these spaces are associated with NPP assessments and serve as key carbon sinks. This makes the ES value in the CC and MYR regions higher than the coastal urban agglomerations. But their SD indicators are lower. It shows that the urban agglomerations in the western regions have made progress in environmental protection, but their economic development still needs further improvement to meet population demands.

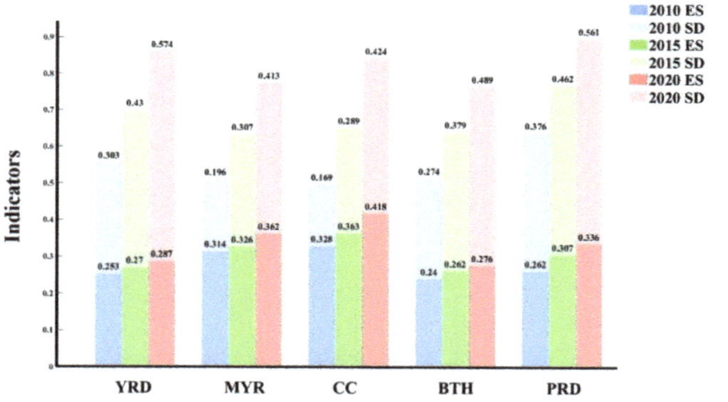

Figure 5. Change in indicators regarding urban livability for the five urban agglomerations in 2010, 2015, and 2020.

4.2. Limitations

The proposed *CDI* and *UACDI* are relative measures rather than absolute values for coordination. They are applied to compare the relative coordination state among the cities in the urban agglomeration and provide an assessment across the five urban agglomerations. They could be much more useful and meaningful in a long-term process for monitoring the development of the urban livability and population dynamic. As the study area includes county-level cities, the collection of certain data (such as fossil fuel usage and disposable income) faced limitations. To ensure data quality, we prioritized official data and public datasets, excluding data from online sources with uncertain quality. This approach led to two issues: first, the difficulty in data collection limited the *ULI* to two essential but constrained dimensions; second, the discontinuity of the data prevented a continuous temporal study of coordination changes. Future research could use more continuous data to describe urban livability, integrate more aspects in the index system, and conduct deeper analyses of livability and population distribution within urban agglomerations, thereby achieving more reliable research results.

5. Conclusions

This study established the Coordination Distance Index to evaluate the coordination level between the urban livability and population distribution within five national urban agglomerations in China. From 2010 to 2020, the coordination levels of the Mid-Yangtze River (MYR), Beijing–Tianjin–Hebei (BTH), Yangtze River Delta (YRD) and Pearl River Delta (PRD) experienced a steady improvement. The Chengdu–Chongqing (CC) region showed initial improvements followed by slight deterioration, and its highest coordination state was in 2015. The Pearl River Delta (PRD) showed significant improvements, maintaining a high coordination state during the study period. By 2020, the PRD and YRD regions reached a high coordination state, ranking first and second, respectively, followed by the BTH, MYR and CC regions, which were in a moderate coordination state. Notably, Beijing in the BTH, Guangzhou and Shenzhen in the PRD demonstrated excellent performance of well-coordinated development, exhibiting both a large population proportion and leading urban livability within their respective agglomerations. In this study, social development (SD) is the main driver influencing urban livability, while the other indicator, ecosystem services (ESs), shows relatively small changes due to the minimal variation in UGS during the study period, and improving urban livability through the enlargement of UGS has great potential to increase the quality of urbanization.

Our results can serve as a basis for policymakers to develop targeted strategies aimed at enhancing the coordination between urban livability and population distribution. The proposed indices highlight the need for monitoring and adaptive management to sustain and improve coordination levels. Cities and regions that have achieved "high coordination" can serve as models for best practices, while those lagging behind can benefit from tailored interventions to address specific imbalances. Furthermore, the insights gained from this study can inform future urban planning initiatives, helping to ensure more sustainable and equitable urban development across China's major urban agglomerations.

Author Contributions: Conceptualization, W.H. and Y.R.; methodology, Y.R.; software, Y.R.; validation, Y.R., C.D., L.Z. and W.H.; formal analysis, J.S. and J.L.; investigation, J.S. and J.L.; resources, L.Z.; data curation, L.Z.; writing—original draft preparation, Y.R. and W.H.; writing—review and editing, W.H. and L.Z.; visualization, Y.R.; supervision, W.H.; project administration, L.Z.; funding acquisition, W.H. All authors have read and agreed to the published version of the manuscript.

Funding: This work was supported by the Fundamental Research Funds for Central Public-Interest Scientific Institution, grant number: AR2419 and AR2421.

Institutional Review Board Statement: Not applicable.

Informed Consent Statement: Not applicable.

Data Availability Statement: The data presented in this study are available on request from the corresponding author.

Conflicts of Interest: The authors declare no conflicts of interest.

References

1. Jiang, L.; O'neill, B.C. Global urbanization projections for the Shared Socioeconomic Pathways. *Glob. Environ. Chang.* **201** *42*, 193–199. [CrossRef]
2. Gu, C. Urbanization: Processes and driving forces. *Sci. China Earth Sci.* **2019**, *62*, 1351–1360. [CrossRef]
3. Sun, L.; Chen, J.; Li, Q.; Huang, D. Dramatic uneven urbanization of large cities throughout the world in recent decades. *Na Commun.* **2020**, *11*, 5366. [CrossRef] [PubMed]
4. Capps, K.A.; Bentsen, C.N.; Ramírez, A. Poverty, urbanization, and environmental degradation: Urban streams in the developin world. *Freshw. Sci.* **2016**, *35*, 429–435. [CrossRef]
5. Ye, C.; Schröder, P.; Yang, D.; Chen, M.; Cui, C.; Zhuang, L. Toward healthy and livable cities: A new framework linking publi health to urbanization. *Environ. Res. Lett.* **2022**, *17*, 064035. [CrossRef]
6. Wang, Y.; Miao, Z. Towards the analysis of urban livability in China: Spatial–temporal changes, regional types, and influencin factors. *Environ. Sci. Pollut. Res.* **2022**, *29*, 60153–60172. [CrossRef]
7. United Nations Department of Economic and Social Affairs (UNDESA). World Urbanization Prospects: The 2018 Revision. 201 Available online: https://www.un.org/en/desa/2018-revision-world-urbanization-prospects (accessed on 24 November 2024
8. Qadeer, M.A. Urbanization by implosion. *Habitat Int.* **2004**, *28*, 1–12. [CrossRef]
9. Liu, Y.; Zhang, X.; Pan, X.; Ma, X.; Tang, M. The spatial integration and coordinated industrial development of urban agglomer tions in the Yangtze River Economic Belt, China. *Cities* **2020**, *104*, 102801. [CrossRef]
10. Tan, R.; Zhou, K.; He, Q.; Xu, H. Analyzing the effects of spatial interaction among city clusters on urban growth—Case of Wuha urban agglomeration. *Sustainability* **2016**, *8*, 759. [CrossRef]
11. United Nations. Department of Economic and Social Affairs, Open Working Group Proposal for Sustainable Development Goal 2015. Available online: https://sdgs.un.org/partnerships (accessed on 24 November 2024).
12. Martins, M.S.; Fundo, P.; Kalil, R.M.L.; Rosa, F.D. Community participation in the identification of neighbourhood sustainabilit indicators in Brazil. *Habitat Int.* **2021**, *113*, 102370. [CrossRef]
13. Valentin, A.; Spangenberg, J.H. A guide to community sustainability indicators. *Environ. Impact Assess. Rev.* **2000**, *20*, 381–39 [CrossRef]
14. Paul, A.; Sen, J. Livability assessment within a metropolis based on the impact of integrated urban geographic factors (IUGFs) o clustering urban centers of Kolkata. *Cities* **2018**, *74*, 142–150. [CrossRef]
15. Paul, A. Developing a methodology for assessing livability potential: An evidence from a metropolitan urban agglomeratio (MUA) in Kolkata, India. *Habitat Int.* **2020**, *105*, 102263. [CrossRef]
16. Parker, J.; Simpson, G.D. Public green infrastructure contributes to city livability: A systematic quantitative review. *Land* **201** *7*, 161. [CrossRef]
17. Martino, N.; Girling, C.; Lu, Y. Urban form and livability: Socioeconomic and built environment indicators. *Build. Cities* **202** *2*, 220–243. [CrossRef]
18. Yang, J.; Wu, T.; Gong, P. Implementation of China's new urbanization strategy requires new thinking. *Sci. Bull.* **2017**, *62*, 81–8 [CrossRef]
19. Ministry of Civil Affairs of the People's Republic of China. Statistical Table of Administrative Divisions of the People's Republi of China. Available online: http://xzqh.mca.gov.cn/statistics/ (accessed on 24 November 2024).
20. Fang, C.; Yu, D. Urban agglomeration: An evolving concept of an emerging phenomenon. *Landsc. Urban Plan.* **2017**, *162*, 126–13 [CrossRef]
21. Douglass, M. From global intercity competition to cooperation for livable cities and economic resilience in Pacific Asia. *Enviro Urban.* **2002**, *14*, 53–68. [CrossRef]
22. Geng, Y.; Fujita, T.; Bleischwitz, R.; Chiu, A.; Sarkis, J. Accelerating the transition to equitable, sustainable, and livable citie Toward post-fossil carbon societies. *J. Clean. Prod.* **2019**, *239*, 118020. [CrossRef]
23. Henderson, J.V. Urbanization and economic development. *Ann. Econ. Financ.* **2003**, *4*, 275–342.
24. Liu, S.; Liao, Q.; Liang, Y.; Li, Z.; Huang, C. Spatio–Temporal Heterogeneity of Urban Expansion and Population Growth in Chin *Int. J. Environ. Res. Public Health* **2021**, *18*, 13031. [CrossRef] [PubMed]
25. Zhang, R.; Jiang, G.; Zhang, Q. Does urbanization always lead to rural hollowing? Assessing the spatio-temporal variations i this relationship at the county level in China 2000–2015. *J. Clean. Prod.* **2019**, *220*, 9–22. [CrossRef]
26. He, Q.; Zeng, C.; Xie, P.; Tan, S.; Wu, J. Comparison of urban growth patterns and changes between three urban agglomeration in China and three metropolises in the USA from 1995 to 2015. *Sustain. Cities Soc.* **2019**, *50*, 101649. [CrossRef]
27. Seto, K.C.; Sánchez-Rodríguez, R.; Fragkias, M. The new geography of contemporary urbanization and the environment. *Ann Rev. Environ. Resour.* **2010**, *35*, 167–194. [CrossRef]
28. Zhan, D.; Kwan, M.-P.; Zhang, W.; Fan, J.; Yu, J.; Dang, Y. Assessment and determinants of satisfaction with urban livability i China. *Cities* **2018**, *79*, 92–101. [CrossRef]

Zhang, Y.; Li, Y.; Chen, Y.; Liu, S.; Yang, Q. Spatiotemporal heterogeneity of urban land expansion and urban population growth under new urbanization: A case study of Chongqing. *Int. J. Environ. Res. Public Health* **2022**, *19*, 7792. [CrossRef]
Hossein, A.; Sarkari, F.; Borhani, M. Coupling coordination analysis between urbanization and ecology in Iran. *Front. Urban Rural. Plan.* **2024**, *2*, 5.
Cai, J.; Li, X.; Liu, L.; Chen, Y.; Wang, X.; Lu, S. Coupling and coordinated development of new urbanization and agro-ecological environment in China. *Sci. Total Environ.* **2021**, *776*, 145837. [CrossRef]
Lv, T.; Wang, L.; Zhang, X.; Xie, H.; Lu, H.; Li, H.; Liu, W.; Zhang, Y. Coupling coordinated development and exploring its influencing factors in Nanchang, China: From the perspectives of land urbanization and population urbanization. *Land* **2019**, *8*, 178. [CrossRef]
Wang, D.; Jiang, D.; Fu, J.; Lin, G.; Zhang, J. Comprehensive Assessment of Production–Living–Ecological Space Based on the Coupling Coordination Degree Model. *Sustainability* **2020**, *12*, 2009. [CrossRef]
Xu, Z.; Yin, Y. Regional Development Quality of Yangtze River Delta: From the Perspective of Urban Population Agglomeration and Ecological Efficiency Coordination. *Sustainability* **2021**, *13*, 12818. [CrossRef]
Zeng, P.; Wei, X.; Duan, Z. Coupling and coordination analysis in urban agglomerations of China: Urbanization and ecological security perspectives. *J. Clean. Prod.* **2022**, *365*, 132730. [CrossRef]
Jiang, S.; Zhang, Z.; Ren, H.; Wei, G.; Xu, M.; Liu, B. Spatiotemporal characteristics of urban land expansion and population growth in Africa from 2001 to 2019: Evidence from population density data. *ISPRS Int. J. Geo-Inf.* **2021**, *10*, 584. [CrossRef]
Regueiro-Ferreira, R.M.; Alonso-Fernández, P. Ecological elasticity, decoupling, and dematerialization: Insights from the EU-15 study (1970–2018). *Ecol. Indic.* **2022**, *140*, 109010. [CrossRef]
Cao, Y.; Zhang, Z.; Fu, J.; Li, H. Coordinated Development of Urban Agglomeration in Central Shanxi. *Sustainability* **2022**, *14*, 9924. [CrossRef]
Hepburn, C.; Qi, Y.; Stern, N.; Ward, B.; Xie, C.; Zenghelis, D. Towards carbon neutrality and China's 14th Five-Year Plan: Clean energy transition, sustainable urban development, and investment priorities. *Environ. Sci. Ecotechnol.* **2021**, *8*, 100130. [CrossRef]
Xiao, R.; Yu, X.; Xiang, T.; Zhang, Z.; Wang, X.; Wu, J. Exploring the coordination between physical space expansion and social space growth of China's urban agglomerations based on hierarchical analysis. *Land Use Policy* **2021**, *109*, 105700. [CrossRef]
Yang, J.; Huang, X. 30 m annual land cover and its dynamics in China from 1990 to 2019. *Earth Syst. Sci. Data* **2021**, *13*, 3907–3925. [CrossRef]
Zhang, Q.; Kong, Q.; Zhang, M.; Huang, H. New-type urbanization and ecological well-being performance: A coupling coordination analysis in the middle reaches of the Yangtze River urban agglomerations, China. *Ecol. Indic.* **2024**, *159*, 111678. [CrossRef]
Li, X.; Lu, Z. Quantitative measurement on urbanization development level in urban Agglomerations: A case of JJJ urban agglomeration. *Ecol. Indic.* **2021**, *133*, 108375. [CrossRef]
Cramer, W.; Field, C.B. Comparing global models of terrestrial net primary productivity (NPP): Introduction. *Glob. Chang. Biol.* **1999**, *5*, 1–15. [CrossRef]
Christopher, G. Potential scale-related problems in estimating the costs of CO_2 mitigation policies. *Clim. Chang.* **2000**, *44*, 331–349.
Jin, J.; Li, M.; Jin, L. Data normalization to accelerate training for linear neural net to predict tropical cyclone tracks. *Math. Probl. Eng.* **2015**, *2015*, 931629. [CrossRef]
Patro, S.G.K.; Sahu, K.K. Normalization: A preprocessing stage. *arXiv* **2015**, arXiv:1503.06462. [CrossRef]
Abdi, H.; Lynne, J. Williams. Normalizing data. *Encycl. Res. Des.* **2010**, *1*, 935–938.
Liu, Y.; Yue, W.; Fan, P.; Zhang, Z.; Huang, J. Assessing the urban environmental quality of mountainous cities: A case study in Chongqing, China. *Ecol. Indic.* **2017**, *81*, 132–145. [CrossRef]
Fu, C.; Zhang, H. Evaluation of Urban Ecological Livability from a Synergistic Perspective: A Case Study of Beijing City, China. *Sustainability* **2023**, *15*, 10476. [CrossRef]
Lv, T.; Geng, C.; Zhang, X.; Hu, H.; Li, Z.; Zhao, Q. Spatiotemporal evolution and influencing factors of urban industrial carbon emission efficiency in the Mid-Yangtze River urban agglomeration of China. *Phys. Chem. Earth Parts A/B/C* **2024**, *135*, 103607. [CrossRef]
Wang, J.; Su, M.; Chen, B.; Chen, S.; Liang, C. A comparative study of Beijing and three global cities: A perspective on urban livability. *Front. Earth Sci.* **2011**, *5*, 323–329. [CrossRef]
Cao, Y.; Li, F.; Xi, X.; van Bilsen, D.J.C.; Xu, L. Urban livability: Agent-based simulation, assessment, and interpretation for the case of Futian District, Shenzhen. *J. Clean. Prod.* **2021**, *320*, 128662. [CrossRef]

sustainability

MDPI

Article

Evaluating Perceived Cultural Ecosystem Services in Urban Green Spaces Using Big Data and Machine Learning: Insights from Fragrance Hill Park in Beijing, China

Lingbo Fu [1] Hongpeng Fu [2,*] and Chengyu Xiong [1]

[1] School of Cultural Industries Management, Communication University of China, Beijing 100024, China; fulingbo@cuc.edu.cn (L.F.); xiongcy@scu.edu.cn (C.X.)
[2] Khoury College of Computer Science, Northeastern University, Seattle, WA 98122, USA
* Correspondence: fu.hongp@northeastern.edu

Abstract: Cultural ecosystem services (CESs) are essential for the sustainable development and management of urban green spaces. However, there remains a gap in leveraging big data and unsupervised machine learning to comprehensively evaluate perceived CESs. This study introduces a hybrid research methodology integrating latent dirichlet allocation (LDA) and importance–performance analysis (IPA) to analyze 20,087 user-generated reviews of Fragrance Hill Park in Beijing from Meituan. The key findings are the following: (1) ten types of CESs were identified, including five related to personal well-being, four to public well-being, and one bridging both categories; (2) the most significant dimensions were "recreational activities", "aesthetic appreciation", "physical well-being", and "mental well-being"; (3) users expressed positive sentiments toward "history and culture", "mental well-being", and "religious engagement", while "social relations" received the most negative feedback; (4) IPA results highlight "recreational activities" and "aesthetic appreciation" as priority areas for improvement. This study provides a scalable, data-driven framework for evaluating CESs in urban green spaces. The insights gained can inform urban green space management and policy decisions to enhance user experiences and promote sustainable urban development.

Keywords: cultural ecosystem services; cultural landscapes; latent dirichlet allocation; importance–performance analysis; sustainable development goals

1. Introduction

In recent decades, international frameworks, such as the United Nations' Sustainable Development Goals (SDGs), emphasize the importance of inclusive and sustainable urban environments (SDG 11) and the conservation of ecosystems (SDG 15). The importance of ecosystem services (ESs) has gained widespread recognition as a key tool for sustainable decision making to improve human well-being [1]. Well-being is a complex and multidimensional concept that encompasses the quality of life of individuals and communities, including material, social, psychological, and health dimensions [2]. Key factors influencing well-being include health, social connections, and environmental quality, as reflected in OECD indicators [3]. Among these, environmental quality, particularly within natural and cultural landscapes, is central to improving individual well-being [4]. ES bridges human well-being and the environment by providing benefits derived from nature, fostering a mutually beneficial relationship between people and ecosystems [5]. Among these, cultural ecosystem services (CESs) stand out for offering intangible benefits such as spiritual enrichment, cognitive growth, recreation, and aesthetic experiences. These services emerge from the dynamic and complex relationship between people and their environments, significantly contributing to human well-being [1,6]. However, compared to the other ESs, the value of CESs remains underexplored in many regions, limiting their integration into global and local policy frameworks. Therefore, evaluating how people perceive CESs is essential for enhancing well-being and improving the management of urban green spaces [7].

Globally, CESs have been recognized as a vital contributor to well-being in an urban context. Previous studies have demonstrated that CESs contribute to stress reduction [8], cognitive enhancement [9], and increased opportunities for physical activities [10]. For instance, McElwee et al. [11] carried out a literature review and found that concepts such as well-being and resilience have been linked to many types of CESs in the Global South. In Russia and China, studies conducted by Dushkova et al. [12] explored the CESs of urban parks in Russia (three public parks in Moscow) and China (six public parks in Xi'an) and found that they enhance quality of life and attract tourists. Additionally, Krsnik and Illán-Fernández [13] examined Murcia, Spain, and highlighted the importance of CES in urban park management and long-term city strategies. Cultural landscapes, characterized by their dynamic evolution and human influence, are increasingly recognized as critical providers of CESs and promote well-being [14]. These landscapes blend natural elements—such as maple, pine, and cypress trees—with cultural features, including traditional architecture and historical relics. This seamless integration of nature and culture creates exemplary cultural landscapes that offer both ecological benefits and recreational opportunities, enriching human experiences and fostering a deeper connection to heritage [15]. Despite the importance of such landscapes, challenges in evaluating CESs arise due to their intangible nature [16,17]. While research has focused on natural landscapes like wetlands [18] and cultural landscapes like urban green spaces [19], studies specifically examining forested urban landscapes that blend natural and cultural attributes remain limited.

Scholars have mainly adopted traditional data sources, such as structured surveys and interviews, to gather data and assess CESs [17,20,21]. For example, Bryce et al. [22] carried out an internet-based survey of 1220 individuals, uncovering several perceived CESs that positively impact the overall well-being of humans. Additionally, Riechers et al. [23] conducted a study that included 558 users of Berlin, discovering that cultural ecosystem services can be perceived through bundles and that those bundles may have a negative influence on each other. Nonetheless, researchers found that these structured surveys and interviews may fail to accurately reflect the values of perceived CESs because they presuppose that all CESs are provided and perceived in a given ecological environment [24,25]. Furthermore, collecting data through traditional data sources has higher costs, offers limited scalability, and is typically restricted to small sample sizes, which hinders their application on a larger scale [26]. In today's digital age, user-generated content (UGC) from platforms like TripAdvisor, Ctrip, and Meituan offers an alternative approach [27–29]. The user-generated content (UGC) of these platforms has become a valuable big data source for Destination Marketing Organizations (DMOs) to gain a bottom-up understanding of how users perceive CESs and to make strategic decisions. However, a comprehensive assessment of perceived CESs from UGC is still underexplored.

The principles of the Data-Intensive Scientific Discovery Paradigm emphasize the use of big data to uncover patterns and generate actionable insights, which provide the potential to extract perceived CESs from UGC. Often referred to as the Fourth Paradigm [30], this framework represents a fundamental shift in scientific inquiry driven by the exponential growth of data and advances in computational power. Unlike traditional approaches that rely on hypothesis-driven experimentation or observational studies, this paradigm prioritizes using large, complex datasets to uncover patterns, generate insights, and enable evidence-based decision making. Its primary benefits include scalability, efficiency, and the ability to analyze diverse data sources, such as user-generated content (UGC), to capture nuanced patterns and perceptions. LDA, an unsupervised machine learning technique, facilitates the extraction of latent themes from large-scale unstructured textual data, capturing the nuanced and multidimensional aspects of CES perception. Compared to widely used supervised classification methods, which face limitations such as the time and cost inefficiencies of annotating large datasets, dependency on prior knowledge that may introduce biases, and a lack of model transparency [24,31], LDA offers a more efficient and unbiased alternative. Researchers in other fields, including transportation [32], information systems [33], and nursing [34], have utilized the unsupervised classification method to investigate how users view specific values to harness the potential of self-generated, volunteer-authored material that distinguishes itself from surveys that pose specific inquiries [35]. IPA complements LDA by translating these themes into actionable strategies, enabling decision-makers to prioritize CES elements that enhance human well-being. This combined approach bridges the gap between data exploration and practical application, offering a robust and scalable framework for CES evaluation. Despite its potential, the integration of LDA and IPA to classify and analyze perceived CESs remains underexplored.

To address the gaps, our study utilizes a combination of latent dirichlet allocation (LDA) and importance–performance analysis (IPA) to analyze perceived CESs from UGC of Fragrance Hill Park from Meituan. Specifically, this study aims to answer the following questions:

(1) What cultural ecosystem services do users perceive in urban green spaces?
(2) How do users evaluate these cultural ecosystem services?
(3) Which cultural ecosystem services should park managers prioritize for improvement?

The insights can help destination managers and environmental organizations develop strategies that balance natural and cultural elements, enhancing user experiences and promoting sustainable urban development [36]. Our study contributes to the literature by applying innovative methods to identify and prioritize CESs, offering practical guidance for managing urban green spaces effectively. Although this study focuses on Fragrance Hill Park in Beijing, the findings have broader implications for urban green spaces worldwide. Many cities, mainly in Europe and the USA, face similar challenges in integrating natural and cultural elements into urban parks.

2. Literature Review

2.1. Perceived Cultural Ecosystem Services

Cultural ecosystem services (CESs) have been defined as intangible and nonmaterial benefits people derive from the ecosystem [1]. CESs are crucial for maintaining cultural identity [37], enhancing human well-being [38], and supporting both physical and mental health [39,40]. Understanding CESs requires exploring and assessing how users perceive these services. This understanding aids in recognizing the value individuals attribute to different aspects of CESs and in identifying ways to enhance services that are less perceived [41]. The perception of CESs represents a dynamic process, where ecosystems offer potential benefits and humans interpret these benefits based on their social norms, experiences, and capabilities [26,42]. Specifically, perceived CESs refers to how individuals feel and interpret the cultural values of their environment, representing a subjective assessment of CESs [43,44]. Perceived CESs emphasizes a people-centric approach by quantifying preferences and perceptions [45,46].

With the knowledge that perceived CESs have an essential influence on people's well-being and the way they view the ecosystem, scholars have been paying attention to assessing perceived CESs from the natural landscape to the cultural landscape to better manage and improve CESs of a place [26,47]. The definition of cultural landscape has been discussed by different agencies, such as UNESCO, The National Park Service, and The Cultural Landscape Foundation, with a focus on human influences on nature [14,48]. The major difference between natural landscape and cultural landscape is whether the landscape is imprinted by human activities. Scholars have explored how users perceive CES in natural landscapes [9,49]. For example, Richards and Friess [50] used MaxEnt 3.4 and content analysis methods to explore users' perceived CESs of four mangroves in Singapore. Zoderer et al. [43] adopted a photo-based survey to identify perceived CESs in the Alpine region and found that land use type, gender, and environmental engagement play important roles. When cultural benefits became important in recent years, scholars' attention has been shifted to how users perceived CESs in the cultural landscape. For example, Langemeyer et al. [51] adopted a survey to determine the perceived CESs in a cultural landscape, and they discovered that a higher emphasis on place values was associated with less intensive management practices, but values connected to tourism were higher in areas that included cultural facilities in their land use. Furthermore, Zhang et al. [19] conducted a comparative study in Huanhuaxi Urban Forest Park to explore the perceived CES between the elderly and the youth. While the importance of balancing nature and culture in tourism sites for destination marketing has been well acknowledged [52], there has been limited focus on a particular landscape that offers both natural and cultural benefits. Nevertheless, there is an immediate need to recognize the perceived value of such a combination of natural and cultural landscapes [48].

Previous research has made contributions to exploring the perceived CES through unidimensional assessment. The formation process of the perception of CESs hinges on the importance of attributes to users, which means that some CESs are placed as more important or useful than other CES values [19]. For example, Stępniewska [53] adopted the survey method and ranked recreational and aesthetic CES values to be the most important perceived CESs for users. Riechers et al. [54] employed a survey methodology and discovered that older users showed a preference for perceived CES values associated with natural experiences, but younger urban dwellers tended to favor perceived CES values that promote social contacts. Furthermore, Bertram and Rehdanz [55] conducted a study in which they examined four cities and discovered that there are resemblances among these towns in terms of their opinions towards perceived CES values. While most methods focus on assessing the importance of different CES values, a multidimensional approach is advocated [24,25]. However, there are not many multidimensional evaluations adopted in the exploration and assessment of perceived CESs.

2.2. Methods to Explore the Perceived Cultural Ecosystem Services

Previous research mainly explored the perceived CESs through traditional data sources while evaluating perceived CESs. Interviews, focus groups, and surveys are the most adopted traditional methods in exploring the perception of CESs [24]. For example, Pleasant et al. [56] adopted surveys and interviews in the landscape management field and found that CES benefits people beyond just the spatial scale. Also, Riechers et al. [23] used problem-centered interviews with experts to find that MEA's standard of CES values should be tailor made to regional contexts to help the policymakers make wiser decisions. Dou et al. [57] also utilized 250 semi-structured interviews and a focus group consisting of 26 local stakeholders, finding that wetlands are seen as having more significant functions than other types of landscapes when it comes to delivering different ecosystem services

(CESs). However, researchers found that traditional data sources such as surveys, interviews, and focus groups yield limited data introducing potential biases and hindering representativeness. As we entered into platform society [58], users now share their experiences and opinions about a destination on digital platforms [16,59]. Big data approaches help researchers cover a wider range of participants and generate more accurate results and insights in environmental design [51,60]. Users convey their opinions, beliefs, and ideas about a place, sharing latent attributes of their perceived CESs online. For instance, Li et al. [61] used social media text data to identify the CESs experienced by citizens and found that five out of nine CESs were identified. Oteros-Rozas et al. [62] performed a comprehensive analysis of a large dataset consisting of 1404 social media photo and their findings revealed that landscapes with wood pastures and grasslands were strongly associated with cultural heritage, as well as social and spiritual qualities. However, a comprehensive assessment of perceived CESs based on big data sources remains limited.

The methods of deriving sub-categories in perceived CESs can be categorized into two types: supervised classification and unsupervised classification. The difference between a supervised classification and an unsupervised classification is whether it involves predefined categories [63]. The supervised classification approach can provide a greater depth of understanding in under-explored areas compared to simply quantitative investigations [23] because they involve a predefined set of knowledge, categories, and datasets. For example, Nowak-Olejnik et al. [49] adopted a survey method in the Pieniny Mountains, southern Poland, and found that different communities vary significantly in terms of perceived CESs based on existing CESs. Also, Csurgó and Smith [10] conducted interviews in two rural areas in Hungary with local stakeholders and found that cultural heritage is the most important perceived CES and is in central relation to social and symbolic values. The unsupervised machine learning approach does not rely on predefined categories or labels, which helps to reduce subjectivity and lower the cost of data collection, making it more flexible in exploring and discovering hidden patterns and structures in big data (Song et al., 2021 [64]). This is particularly valuable when working with large amounts of unlabeled text data or in research areas that are not yet fully explored with enough datasets. There is a lack of research on the use of unsupervised machine learning in understanding perceptions and preferences. For instance, Song et al. [64] gathered 20,476 online evaluations from 18,387 consumers of TripAdvisor between 2008 and 2019 and they identified 30 separate subjects that were connected to the Strip in Las Vegas. However, there is limited research that employs unsupervised machine learning methods to categorize and examine the perceived CESs for the natural–cultural landscape.

3. Materials and Methods

3.1. Study Area

Fragrance Hill Park (https://xiangshanpark.com/en/, accessed on 10 December 2024), located in Beijing's Haidian District at the foothills of the Western Hills, spans 1,880,000 square meters (Figure 1). Renowned for its seamless blend of natural beauty and cultural heritage, the park is one of Beijing's most iconic scenic attractions. It features diverse vegetation, including forests of maple, pine, and cypress trees, which create a breathtaking display of red leaves each autumn. The park also hosts over 5800 rare plant species, adding to its ecological and botanical significance.

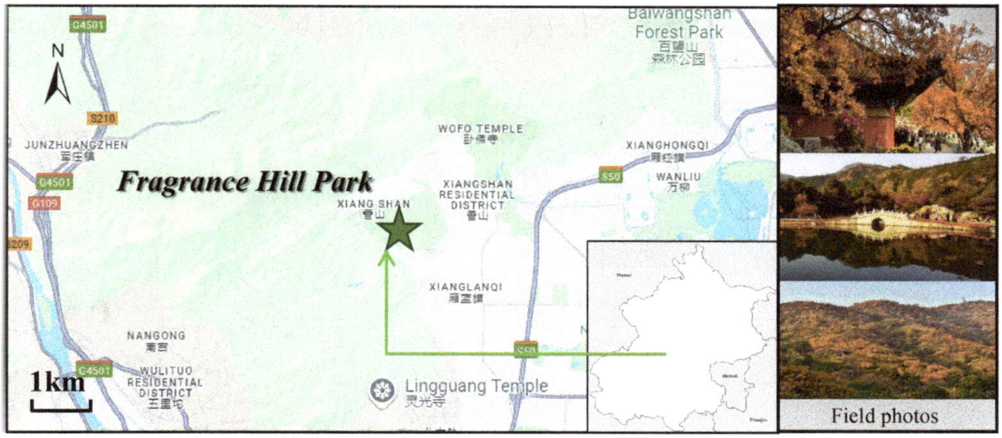

Figure 1. Location of Fragrance Hill Park.

The history of Fragrance Hill Park dates back to the Tang Dynasty (618–907), when it was initially established as an imperial garden. It was later expanded during the Jin (1115–1234), Yuan (1271–1368), and Ming (1368–1644) Dynasties, with significant development occurring during the reign of Emperor Qianlong of the Qing Dynasty (1735–1796). During this period, many of the park's notable structures were built, including the Fragrant Temple (Xiangshan Temple) and the Bright Temple (Zhao Temple), showcasing exquisite traditional Chinese architecture.

Opened to the public in 1956, the park is now equipped with a variety of facilities catering to different user needs. Hiking enthusiasts can explore trails leading to Xianglu Peak, the park's highest point at 557 m, offering stunning panoramic views of Beijing. Scenic cable cars provide a more relaxed way to ascend the peak. Families and casual users can enjoy picnic areas, teahouses, and tranquil gardens. Educational resources include botanical gardens and an information center detailing the park's ecological and historical significance. Additionally, the park hosts the annual Red Leaves Festival, drawing large crowds to experience the vibrant autumn scenery. Modern amenities ensure a comfortable visit for over 5 million annual guests, with visitor centers, accessible pathways, and guided tours. In 2001, Fragrance Hill Park was designated as an AAAA-level scenic area by the National Tourism Administration of China. It is also a proud member of the World-Famous Mountain Association, affirming its global recognition.

To understand how users perceive Fragrance Hill Park's CESs, this study used the LDA and the IPA approach and collected 20,087 travel reviews from Meituan from 2010 to 2020. The LDA method was applied to extract perceived CES topics from these reviews. These topics were then classified from a cultural ecosystem services perspective. Subsequently, the study conducted frequency and sentiment analysis to further assess these topics. Finally, the IPA was used to evaluate each topic, aiming to provide actionable implications to improve the users' experience at the urban green space (Figure 2).

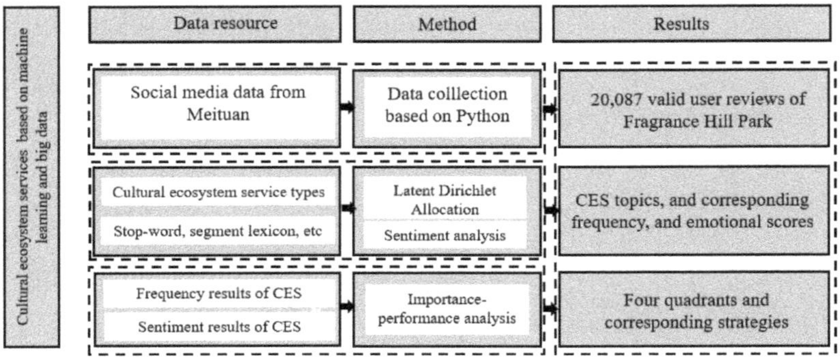

Figure 2. Research Framework.

3.2. Data Collection

User-generated content reviews have become readily accessible on the internet, particularly in an era defined by the pervasive sharing of experiences online [58]. In this study, user reviews from Meituan.com, China's largest online travel platform [65], were chosen as the primary data source. Meituan (http://www.meituan.com, accessed on 10 December 2024) is a platform renowned as the largest mainland China-based travel website, enabling its members to share their travel experiences and insights [66]. It provides abundant travel information and tips, convenient booking services, user evaluation, and communication functions. The registered users have reached 470 million, with daily traffic exceeding 10 million times. Therefore, collecting online review data of a natural-cultural landscape with this website can fully ensure the reliability and representativeness of the data from the sample sources.

With the help of Python 3.5 crawl, user-generated reviews are crawled and stored in a text file. To ensure data quality and integrity, we meticulously filtered out duplicate entries and irrelevant data, including those with empty content, advertisements, and other non-contributory content. This curation process culminated in acquiring a robust dataset comprising 20,087 valid user reviews, forming the foundation of our subsequent research. This study relies on user-generated content from social media platforms, so our review data reflect the perceptions of all users engaging with the park's CES.

3.3. Cultural Ecosystem Services Extraction Based on Latent Dirichlet Allocation

The Latent Dirichlet Allocation model is a widely used topic modeling method for uncovering latent topics in a collection of documents [67], which has been commonly used to extract the important dimensions of the products or services from online reviews [59,68]. In the context of this study, users express their beliefs, ideas, and travel experiences through online blogs, effectively conveying their perceptions of CESs. Each UGC review represents a composite of latent dimensions associated with CESs, captured through the unique vocabulary chosen by users [69]. In our context, the dimensions of CESs can be referred to as "topics", which are derived from the LDA model. The implementation of LDA to extract CES dimensions involves two steps, including preprocessing online reviews and extracting topics using LDA.

First, data preprocessing serves as a pivotal step in ensuring data cleanliness and coherence, encompassing several key processes. These include Chinese word segmentation, the elimination of stop words, both high and low-frequency words, as well as the consolidation of synonyms. To initiate the process, we employed Jieba for Chinese word segmentation. Subsequently, an initial stop-word list was compiled, combining both Baidu's stop-word list and a Chinese stop-word list. Following this, a word frequency analysis was performed, leading to the removal of insignificant, high-frequency words. Additionally, low-frequency words, specifically those identified within the context of perception dimensions, were also added to the final stop words list. In the last step, we conducted a comprehensive analysis of words with similar meanings within a thesaurus, facilitating semantic deduplication and integration. This meticulous data processing regimen ensured the elimination of noise and enhanced the quality of the 711 final dataset.

Second, this study used the gensim toolkit of Python to conduct the LDA model. Determining the optimal number of LDA dimensions is a crucial step. Following the previous studies, we adopted a method based on coherence and perplexity metrics [70]. The coherence metric emphasizes the statistical similarity among terms within a topic, reflecting the uncertainty of assigning a text piece to a specific topic. On the other hand, perplexity gauges the model's ability to fit new data [71]. Previous studies suggest that coherence scores in the range of 0.4 to 0.6 generally indicate meaningful and interpretable topics [72]. To attain the optimum number of topics, this study experimented with different quantities, ranging from 1 to 20, ultimately settling on a 10-topic model backed by a 0.56 coherence score and low perplexity, which aligned with the state-of-the-art performance in previous topic modeling research [72]. We employed Python's data visualization library, Matplotlib 3.8, to create a line graph illustrating the relationship between the number of dimensions and the confusion level. This graph enabled us to discern the inflection point and identify the optimal number of dimensions for the LDA model. Each extracted topic was assigned a meaningful name based on its representative words. Subsequently, the chosen number of dimensions was meticulously tested to ensure that there was no overlap or ambiguity among various topics.

3.4. Cultural Ecosystem Services Evaluation Based on Importance–Performance Analysis

Importance–performance analysis (IPA) is a widely recognized analytical method designed to evaluate the significance of performance-related factors within the context of specific products or services [73]. IPA serves as a valuable tool for identifying dimensions requiring improvement or offering opportunities for cost-saving measures, all while minimizing the impact on quality [74,75]. It has found extensive application across marketing, tourism, and management domains [68,76]. This study applied the IPA framework to evaluate the perceived CES dimensions derived from the LDA model, following a three-step process: calculating the importance score, calculating the performance score, and constructing the IPA matrix.

First, importance scores were derived from word frequency analysis, a widely used method for quantifying public interest in textual data. Frequent terms reflect tangible public focus, making word frequency an appropriate proxy for importance. Following previous studies, this study utilized Python to conduct the word frequency of each topic as the importance indicator [77,78]. With the results of LDA dimension classification, the perception dimension corresponding to each review can be calculated. The score is obtained by summing the probability of the CES appearing in each review and then dividing by the total number of reviews:

$$I_j = \frac{\sum_{i=1}^{n} P_{j_i}}{n} \quad (1)$$

In Formula (1), $\sum_{i=1}^{n} P_{j_i}$ represents the sum of the probabilities of the CES j appearing in all reviews, and n is the total number of reviews.

Second, this study estimates the performance of topics through sentiment analysis, which evaluates users' subjective opinions of each CES dimension [68]. Following previous studies [77,78], by calculating the topics from LDA and its corresponding emotional value for each comment, we can derive the average emotional value for each CES type as the performance score. As for the sentiment analysis, this study employs the authoritative

Simplified Chinese Sentiment Lexicon Ontology from Mainland China. Developed and released by the Information Retrieval Laboratory at Dalian University of Technology [79], this lexicon serves as a comprehensive dictionary for Simplified Chinese emotional vocabulary in mainland China's social media. Based on Ekman's emotions, the lexicon is categorized into 21 subcategories, with each emotional word marked for polarity, part of speech, and emotional intensity. The part of speech is categorized into seven types: noun, verb, adjective, adverb, online slang, idiom, and prepositional phrase. Each word is assigned a value: positive words are marked as 1, neutral words as 0, and negative words as −1 [65]. The emotional intensity labels range from 1, 3, 5, 7, to 9, with level 1 representing the lowest intensity and level 9 the highest. The lexicon encompasses 27,466 emotional words, and using this vocabulary, the emotional tendency of a text is calculated. This structured approach allows for a precise and methodologically defined classification of "positive" and "negative" emotions in the analysis. The calculation formula is as follows:

$$\text{Sentiment}(d) = \begin{cases} \text{positive}\sum_i^n \text{Sentiment}_{\text{positive}}(i) + \sum_i^n \text{Sentiment}_{\text{neutral}}(i) + \sum_i^n \text{Sentiment}_{\text{negative}}(i) > 0 \\ \text{positive}\sum_i^n \text{Sentiment}_{\text{positive}}(i) + \sum_i^n \text{Sentiment}_{\text{neutral}}(i) + \sum_i^n \text{Sentiment}_{\text{negative}}(i) = 0 \\ \text{positive}\sum_i^n \text{Sentiment}_{\text{positive}}(i) + \sum_i^n \text{Sentiment}_{\text{neutral}}(i) + \sum_i^n \text{Sentiment}_{\text{negative}}(i) < 0 \end{cases} \quad (2$$

In Formula (2), d is the text d, i is word i. Sentiment(d) is the sentiment score of the text d. $\sum_i^n \text{Sentiment}_{\text{positive}}(i)$ is the accumulative value of n positive emotion words. $\sum_i^n \text{Sentiment}_{\text{neutral}}(i)$ is the accumulative value of n neutral emotion words. $\sum_i^n \text{Sentiment}_{\text{negative}}(i)$ is the accumulative value of n negative emotion words.

Finally, the IPA framework enables the classification of CES topics into four distinct quadrants based on their importance and performance levels. Each CES dimension was mapped onto the matrix based on its importance and performance scores. Following the previous studies, the four-quadrant matrix of importance–performance was constructed using mean values of all importance and performance values of online reviews as the coordinate origin [80,81]. As shown in Figure 3, the vertical coordinate represents performance and the horizontal coordinate represents importance. The four quadrant matrices are quadrant I, "high importance–high performance" (advantage), quadrant II, "low importance–high performance" (possibly overkill), quadrant III, "low importance–low performance" (low priority), quadrant IV, "high importance–low performance" (concentrate here).

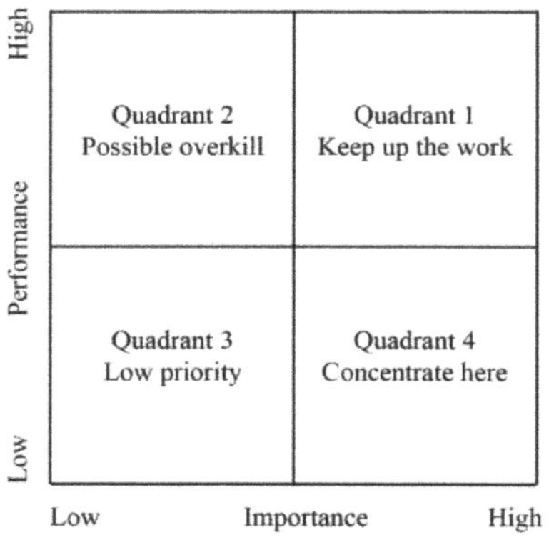

Figure 3. Four quadrants of importance–performance analysis.

4. Results

4.1. Perceived Cultural Ecosystem Services of Users

The results of the LDA model show that users mainly perceived ten dimensions of cultural ecosystem services (Table 1). Users mainly perceived five CES-related personal well-being and four CES-related public well-being dimensions. As a CES related to both social well-being and personal well-being, social relations are also perceived. Five dimensions associated with personal well-being, encompassing aspects such as "religious engagement", "recreational activities", "mental wellness", "physical health", and "aesthetic appreciation", are mostly perceived by users. In addition, four dimensions pertain to public well-being, covering air quality, habitat and biodiversity, climate regulation, and historical and cultural aspects. Although these dimensions are mentioned less frequently, they underscore the vital role of Fragrant Hill Park in fostering social well-being. "Social relations" straddles the domains of both personal and public well-being. Despite its lower frequency of mention, this dimension reveals the social needs of users in facilitating interaction opportunities among users and between users and the environment.

Table 1. Results of destination image of Fragrance Hill Park.

No.	Topics	Relation to Human Well-Being	Representative Keywords
Topic 1	Religion	Personal well-being	Incense burner (0.028), Fragrant Hills Temple (0.009), Azure Cloud Temple (0.008)
Topic 2	Recreational activities	Personal well-being	Climb down the mountain (0.03), cable car (0.029), Climb the mountain (0.024), climb up the mountain (0.02)
Topic 3	Mental well-being	Personal well-being	Mood (0.03), relaxation (0.016), nature (0.012), comfortable (0.008)
Topic 4	Air quality	Social well-being	Breathing (0.026), fresh air (0.021)
Topic 5	Habitat and biodiversity	Social well-being	Yellow horn (0.011), maple tree (0.006), leaves (0.006), frost leaves (0.006), maple leaves (0.005), Ginkgo (0.005)
Topic 6	Climate regulation	Social well-being	Ancient trees (0.009), climate (0.008), refreshing (0.005), summer retreat (0.005)
Topic 7	Physical well-being	Personal well-being	Mountain climbing (0.072), exercise (0.052), sports (0.018)
Topic 8	Social relations	Social well-being/ personal well-being	Colleagues (0.005), convenience (0.005), best friends (0.005)
Topic 9	Aesthetic appreciation	Personal well-being	Red leaves (0.2), maple leaves (0.029), viewing (0.012), beautiful (0.011), gorgeous (0.008)
Topic 10	History and culture	Social well-being	Architecture (0.01), Diligent Governance Hall (0.007), history (0.012), Glass Pagoda (0.005)

"Recreational activities", and "aesthetic appreciation" are the most perceived CESs by users, while "air quality", "climate regulation", and "social relations" are the least perceived CESs. "Recreational activities" emerged as the most frequently referenced dimension, occupying a significant 52% of all extracted dimensions. This category was further differentiated into four attributes, namely, "climb down the mountain", "cable car", "climb the mountain", and "climb up the mountain". It is evident that users predominantly visit Fragrance Hill Park for recreational purposes. "Aesthetic appreciation" represents the second most important dimension of Fragrance Hill Park's CESs, constituting 21% of all dimensions. This dimension consists of "red leaves", "maple leaves", "viewing", "beautiful", and "gorgeous". It reflects the motivation of users to appreciate the natural beauty of Fragrance Hill Park, particularly its renowned maple trees in the fall. Conversely, "air quality", "climate regulation", and "social relations" are the least perceived CESs of Fragrance Hill Park, each garnering relatively less attention from users.

4.2. Importance and Performance Assessment of Cultural Ecosystem Services

Users perceived highly on CES related to personal well-beings, but low on CES related to public well-beings. The results indicated that "recreational activities" (37.2%), "aesthetic appreciation" (19.3%), "physical well-being" (10.0%), and "mental well-being" (9.5%) as the top four most crucial dimensions (Figure 3), which are all related to person well-beings (Figure 4). However, "air quality" (4.1%), "climate regulation" (3.0%), and "social relations" (2.8%) are perceived the least important dimensions, which are all related to public well-beings.

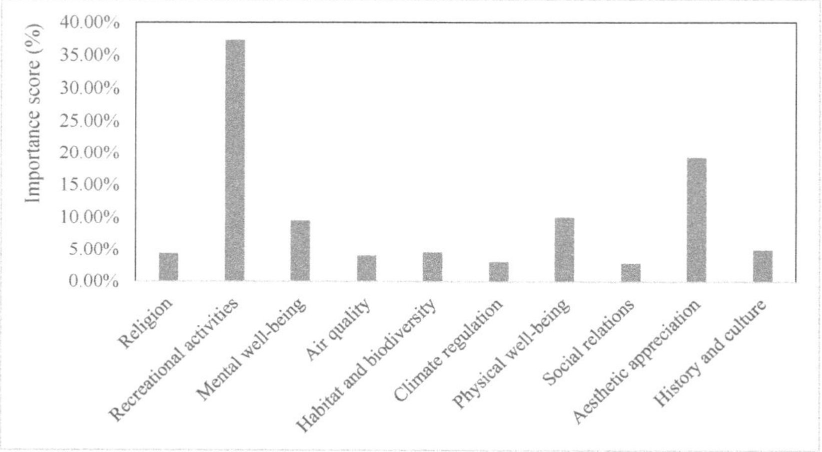

Figure 4. Importance scores of CES.

As for the performance assessment (Figure 5), "mental well-being" and "history and culture" are identified as the most positively perceived dimensions, whereas "recreational activities", "social relations", and "aesthetic appreciation" are the most negatively perceived CESs. "History and culture" stands out with the highest proportion of positive emotion, proving that users hold a significant appreciation for the CES value of Fragrance Hill Park. In addition, "mental well-being", "religions", "air quality", "habitat and biodiversity", and "climate regulation" all exhibit positive proportions exceeding 70.0%, which indicates that users not only have mental needs met from the trip but find solace and enjoyment in the natural surroundings while exploring Fragrance Hill. In contrast, "social relations" emerges as the dimension with the highest proportion of negative emotions, underscoring the influence of user-to-user interactions on users' attitudes. Analysis of reviews displaying a higher proportion of negative emotions suggests that dimensions leading to low user satisfaction might be attributed to pessimistic social relations during the trip. Notably, "recreational activities" follows with the second-highest proportion of negative emotions, accounting for 16.4%. This finding points to specific aspects within this dimension that might require attention to enhance the overall user experience.

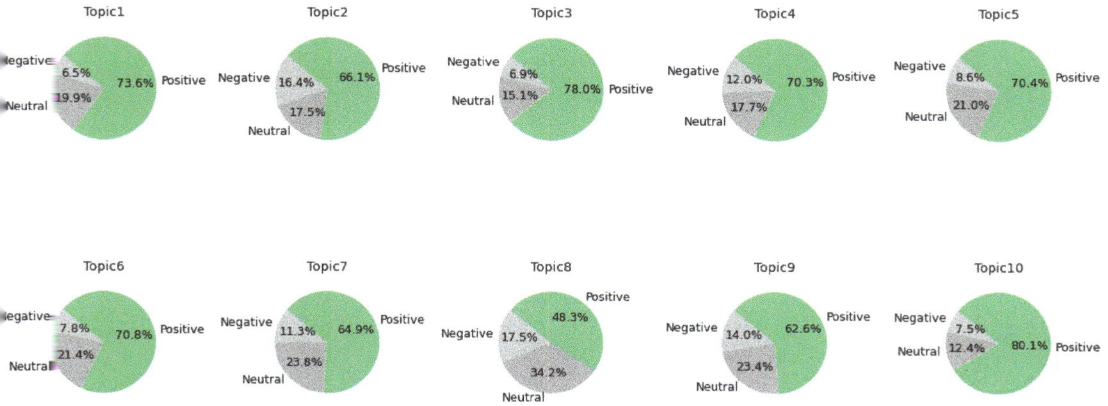

Figure 5. Performance scores of CESs.

When considering the optimization priorities based on importance or satisfaction, divergent focal points emerge for Fragrance Hill Park's enhancement. Focusing solely on importance suggests prioritizing improvements in "recreational activities", "aesthetic appreciation", "physical well-being", and "mental well-being", as these dimensions are integral to the park's core offerings and crucial for the user's holistic experience. On the flip side, when only satisfaction levels are considered, "recreational activities", "social relations", and "aesthetic appreciation" demand immediate attention due to their lower satisfaction scores, despite their significant impact on the overall user sentiment. This discrepancy points to the need for a balanced approach that weighs both the importance of each dimension to the user's experience and the current level of satisfaction, ensuring that enhancements are both impactful and responsive to user feedback.

4.3. Importance–Performance Analysis of Perceived Cultural Ecosystem Services Topics

Six dimensions, including all of the CESs related to public welfare and two CESs related to personal welfare, are perceived in lower importance and high performance and fall into the "possible overkill" quadrant (Figure 6). These dimensions, namely "history and culture", "mental well-being", "religion", "climate regulation", "habitat and biodiversity", "air quality", and "physical well-being", exhibit a descending performance score order. This suggests that the destination management organization of Fragrance Hill Park has invested considerable effort into enhancing these aspects of their services, resulting in higher performance levels. However, the majority of users do not perceive these factors as relatively important in shaping their destination image. Hence, these dimensions should receive less attention in future efforts.

Quadrant IV encompasses two dimensions, which are "recreational activities" and "aesthetic appreciation". users consider these dimensions as highly important, but their performance levels fall below satisfaction levels. These dimensions serve as critical areas for enriching and enhancing the CESs of Fragrance Hill Park, offering potential to increase user satisfaction and foster positive emotional connections with the destination. The unique mountain-specific resources, characterized by diversity, access difficulty, fragility, and aesthetic qualities call for a harmonious fusion of both natural and cultural elements to provide exceptional recreational activities and aesthetic appreciation opportunities.

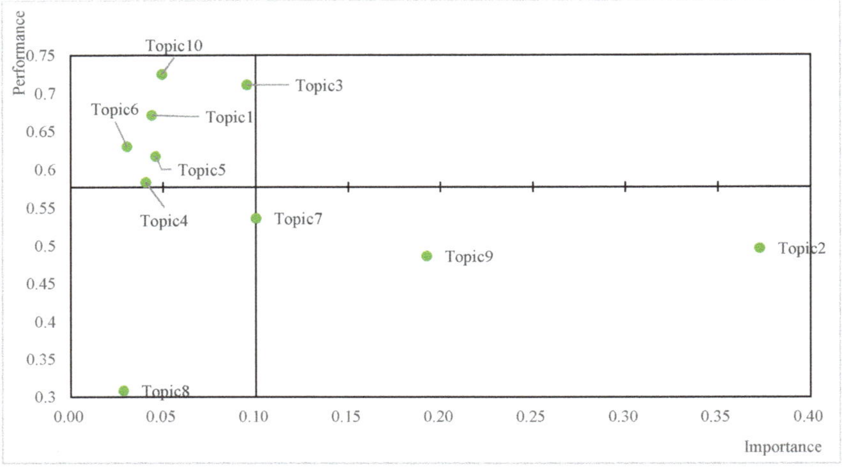

Figure 6. Importance–performance analysis of CES.

Quadrant III features two dimensions, which are "physical well-being" and "social relations". Users consider these dimensions as low importance and performance. These dimensions indicate that users initially visit Fragrance Hill Park for physical exercise purposes, making these dimensions relatively less important in the overall perception of the destination.

Lastly, the absence of any dimension in quadrant I implies that there is still room for improvement within the domain of destination management organization, highlighting key dimensions that warrant continued attention and maintenance.

5. Discussion

5.1. Perceived Cultural Ecosystem Services of Natural-Cultural Heritage

The results of this study demonstrate the multifaceted nature of perceived cultural ecosystem services (CES) in Fragrance Hill Park, an urban forest landscape that blends natural and cultural elements. The analysis of user-generated content (UGC) identified ten dimensions of CESs, with "recreational activities", "aesthetic appreciation", and "physical well-being" emerging as the most prominent. These findings align with existing literature emphasizing the importance of recreation, aesthetic experiences, and physical health benefits in urban green spaces [82]. The prominence of "recreational activities" suggests that the diversity and quality of these activities hold significant importance for users [53]. This dimension serves as a measure of the utility users derive from the destination, influenced by a multitude of factors, including climate conditions and natural resources [83]. Recreational activities, such as climbing down/up the mountain, taking photos, and visiting historical sites, are usually spatially diverse due to the park's natural conditions. Similarly, "aesthetic appreciation" signifies the substantial attention users attribute to the natural scenery of Fragrance Hill Park. Words like "red leaves", "maple leaves", "viewing", "beautiful", and "gorgeous" underscore the emphasis on the aesthetic aspect of the destination. This aligns with research suggesting that aesthetic experiences in natural environments foster emotional well-being and positive attitudes [84]. Fragrance Hill Park is famous for its red maples in autumn and has been sustained by hosting the Red Maple Festival between October and November for several decades, a successful event that draws a multitude of excited users annually. For the "physical well-being" dimension, keywords are "mountain climbing", "exercise", "sports", and so on. Previous research has consistently affirmed the significant role of physical exercise in improving well-being and health [85,86]. It is, therefore, unsurprising that users frequent this destination for physical exercise. Fulfilling physical well-being needs is an essential motivation for users in mountain-based tourism, and it aligns with their desire for health-promoting activities [87,88].

There are small variations in users' emotions towards different dimensions of Fragrance Hill Park. The "history and culture" dimension shows the highest proportion of positive emotions, indicating the predominantly positive sentiments users express towards the architectural, historical, and cultural elements of Fragrance Hill. This positive disposition aligns with the established acknowledgment that users tend to view historical and architectural sites with enthusiasm and appreciation [89]. The "mental well-being" dimension emerges as the second-highest positive emotion, characterized by keywords like "mood", "relaxation", "nature", and "comfortable", which indicates that users often seek refuge in natural environments to escape the demands of daily life and endeavor to lead a healthy, chill, and refresh lifestyle [90–92]. In contrast, "social relations" stands out with the highest proportion of negative emotions, underscoring the essential role of user-to-user interactions in shaping the quality of the trip [47,93]. This also could reflect the challenges associated with interactions among users in the park. Possible reasons for this could include overcrowding, lack of social spaces, or the diversity of user groups with differing expectations and behaviors. Overcrowding, lack of seating areas for socializing, and misalignment between users' social preferences and park design could have contributed to the negative perceptions. To improve this aspect, park management could focus on creating more designated social spaces, improving signage for group activities, and fostering community-driven events. These efforts could enhance the quality of user interactions and reduce negative emotional responses associated with social relations in the park. Also, "recreational activities" ranked the second highest negative emotions, which indicates that recreational activities are what make a memorable outdoor experience, echoing back on research on how experience affects user experiences and emotions [88,94,95]. Moreover, "aesthetic appreciation" ranked as the third negative emotion, indicating that users varied in aesthetic appreciation based on their individual experiences, aesthetic literacy, emotions, etc. [96].

In perceived CES assessment, there is an ongoing debate regarding the application of different evaluative dimensions. Recent studies underscore that user satisfaction and perceived importance are not always aligned, which presents a complex scenario for management decisions. This study highlights the inconsistencies in CES evaluation results when different dimensions are considered separately. For instance, the relevance of physical and mental well-being in urban green spaces suggests that these attributes are of paramount importance for users [91]. In contrast, research by Langemeyer et al. [51] suggests that while importance is a critical driver of user experience, satisfaction—or the lack thereof—can significantly influence the perceived quality and value of CES.

5.2. Utilizing Unsupervised Machine Learning from User-Generated Text to Assess Perceived Cultural Ecosystem Services

The increasing use of mixed methodologies in CES research is essential for capturing the multifaceted nature of these services. Recent studies have emphasized the benefits of combining participatory methods with spatial analysis or environmental modeling to better understand the complex interplay between humans and their environment. It acknowledges the multifaceted nature of CESs, capturing both tangible and intangible benefits derived from ecosystems. Mixed-methods research offers a comprehensive framework to evaluate perceived CESs. For example, Plieninger et al. [97] advocate for an approach emphasizing that it allows for the triangulation of data, thereby enhancing the validity of the results. More recent studies have applied this notion by integrating participatory methods, such as stakeholder interviews, with spatial analysis or environmental modeling, to gain a multidimensional understanding of perceived CESs. Furthermore, advances in machine learning and natural language processing have also opened new possibilities for using large datasets. Integrating these cutting-edge computational methods with traditional survey data can yield insights into the patterns and drivers of CES values across different landscapes [7]. The notion of "hybrid methods" that combine econometric techniques with participatory GIS is highlighted as a means to capture the economic valuation of CES while also considering spatial and social dimensions [18]. This blend of approaches can help policymakers and managers prioritize efforts based on a more comprehensive valuation of CESs [98]. The study builds on these foundations, leveraging the combined strengths of mixed methodologies to offer an assessment of CESs at Fragrance Hill Park. By integrating different methods, the study could more accurately capture the wide array of benefits that the urban green space provides to individuals and the community, spanning from mental well-being and social relations to physical health and recreational opportunities.

The unsupervised machine learning based on user-generated data is beneficial to clear the articulation and assessment of CESs considering the inherent intangibility, multidimensionality, and experiential nature of CESs. User-generated data from social media data are beneficial for precisely characterizing user evaluations. Social media data are a valuable source of information about today's conceptualization of information sharing and the cultural creation process relative to the natural environment [99], such as the Weibo tags we often use, and reviews from the public. Sharing personal opinions and pictures on social media platforms reflects the collective values of society, helping researchers to understand the shared image cognition of people towards the ecosystem, and is powerful evidence of the continuous existence or extinction of ecosystem services. At the same time, the openness of social media platforms may also encourage people to construct the same values in different ways [100]. Traditional surveys also have a strong guiding nature. For example, when being asked about the purpose of going to the park, respondents may pay more attention to purposeful options, and services that subtly happen like mental relaxation may be ignored. As social media platforms become more and more popular, free expression breaks through the limitations of traditional respondents, providing a place for people of different social classes to express their feelings. The complexities of quantifying perceived CESs have driven researchers to explore innovative approaches beyond traditional surveys and statistical regression. In addition, the use of unsupervised machine learning techniques offers an objective way to quantify qualitative aspects of CES by discerning themes within textual data. The robustness of LDA is confirmed by metrics that reflect the degree to which a model meaningfully extracts themes from unstructured big data. For instance, McPhearson et al. [101] utilized LDA in their analysis of urban green space and found that it effectively identified the variety of services and values that communities associate with these areas. Additionally, Jiang et al. [19] leveraged unsupervised machine learning to explore social media content, revealing public perceptions and values related to urban green space, indicating the method's capability to reduce subjective biases. The reported coherence measure (CV) in this context would indicate a moderate level of topic extraction success, aligning with benchmarks suggested in studies [102]. This reflects the potential of unsupervised learning in CESs in the future.

The integration of the LDA approach and the IPA approach is also beneficial for cost effectiveness. By categorizing identified CESs into four distinct quadrants that reflect varying evaluation situations and optimization priorities, researchers can effectively target improvements where they are most needed. Similarly, Brown and MacLeod [37] used IPA to map and evaluate the performance of ESs, highlighting the importance of considering both local community priorities and the actual state of ecosystem service provision. Our framework is particularly adept at informing the marketing and management strategies of natural–cultural landscapes such as heritage sites and urban green space, as it not only identifies the gaps between users' expectations and their experiences but also suggests the most cost-effective areas for improvement.

5.3. Managerial and Policy Implication for the Cultural Ecosystem Services of Natural Cultural Heritage

Based on the latent dirichlet allocation (LDA) and importance–performance analysis (IPA) of Fragrance Hill Park's perceived cultural ecosystem services (CESs), this study offers actionable recommendations for improving the management of urban green spaces. The findings are not only applicable to Fragrance Hill Park but also transferable to other urban parks worldwide, particularly as cities increasingly seek to balance natural preservation with recreational and cultural offerings.

From a marketing perspective, it is imperative to make good use of a range of marketing applications to construct a comprehensive CES that transcends the dichotomy between nature and culture [36,103]. Since recreational activities emerged as one of the most important CES, it is crucial for Destination Marketing Organizations (DMOs) and environmental service providers to prioritize the enhancement of recreational offerings. The study suggests that, beyond activities like hill climbing, soft adventure programs should be introduced to cater to diverse user preferences [92]. This could include the development of themed hiking and trekking trails that highlight the park's cultural and natural heritage, providing both physical exercise and educational opportunities. Additionally, fitness programs, such as yoga in scenic areas or guided walks, should be offered to further diversify the recreational experiences available to park users [91].

Aesthetic experiences, another highly valued CES, should also be prioritized. Managers must preserve and enhance the park's visual appeal through both seasonal events and artistic interventions. Continuing the Red Maple Festival is one example of maintaining the seasonal beauty that attracts users. Expanding these events to include art exhibitions or cultural performances that reflect the park's historical significance can further enrich the user experience. Moreover, creating interactive art installations or outdoor galleries within the park would deepen users' engagement, allowing them to connect emotionally and intellectually with the landscape [96]. This approach would also foster aesthetic literacy, enhancing users' appreciation of the park's unique combination of natural and cultural features. In addition to these initiatives, DMOs should leverage social media platforms to create a dynamic, interactive community around the park. This can foster engagement, allowing users to share experiences, exchange tips, and provide feedback. Social media can also serve as a powerful tool for promoting events and activities, thereby increasing awareness and participation. By regularly monitoring social media and user-generated content (UGC), park managers can adapt marketing strategies and update CES offerings to maintain relevance and meet evolving user expectations [65].

Moreover, establishing a continuous feedback loop is essential for ensuring that the park's services remain aligned with user needs. Annual surveys, real-time feedback mechanisms, and UGC analysis can provide valuable insights into user satisfaction and highlight areas for improvement. For instance, users could be encouraged to share feedback about specific recreational offerings or aesthetic experiences, allowing for targeted adjustments in park management. This adaptive management approach will help ensure that the park remains a vibrant and engaging space for both locals and tourists, while continuously improving based on real-time data.

By focusing on these actionable recommendations, urban green space managers can enhance CES delivery, creating spaces that not only meet recreational needs but also foster emotional and intellectual connections with users. These strategies are designed to improve the overall quality of urban green spaces, ensuring their long-term sustainability and enhancing the well-being of users.

6. Conclusions

Taking Fragrance Hill Park as an example, this study provides a comprehensive analysis of the perceived CESs in cultural landscapes, including exploring the dimensions of CESs, understanding how users perceive these services, and assessing their relative importance and performance. By applying innovative methodologies such as LDA and IPA based on the UGC, this research contributes to the growing body of literature on urban green spaces and CESs, offering new insights into the CESs in cultural landscapes.

This study integrates the IPA framework with the LDA method, addressing the principles of the Data-Intensive Scientific Discovery Paradigm. Through mining UGC, the research identified ten types of CESs, including five related to personal well-being, four to public well-being, and one bridging both categories, reflecting public perceptions of these ten CES types. The use of LDA for CES identification ensures objectivity, avoiding the subjective biases often introduced by predefined CES categories. IPA results highlight "recreational activities" and "aesthetic appreciation" as priority areas for improvement. Six dimensions, including all of the CESs related to public welfare and two CESs related to personal welfare, are perceived as lower importance and high performance and fall into the "possible overkill" quadrant.

This study also extracts the different dimensions of CESs from both importance and performance. Our findings underscore the significance of recreational activities and aesthetic appreciation as the most valued CESs by users, highlighting areas where performance improvements are necessary to meet user expectations. We also identified that dimensions such as history and culture, mental well-being, and religions predominantly evoke positive emotions, whereas social relations generated the highest levels of negative emotion, signaling potential areas for intervention. These insights are critical for improving both the user experience and the management of urban forests, where balancing natural preservation with the delivery of cultural services is essential for sustainability.

The practical implications of this study are wide-ranging. For Destination Marketing Organizations (DMOs) and environmental service providers, our findings offer valuable guidance in crafting targeted marketing strategies and tourism experiences that enhance user satisfaction and promote sustainable tourism. Specifically, park managers can focus on developing and expanding recreational facilities such as hiking trails and themed cultural events, ensuring that these experiences align with the park's natural environment and cultural heritage. This approach will not only improve both the quantity and quality of recreational opportunities available to park users. Moreover, understanding the emotional impacts of CESs can assist park managers in adapting their strategies to address both the positive and negative emotional responses identified in the study. Also, investments should be made into aesthetic improvements by preserving the park's natural beauty, enhancing visual experiences, and promoting cultural heritage. Initiatives like maintaining seasonal festivals, planting aesthetically pleasing flora, and integrating artistic and cultural elements into the park's design can enhance the aesthetic appeal and users' emotional connections to the space. A dynamic and responsive marketing strategy will help monitor evolving user perceptions, ensuring that Fragrance Hill Park remains a valued and evolving urban forest landscape.

Despite its contributions, this study has certain limitations. First, the CES dimensions are based on a single dataset at a specific point in time, which may not fully capture evolving user perceptions. Long-term data collection is needed to track changes over time. Second, the reliance on Meituan.com introduces platform bias, as its user base may not represent diverse demographics. Future research should incorporate data from multiple platforms, such as TripAdvisor and C-trip, to ensure a more comprehensive and balanced dataset. Third, the absence of user demographic information limits the ability to differentiate CES perceptions among groups based on factors like age or socioeconomic status. Future studies should explore these differences using survey data or advanced machine learning methods for deeper insights. Finally, the use of big data and machine learning may raise ethical concerns, including data privacy, informed consent, and algorithmic bias. Future research should adopt ethical data practices and fairness-aware algorithms to ensure transparency and inclusivity.

Author Contributions: Conceptualization, L.F. and H.F.; methodology, H.F.; formal analysis, L.F. and H.F.; investigation, L.F. and H.F.; writing—original draft preparation, L.F. and H.F.; writing—review and editing, L.F. and H.F.; supervision, C.X.; funding acquisition, L.F. All authors have read and agreed to the published version of the manuscript.

Funding: This research was supported by "the Fundamental Research Funds for the Central Universities", grant number CUC230D031.

Institutional Review Board Statement: Not applicable.

Informed Consent Statement: Not applicable.

Data Availability Statement: Data supporting conclusions of this article will be made available by authors on request.

Conflicts of Interest: The authors declare no conflicts of interest.

References

MEA. *Ecosystems and Human Well-Being*; Island Press: Washington, DC, USA, 2005; Volume 5.

Uysal, M.; Sirgy, M.J.; Woo, E.; Kim, H.L. Quality of Life (QOL) and Well-Being Research in Tourism. *Tour. Manag.* **2016**, *53*, 244–261. [CrossRef]

Shek, D.T. Protests in Hong Kong (2019–2020): A Perspective Based on Quality of Life and Well-Being. *Appl. Res. Qual. Life* **2020**, *15*, 619–635. [CrossRef]

Wójcik-Raźniak, A.; Raźniak, P. Are Pueblos Mágicos Really Magic? Tourism Development Program in the Context of the Quality of Life of Town Residents. *Land* **2021**, *10*, 1342. [CrossRef]

Chan, K.M.A.; Satterfield, T.; Goldstein, J. Rethinking Ecosystem Services to Better Address and Navigate Cultural Values. *Ecol. Econ.* **2012**, *74*, 8–18. [CrossRef]

Kosanic, A.; Petzold, J. A Systematic Review of Cultural Ecosystem Services and Human Wellbeing. *Ecosyst. Serv.* **2020**, *45*, 101168. [CrossRef]

Bieling, C.; Plieninger, T. Recording Manifestations of Cultural Ecosystem Services in the Landscape. *Landsc. Res.* **2013**, *38*, 649–667. [CrossRef]

Grahn, P.; Stigsdotter, U.A. Landscape Planning and Stress. *Urban For. Urban Green.* **2003**, *2*, 1–18. [CrossRef]

Gidlow, C.J.; Jones, M.V.; Hurst, G.; Masterson, D.; Clark-Carter, D.; Tarvainen, M.P.; Smith, G.; Nieuwenhuijsen, M. Where to Put Your Best Foot Forward: Psycho-Physiological Responses to Walking in Natural and Urban Environments. *J. Environ. Psychol.* **2016**, *45*, 22–29. [CrossRef]

Csurgó, B.; Smith, M.K. The Value of Cultural Ecosystem Services in a Rural Landscape Context. *J. Rural Stud.* **2021**, *86*, 76–86. [CrossRef]

McElwee, P.; He, J.; Hsu, M. Challenges to Understanding and Managing Cultural Ecosystem Services in the Global South. *Ecol. Soc.* **2022**, *27*, 23. [CrossRef]

Dushkova, D.; Ignatieva, M.; Konstantinova, A.; Yang, F. Cultural Ecosystem Services of Urban Green Spaces. How and What People Value in Urban Nature? In *Advanced Technologies for Sustainable Development of Urban Green Infrastructure, Proceedings of Smart and Sustainable Cities 2020, Moscow, Russia, 8–10 July 2020*; Vasenev, V., Dovletyarova, E., Valentini, R., Cheng, Z., Calfapietra, C., Inostroza, L., Leuchner, M., Eds.; Springer International Publishing: Cham, Switzerland, 2021; pp. 292–318.

Kosnik, G.; Illán-Fernández, E.J. Assessing Indicators and Preferences of Cultural Ecosystem Services in Urban Areas: A Case Study of Murcia, Spain. *Landsc. Ecol.* **2024**, *39*, 190. [CrossRef]

14. Jones, M. The Concept of Cultural Landscape: Discourse and Narratives. In *Landscape Interfaces: Cultural Heritage in Changing Lan scapes*; Palang, H., Fry, G., Eds.; Landscape Series; Springer: Dordrecht, The Netherlands, 2003; pp. 21–51. ISBN 978-94-017-0189-
15. Di, S.; Chen, Z.; Ren, Z.; Ding, T.; Zhao, Z.; Hou, Y.; Chen, Z. Transformation of Modern Urban Park Based on User's Spatia Perceived Preferences: A Case Study of Kowloon Walled City Park in Hong Kong. *Forests* **2024**, *15*, 1637. [CrossRef]
16. Dai, P.; Zhang, S.; Chen, Z.; Gong, Y.; Hou, H. Perceptions of Cultural Ecosystem Services in Urban Parks Based on Social Networ Data. *Sustainability* **2019**, *11*, 5386. [CrossRef]
17. Wang, W.; Wu, C.; Fang, Q.; Harrison, O.I. Cultural Ecosystem Services Evaluation in a Coastal City of China Using Social Medi Data. *Ocean Coast. Manag.* **2023**, *242*, 106693. [CrossRef]
18. Chen, M.; Lin, G. How Perceived Sensory Dimensions of Urban Green Spaces Affect Cultural Ecosystem Benefits: A Study o Haizhu Wetland Park, China. *Urban For. Urban Green.* **2023**, *86*, 127983. [CrossRef]
19. Zhang, K.; Tang, X.; Zhao, Y.; Huang, B.; Huang, L.; Liu, M.; Luo, E.; Li, Y.; Jiang, T.; Zhang, L.; et al. Differing Perceptions of th Youth and the Elderly Regarding Cultural Ecosystem Services in Urban Parks: An Exploration of the Tour Experience. *Sci. Tot Environ.* **2022**, *821*, 153388. [CrossRef] [PubMed]
20. Helmer, M.; Lipton, J.; Snitker, G.; Hackenberger, S.; Triplett, M.; Cerveny, L. Mapping Heritage Ecosystem Services in Ecologic Restoration Areas: A Case Study from the East Cascades, Washington. *J. Outdoor Recreat. Tour.* **2020**, *31*, 100314. [CrossRef]
21. Grzyb, T. Mapping Cultural Ecosystem Services of the Urban Riverscapes: The Case of the Vistula River in Warsaw, Polan *Ecosyst. Serv.* **2024**, *65*, 101584. [CrossRef]
22. Bryce, R.; Irvine, K.N.; Church, A.; Fish, R.; Ranger, S.; Kenter, J.O. Subjective Well-Being Indicators for Large-Scale Assessment c Cultural Ecosystem Services. *Ecosyst. Serv.* **2016**, *21*, 258–269. [CrossRef]
23. Riechers, M.; Barkmann, J.; Tscharntke, T. Perceptions of Cultural Ecosystem Services from Urban Green. *Ecosyst. Serv.* **2016**, *1* 33–39. [CrossRef]
24. Cheng, X.; Van Damme, S.; Li, L.; Uyttenhove, P. Evaluation of Cultural Ecosystem Services: A Review of Methods. *Ecosyst. Ser* **2019**, *37*, 100925. [CrossRef]
25. Tian, T.; Sun, L.; Peng, S.; Sun, F.; Che, Y. Understanding the Process from Perception to Cultural Ecosystem Services Assessmen by Comparing Valuation Methods. *Urban For. Urban Green.* **2021**, *57*, 126945. [CrossRef]
26. Kumar, M.; Kumar, P. Valuation of the Ecosystem Services: A Psycho-Cultural Perspective. *Ecol. Econ.* **2008**, *64*, 808–81 [CrossRef]
27. Pike, S.; Kotsi, F. Stopover Destination Image—Perceptions of Dubai, United Arab Emirates, among French and Australia Travellers. *J. Travel Tour. Mark.* **2018**, *35*, 1160–1174. [CrossRef]
28. Ye, B.H.; Luo, J.M.; Vu, H.Q. Spatial and Temporal Analysis of Accommodation Preference Based on Online Reviews. *J. Desti Mark. Manag.* **2018**, *9*, 288–299. [CrossRef]
29. Afshardoost, M.; Eshaghi, M.S. Destination Image and Tourist Behavioural Intentions: A Meta-Analysis. *Tour. Manag.* **2020**, *8* 104154. [CrossRef]
30. Hey, T.; Tansley, S.; Tolle, K.M. *The Fourth Paradigm: Data-Intensive Scientific Discovery*; Microsoft Research: Redmond, WA, US 2009; Volume 1.
31. Schirpke, U.; Scolozzi, R.; Dean, G.; Haller, A.; Jäger, H.; Kister, J.; Kovács, B.; Sarmiento, F.O.; Sattler, B.; Schleyer, C. Cultura Ecosystem Services in Mountain Regions: Conceptualising Conflicts Among Users and Limitations of Use. *Ecosyst. Serv.* **2020**, *4* 101210. [CrossRef]
32. Das, S.; Sheykhfard, A.; Liu, J.; Nasim Khan, M. Understanding Non-Motorists' Views on Automated Vehicle Safety throug Bayesian Network Analysis and Latent Dirichlet Allocation. *Int. J. Transp. Sci. Technol.* **2023**, *14*, 289–304. [CrossRef]
33. Zhang, P.; Gu, H.; Gartrell, M.; Lu, T.; Yang, D.; Ding, X.; Gu, N. Group-Based Latent Dirichlet Allocation (Group-LDA): Effectiv Audience Detection for Books in Online Social Media. *Knowl.-Based Syst.* **2016**, *105*, 134–146. [CrossRef]
34. Zhou, Q.; Xu, Y.; Yang, L.; Menhas, R. Attitudes of the Public and Medical Professionals toward Nurse Prescribing: A Text-Minin Study Based on Social Medias. *Int. J. Nurs. Sci.* **2023**, *11*, 99–105. [CrossRef]
35. Wilson, A.; Murphy, H.; Fierro, J.C. Hospitality and Travel: The Nature and Implications of User-Generated Content. *Cornell Hos Q.* **2012**, *53*, 220–228. [CrossRef]
36. Wolski, J. The Landscape of Abandoned Villages in the Western Bieszczady: The Problem of Definition and Classification. *Geog Pol.* **2016**, *89*, 371–387. [CrossRef]
37. Brown, J.; MacLeod, N. A Site-Based Approach to Delivering Rangeland Ecosystem Services. *Rangel. J.* **2011**, *33*, 99–10 [CrossRef]
38. Hirons, M.; Comberti, C.; Dunford, R. Valuing Cultural Ecosystem Services. *Annu. Rev. Environ. Resour.* **2016**, *41*, 545–57 [CrossRef]
39. de Vries, S.; Goossen, M. Modelling Recreational Visits to Forests and Nature Areas. *Urban For. Urban Green.* **2002**, *1*, 5–1 [CrossRef]

Kim, J.; Kaplan, R. Physical and Psychological Factors in Sense of Community: New Urbanist Kentlands and Nearby Orchard Village. *Environ. Behav.* **2004**, *36*, 313–340. [CrossRef]

Fish, R.; Church, A.; Winter, M. Conceptualising Cultural Ecosystem Services: A Novel Framework for Research and Critical Engagement. *Ecosyst. Serv.* **2016**, *21*, 208–217. [CrossRef]

Polishchuk, Y.; Rauschmayer, F. Beyond "Benefits"? Looking at Ecosystem Services through the Capability Approach. *Ecol. Econ.* **2012**, *81*, 103–111. [CrossRef]

Zoderer, B.M.; Tasser, E.; Erb, K.-H.; Lupo Stanghellini, P.S.; Tappeiner, U. Identifying and Mapping the Tourists' Perception of Cultural Ecosystem Services: A Case Study from an Alpine Region. *Land Use Policy* **2016**, *56*, 251–261. [CrossRef]

Ral, E.; Bieling, C.; Zytynska, S.; Haase, D. Exploring City-Wide Patterns of Cultural Ecosystem Service Perceptions and Use. *Ecol. Indic.* **2017**, *77*, 80–95. [CrossRef]

Conrad, E.; Christie, M.; Fazey, I. Understanding Public Perceptions of Landscape: A Case Study from Gozo, Malta. *Appl. Geogr.* **2011**, *31*, 159–170. [CrossRef]

Soini, K.; Vaarala, H.; Pouta, E. Residents' Sense of Place and Landscape Perceptions at the Rural–Urban Interface. *Landsc. Urban Plan.* **2012**, *104*, 124–134. [CrossRef]

Prayag, G.; Hosany, S.; Muskat, B.; Del Chiappa, G. Understanding the Relationships between Tourists' Emotional Experiences, Perceived Overall Image, Satisfaction, and Intention to Recommend. *J. Travel Res.* **2017**, *56*, 41–54. [CrossRef]

Plumwood, V. The Concept of a Cultural Landscape: Nature, Culture and Agency in the Land. *Ethics Environ.* **2006**, *11*, 115–150. [CrossRef]

Nowak-Olejnik, A.; Mocior, E.; Hibner, J.; Tokarczyk, N. Human Perceptions of Cultural Ecosystem Services of Semi-Natural Grasslands: The Influence of Plant Communities. *Ecosyst. Serv.* **2020**, *46*, 101208. [CrossRef]

Richards, D.R.; Friess, D.A. A Rapid Indicator of Cultural Ecosystem Service Usage at a Fine Spatial Scale: Content Analysis of Social Media Photographs. *Ecol. Indic.* **2015**, *53*, 187–195. [CrossRef]

Langemeyer, J.; Calcagni, F.; Baró, F. Mapping the Intangible: Using Geolocated Social Media Data to Examine Landscape Aesthetics. *Land Use Policy* **2018**, *77*, 542–552. [CrossRef]

Pike, S. *Destination Marketing Organisations*; Routledge: London, UK, 2007.

Stępniewska, M. The Capacity of Urban Parks for Providing Regulating and Cultural Ecosystem Services versus Their Social Perception. *Land Use Policy* **2021**, *111*, 105778. [CrossRef]

Riechers, M.; Barkmann, J.; Tscharntke, T. Diverging Perceptions by Social Groups on Cultural Ecosystem Services Provided by Urban Green. *Landsc. Urban Plan.* **2018**, *175*, 161–168. [CrossRef]

Bertram, C.; Rehdanz, K. Preferences for Cultural Urban Ecosystem Services: Comparing Attitudes, Perception, and Use. *Ecosyst. Serv.* **2015**, *12*, 187–199. [CrossRef]

Pleasant, M.M.; Gray, S.A.; Lepczyk, C.; Fernandes, A.; Hunter, N.; Ford, D. Managing Cultural Ecosystem Services. *Ecosyst. Serv.* **2014**, *8*, 141–147. [CrossRef]

Dou, Y.; Yu, X.; Bakker, M.; De Groot, R.; Carsjens, G.J.; Duan, H.; Huang, C. Analysis of the Relationship between Cross-Cultural Perceptions of Landscapes and Cultural Ecosystem Services in Genheyuan Region, Northeast China. *Ecosyst. Serv.* **2020**, *43*, 101112. [CrossRef]

van Dijck, J.; Poell, T.; de Waal, M. *The Platform Society: Public Values in a Connective World*; Oxford University Press: Oxford, UK, 2018; ISBN 978-0-19-088978-4.

Guo, Y.; Barnes, S.J.; Jia, Q. Mining Meaning from Online Ratings and Reviews: Tourist Satisfaction Analysis Using Latent Dirichlet Allocation. *Tour. Manag.* **2017**, *59*, 467–483. [CrossRef]

Manley, K.; Nyelele, C.; Egoh, B.N. A Review of Machine Learning and Big Data Applications in Addressing Ecosystem Service Research Gaps. *Ecosyst. Serv.* **2022**, *57*, 101478. [CrossRef]

Li, J.; Gao, J.; Zhang, Z.; Fu, J.; Shao, G.; Zhao, Z.; Yang, P. Insights into Citizens' Experiences of Cultural Ecosystem Services in Urban Green Spaces Based on Social Media Analytics. *Landsc. Urban Plan.* **2024**, *244*, 104999. [CrossRef]

Oteros-Rozas, E.; Martín-López, B.; Fagerholm, N.; Bieling, C.; Plieninger, T. Using Social Media Photos to Explore the Relation between Cultural Ecosystem Services and Landscape Features across Five European Sites. *Ecol. Indic.* **2018**, *94*, 74–86. [CrossRef]

Love, B.C. Comparing Supervised and Unsupervised Category Learning. *Psychon. Bull. Rev.* **2002**, *9*, 829–835. [CrossRef]

Song, Y.; Wang, R.; Fernandez, J.; Li, D. Investigating Sense of Place of the Las Vegas Strip Using Online Reviews and Machine Learning Approaches. *Landsc. Urban Plan.* **2021**, *205*, 103956. [CrossRef]

Chen, X.; Li, J.; Han, W.; Liu, S. Urban Tourism Destination Image Perception Based on LDA Integrating Social Network and Emotion Analysis: The Example of Wuhan. *Sustainability* **2022**, *14*, 12. [CrossRef]

Leung, D.; Law, R.; Lee, H.A. The Perceived Destination Image of Hong Kong on Ctrip.Com. *Int. J. Tour. Res.* **2011**, *13*, 124–140. [CrossRef]

Blei, D.M.; Ng, A.Y.; Jordan, M.I. Latent Dirichlet Allocation. *J. Mach. Learn. Res.* **2003**, *3*, 993–1022.

68. Bi, J.-W.; Liu, Y.; Fan, Z.-P.; Zhang, J. Wisdom of Crowds: Conducting Importance-Performance Analysis (IPA) through Onlin Reviews. *Tour. Manag.* **2019**, *70*, 460–478. [CrossRef]
69. Luo, Y.; He, J.; Mou, Y.; Wang, J.; Liu, T. Exploring China's 5A Global Geoparks through Online Tourism Reviews: A Minin Model Based on Machine Learning Approach. *Tour. Manag. Perspect.* **2021**, *37*, 100769. [CrossRef]
70. Peng, G.; Yuefen, W. Research on the Method of Determining the Optimal Topic Number of LDA Topic Model in Scientific an Technological Information Analysis. *Mod. Libr. Inf. Technol.* **2016**, *32*, 42–50.
71. Lee, J.W.; Kim, Y.; Han, D.H. LDA-Based Topic Modeling for COVID-19-Related Sports Research Trends. *Front. Psychol.* **2022**, 1 1033872. [CrossRef] [PubMed]
72. Röder, M.; Both, A.; Hinneburg, A. Exploring the Space of Topic Coherence Measures. In Proceedings of the Eighth ACM International Conference on Web Search and Data Mining, Shanghai, China, 2–6 February 2015; ACM: New York, NY, USA, 201 pp. 399–408.
73. Martilla, J.A.; James, J.C. Importance-Performance Analysis. *J. Mark.* **1977**, *41*, 77–79. [CrossRef]
74. Abalo, J.; Varela, J.; Manzano, V. Importance Values for Importance–Performance Analysis: A Formula for Spreading out Value Derived from Preference Rankings. *J. Bus. Res.* **2007**, *60*, 115–121. [CrossRef]
75. Deng, W. Using a Revised Importance–Performance Analysis Approach: The Case of Taiwanese Hot Springs Tourism. *Tou Manag.* **2007**, *28*, 1274–1284. [CrossRef]
76. Boley, B.B.; McGehee, N.G.; Tom Hammett, A.L. Importance-Performance Analysis (IPA) of Sustainable Tourism Initiatives: Th Resident Perspective. *Tour. Manag.* **2017**, *58*, 66–77. [CrossRef]
77. Havinga, I.; Marcos, D.; Bogaart, P.; Tuia, D.; Hein, L. Understanding the Sentiment Associated with Cultural Ecosystem Service Using Images and Text from Social Media. *Ecosyst. Serv.* **2024**, *65*, 101581. [CrossRef]
78. Zang, Z.; Fu, H.; Cheng, J.; Raza, H.; Fang, D. Digital Threads of Architectural Heritage: Navigating Tourism Destination Imag through Social Media Reviews and Machine Learning Insights. *J. Asian Archit. Build. Eng.* **2024**, 1–18. [CrossRef]
79. Xu, L.; Lin, H.; Pan, Y.; Ren, H.; Chen, J. Constructing the Affective Lexicon Ontology. *J. China Soc. Sci. Tech. Inf.* **2008**, *27*, 180–18
80. Hua, J.; Chen, W.Y. Prioritizing Urban Rivers' Ecosystem Services: An Importance-Performance Analysis. *Cities* **2019**, *94*, 11–2 [CrossRef]
81. Zhang, J.; Cheng, Y.; Zhao, B. Assessing the Inequities in Access to Peri-Urban Parks at the Regional Level: A Case Study i China's Largest Urban Agglomeration. *Urban For. Urban Green.* **2021**, *65*, 127334. [CrossRef]
82. Stålhammar, S.; Pedersen, E. Recreational Cultural Ecosystem Services: How Do People Describe the Value? *Ecosyst. Serv.* **201** *26*, 1–9. [CrossRef]
83. Hall, C.M.; Higham, J. *Tourism, Recreation and Climate Change*; Channel View Publications: Bristol, UK, 2005; ISBN 978-1-8454 314-9.
84. Knudsen, D.C.; Metro-Roland, M.M.; Rickly, J.M. Tourism, Aesthetics, and Touristic Judgment. *Tour. Rev. Int.* **2015**, *19*, 179–19 [CrossRef]
85. Laesser, C. Health Travel Motivation and Activities: Insights from a Mature Market—Switzerland. *Tour. Rev.* **2011**, *66*, 83–8 [CrossRef]
86. Han, H.; Lho, L.H.; Al-Ansi, A.; Yu, J. Cycling Tourism: A Perspective Article. *Tour. Rev.* **2020**, *75*, 162–164. [CrossRef]
87. Butler, R. Geographical Research on Tourism, Recreation and Leisure: Origins, Eras and Directions. *Tour. Geogr.* **2004**, *6*, 143–16 [CrossRef]
88. Williams, A.M.; Shaw, G. Future Play: Tourism, Recreation and Land Use. *Land Use Policy* **2009**, *26*, S326–S335. [CrossRef]
89. Bhuiyan, M.A.H.; Darda, M.A. Tourists' Satisfaction on Heritage Sites of Dhaka City in Bangladesh. *Int. J. Cult. Tour. Hosp. Re* **2019**, *14*, 34–43. [CrossRef]
90. Kim, H.; Lee, S.; Uysal, M.; Kim, J.; Ahn, K. Nature-Based Tourism: Motivation and Subjective Well-Being. *J. Travel Tour. Mar* **2015**, *32*, S76–S96. [CrossRef]
91. Nordbø, I.; Prebensen, N.K. Hiking as Mental and Physical Experience. In *Advances in Hospitality and Leisure*; Emerald Grou Publishing Limited: Bingley, UK, 2015; Volume 11, pp. 169–186. ISBN 978-1-78560-271-9.
92. Bichler, B.F.; Peters, M. Soft Adventure Motivation: An Exploratory Study of Hiking Tourism. *Tour. Rev.* **2020**, *76*, 473–48 [CrossRef]
93. Lin, H.; Zhang, M.; Gursoy, D.; Fu, X. Impact of Tourist-to-Tourist Interaction on Tourism Experience: The Mediating Role c Cohesion and Intimacy. *Ann. Tour. Res.* **2019**, *76*, 153–167. [CrossRef]
94. Andrews, H. Tourism as a 'Moment of Being. *Suom. Antropol. J. Finn. Anthropol. Soc.* **2009**, *34*, 5–21. [CrossRef]
95. Knobloch, U.; Robertson, K.; Aitken, R. Experience, Emotion, and Eudaimonia: A Consideration of Tourist Experiences an Well-Being. *J. Travel Res.* **2017**, *56*, 651–662. [CrossRef]
96. Kirillova, K.; Fu, X.; Lehto, X.; Cai, L. What Makes a Destination Beautiful? Dimensions of Tourist Aesthetic Judgment. *Tou Manag.* **2014**, *42*, 282–293. [CrossRef]

. Plieninger, T.; Dijks, S.; Oteros-Rozas, E.; Bieling, C. Assessing, Mapping, and Quantifying Cultural Ecosystem Services at Community Level. *Land Use Policy* **2013**, *33*, 118–129. [CrossRef]
. Raymond, C.M.; Kyttä, M.; Stedman, R. Sense of Place, Fast and Slow: The Potential Contributions of Affordance Theory to Sense of Place. *Front. Psychol.* **2017**, *8*, 1674. [CrossRef]
. Calcagni, F.; Amorim Maia, A.T.; Connolly, J.J.T.; Langemeyer, J. Digital Co-Construction of Relational Values: Understanding the Role of Social Media for Sustainability. *Sustain. Sci.* **2019**, *14*, 1309–1321. [CrossRef]
0. Cord, A.F.; Roeßiger, F.; Schwarz, N. Geocaching Data as an Indicator for Recreational Ecosystem Services in Urban Areas: Exploring Spatial Gradients, Preferences, and Motivations. *Landsc. Urban Plan.* **2015**, *144*, 151–162. [CrossRef]
1. McPhearson, T.; Hamstead, Z.A.; Kremer, P. Urban Ecosystem Services for Resilience Planning and Management in New York City. *Ambio* **2014**, *43*, 502–515. [CrossRef] [PubMed]
2. Gordon, N.; Tsuchiya, N.; Koenig-Robert, R.; Hohwy, J. Expectation and Attention Increase the Integration of Top-down and Bottom-up Signals in Perception through Different Pathways. *PLoS Biol.* **2019**, *17*, e3000233. [CrossRef]
3. Fu, H.; Wang, Z.; Jie, H.; Wang, L. Emotional Characteristics and Influencing Factors of Urban Park Users: A Case Study of South China Botanical Garden and Yuexiu Park. *Beijing Da Xue Xue Bao* **2021**, *57*, 1108–1120.

sustainability

MDPI

Article

Spatiotemporal Evolution and Driving Factors of the Cooling Capacity of Urban Green Spaces in Beijing over the Past Four Decades

Chao Wang [1], Chaobin Yang [1,2,*], Huaiqing Wang [1] and Lilong Yang [1]

1 School of Civil Engineering and Geomatics, Shandong University of Technology, Zibo 255000, China; 23507020861@stumail.sdut.edu.cn (C.W.); 24407010857@stumail.sdut.edu.cn (H.W.); 22407010007@stumail.sdut.edu.cn (L.Y.)

2 State Key Laboratory of Regional and Urban Ecology, Research Center for Eco-Environmental Sciences, Chinese Academy of Sciences, Beijing 100085, China

* Correspondence: yangchaobin@sdut.edu.cn

Abstract

Urban green spaces (UGS) are crucial for mitigating rising urban land surface temperatures (LST). Rapid urbanization presents unresolved questions regarding (a) seasonal variations in the spatial co-distribution of UGS and LST, (b) the temporal and spatial changes in UGS cooling, and (c) the dominant factors driving cooling effects during different periods. This study focuses on Beijing's Fifth Ring Road area, utilizing nearly 40 years of Landsat remote sensing imagery and land cover data. We propose a novel nine-square grid spatial analysis approach that integrates LST retrieval, profile line analysis, and the XGBoost algorithm to investigate the long-term spatiotemporal evolution of UGS cooling capacity and its driving mechanisms. The results demonstrate three key findings: (1) Strong seasonal divergence in UGS-LST correlation: A significant negative correlation dominates during summer months (June–August), whereas winter (December–February) exhibits marked weakening of this relationship, with localized positive correlations indicating thermal inversion effects. (2) Dynamic evolution of cooling capacity under urbanization: Urban expansion has reconfigured UGS spatial patterns, with a cooling capacity of UGS showing an "enhancement–decline–enhancement" trend over time. Analysis through machine learning on the significance of landscape metrics revealed that scale-related metrics play a dominant role in the early stage of urbanization, while the focus shifts to quality-related metrics in the later phase. (3) Optimal cooling efficiency threshold: Maximum per-unit-area cooling intensity occurs at 10–20% UGS coverage, yielding an average LST reduction of approximately 1 °C relative to non-vegetated surfaces. This study elucidates the spatiotemporal evolution of UGS cooling effects during urbanization, establishing a robust scientific foundation for optimizing green space configuration and enhancing urban climate resilience.

Keywords: cooling capacity; land surface temperature; spatial patterns; urban green space; XGBoost

1. Introduction

With rapid socio-economic advancement, urbanization is progressing at an unprecedented pace. This process has triggered a sharp increase in urban populations [1,2], the continuous expansion of impervious surfaces (e.g., buildings and roads), and the gradual replacement of traditional low-rise buildings with mid-to-high-rise structures [3–5]. These impervious surfaces absorb and retain substantial heat, significantly elevating urban temperatures above those in surrounding rural areas—a phenomenon known as the urban heat island (UHI) effect [6–8]. In China, major cities including Beijing, Shanghai, and Nanjing exhibit pronounced UHI intensity, a critical environmental challenge in contemporary urban development [9–12].

Sustainable urban development has become a globally recognized objective, with particular urgency for rapidly urbanizing nations such as China. Mitigating urbanization's adverse impacts and enhancing rural-urban environmental linkages now constitute critical research priorities. Consequently, diverse mitigation strategies have been adopted—including improved energy efficiency, greenhouse gas reduction, and green infrastructure development—to foster climate-resilient urbanization [13–16]. Among these strategies, UGS plays a critical role [17–19]. Through transpirative cooling and shading effects, UGS significantly reduces local LST [20–22]. Compared to alternative UHI mitigation measures, UGS provides more practical and scalable solutions for enhancing urban thermal comfort [23–25]. Consequently, UGS expansion and protection are globally endorsed as effective UHI countermeasures [26,27]. These green spaces—varying in size and distribution—generate localized cooling not only within their boundaries but also across adjacent areas, establishing "cooling islands" in urban landscapes [28–33].

A comprehensive understanding of UGS cooling effects critically informs UHI mitigation, microclimate improvement, and thermal comfort enhancement. Such studies provide scientific foundations for climate adaptation strategies, advancing green city development and facilitating synergistic optimization of ecological-human habitat systems [34–36]. In recent years, research on the cooling effects of UGS has advanced significantly in data acquisition, methodological innovation, and result interpretation. Studies have progressed from early field measurements to extensively utilizing remote sensing imagery—such as MODIS, Landsat, and Sentinel-2—for extracting LST and vegetation indices [37–40]. Methodologically, the integration of statistical analysis, spatial modeling, machine learning, and computational fluid dynamics (CFD) simulations has significantly advanced our understanding of UGS cooling mechanisms. A growing body of evidence confirms that UGS exert substantial cooling effects, with intensity modulated by factors such as green space area, shape, vegetation type, and surrounding environment [41–44]. Composite green spaces—such as forest-water complexes—demonstrate enhanced cooling efficiency. Typically, the cooling effect extends up to 500 m from green space boundaries, though significant spatial heterogeneity occurs across climatic regions [28,42,45].

Despite considerable progress, current research remains limited. Temporally, studies are often restricted to short time frames and suffer from a scarcity of long-term time series data, impeding comprehensive analyses of UGS cooling effect dynamics [46–48]. Conventional studies have predominantly focused on individual parks or specific urban blocks, thus failing to capture a holistic perspective of the thermal landscape across broader urban areas. Furthermore, their methodological approaches remain relatively conventional and have not adequately integrated cutting-edge technologies such as machine learning [49].

To address these gaps, this study focuses on the area within Beijing's Fifth Ring Road, employing granular spatial indicators and advanced methodologies—including machine learning modeling. Specifically, it aims to (1) investigate the spatiotemporal evolution of UGS–LST relationships; (2) analyze the cooling capacity dynamics of UGS across space and time; and (3) identify key drivers of UGS cooling effects. These findings will complement the research gap on the relationship between green space configuration and thermal regulation over multiple decades at the urban scale, a field that has not been adequately explored in previous studies, whether in Beijing or other metropolitan areas.

It is anticipated that these findings will provide empirical evidence for optimizing UGS design and support urban planning strategies aimed at maximizing the potential for UHI.

2. Materials and Methods

2.1. Materials

2.1.1. Study Area

Beijing (39°56′ N, 116°20′ E) is situated in northern China along the northern margin of the North China Plain, bordering Tianjin Municipality to the east and surrounded by Hebei Province elsewhere [50]. As one of China's most urbanized cities, Beijing exhibits distinct topography and a temperate-monsoon climate. The terrain descends from the northwestern highlands to the southeastern lowlands, with mountains encircling the western, northern, and northeastern sectors. The southeastern slopes transition into a broad alluvial plain extending toward the Bohai Sea. Climatically, summers are hot-humid and winters cold-dry, with abbreviated spring and autumn seasons. Rapid urbanization has intensified summer high-temperature events and amplified the UHI effect, positioning Beijing as an exemplary study area for evaluating UGS cooling efficacy. This research concentrates on the zone between the Inner City and Fifth Ring Road—encompassing Beijing's primary transportation hub and core development corridor. This densely populated region plays a pivotal role in the city's spatial-functional structure (study area location shown in Figure 1.

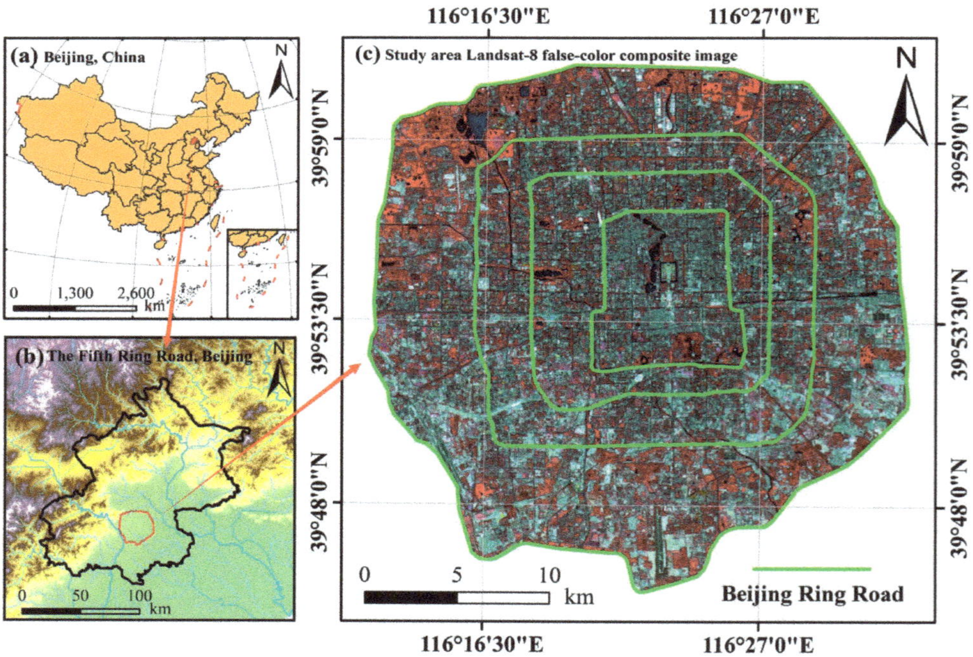

Figure 1. Location of the study area (Landsat-8 false-color composite image).

2.1.2. Data Sources

This study utilized Landsat TM-5 and Landsat 8 OLI/TIRS imagery (30 m spatial resolution), acquired from the China Geospatial Data Cloud (http://www.gscloud.cn). To capture long-term trends, summer (June–August) and winter (December–February) scenes were selected at decadal intervals from 1984 to 2021, ensuring representation of key variables under contrasting climatic conditions. For enhanced temporal reliability, seasonal composites were generated from the mean of three high-quality, cloud-free images. Where intra-seasonal data gaps existed, imagery from adjacent years within the same season was substituted to maintain temporal continuity in UGS cooling analysis. Land cover data (CLCD; 30 m resolution) were sourced from Zenodo (https://zenodo.org/records/8176941, Accessed on 9 December 2024). Auxiliary datasets included meteorological station records, Beijing administrative boundaries, a digital elevation model (DEM), and Google Earth high-resolution imagery. Detailed image metadata is provided in Table 1.

Table 1. Acquisition dates of Landsat imagery used in this study.

Season	1980s	1990s	2000s	2010s	2020s
Summer	1984-06-29	1993-06-22	2004-07-06	2013-07-31	2021-06-03
	1984-08-16	1993-08-25	2005-06-23	2014-08-19	2021-06-19
	1985-07-18	1996-06-30	2005-07-09	2017-07-10	2021-08-06
Winter	1984-12-06	1994-12-02	2003-12-11	2014-12-25	2020-12-25
	1985-01-07	1995-12-05	2003-12-27	2015-01-10	2021-12-12
	1986-02-27	1996-02-07	2006-02-18	2016-12-14	2021-12-28

2.2. Methods

The technical workflow of this study is illustrated in Figure 2. Landsat-derived LST, Normalized Difference Vegetation Index (NDVI), and Percentage of Vegetation (PV) were integrated with CLCD land cover data. Regression analysis, machine learning, profile-based assessment, and multi-scale spatial analysis were employed to examine: (1) UGS spatial pattern evolution, (2) LST heterogeneity, and (3) spatiotemporal dynamics and drivers of UGS cooling effects.

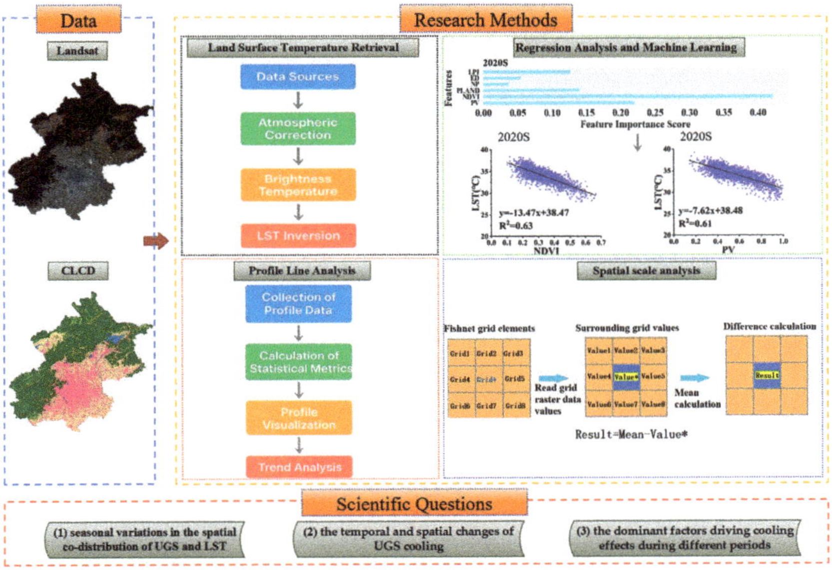

Figure 2. Technical workflow of the study. Different colors represent different types of land. "*" It is only used to mark the computational grid, with no other special purpose.

2.2.1. LST Retrieval

To ensure accurate LST retrieval, this study employed the atmospheric correction method [51]. Brightness temperature—the radiatively derived temperature recorded by satellite sensors—was corrected to retrieve LST. The relationship between brightness temperature and LST is expressed as follows:

$$L_\lambda = \frac{\left(B_\lambda\left(T_{surf}\right) - \varepsilon_\lambda B_\lambda(T_{atm})\right)}{1-\varepsilon_\lambda} \quad (1)$$

In this context, L_λ represents the brightness temperature (K) obtained from remote sensing imagery; $B_\lambda\left(T_{surf}\right)$ denotes the blackbody radiance of the land surface ($W/m^2 \cdot sr \cdot \mu m$), which is directly associated with LST(T_{surf}); ε_λ is the land surface emissivity, typically determined by surface material characteristics; and T_{atm} indicates the atmospheric temperature (K).

2.2.2. Profile Line Analysis

Profile-line analysis—a key spatial technique in urban thermal environment studies—elucidates temperature gradients across land cover transitions [52,53]. This method quantifies thermal variations from urban cores to peripheries or between green spaces and built-up areas. Reflecting Beijing's axial-radial spatial structure, north–south and east–west transects were established centered on the Forbidden City. As Figure 3 illustrates, the east–west transect spans the Fifth Ring Road's western to eastern extents, while the north–south transect connects its northern and southern boundaries.

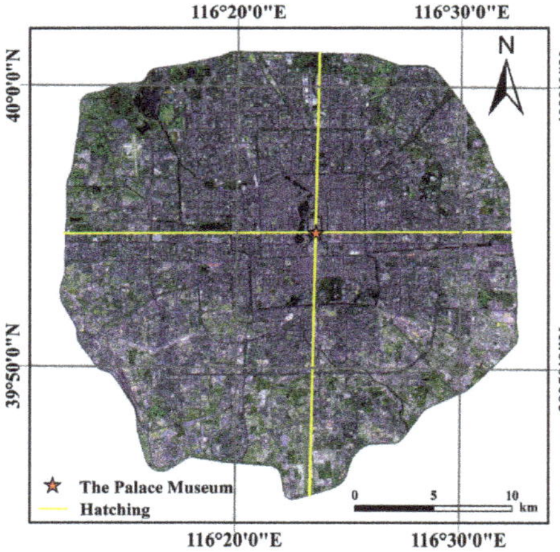

Figure 3. Schematic diagram of transect lines in the study area.

2.2.3. Spatial Scale Analysis

Spatial scale selection critically influences the evaluation of UGS cooling effects and the interpretation. Local-scale analyses reveal direct temperature reduction mechanisms, while macro-scale analyses illuminate integrated UGS roles in UHI mitigation. Therefore, multi-scale integrated analyses are essential to comprehensively assess UGS cooling potential [54].

To rigorously characterize the spatial distribution of UGS cooling effects, this study implemented high-resolution spatial analysis. The study area was partitioned into 1 km × 1 km grid units, with thermal adjacency effects quantified for each cell relative to neighboring grids (Figure 4). Grid cells were statistically clustered based on cooling capacity. Comparative temporal analysis then assessed cooling intensity variations, while inter-period grid-level differences systematically revealed interannual evolution of UGS cooling effects.

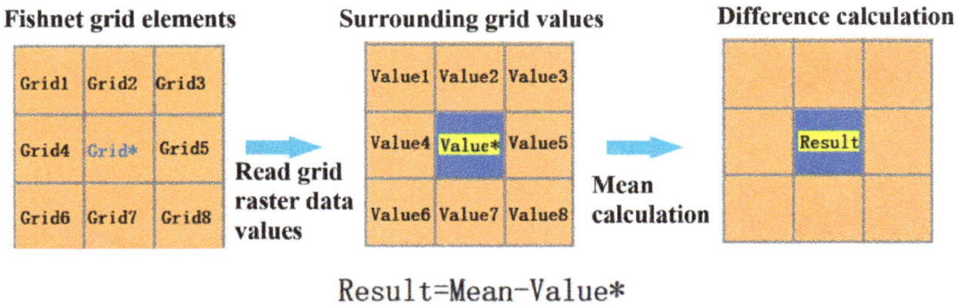

Figure 4. Schematic diagram illustrating the calculation method of green space cooling intensity between adjacent grid cells. "*" It is only used to mark the computational grid, with no other special purpose.

2.2.4. Machine Learning and Regression Analysis

Machine learning has advanced the study of UGS cooling effects in recent years. Integrating machine learning with landscape pattern analysis helps determine critical landscape determinants of cooling efficacy. This study computed class-level landscape metrics using Fragstats 4.2, selecting four representative indices for analysis (Table 2): Percentage of Landscape (PLAND), number of patches (NP), largest patch index (LPI), and edge density (ED). Then, use the XGBoost model to evaluate the relative importance of six indicators, including NDVI and PV. The dataset was divided into a training set (70%), a test set (20%), and a validation set (10%) for model tuning and accuracy assessment. The model was trained and optimized using the training and validation sets, and its predictive performance was evaluated on the test set using the coefficient of determination (R^2) and root mean square error (RMSE). The XGBoost model demonstrated satisfactory accuracy across different decades (1980s–2020s).

Table 2. Green space landscape pattern index.

Metric Name	Calculation Formula	Meaning
Percent of Landscape (PLAND)	$PLAND = \left(\frac{\sum_{j=1}^{n} a_{ij}}{A}\right) \times 100$	a_{ij} : The area of the j-th patch belonging to the i-th class (e.g., green space), measured in square meters or number of pixels. n : The total number of plaques in that category. A: The total area of the entire landscape. $\sum_{j=1}^{n} a_{ij}$: The total area of all patches in that category (e.g., green space).
Number of Patches (NP)	$NP = \text{n}$	n: Refers to the total number of patches in a given landscape type (e.g., green space, forest land, construction land, etc.).
Largest Patch Index (LPI)	$LPI = \left(\frac{max(a_{ij})}{A}\right) \times 100$	$\max(a_{ij})$: The largest patch in category i (e.g., green space). A: The total area of the entire landscape.
Edge Density (ED)	$ED = \left(\frac{\sum e_{ik}}{A}\right) \times 10,000$	$\sum e_{ik}$: The total length (in meters) of all patch boundaries in category i (e.g., green space). A: The total area of the entire landscape.

This study established a regression model to quantify relationships between LST and predictors, evaluating UGS cooling effects [55]. LST served as the dependent variable, with

PV, NDVI, and green space area as independent variables. Linear regression analyzed temporal variations in LST-predictor correlations.

3. Results

3.1. Spatial and Temporal Pattern Evolution of UGS

3.1.1. Temporal and Spatial Changes in UGS

This study examines the evolution of UGS distribution in Beijing from the 1980s to the 2020s using summer and winter PV data. Figure 5 shows limited UGS coverage in the city center during the 1980s, mainly in major parks such as Tiantan [Temple of Heaven] and Ditan [Temple of Earth], while high vegetation indices were mainly along the Fourth–Fifth Ring Road interzone. In the 1990s and 2000s, central UGS coverage increased modestly but decreased significantly between the Fourth and Fifth Ring Roads. Since the 2010s, UGS distribution has become more decentralized, with a notable recovery in the Fourth–Fifth Ring Road corridor. Overall, UGS within Beijing's Fifth Ring Road experienced two main phases: (1) a decline from the 1980s to the 2000s driven by rapid urbanization replacing UGS with impervious surfaces; and (2) a recovery from the 2010s to the 2020s linked to urban stabilization, functional improvements, and better public infrastructure. This resurgence is especially strong in Chaoyang, Haidian, and Dongcheng districts—key areas for modern service industries. The pre-Olympic greening initiatives in 2008 further significantly expanded urban green coverage.

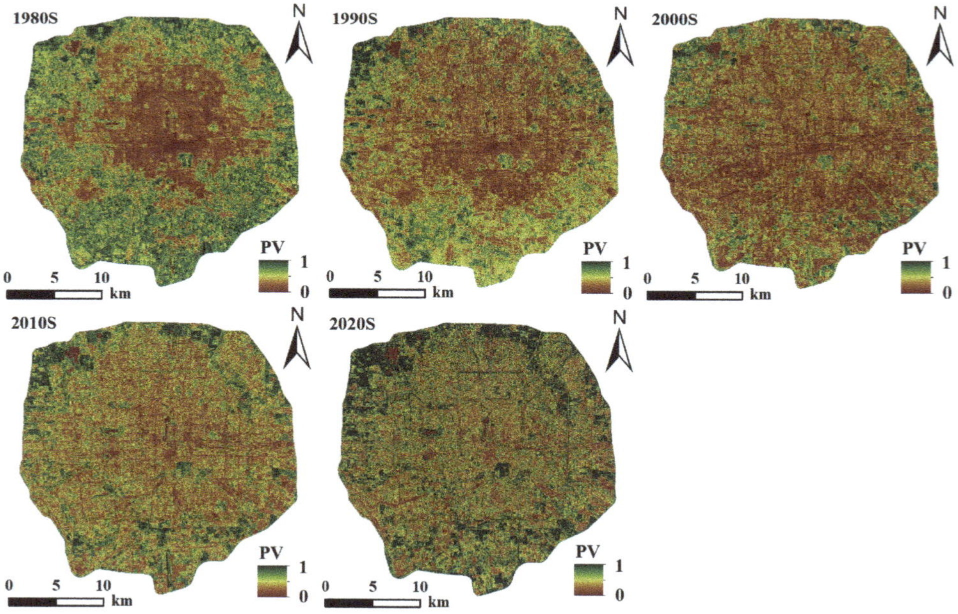

Figure 5. Spatial distribution of summer PV in the study area.

Figure 6 shows lower winter vegetation indices compared to summer values, due to the dominance of deciduous vegetation in northern temperate climates. However, the addition of evergreen species and improved UGS management has greatly increased winter vegetation indices in historical and cultural parks, showing better urban ecological sustainability.

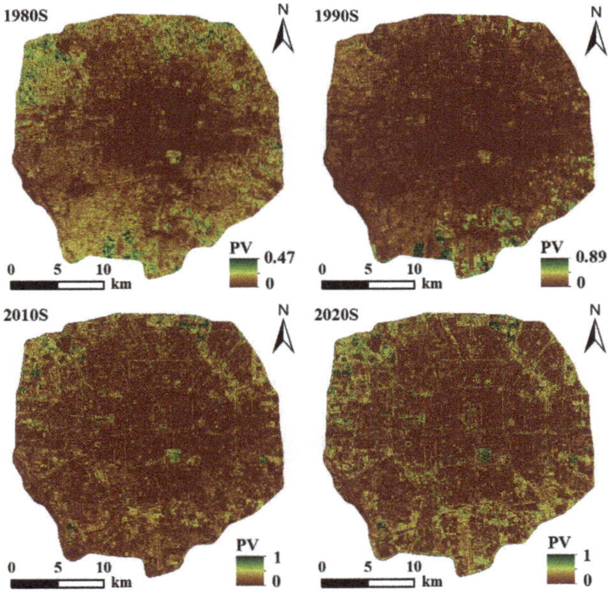

Figure 6. Spatial distribution of winter PV in the study area.

3.1.2. Evolution of UGS Landscape Patterns

Analysis of green space landscape metrics at a 500 m spatial scale (Figure 7) reveals distinct spatiotemporal variations in four key indices, signifying dynamic pattern changes during rapid urbanization. The green landscape followed an "initial fragmentation → gradual restoration" trajectory. From the 1980s to 2000s, urban core PLAND declined continuously while NP increased sharply, indicating intensified fragmentation. Concurrently, decreasing LPI and fluctuating ED (initial increase followed by a decrease) signified greater patch segmentation and structural complexity. Since the 2010s, PLAND recovery with concurrent NP, ED, and LPI increases reflects landscape expansion, reconnection, and integration. Green infrastructure development during this period enhanced ecological spatial structure, improving connectivity and stability.

A detailed stage-wise analysis indicates that during the 1980s, UGS were extensively distributed and spatially concentrated, exhibiting high ecological integrity. Beginning in the 1990s, the UGS area declined markedly, fragmentation intensified, and dominant patches diminished. By the 2000s, the landscape exhibited peak fragmentation and dispersion, with dominant patches disappearing and ecosystem vulnerability escalating sharply. Although the UGS area partially recovered in the 2010s, fragmentation persisted, evidenced by heightened ED. In the 2020s, green space continued to expand, yet patch aggregation and boundary complexity had not reached optimal levels, and spatial integrity still requires enhancement. In summary, the UGS system is transitioning from a development-led model toward ecological restoration prioritization, though spatial optimization remains ongoing.

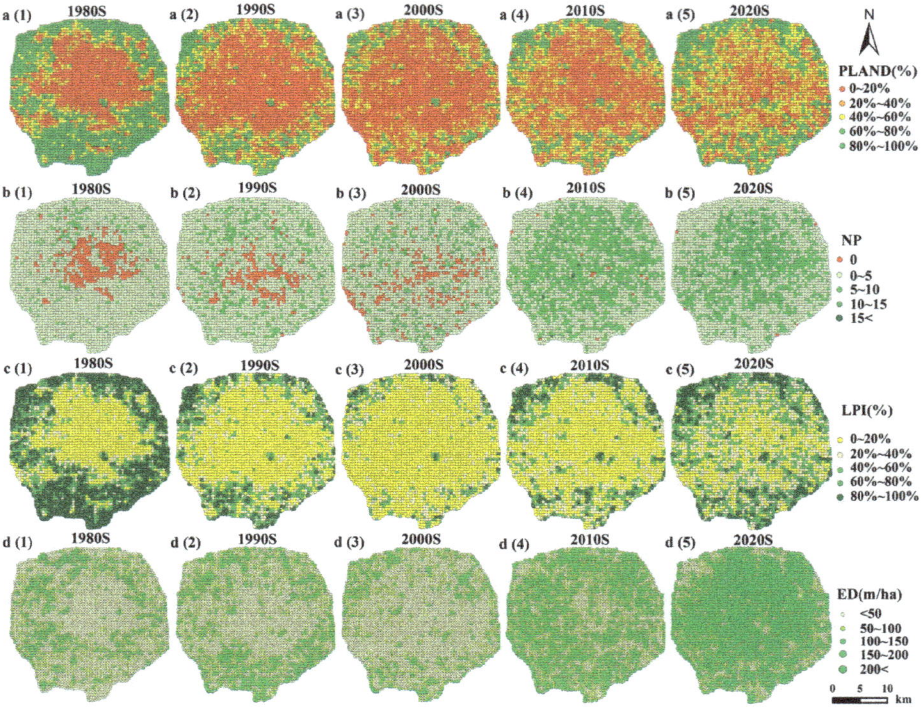

Figure 7. Schematic diagram of the distribution of landscape pattern in a unit area.

3.2. Spatiotemporal Evolution of LST

3.2.1. Accuracy Validation of LST

To improve data accuracy and enhance the scientific rigor of subsequent analyses, the retrieved LST data were validated against ground-based observations. Meteorological data were obtained from six national standard meteorological stations within North China, including sites in Beijing, Tangshan, Baoding, Chengde, and surrounding regions. As part of China's national meteorological observation network, these baseline stations feature long-term monitoring records and stable operational protocols, ensuring high data accuracy. Consequently, their observations serve as reliable references for evaluating the accuracy of LST retrievals.

Figure 8 compares retrieved LST values with station-based air temperature observations. The results indicate limited deviations between retrieved and observed values, with consistent temporal trends. The error remains within acceptable thresholds, confirming that the adopted LST retrieval method is robust and regionally appropriate. This approach accurately reflects land surface thermal characteristics and thus provides a reliable data source for subsequent research.

3.2.2. Temporal Trends of LST Across Beijing's Ring Roads

Figure 9 presents the temporal trends of mean LST during summer and winter across distinct ring road zones in Beijing (specifically, the Fifth, Fourth, Third, and Second Ring Roads). Over the past four decades, summer LST in all zones exhibited a generally increasing trend with fluctuations, rising overall by approximately 6–8 °C. The most rapid increase occurred from the 1980s to the 2000s, followed by a reduced rate of warming between the 2000s and 2010s. Conversely, winter LST displayed a more complex pattern, characterized by a three-phase trend: initial increase, subsequent stabilization, and eventual slight decline. Notably, temperature differences among zones during winter were smaller than those observed in summer.

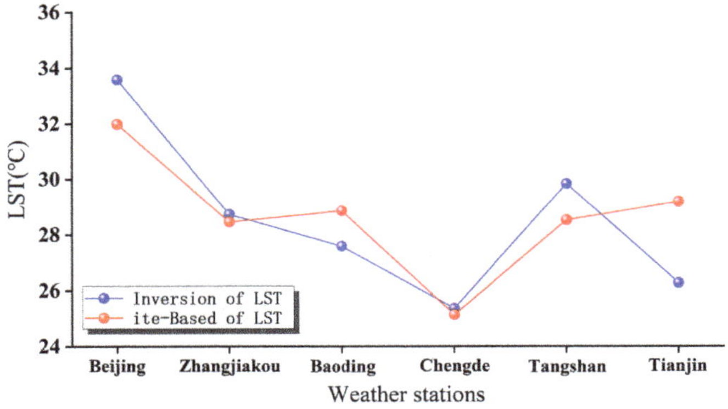

Figure 8. Temporal Trends of LST across Beijing's Ring Roads.

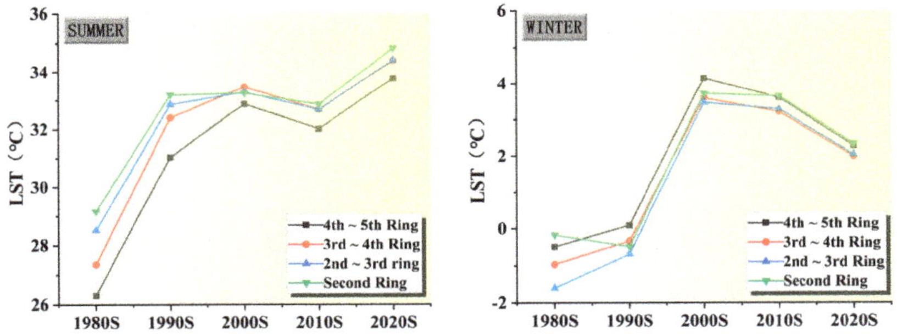

Figure 9. Temporal trends of LST across different ring roads in Beijing.

This seasonal disparity in LST trends may partially reflect changes in urban architectural morphology. Since the 2000s, the proliferation of high-rise buildings has increased shadow coverage, reducing ground-level direct solar radiation. This effect likely explains the suppressed winter LST observed in urban areas.

Spatially, the temperature gradient among ring road zones has undergone significant temporal changes. During the 1980s, marked LST differences existed across zones. By the 2020s, however, these differences had substantially decreased. This thermal homogenization likely results from accelerated urbanization, climate change, improved urban greening, and heating system evolution—collectively reducing LST variability across the urban landscape.

3.2.3. Spatiotemporal Heterogeneity of LST

Figure 10 illustrates the marked spatiotemporal heterogeneity and dynamic evolution of Beijing's thermal environment over the past four decades. Temporally, LST exhibited continuous warming from the 1980s to the 2020s. Temperature peaks in high-temperature zones consistently exceeded 40 °C since the 1990s, reaching a maximum value of 44.23 °C in the 2020s. The warming rate varied interdecadally: rapid growth during the 1980s–1990s coincided with substantial expansion of high-temperature areas; moderated growth in the 2000s; a slight decline in the 2010s; and a resurgence in the 2020s. Spatially, the 1980s featured concentrated, contiguous high-temperature zones primarily within the Third Ring Road, forming a single thermal core. These zones subsequently expanded outward with decreasing spatial concentration. By the 2020s, high-temperature areas exhibited a polycentric, dispersed pattern featuring independent thermal patches in both central urban areas and suburban development zones. This transition from centralized to dispersed UHI patterns reflects urban expansion and polycentric land-use development.

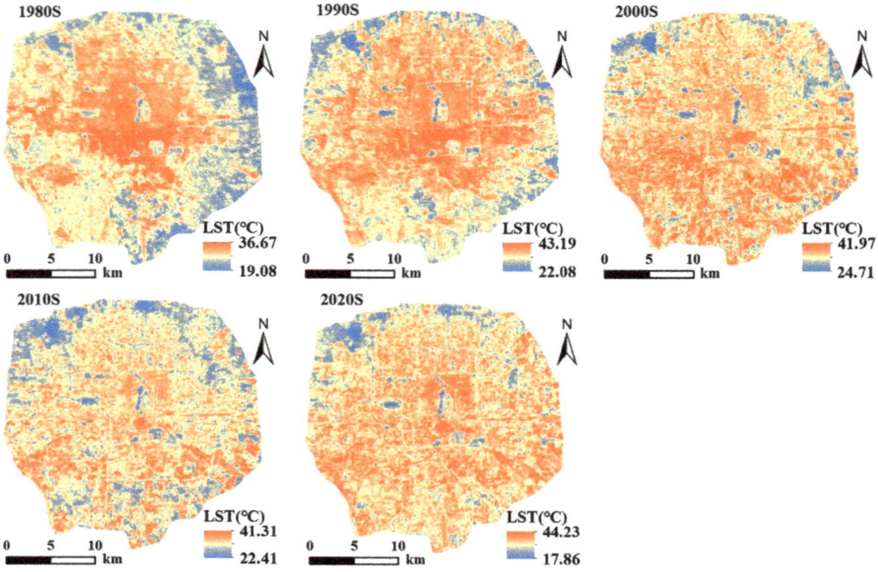

Figure 10. Spatial distribution characteristics of summer LST within the study area.

To evaluate the winter temperature regulation capacity of UGS within Beijing's Fifth Ring Road, LST was analyzed for winter months (December–February). Figure 11 presents winter LST temporal variations. Analysis reveals a spatial distribution pattern similar to summer, but with generally higher temperatures at the study area's periphery than in the central region. This "Inverse Urban Heat Island" (IUHI) phenomenon primarily results from: (1) industrial zone concentration between the Fourth and Fifth Ring Roads, and (2) frequent temperature inversion events characteristic of northern winter climates. Notably, large UGS containing evergreen vegetation (e.g., Tiantan Park) maintain significant cooling effects during winter.

3.3. Spatiotemporal Evolution of UGS Cooling Capacity

3.3.1. Spatiotemporal Variation in LST and UGS Along Profile Lines in Urban Areas

Figure 12 displays the spatiotemporal variations in LST and PV. A distinct inverse relationship exists between LST and PV: areas with high PV values (dense vegetation) exhibit lower surface temperatures, while areas with low PV (built-up/bare surfaces) display significantly elevated temperatures, demonstrating UGS cooling capacity. Temporally, LST shows a general increasing trend with growing thermal fluctuations over decades. PV values underwent more pronounced changes, reaching near-maximum values (PV ≈ 1) in numerous areas during the 2020s. This indicates that despite increased fragmentation of green spaces, enhanced vegetation vigor within individual patches demonstrates effective urban greening implementation with robust thermal regulation capabilities. This study confirms a robust inverse correlation between UGS and LST during summer, demonstrating substantial spatiotemporal UGS evolution impacts on urban thermal environments.

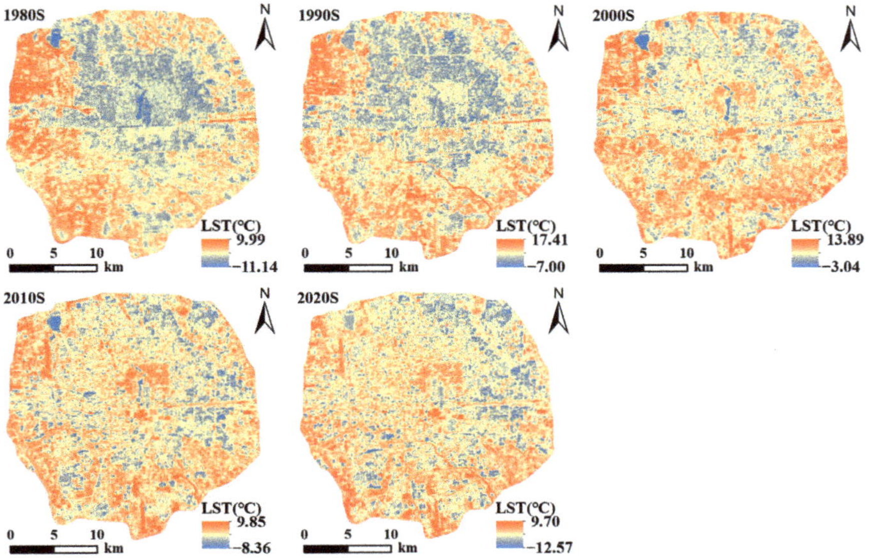

Figure 11. Spatial distribution of winter LST within the study area.

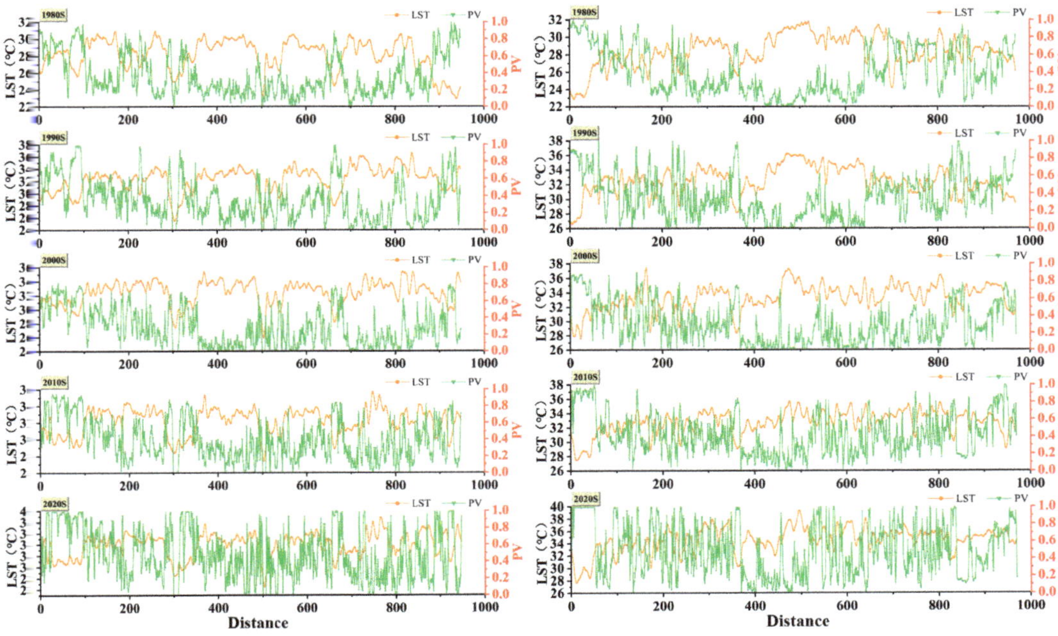

Figure 12. Left: Horizontal (east–west) profiles of summer LST and UGS in the study area; Right: Vertical (north–south) profiles of summer LST and UGS in the study area; The distance between two points on the X-axis is 30 m.

3.3.2. Spatial Pattern Changes in UGS Cooling Capacity

To quantify spatiotemporal variations in UGS cooling effects, this study analyzed the influence of individual 1 km × 1 km grid cells on LST in their eight adjacent neighbors. As Figure 13 demonstrates, grid-level cooling effects exhibited significant spatiotemporal heterogeneity. Across study periods, grid cells with pronounced warming effects clustered predominantly in the study area's southeastern sector. ArcGIS 10.7 Online comparison with reference basemaps revealed this region's characteristics: relatively low UGS coverage, extensive impervious surfaces, and dense industrial facilities—collectively elevating summer LST above surrounding areas.

Compared to the 1980s, more grid cells exhibited pronounced cooling capacity in the 2020s, with increased spatial homogeneity in their distribution. Analysis of summer vegetation coverage (Figure 5) reveals declining and increasingly fragmented green spaces at urban fringes over four decades, yet overall distribution has trended toward greater uniformity. This implies that large UGS patches alone are insufficient for optimal cooling. An integrated approach considering spatial configuration, adjacent land-use, and anthropogenic activities is essential. As shown in Figure 14, among the 100 randomly sampled points, most grid cells maintain a stable cooling intensity throughout all periods, within a range of ±1 °C, with only a few exceptions showing significant warming. Crucially, cooling peaks consistently outnumbered warming peaks, confirming UGS predominantly cooled local environments rather than contributing to warming. This provides robust evidence of UGS's sustained role in mitigating summer UHI effects.

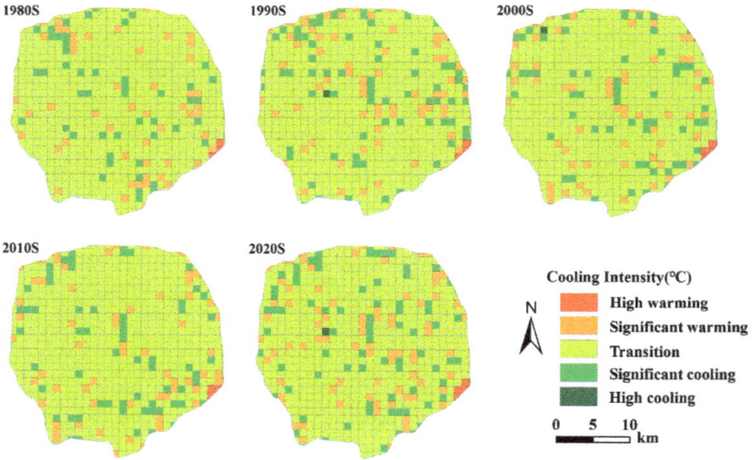

Figure 13. Spatial distribution patterns of UGS cooling capacity across different periods.

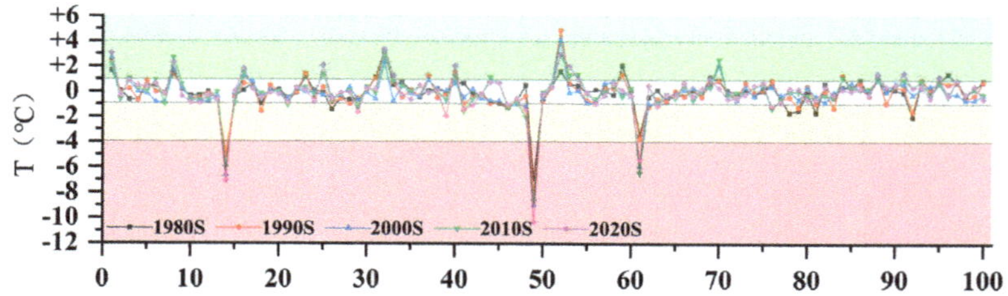

Figure 14. Line chart of crid-based cooling intensity ("+" for Cooling, "−" for Warming).

In summary, urban planning and ecological development strategies must recognize that solely augmenting green space area is insufficient to fully realize UGS cooling potential. Maximizing thermal regulation performance requires integrated consideration of UGS spatial distribution, structural composition, and local environmental contexts.

3.3.3. Interannual Variation Patterns of UGS Cooling Capacity

To assess interannual variation in UGS cooling effects, this study performed a cross-decadal comparative analysis of thermal regulation performance within fixed 1 km × 1 km grid cells (Table 3). Findings indicate short-term changes were predominantly 'transition' type (>50% per interval), peaking at 76.9% during 2010s–2020s. This demonstrates moderate inter-period stability in cooling capacity for most grids. 'Enhancement' grids remained limited in short-term intervals (typically 17–21%), contrasting with the 1980s–2020s long-term proportion of 25.4%. This suggests minimal short-term improvements but moderate long-term enhancement. 'Weakening' grids maintained a low proportion (minimum of 10.7% during 2010s–2020s), suggesting either a relatively stable UGS structure throughout this phase or a limited impact of urban expansion on the cooling capacity of UGS.

Table 3. The temporal distribution of cooling capacity intensity across grid cells.

Cooling Ability	1990s–1980s	2000s–1990s	2010s–2000s	2020s–2010s	2020s–1980s
Weakened	18.4%	21.0%	18.4%	10.7%	15.1%
Transition	63.6%	58.50%	64.2%	76.9%	59.5%
Enhanced	18.0%	20.50%	17.4%	12.4%	25.4%

Note: Weakened: $T \leq -0.5$. Transition: $-0.5\ °C < T < 0.5\ °C$. Enhanced: $0.5 \leq T$.

Overall, the period 2010s–2020s exhibited the highest thermal stability in the urban environment and was the only decade where "Enhanced" grids outnumbered "Weakened" grids. This can be attributed to the near-completion of urbanization within Beijing's Fifth Ring Road, during which the development model shifted from extensive expansion to intensive renewal. From a long-term perspective (1980s–2020s), the proportion of "Enhanced" grids was significantly higher than that of "Weakened" grids, indicating sustained improvement in the cooling capacity of UGS across specific areas.

3.4. Analysis of the Driving Factors of UGS Cooling Effects

3.4.1. Analysis of Feature Importance for Driving Factors

To systematically elucidate the relationship between LST variation and landscape patterns, feature importance analysis was conducted using XGBoost models across distinct temporal periods. As illustrated in Figure 15, the PLAND consistently demonstrated the highest feature importance from the 1990s to the 2010s, indicating that UGS proportion was the decisive predictor of LST variations during these decades. LPI was also relatively prominent only in the 1980s, highlighting the dominant role of large green patches in the early urban landscape configuration. By the 2020s, the importance of the NDVI and PV substantially increased, reflecting the growing influence of vegetation quality on environmental regulation. This shift signifies decreasing reliance on single-metric area indicators and a transition toward integrated consideration of vegetation condition and spatial configuration. Such evolution likely corresponds with advances in refined UGS management and ecological restoration technologies. Collectively, these findings reveal a clear transition in urban ecosystems: from a primary focus on UGS quantity toward an emphasis on UGS quality.

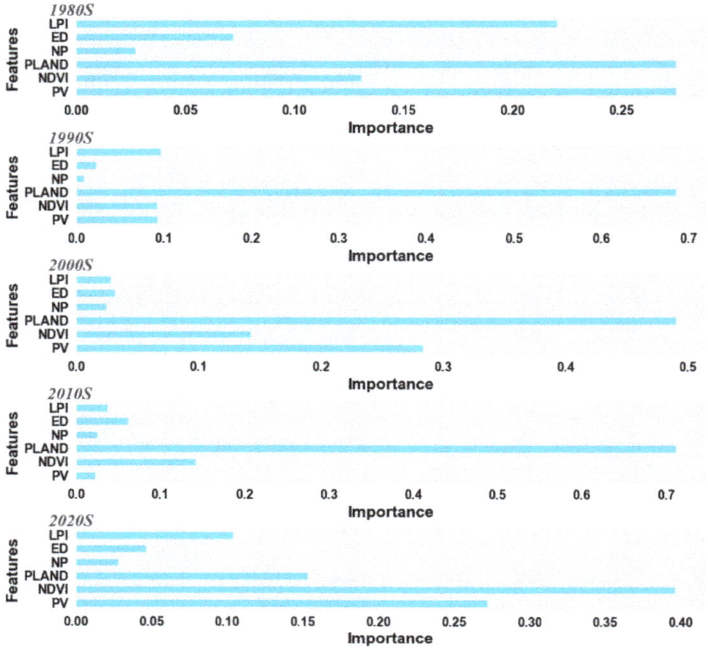

Figure 15. Feature importance of UGS cooling drivers across decades.

3.4.2. Correlation Between NDVI and LST

NDVI is a key metric for characterizing vegetation coverage on the land surface. Higher NDVI values correspond to denser vegetation, enhancing evapotranspiration (including evaporation and transpiration) and consequently reducing LST. Conversely, lower NDVI values indicate sparse vegetation and reduced evapotranspiration, typically leading to elevated LST. This study quantified the linear relationship between NDVI and LST across four decades (1980s–2020s), as presented in Figure 16.

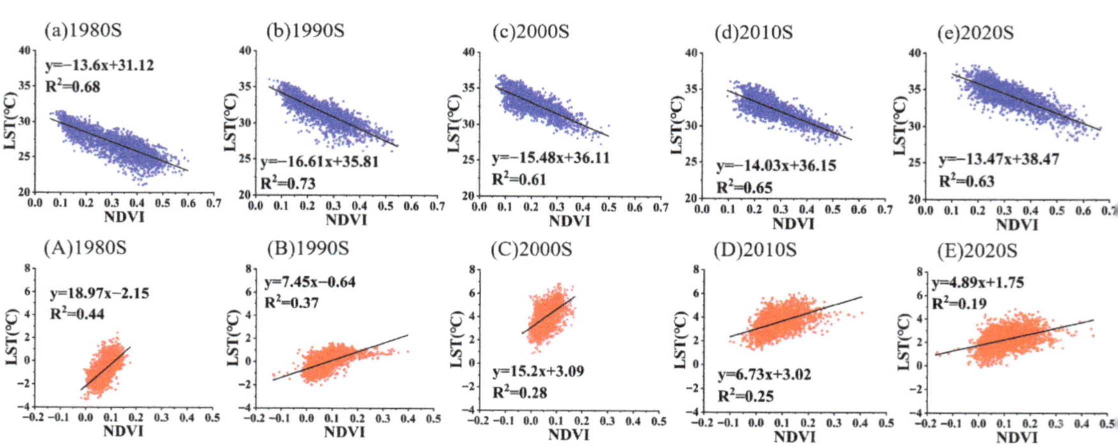

Figure 16. Linear relationships between NDVI and LST in different periods. (**a–e**) represent the summer NDVI–LST linear relationships for each period; (**A–E**) represent the winter NDVI–LST linear relationships for each period.

During summer, NDVI and LST exhibited a significant negative correlation (correlation coefficients > 0.6), indicating strong coupling. The steepest linear regression slope occurred in the 1990s, reflecting the highest cooling efficiency per unit vegetation increase during that decade. Comparison of 1980s and 2020s summer data reveals that although localized LST increases occurred due to rapid urbanization within Beijing's Fifth Ring Road, NDVI values concurrently increased in certain areas. This demonstrates that UGS has effectively moderated the urban thermal environment over the past four decades. Conversely, winter exhibited a positive NDVI-LST correlation with a weaker linear fit (R^2), and both correlation strength and regression slopes diminished progressively over the study period. Winter NDVI values were significantly lower than summer values, consistent with limited evergreen vegetation coverage and reduced physiological activity characteristic of temperate cities during cold seasons. These findings indicate that vegetation's influence on LST within the Fifth Ring Road is substantially diminished in winter.

3.4.3. Correlation Between PV and LST

As presented in Figure 17, linear relationships between PV and LST were analyzed from the 1980s to the 2020s. Results demonstrate a consistently significant negative PV-LST correlation during summer across all periods ($R^2 > 0.6$), confirming vegetation coverage's critical role in regulating urban thermal environments. The strongest association occurred in the 1990s ($R^2 = 0.72$, steepest slope), indicating peak cooling efficiency per unit vegetation increase. These findings underscore the essential function of summer heat mitigation. Conversely, winter exhibited positive PV-LST correlations with progressively diminishing explanatory power (all $R^2 < 0.4$). By the 2000s–2020s, R^2 declined to approximately 0.1, demonstrating vegetation's negligible role in winter LST regulation within Beijing's Fifth Ring Road. Given that vegetation cooling primarily depends on evapotranspiration and shading processes substantially inhibited during winter in temperate cities, these results align with expected seasonal limitations in ecological service functions.

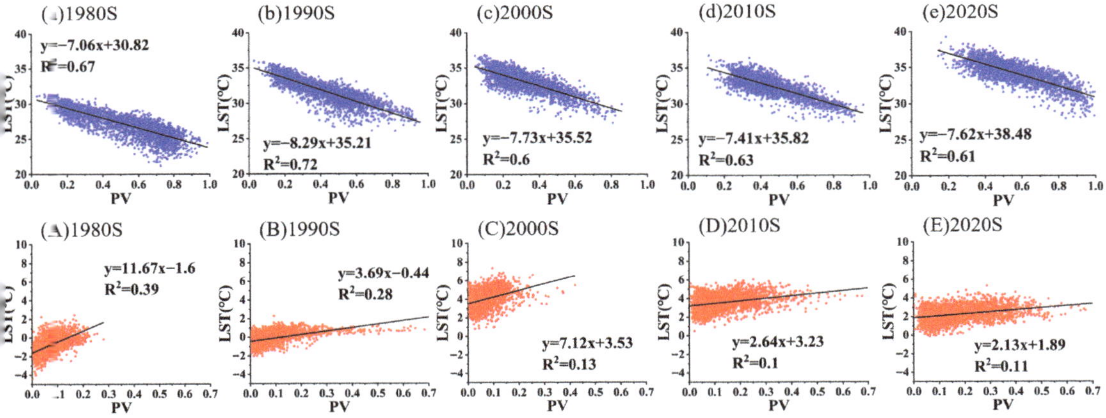

Figure 17. Linear relationships between PV and LST during different periods. Panels (**a–e**) show the summer correlations for each respective period, while panels (**A–E**) display the corresponding winter correlations.

3.4.4. Effect of UGS Area on LST

To identify optimal UHI mitigation strategies through UGS, this study analyzed UGS-LST relationships at a 500 × 500 m grid unit scale (Figure 18). Grid classification based on UGS proportion revealed decade-specific thermal regulation patterns: Maximum cooling efficiency occurred at 10–20% UGS coverage throughout 1980–2020, achieving approximately 1 °C reduction with favorable ecological efficiency. Notably, during the 1990s, LST decreased monotonically with increasing UGS, peaking at 80–90% coverage. In contrast, the 2010s exhibited an inflection point at 50–60% coverage while maintaining cooling benefits beyond this threshold. Throughout the 1980s, 2000s, and 2020s, LST declined with UGS increase but demonstrated marginal thermal reductions beyond 80% coverage. These findings indicate that while UGS expansion effectively mitigates urban heat, diminishing returns occur beyond coverage thresholds. Consequently, optimizing thermal regulation requires holistic strategies integrating socioeconomic constraints with ecological objectives.

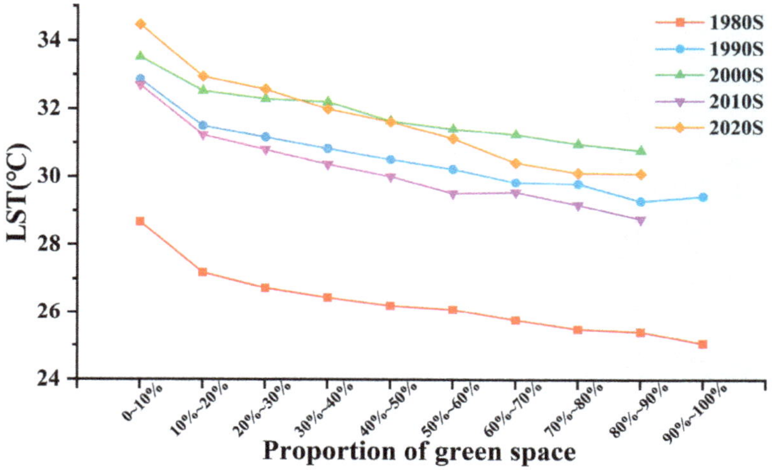

Figure 18. Temporal variation in the relationship between UGS area and LST over the past 40 years.

Since the 2000s, grid cells exceeding 90% UGS coverage have largely disappeared within Beijing's Fifth Ring Road—a consequence of rapid urbanization converting green areas into built-up land. This transformation, driven by growing demands for transportation, residential, commercial, and public infrastructure, is particularly pronounced in the Central Business District (CBD) and Financial Street, where intensive land development pressures have precipitated significant UGS decline.

4. Discussion

4.1. Temporal Changes in Dominant Factors Influencing Cooling Effects of UGS

Multi-temporal XGBoost-based feature importance analysis reveals significant temporal shifts in primary drivers of UGS cooling capacity [56,57]. During the 1980s, higher LPI and PLAND values reflected that large continuous green spaces in the early UGS configuration played a dominant role in urban thermal regulation. From the 1990s to the 2010s, the PLAND consistently demonstrated the highest influence, confirming the total green coverage's decisive role in urban heat mitigation. By the 2020s, vegetation quality indicators—particularly the NDVI and PV—exhibited substantially increased importance. This progression reflects a strategic transition from UGS quantity toward quality optimization [58–60], highlighting advancements in precision greening and ecological restoration technologies. Methodologically, temporal replication of XGBoost modeling effectively disentangles diverse drivers' influence mechanisms on urban thermal environments [61].

Furthermore, spatially constraining the study area within Beijing's Fifth Ring Road ensures representative specificity, enhancing the findings' robustness and contextual relevance.

4.2. Evolution of Cooling Mechanisms and Optimal UGS Configuration

The overall findings of this study align closely with previous research, such as Hamada (2010) [62], which identified a significant negative correlation between vegetation cover and LST [63,64]. Specifically, the regulatory function of vegetation in reducing surface temperature—primarily mediated through transpiration and shading mechanisms—was robustly validated during summer. However, unlike earlier studies that predominantly employed static, single-time-point analyses, this research utilized multi-year, multi-temporal remote sensing data to systematically reveal the dynamic evolution of the UGS cooling effect across both temporal and seasonal dimensions in Beijing. This underscores a pivotal shift in urban thermal environment management: from prioritizing UGS quantity to emphasizing vegetation quality. Consequently, our work expands traditional frameworks in temporal scale and regulatory mechanisms. Significant seasonal and temporal variations in UGS cooling efficacy were observed. Cooling effects were most pronounced in summer, whereas the vegetation–LST correlation weakened considerably in winter. This finding corroborates Zhao (2021) [48], though our study extends spatial coverage to the entire area within Beijing's Fifth Ring Road, yielding broader and more representative spatial insights [62,65–67]. Furthermore, an optimal UGS proportion threshold was identified: maximum cooling benefits occurred when vegetated areas comprised approximately 20% of the surrounding environment. This result closely aligns with the optimal configuration range proposed by Zeng (2025) [68]. Such evidence provides an operational, quantitative foundation for UGS planning and thermal environment optimization, offering substantive policy relevance and practical application value.

4.3. Optimization Strategies for UGS Cooling

Amid intensifying global climate warming, the UHI effect has become increasingly pronounced, posing significant challenges to urban ecological systems and public health. Studies demonstrate that merely expanding the UGS area is insufficient for UHI mitigation, as cooling benefits exhibit diminishing marginal returns [69]. Consequently, enhancing vegetation quality and vitality has emerged as essential for alleviating urban thermal stress and strengthening ecological resilience [58,69]. Urban planning must therefore transition from a quantitative to a qualitative paradigm, emphasizing diverse species composition, optimized ecosystem structures, and science-based ecological restoration techniques. Such approaches can elevate the holistic ecological service functions and microclimate regulation capacity of UGS. Focusing on Beijing, this study employs multi-source remote sensing data integrated with a nine-square grid spatial analysis framework to systematically investigate spatiotemporal dynamics of UGS patterns across regions and temporal scales. A notable example of policy-driven UGS change occurred during the Beijing Olympics, when a significant increase in green space resulted from targeted interventions and urban planning measures. These actions altered the urban landscape and, in turn, influenced the distribution of UGS and their associated cooling effects. Results reveal significant regional disparities in UGS distribution, driven by urban expansion, land-use transitions, and policy interventions. These findings establish a scientific foundation for optimized resource allocation and zonal management of green infrastructure, enabling spatially tailored governance strategies. This approach facilitates more efficient regulation of the urban thermal environment while advancing urban resilience and socio-ecological livability.

4.4. Limitations and Future Perspectives

This study had some limitations. First, although machine learning methods can effectively identify potential key predictors, they primarily reveal statistical associations rather than causal relationships. Therefore, to enhance the scientific rigor and interpretability of future research, it is essential to integrate causal inference models and spatial statistical approaches and to explore the multi-scale dynamics of underground space cooling effects. Second, although this study reveals long-term evolution patterns of UGS cooling effects within Beijing's Fifth Ring Road, the scope remains confined to the urban core. This exclusion of rapidly expanding suburban areas due to urban sprawl fails to comprehensively characterize urbanization impacts on metropolitan-scale cooling effects. Additionally, the decadal temporal resolution may obscure short-term variation patterns. Third, vegetation-type differentiation in cooling efficacy was not addressed. Future work should incorporate species composition and canopy structure parameters for granular analysis [70]. Furthermore, multidimensional influences of urban morphology, socio-cultural dynamics, and socioeconomic factors require more systematic consideration. Subsequent research should integrate diverse data sources to enable comprehensive multifactorial analysis.

5. Conclusions

This study systematically analyzes the spatiotemporal evolution of the cooling capacity of UGS within the Beijing Fifth Ring Road area over the past four decades and its key driving factors. The main findings are as follows:

(1) Pronounced seasonal differences: A significant negative correlation was observed between UGS and LST in summer (mean $R^2 > 0.64$), strongly confirming the critical role of vegetation in regulating the urban thermal environment and mitigating the heat island effect. In contrast, the correlation weakened considerably in winter (mean $R^2 < 0.31$), indicating a notable decline in the cooling capacity of vegetation during the cold season.

(2) Phased fluctuations in cooling effectiveness: Accelerated urbanization has intensified the fragmentation of UGS inside the Fifth Ring Road, with a clear shift in spatial configuration from continuous patches to fragmented and isolated units. Concurrently, the cooling capacity exhibited distinct phased fluctuations, characterized by an "enhancement–decline–re-enhancement" trajectory. In the temporal comparison, the proportion of grids with an enhanced cooling effect consistently remained over 10% higher than in the preceding period. This trend was notably pronounced during the 1990s–2000s, with the proportion reaching 20.5%.

(3) Spatiotemporal dynamics of landscape pattern: Urbanization has not only altered the spatial distribution of UGS but also reshaped its landscape pattern characteristics. Importance analysis based on the XGBoost model revealed significant temporal heterogeneity in the key indicators governing the cooling effect of UGS: in the 1980s, landscape-scale metrics (e.g., PLAND, LPI) were dominant, whereas by the 2020s, quality-based metrics (e.g., NDVI) became more influential. This shift indicates that the underlying influencing mechanism has transitioned from "scale-dominated" to "quality-dominated." Therefore, future urban greening strategies should shift from merely pursuing green space expansion to optimizing spatial configuration and enhancing the ecological quality of vegetation.

(4) Existence of an optimal cooling efficiency threshold: Investigation into the relationship between UGS area and LST over the past 40 years revealed that when the UGS coverage ratio increased from 0–10% to 10–20%, the cooling efficiency reached its maximum, achieving a temperature reduction of approximately 1 °C.

Author Contributions: Conceptualization, C.Y. and C.W.; methodology, C.W. and H.W.; software, L.Y.; validation, C.W. and H.W.; formal analysis, L.Y.; investigation, C.W.; resources, C.Y.; data curation, C.W. and L.Y.; writing—original draft preparation, C.W.; writing—review and editing, C.Y.; visualization, C.W.; project administration, C.Y.; funding acquisition, C.Y. All authors have read and agreed to the published version of the manuscript.

Funding: This study was funded by the National Natural Science Foundation of China (grant number 32471645, 42201123) and the Shandong Provincial Natural Science Foundation, China (grant number ZR2024MD024). This study was also supported by Shandong Provincial College and University Youth Innovation Technology Program (2024KJG015).

Institutional Review Board Statement: Not applicable.

Informed Consent Statement: Not applicable.

Data Availability Statement: The data presented in this study are available on request from the corresponding author. The data are not publicly available due to privacy.

Acknowledgments: The authors sincerely thank all contributors to this work. Special appreciation goes to C.Y. for her insightful guidance and project supervision.

Conflicts of Interest: The authors declare no conflicts of interest.

References

Hamstead, Z.A.; Kremer, P.; Larondelle, N.; McPhearson, T.; Haase, D. Classification of the heterogeneous structure of urban landscapes (STURLA) as an indicator of landscape function applied to surface temperature in New York City. *Ecol. Indic.* **2016**, *70*, 57[illegible]–585. [CrossRef]

Bai, Y.; Jia, R.X. Elite Recruitment and Political Stability: The Impact of the Abolition of China's Civil Service Exam. *Econometrica* **2016**, *84*, 677–733. [CrossRef]

Rajagopal, P.; Priya, R.S.; Senthil, R. A review of recent developments in the impact of environmental measures on urban heat island. *Sustain. Cities Soc.* **2023**, *88*, 104279. [CrossRef]

Zhang, W.; Li, W.; Zhang, C.; Hanink, D.M.; Liu, Y.; Zhai, R. Analyzing horizontal and vertical urban expansions in three East Asian megacities with the SS-coMCRF model. *Landsc. Urban Plan.* **2018**, *177*, 114–127. [CrossRef]

Yu, B.; Liu, H.; Wu, J.; Hu, Y.; Zhang, L. Automated derivation of urban building density information using airborne LiDAR data and object-based method. *Landsc. Urban Plan.* **2010**, *98*, 210–219. [CrossRef]

Stumpe, B.; Bechtel, B.; Heil, J.; Jörges, C.; Jostmeier, A.; Kalks, F.; Schwarz, K.; Marschner, B. Soil texture mediates the surface cooling effect of urban and peri-urban green spaces during a drought period in the city area of Hamburg (Germany). *Sci. Total Environ.* **2023**, *897*, 165228. [CrossRef] [PubMed]

Yin, H.; Zhao, X. Urban heat island analysis based on high resolution measurement data: A case study in Beijing. *Sustain. Cities Soc.* **2024**, *106*, 105389. [CrossRef]

Yuan, F.; Bauer, M.E. Comparison of impervious surface area and normalized difference vegetation index as indicators of surface urban heat island effects in Landsat imagery. *Remote Sens. Environ.* **2007**, *106*, 375–386. [CrossRef]

Leal, W.; Wolf, F.; Castro-Díaz, R.; Li, C.L.; Ojeh, V.N.; Gutiérrez, N.; Nagy, G.J.; Savic, S.; Natenzon, C.E.; Al-Amin, A.Q.; et al. Addressing the Urban Heat Islands Effect: A Cross-Country Assessment of the Role of Green Infrastructure. *Sustainability* **2021**, *13*, 753. [CrossRef]

Gago, E.J.; Roldan, J.; Pacheco-Torres, R.; Ordóñez, J. The city and urban heat islands: A review of strategies to mitigate adverse effects. *Renew. Sustain. Energy Rev.* **2013**, *25*, 749–758. [CrossRef]

Leal Filho, W.; Echevarria Icaza, L.; Neht, A.; Klavins, M.; Morgan, E.A. Coping with the impacts of urban heat islands. A literature based study on understanding urban heat vulnerability and the need for resilience in cities in a global climate change context. *J. Clean. Prod.* **2018**, *171*, 1140–1149. [CrossRef]

Santamouris, M. Recent progress on urban overheating and heat island research. Integrated assessment of the energy, environmental, vulnerability and health impact. Synergies with the global climate change. *Energy Build.* **2020**, *207*, 109482. [CrossRef]

Liu, R.; Dong, X.; Wang, X.; Zhang, P.; Liu, M.; Zhang, Y. Relationship and driving factors between urbanization and natural ecosystem health in China. *Ecol. Indic.* **2023**, *147*, 109972. [CrossRef]

Akdag, S.; Yıldırım, H. Toward a sustainable mitigation approach of energy efficiency to greenhouse gas emissions in the European countries. *Heliyon* **2020**, *6*, e03396. [CrossRef] [PubMed]

15. Skillington, K.; Crawford, R.H.; Warren-Myers, G.; Davidson, K. A review of existing policy for reducing embodied energy an greenhouse gas emissions of buildings. *Energy Policy* **2022**, *168*, 112920. [CrossRef]
16. Lin, T.-Y.; Chiu, Y.-h.; Chen, C.-H.; Ji, L. Renewable energy consumption efficiency, greenhouse gas emission efficiency, an climate change in Europe. *Geoenergy Sci. Eng.* **2025**, *247*, 213665. [CrossRef]
17. Wu, Z.; Zhou, Y.; Ren, Y. Green space-building integration for Urban Heat Island mitigation: Insights from Beijing's fifth rin road district. *Sustain. Cities Soc.* **2024**, *116*, 105917. [CrossRef]
18. Wu, C.; Li, J.; Wang, C.; Song, C.; Chen, Y.; Finka, M.; La Rosa, D. Understanding the relationship between urban blu infrastructure and land surface temperature. *Sci. Total Environ.* **2019**, *694*, 133742. [CrossRef]
19. Furlan, R.; Sinclair, B.R. Planning for a neighborhood and city-scale green network system in Qatar: The case of MIA Par *Environ. Dev. Sustain.* **2021**, *23*, 14933–14957. [CrossRef]
20. Huang, J.; Kong, F.; Yin, H.; Middel, A.; Liu, H.; Zheng, X.; Wen, Z.; Wang, D. Transpirational cooling and physiological response of trees to heat. *Agric. For. Meteorol.* **2022**, *320*, 108940. [CrossRef]
21. Su, Y.; Wu, J.; Zhang, C.; Wu, X.; Li, Q.; Liu, L.; Bi, C.; Zhang, H.; Lafortezza, R.; Chen, X. Estimating the cooling effect magnitud of urban vegetation in different climate zones using multi-source remote sensing. *Urban Clim.* **2022**, *43*, 101155. [CrossRef]
22. Armson, D.; Stringer, P.; Ennos, A.R. The effect of tree shade and grass on surface and globe temperatures in an urban area. *Urba For. Urban Green.* **2012**, *11*, 245–255. [CrossRef]
23. Li, P.; Wu, W.; Yin, Y. Enhanced accessibility to park cooling services in developed areas: Experimental insights on the walkabilit in large urban agglomerations. *Build. Environ.* **2025**, *272*, 112665. [CrossRef]
24. Gao, Z.; Zaitchik, B.F.; Hou, Y.; Chen, W. Toward park design optimization to mitigate the urban heat Island: Assessment of th cooling effect in five U.S. cities. *Sustain. Cities Soc.* **2022**, *81*, 103870. [CrossRef]
25. Yang, F.; Yousefpour, R.; Zhang, Y.; Wang, H. The assessment of cooling capacity of blue-green spaces in rapidly developing citie A case study of Tianjin's central urban area. *Sustain. Cities Soc.* **2023**, *99*, 104918. [CrossRef]
26. Zhou, W.; Yu, W.; Zhang, Z.; Cao, W.; Wu, T. How can urban green spaces be planned to mitigate urban heat island effect unde different climatic backgrounds? A threshold-based perspective. *Sci. Total Environ.* **2023**, *890*, 164422. [CrossRef]
27. Das, M.; Das, A.; Momin, S. Quantifying the cooling effect of urban green space: A case from urban parks in a tropical meg metropolitan area (India). *Sustain. Cities Soc.* **2022**, *87*, 104062. [CrossRef]
28. Liu, Q.; Dong, Q.; Zhang, L.; Sun, C. Summer cooling island effects of blue-green spaces in severe cold regions: A case study o harbin, China. *Build. Environ.* **2024**, *257*, 111539. [CrossRef]
29. Zhang, M.; Huang, L.; Jiang, W.; Fan, M.; Dai, W.; Han, W.; Xu, Q.; Dong, Y. Urban core greening and industrial decentralizatio lead to contrasting trends in surface urban heat islands in a metropolitan area in China. *J. Environ. Manag.* **2025**, *374*, 12404 [CrossRef] [PubMed]
30. Peng, J.; Dan, Y.; Yu, X.; Xu, D.; Yang, Z.; Wang, Q. Response of urban green space cooling effect to urbanization in the Three Rin Road area of Changsha City. *Sustain. Cities Soc.* **2024**, *109*, 105534. [CrossRef]
31. Wang, Y.; Meng, D.; Zhu, Y.; Zhang, F. Impacts of regional urbanization development on plant diversity within boundary o built-up areas of different settlement categories in Jinzhong Basin, China. *Landsc. Urban Plan.* **2009**, *91*, 212–218. [CrossRef]
32. Xiao, X.D.; Dong, L.; Yan, H.; Yang, N.; Xiong, Y. The influence of the spatial characteristics of urban green space on the urba heat island effect in Suzhou Industrial Park. *Sustain. Cities Soc.* **2018**, *40*, 428–439. [CrossRef]
33. Shah, A.; Garg, A.; Mishra, V. Quantifying the local cooling effects of urban green spaces: Evidence from Bengaluru, India. *Lands Urban Plan.* **2021**, *209*, 104043. [CrossRef]
34. Kabisch, N.; Frantzeskaki, N.; Pauleit, S.; Naumann, S.; Davis, M.; Artmann, M.; Haase, D.; Knapp, S.; Korn, H.; Stadler, J.; et a Nature-based solutions to climate change mitigation and adaptation in urban areas: Perspectives on indicators, knowledge gap barriers, and opportunities for action. *Ecol. Soc.* **2016**, *21*, 39. [CrossRef]
35. Fedele, G.; Donatti, C.I.; Harvey, C.A.; Hannah, L.; Hole, D.G. Transformative adaptation to climate change for sustainabl social-ecological systems. *Environ. Sci. Policy* **2019**, *101*, 116–125. [CrossRef]
36. Wolch, J.R.; Byrne, J.; Newell, J.P. Urban green space, public health, and environmental justice: The challenge of making citie 'just green enough'. *Landsc. Urban Plan.* **2014**, *125*, 234–244. [CrossRef]
37. Rasul, A.; Balzter, H.; Smith, C.; Remedios, J.; Adamu, B.; Sobrino, J.A.; Srivanit, M.; Weng, Q.; Millington, A.; Nagendra, H.; et a A Review on Remote Sensing of Urban Heat and Cool Islands. *Land* **2017**, *6*, 38. [CrossRef]
38. Pritipadmaja; Garg, R.D.; Sharma, A.K. Assessing the Cooling Effect of Blue-Green Spaces: Implications for Urban Heat Islanc Mitigation. *Water* **2023**, *15*, 2983. [CrossRef]
39. Budzik, G.; Sylla, M.; Kowalczyk, T. Understanding Urban Cooling of Blue-Green Infrastructure: A Review of Spatial Data anc Sustainable Planning Optimization Methods for Mitigating Urban Heat Islands. *Sustainability* **2025**, *17*, 142. [CrossRef]
40. Zhao, Z.; Zhang, G.; Chen, Q.; Cai, D.; Meng, F.; Long, X.; Zhang, T.; Wang, Y.; Xu, T.; Yang, H.; et al. Gold exploration usin multi-source remote sensing data in the northern part of the Wa State, Myanmar. *Ore Geol. Rev.* **2025**, *183*, 106703. [CrossRef]

Bowler, D.E.; Buyung-Ali, L.; Knight, T.M.; Pullin, A.S. Urban greening to cool towns and cities: A systematic review of the empirical evidence. *Landsc. Urban Plan.* **2010**, *97*, 147–155. [CrossRef]

Zhou, W.; Cao, W.; Wu, T.; Zhang, T. The win-win interaction between integrated blue and green space on urban cooling. *Sci. Total Environ.* **2023**, *863*, 160712. [CrossRef]

Wang, M.; Song, H.Y.; Zhu, W.; Wang, Y.C. The Cooling Effects of Landscape Configurations of Green-Blue Spaces in Urban Waterfront Community. *Atmosphere* **2023**, *14*, 833. [CrossRef]

Lin, B.-S.; Tsai, Y.-H.; Chang, H.-C.; Yu, C.-C.; Hsieh, C.-I. Building with nature: Morphological spatial pattern of green infrastructure in urban heat mitigation. *Build. Environ.* **2025**, *279*, 113087. [CrossRef]

Pan, Z.W.; Xie, Z.Y.; Wu, L.Y.; Pan, Y.; Ding, N.; Liang, Q.S.; Qin, F. Simulation of Cooling Island Effect in Blue-Green Space Based on Multi-Scale Coupling Model. *Remote Sens.* **2023**, *15*, 2093. [CrossRef]

Yu, Z.; Yang, G.; Zuo, S.; Jørgensen, G.; Koga, M.; Vejre, H. Critical review on the cooling effect of urban blue-green space: A threshold-size perspective. *Urban For. Urban Green.* **2020**, *49*, 126630. [CrossRef]

Gomez-Baggethun, E.; Barton, D.N. Classifying and valuing ecosystem services for urban planning. *Ecol. Econ.* **2013**, *86*, 235–245. [CrossRef]

Zhao, J.C.; Zhao, X.; Liang, S.L.; Wang, H.Y.; Liu, N.J.; Liu, P.; Wu, D.H. Dynamic Cooling Effects of Permanent Urban Green Spaces in Beijing, China. *Remote Sens.* **2021**, *13*, 3282. [CrossRef]

Yan, D.Y.; Xu, L.; Wang, Q.F.; Feng, J.; Wu, X.X. Quantifying Landscape Effects on Urban Park Thermal Environments Using ENVI-Met and 3D Grid Profile Analysis. *Forests* **2025**, *16*, 1085. [CrossRef]

Pan, T.; Kuang, W.H.; Pan, R.Y.; Niu, Z.G.; Dou, Y.Y. Hierarchical Urban Land Mappings and Their Distribution with Physical Medium Environments Using Time Series of Land Resource Images in Beijing, China (1981–2021). *Remote Sens.* **2022**, *14*, 580. [CrossRef]

Kiener, D.; Láska, K.; Stachoň, Z. Assessment and validation of Land Surface Temperature retrieval algorithms using Landsat 8 TIRS data in Antarctic ice-free areas. *Polar Sci.* **2024**, *42*, 101127. [CrossRef]

Alavipanah, S.; Haase, D.; Makki, M.; Nizamani, M.M.; Qureshi, S. On the Spatial Patterns of Urban Thermal Conditions Using Indoor and Outdoor Temperatures. *Remote Sens.* **2021**, *13*, 640. [CrossRef]

Rodríguez, L.R.; Ramos, J.S.; de la Flor, F.J.S.; Domínguez, S.A. Analyzing the urban heat Island: Comprehensive methodology for data gathering and optimal design of mobile transects. *Sustain. Cities Soc.* **2020**, *55*. [CrossRef]

Jeon, G.; Park, Y.; Guldmann, J.M. Impacts of Urban Morphology on Seasonal Land Surface Temperatures: Comparing Grid- and Block-Based Approaches. *Isprs Int. J. Geo-Inf.* **2023**, *12*, 482. [CrossRef]

Yang, C.; Kui, T.; Zhou, W.; Fan, J.; Pan, L.; Wu, W.; Liu, M. Impact of refined 2D/3D urban morphology on hourly air temperature across different spatial scales in a snow climate city. *Urban Clim.* **2023**, *47*, 101404. [CrossRef]

Huang, J.D.; Lu, X.H.; Wang, Y.C. Spatio-Temporal Changes and Key Driving Factors of Urban Green Space Configuration on Land Surface Temperature. *Forests* **2024**, *15*, 812. [CrossRef]

Bhattarai, S.; Banjara, P.; Pandey, V.P.; Aryal, A.; Pradhan, P.; Al-Douri, F.; Pradhan, N.R.; Talchabhadel, R. Quantifying the cooling effects of blue-green spaces across urban landscapes: A case study of Kathmandu Valley, Nepal. *Urban Clim.* **2025**, *61*, 102493. [CrossRef]

Knobel, P.; Dadvand, P.; Alonso, L.; Costa, L.; Español, M.; Maneja, R. Development of the urban green space quality assessment tool (RECITAL). *Urban For. Urban Green.* **2021**, *57*, 126895. [CrossRef]

Gundersen, V.; Frivold, L.H.; Löfström, I.; Jørgensen, B.B.; Falck, J.; Øyen, B.-H. Urban woodland management—The case of 13 major Nordic cities. *Urban For. Urban Green.* **2005**, *3*, 189–202. [CrossRef]

Acdas, A. Influence of Urban Green Spaces on Quality of Life and Health with Smart City Design. *Land* **2023**, *12*, 960. [CrossRef]

Wei, J.Q.; Li, Y.H.; Jia, L.Q.; Liu, B.T.; Jiang, Y.H. The Impact of Spatiotemporal Effect and Relevant Factors on the Urban Thermal Environment Through the XGBoost-SHAP Model. *Land* **2025**, *14*, 394. [CrossRef]

Hamada, S.; Ohta, T. Seasonal variations in the cooling effect of urban green areas on surrounding urban areas. *Urban For. Urban Green.* **2010**, *9*, 15–24. [CrossRef]

Zhong, Y.; Li, S.; Liang, X.; Guan, Q. Causal inference of urban heat island effect and its spatial heterogeneity: A case study of Wuhan, China. *Sustain. Cities Soc.* **2024**, *115*, 105850. [CrossRef]

Imhoff, M.L.; Zhang, P.; Wolfe, R.E.; Bounoua, L. Remote sensing of the urban heat island effect across biomes in the continental USA. *Remote Sens. Environ.* **2010**, *114*, 504–513. [CrossRef]

Yang, C.B.; He, X.Y.; Wang, R.H.; Yan, F.Q.; Yu, L.X.; Bu, K.; Yang, J.C.; Chang, L.P.; Zhang, S.W. The Effect of Urban Green Spaces on the Urban Thermal Environment and Its Seasonal Variations. *Forests* **2017**, *8*, 153. [CrossRef]

Lin, W.; Zhao, H.Y.; Sun, S.B.; Xu, X.Y.; Huang, T.T.; Zhu, J.N. Green Space Cooling Effect and Contribution to Mitigate Heat Island Effect of Surrounding Communities in Beijing Metropolitan Area. *Front. Public Health* **2022**, *10*, 870403. [CrossRef]

Aram, F.; Higueras García, E.; Solgi, E.; Mansournia, S. Urban green space cooling effect in cities. *Heliyon* **2019**, *5*, e01339. [CrossRef]

68. Zeng, W.X.; Yang, K.; Zhang, S.H.; Bi, C.Y.; Liu, J.; Yang, X.F.; Rao, Y.; Ma, Y. Configuration of Green-Blue-Grey Spaces for Efficier Cooling of Urban Physical and Perceptual Thermal Environments. *Land* **2025**, *14*, 645. [CrossRef]
69. Xu, Z.; Zhao, S. Scale dependence of urban green space cooling efficiency: A case study in Beijing metropolitan area. *Sci. Tot Environ.* **2023**, *898*, 165563. [CrossRef]
70. Wang, X.; Dallimer, M.; Scott, C.E.; Shi, W.; Gao, J. Tree species richness and diversity predicts the magnitude of urban heat islan mitigation effects of greenspaces. *Sci. Total Environ.* **2021**, *770*, 145211. [CrossRef]

sustainability

MDPI

Article

Ecological Security Assessment Based on Sensitivity, Connectivity, and Ecosystem Service Value and Pattern Construction: A Case Study of Chengmai County, China

oyao Zhao, Yuan Feng, Qing Liu, Yixian Mo, Shuhai Zhuo and Peng Zhou *

College of Tropical Agriculture and Forestry, Hainan University, Haikou 570228, China;
23220953000049@hainanu.edu.cn (Y.Z.); lancer0918@163.com (Y.F.); 23210834000006@hainanu.edu.cn (Q.L.);
yixianmo@hainanu.edu.cn (Y.M.); hilario311@hainanu.edu.cn (S.Z.)
* Correspondence: zp85611@163.com

Abstract

Against the backdrop of continuous natural space loss and accelerated urbanization, considerable attention has been directed toward balancing economic development demands with the protection of fragile ecosystems within limited spatial boundaries to achieve regional sustainable development. This study therefore focuses on Chengmai County, a small-scale region prioritizing both green development and ecological conservation. Land-use changes and trends in ecosystem services value (ESV) from 2000 to 2020 were analyzed. An ecological security assessment model was developed, integrating ecosystem services, ecological sensitivity, and landscape connectivity, which enabled the identification of areas with high ecological security value as ecological sources. Ecological corridors and nodes were extracted using the minimum cumulative resistance model and the gravity model, culminating in the construction of Chengmai County's ecological security pattern through overlay analysis. The main findings are summarized as follows: (1) Construction land expanded rapidly between 2000 and 2020. The ecological sensitivity of Chengmai County displayed a spatial distribution pattern of "high in the south, low in the north," while ESV exhibited a pattern of "high in the central-south and low in the northeast," showing an overall increasing trend. (2) The overall ecological security status was relatively favorable. A total of 10 ecological nodes and 45 ecological corridors were identified, including 16 core corridors. (3) Based on these analyses, an ecological security pattern described as "one axis, two belts, and three zones" was established for Chengmai County. This study provides a practical spatial strategy for ecological conservation and sustainable development in Chengmai County and offers a transferable methodological framework for similar coastal regions facing development pressures.

Keywords: ecological security pattern; ecosystem services; ecosystem sensitivity; landscape connectivity

1. Introduction

The ongoing loss of natural spaces, coupled with intensified urbanization, presents a critical challenge for many regions in striking a balance between economic development and ecological conservation [1]. Coastal areas, characterized by dense populations [2,3], robust economic activity, and high ecological sensitivity, host critical ecosystems such as mangroves and coral reefs [4]. Simultaneously, these regions are subjected to multiple

 https://doi.org/10.3390/su172310724

pressures, including land-based pollution, coastal erosion, and sea-level rise, which render their ecological security issues particularly pronounced and urgent [5,6].

Within this context, Chengmai County in Hainan Province is selected as a representative case of a developing coastal region confronting distinct and pressing ecological security pressures. As an integral component of the Haikou Economic Circle and a strategic reserve for the development of the free trade port, Chengmai County is undergoing a critical phase of accelerated urbanization and industrialization. This growth is manifested in the rapid expansion of construction land, which increasingly encroaches upon ecological spaces. Concurrently, long-term dependence on coastal resources for developing resource-intensive industries has triggered localized issues, including mangrove degradation, the weakening of coastal ecological functions, and biodiversity impairment. The conflict between anticipated rapid economic growth and the carrying capacity of its fragile ecosystems poses a practical challenge for optimizing the local ecological spatial structure, specifically in curbing ecological fragmentation and maintaining the continuity of key ecological processes. Furthermore, China's heightened emphasis on ecological and environmental protection is demonstrated by its endorsement of Hainan Island's designation as a National Ecological Civilization Pilot Zone, which prioritizes ecological conservation and green, low-carbon development. This national policy framework provides a clear mandate and urgent direction for Chengmai County to address its development challenges by establishing an ecological security framework.

The ESP is defined as a spatial network structure established through the identification and protection of key ecological elements to maintain regional ecosystem stability, safeguard ecosystem service functions, and preserve biodiversity [7,8]. This concept originated from the "patch-corridor-matrix" model in landscape ecology and the theory of island biogeography [9]. In 1995, Kongjian Yu introduced the concept of the "landscape security pattern," which operationalized the security thresholds of ecological processes, thereby laying the foundation for spatial identification and planning. Currently, a mainstream framework has been established in ESP research, focusing on three core components: ecological source identification, resistance surface construction, and ecological corridor extraction [10,11]. Methodologically, ecological sources are typically identified using MSPA and dPC [12,13]. The regional contribution of ecological functions is quantified through an ESV assessment [14], while ecologically sensitive and vulnerable areas are delineated via an ecological sensitivity assessment [15]. Furthermore, ecological flow paths are simulated, and corridors are extracted through the widespread application of the MCR model and circuit theory [16]. However, significant limitations are commonly encountered when these methods are applied individually to small-scale coastal counties. MSPA and dPC are highly dependent on land use classification accuracy, primarily reflecting landscape structural connectivity, while inadequately capturing connectivity in ecological processes [17,18]. ESV assessments, which often rely on static value coefficients [19], exhibit limited responsiveness to internal structural and qualitative differences within local ecosystems [20,21]. For ecological sensitivity modeling, the selection of factors and assignment of their weights involve a degree of subjectivity [22], which potentially compromises the accuracy of the results [23]. Furthermore, the traditional 'generalized' resistance surface used in the MCR model fails to account for functional differences among species, thus inaccurately characterizing functional connectivity.

However, the application of these ESP methodologies to small-scale coastal counties is particularly complicated by issues of parameter calibration and scale compatibility [24,25]. This complication arises due to the convergence of inherent ecological fragility, intense development pressure, and complex spatial heterogeneity in such regions. Consequently, traditional ESP models developed at macro scales demonstrate significant limitations when directly applied, specifically in ecological source identification, comprehensive assessment, and corridor prioritization.

Furthermore, a significant imbalance is consistently observed in the existing literature about both research scale and methodological application. Regarding research scale, studies are predominantly concentrated in developed regions [26], large urban agglomerations, or ecologically fragile zones [27,28]. At the same time, insufficient attention has been given to small-scale coastal counties, which are characterized by high ecological sensitivity and substantial development pressure [29]. Methodologically, many studies still emphasize single-method approaches or isolated factor analyses, lacking integrated ESP construction that incorporates multiple methods and indicators [30–32]. Case studies of small-scale counties remain relatively scarce, particularly those that systematically integrate landscape connectivity [33,34], ecosystem services, and ecological sensitivity for comprehensive ecological source identification [35]. These regions are typically characterized by limited spatial extent, where the conflict between ecological conservation and economic development is particularly acute. Nevertheless, refined ESP construction methodologies and practical case studies specifically tailored to their scale-specific characteristics remain a notably understudied area in the literature, both domestically and internationally [36]. Therefore, this study is designed to address these research gaps by adopting Chengmai County as a typical case study. It seeks to develop a comprehensive pathway for constructing an ESP that integrates multi-source data and synthesized methodologies. The aim is to provide a transferable framework and offer a reference for refined management solutions to support the sustainable development of analogous coastal areas.

Based on this research background, Chengmai County was selected as a typical case study to construct an ESP that integrates multi-source data and comprehensive methods, clarifying the spatial mechanisms for the synergistic optimization of ecological conservation and economic development. The study integrated three assessment methods—ecological sensitivity, ESV, and landscape connectivity—to collectively identify ecological sources, thereby mitigating biases associated with single-method approaches. An integrated resistance surface was subsequently constructed, and multi-level ecological corridors and nodes were extracted using a combination of the MCR model and the gravity model, ultimately forming an integrated ESP of "sources-corridors-nodes." The specific objectives of this study were defined as follows: (1) To elucidate the spatiotemporal evolution characteristics of land use and ecosystem service value in Chengmai County from 2000 to 2020, and to reveal the spatial differentiation patterns and underlying causes of ecological sensitivity. (2) To identify core ecological sources through the integrated "sensitivity-service-connectivity" framework, construct a comprehensive resistance surface, and extract key corridors and nodes. (3) To establish an integrated "sources-corridors-nodes" ESP for Chengmai County and propose targeted spatial optimization strategies. This study provides a scientific basis for territorial spatial planning and ecological restoration in Chengmai County, while also offering a methodological reference for coordinating conservation and development practices in similar coastal regions.

2. Materials and Methods

2.1. Study Area

Chengmai County, located at 19°23′–20°01′ N and 109°45′–110°15′ E, lies in the northwestern part of Hainan Island (Figure 1). It borders the Qiongzhou Strait to the north, Haikou City, the provincial capital of Hainan, to the east, and Danzhou City, a prefecture-level city, to the west. The county spans 56.25 km from east to west and 70 km from north to south, encompassing a total land area of 2072.97 km^2 and a marine area of 470.53 km^2. The region features a tropical monsoon climate, with forest land as the dominant land cover. Topographically, the county is higher in the northwest and lower in the southeast, and it is traversed by several water systems, including the Nandu River, Wencai Stream, and Jiale Stream. With the continuous acceleration of urbanization, the area of construction land increased by 2.89 times between 2000 and 2020, while natural areas have become increasingly fragmented. As a region with pending economic development, Chengmai County faces potential conflicts between high-quality economic growth and high-level ecological conservation [37]. Therefore, the scientific construction of an ESP is essential to balance ecosystem health with sustainable economic development.

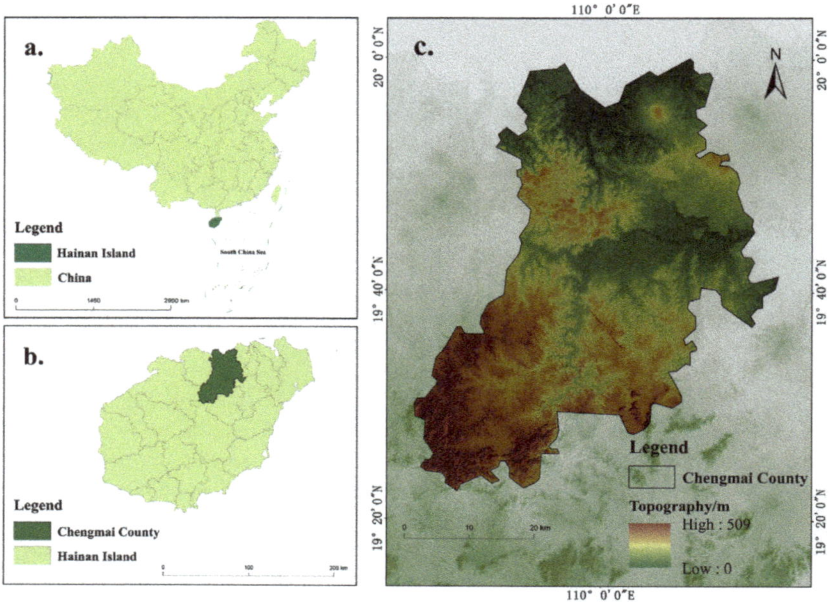

Figure 1. Location and characteristics of the study area: (**a**) China; (**b**) Hainan Island; (**c**) Chengmai County.

2.2. Data Source

To ensure consistent spatial resolution for overlay analysis, all lower-resolution datasets, such as NDVI and rainfall data, were uniformly resampled to a 30-m resolution, aligning with the highest precision land-use and DEM data (Table 1). Resampling techniques, including bilinear interpolation and nearest-neighbor assignment, were employed [29,38]. This preprocessing approach optimally preserved the spatial details of high-precision data, thereby ensuring the accuracy of Ecological Source identification and corridor extraction [39,40].

2.2.1. Land Use Data Preprocessing

Based on the National Land Use Remote Sensing Monitoring Classification System, land use data were categorized into six primary types: cultivated land, forest land, grassland, water bodies, construction land, and unutilized land [41]. Following established methodologies, the land use data for Chengmai County for 2000, 2010, and 2020 were divided into 2243 grid cells, each measuring 1 km × 1 km, using the ArcGIS 10.8 Fishnet tool [42]. The area of each land use type within every grid cell was then calculated [43].

Table 1. Data sources and information.

Type	Resolution (m)	Source	Year
Land-Use Data	30	National Catalogue Service For Geographic Information (https://www.webmap.cn/ (accessed on 16 June 2025))	2000, 2010, 2020
DEM	30	Geospatial Data Cloud (http://www.gscloud.cn/ (accessed on 17 June 2025))	2020
Soil	1000	Soil Dataset (v2.0) from the World Soil Database (HWSD) of the Cold and Arid Regions Scientific Data Centre	2023
Precipitation	1000	Geospatial Data Cloud (http://www.gscloud.cn/ (accessed on 1 June 2025))	2022
Population density data	90	(https://landscan.ornl.gov/ (accessed on 3 June 2025))	2022
NDVI	1000	Geospatial Data Cloud (http://www.Gscloud.cn (accessed on 1 June 2025))	2021
Road and water data body buffer data	-	National Catalogue Service For Geographic Information (https://www.webmap.cn/ (accessed on 20 July 2025))	2021
Socioeconomic data	-	The Hainan Statistical Yearbook, China Statistical Yearbook, and National Agricultural Product Compilation.	2000–2020

2.3. Methods

The technical framework of this research is presented in Figure 2.

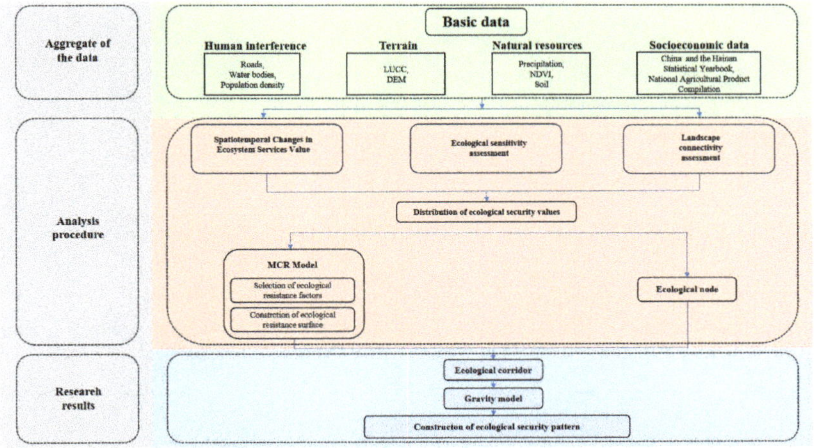

Figure 2. Technology framework.

2.3.1. Ecosystem Sensitivity Assessment

Ecological sensitivity serves as a comprehensive metric for assessing ecological quality and the rationality of land use, reflecting an ecosystem's self-regulatory capacity in response to anthropogenic disturbances and natural environmental changes [44]. Lower ecosystem sensitivity indicates a less stable ESP. At the core of ecological sensitivity assessment is the development of an evaluation indicator system [45]. Due to the multitude of potential sensitivity factors and their variability across regions and objectives, no universally standardized protocol exists; therefore, the system must be tailored to the specific conditions of the study area [46,47]. Building upon existing research and considering the natural environmental conditions, data availability, and temporal relevance for Chengmai County, an evaluation indicator system for ecosystem sensitivity was constructed (Table 2). This system was developed with reference to the Guidelines for Resource and Environmental Carrying Capacity Evaluation and Territorial Spatial Suitability Assessment (Trial) issued by the Ministry of Natural Resources in January 2020. The framework encompasses three main dimensions: soil erosion, human disturbance, and habitat quality [48]. The study employed the natural breaks method to classify the ecological sensitivity index, categorizing the sensitivity levels of factors into four grades: Low Sensitivity, Moderate Sensitivity, High Sensitivity, and Extreme Sensitivity. This approach effectively maximizes the differentiation between categories, ensuring the objectivity and rationality of the classification results [49]. The specific formula for calculating the sensitivity model is presented as follows:

$$S = \sum_{i=1}^{n} (W_i \times C_i) \quad (1)$$

In this model, S represents the integrated ecological sensitivity assessment result for the watershed; i denotes the index of an individual evaluation factor; n is the total number of factors considered; W_i is the weight assigned to the i-th evaluation factor; and C_i is the standardized assessment value of the i-th factor.

Table 2. Ecological Environment Sensitivity Evaluation System.

Evaluation Factors		Low Sensitivity	Moderate Sensitivity	High Sensitivity	Extreme Sensitivity
Soil Erosion	Rainfall Erosivity/(MJ·mm/hm^2·h)	<14,074.39	[14,074.39, 14,696.10)	[14,696.10, 15,202.60)	≥15,202.60
	Soil Erodibility/[t·hm^2·h/(hm^2·MJ·mm^2)]	<0.010	[0.010, 0.014)	[0.014, 0.018)	≥0.018
	Topographic Roughness/m	<26	[26, 53)	[53, 105)	[105, 272]
Human Disturbance	Population Density	≥240	[50, 240)	[14, 50)	<14
	Road Buffer Zone/m	<100	[100, 150)	[150, 200)	≥200
Habitat Quality	Land Use Type	Construction Land	Unutilized Land	Cultivated Land, Grassland	Forest Land, Water Bodies
	Elevation/m	[0, 62)	[62, 117)	[117, 213)	≥213
	Slope/(°)	[0, 4)	[4, 9)	[9, 17)	≥17
	Aspect	Flat terrain, due north, northwest	Northeast, Due West	East	South, Southeast, Southwest
	Water Body Buffer Zone/m	Buffer distance > 500	Buffer distance 200–500	Buffer distance 50–200	Buffer distance < 50
	NDVI	[0, 0.23)	[0.23, 0.51)	[0.51, 0.74)	[0.74, 1]

Subjective weighting methods are often characterized by significant subjectivity and cognitive bias. Although objective weighting methods avoid human subjective bias, they may assign inappropriately low weights to important indicators with small data dispersion, resulting in substantial deviations from actual conditions. To overcome the limitations of single-weighting approaches, a combined weighting method integrating the AHP and entropy weight methods was employed in this study, as detailed in Table 3. The multiplicative synthesis formula was utilized for combination scoring. Within the AHP framework, ten domain experts were invited to evaluate the relative importance of factors at the same hierarchy level using the 1–9 scale method. Expert evaluations were integrated through the geometric mean method to construct a comprehensive judgment matrix, from which initial factor weights were calculated and consistency was verified. The multiplicative synthesis formula is presented below:

$$w_j = \frac{W_{s,j} \cdot W_{e,j}}{\sum_{k=1}^{11} W_{s,k} \cdot W_{e,k}} \quad (2)$$

In the formula, wj denotes the combined weight of the j-th indicator; Ws,j represents the subjective weight of the j-th indicator, derived from the AHP; We,j indicates the objective weight of the j-th indicator, obtained through the Entropy Weight Method; and k is the total number of indicators.

Table 3. Weights of the evaluation indicators. Note: w_{Ai} represents the subjective weight determined by the AHP; w_{Bi} represents the objective weight determined by the Entropy Weight Method; w_i represents the final combined weight calculated via the multiplicative integration method.

Weight Type	DEM	Slope Gradient	Slope Aspect	Water Body Buffer Zone	NDVI	Land Use Type	Rainfall Erosive Power	Soil Erodibility	Terrain Undulation	Population Density	Road Buffer Zone
w_{Ai}	0.06	0.07	0.05	0.09	0.22	0.08	0.08	0.12	0.08	0.10	0.05
w_{Bi}	0.09	0.09	0.09	0.09	0.09	0.09	0.08	0.08	0.09	0.12	0.08
w_i	0.06	0.07	0.05	0.09	0.22	0.08	0.07	0.11	0.08	0.13	0.04

2.3.2. Ecosystem Service Value Assessment

(1) Ecosystem Service Value Assessment

Based on recent studies on ESV and the equivalent value table for Chinese terrestrial ecosystems established by Xie Gaodi et al. [50] the economic value of one standard equivalent factor for ecosystems in China was determined to be 3406.50 yuan per hectare. The average actual grain yield in Chengmai County during the study period was 5192.26 kg/ha, compared to the national average of 5578.71 kg/ha [43]. This ratio yielded a correction coefficient of 0.9307 for the ecosystem service equivalent value in the study area. Accordingly, the economic value of one standard equivalent factor was adjusted to 3077.36 yuan/ha for Chengmai County. Furthermore, by incorporating insights from studies on specific regions of Hainan Island conducted by Lei Jinrui et al. [43], the ESV equivalents for Chengmai County were finalized. The value equivalent for built-up land was set to zero. The ESV coefficient for the study area was derived based on these adjusted value equivalents and the spatial distribution of each land use type (Table 4). The total ESV was calculated using the following formula:

$$ESV = \sum_{i=1}^{n} A_i \times VC_i \quad (3)$$

$$ESV_f = \sum_{i=1}^{n} A_i \times VC_{fi} \quad (4)$$

$$VC_i = \sum_{j=1}^{k} EC_{ij} \times E_a \quad (5)$$

In the equation, ESV represents the total value of ecosystem services; A_i is the area of the i-th land use type; VC_i is the ESV coefficient for the i-th land use type; ESV_f denotes the value of the f-th ecosystem service; VC_{fi} is the ESV coefficient for the f-th ecosystem service of the i-th land use type; EC_f represents the value equivalent of the f-th ecosystem service for a given land use type; and Ea is the standard equivalent ESV for the study area, which is 3077.36 yuan/hm^2.

Table 4. Ecosystem service value coefficients per unit area for land-use types (yuan/hm^2).

Ecosystem Service	Secondary Type	Cultivated Land	Forest Land	Grassland	Water Bodies	Construction Land	Unutilized Land
Provisioning Services	Food Production	3169.77	800.14	892.46	2031.12	0	30.77
	Raw Material Production	892.46	1846.47	1323.3	1138.66	0	92.32
	Water Supply	−2800.48	954.01	738.59	16,741.33	0	61.55
Regulating Services	Gas Regulation	2523.51	6093.35	4646.95	4123.78	0	307.75
	Climate Regulation	1323.3	18,218.5	12,309.8	9078.48	0	276.97
	Purification of the Environment	369.29	5323.99	4062.23	14,094.72	0	892.46
	Hydrological Regulation	3477.52	11,909.73	9016.93	194,617.94	0	584.72
Supporting Services	Soil Conservation	2061.89	7416.65	5662.51	4985.47	0	369.29
	Nutrient Cycling Maintenance	430.84	553.94	430.84	400.07	0	30.77
	Biodiversity	492.39	6739.62	5139.34	16,033.51	0	338.52
Cultural Services	Aesthetic Landscape	215.42	2954.35	2277.31	10,186.36	0	153.87

(2) Ecosystem Service Change Index (ESCI)

The ESCI was employed to quantify relative gains or losses in ecosystem services. To comply with measurement accuracy and presentation standards, all ESCI values utilized for mapping and analysis were uniformly rounded to two decimal places. The calculation formula is as follows:

$$ESCI_x = \frac{ES_{CUR_x} - ES_{HIS_x}}{ES_{HIS_x}} \quad (6)$$

In the equation, $ESCI_x$ represents the change index of a single ecosystem service; ES_{CURx} denotes the ESV at the final state; and ES_{HISx} indicates the ESV at the initial state.

2.3.3. Landscape Connectivity Assessment

MSPA is an image-processing method based on mathematical morphology principles, designed to evaluate the morphological connectivity of landscapes. This method identifies patches with significant ecological functions within land use types by measuring, segmenting, and analyzing raster images. Through this process, seven landscape elements are distinguished: core, edge, bridge, islet, perforation, loop, and branch, which clarify the spatial topological relationships between target pixel sets and structural elements. MSPA thus enables the identification of ecologically functional patches and individual pixel units, providing a more scientific basis for selecting ecological sources [51].

Landscape connectivity refers to the degree to which a landscape facilitates or impedes movement between patches. Stronger connectivity supports ecosystem stability. A commonly used metric for assessing connectivity is the integral connectivity index, dPC, which is calculated as follows:

$$dPC = \frac{PC - PC_{i\_remove}}{PC} \quad (7)$$

In the formula, dPC represents the patch importance index. Pc_{remove} denotes the likelihood of connectivity in the study area after the specific patch and all other corresponding data are removed.

In this study, areas of high ecological value, such as forest land, grassland, and water bodies, were defined as foreground data, while all other areas were classified as background [52]. Foreground connectivity was assigned a value of 1. Using the Guidos

Toolbox with an eight-neighbor analysis method, seven landscape types were delineated: core, bridge, edge, loop, branch, islet, and perforation, as shown in Section 3.4.2. Core areas were emphasized due to their high morphological connectivity within the foreground and were subsequently used as the primary basis for selecting potential ecological sources [53]. Patches larger than 1 km^2 within the core areas were then selected. Connectivity distance was set to 2 km, and the connectivity probability was set to 0.5. Inter-patch connectivity levels were calculated using Conefor 2.6. Finally, the landscape connectivity of Chengmai County was categorized into three levels, low, medium, and high, based on the resulting dPC values.

2.3.4. Ecological Security Pattern Construction

(1) Resistance Surface Construction

A resistance surface represents the cumulative resistance that ecological elements must overcome when moving between sources and serves as a fundamental component in constructing an ESP. Higher resistance coefficients indicate greater resistance, thereby increasing the "cost" of ecological land expansion. Based on relevant studies and considering the unique physiographical characteristics of the study area, six factors were selected as resistance coefficients: elevation, slope, NDVI, land use type, distance from water bodies, and distance from roads. These coefficients were assigned values ranging from 1 to 9, with higher values corresponding to greater resistance levels (Table 5). The resistance surface indicators were allocated using ArcGIS 10.8, and the Analytic Hierarchy Process was implemented. In this process, five domain experts were invited to assess the relative importance of factors within the same hierarchy level using the 1–9 scale method. The expert evaluations were subsequently integrated through the geometric mean method to construct a comprehensive judgment matrix. Initial weights for each factor were then calculated and subjected to consistency verification. Based on the determined weights of each indicator, a comprehensive resistance surface for Chengmai County was generated [54].

Table 5. Ecological resistance surface evaluation system.

Resistance Type	Resistance Value					Weight
	1	3	5	7	9	
Land Use Type	Forest land, Water bodies	Grassland	Cultivated land	Unutilized land	Construction land	0.43
Elevation/m	<47	47–87	87–135	135–227	>227	0.06
Slope/°	<3	3–7	7–12	12–20	>20	0.11
NDVI	>0.45	0.36–0.45	0.24–0.36	0.08–0.24	0–0.08	0.17
Distance from Road/m	>5867	3224–5867	1881–3224	850–1881	<850	0.12
Distance from Water Dodies/m	<50	50–200	200–500	500–800	>800	0.11

(2) Potential Corridor Extraction

The MCR model is a GIS-based spatial analysis tool that calculates the minimum cost distance for landscape elements moving from a source to a target. This model identifies optimal pathways with the least resistance or lowest cost between ecological elements, thereby facilitating the identification of potential points of ecological connectivity [55]. The formula is expressed as follows:

$$MCR = f_{min} \sum_{j=n}^{i=m} D_{ij} \times R_i \quad (8)$$

In the formula, D_{ij} denotes the spatial distance between ecological source i and ecological source j; R_i represents the resistance coefficient of ecological source i to species movement; and m and n are the numbers of ecological sources i and j, respectively.

(3) Significant Corridor Identification

The gravity model assesses the relative importance of ecological corridors by quantifying the intensity of interaction between different ecological patches. Corridors with higher interaction forces are considered more important. Identifying key corridors using this model provides valuable guidance for establishing conservation priorities [56]. The equation for the gravity model is presented below:

$$G_{ab} = \frac{N_a N_b}{D_{ab}^2} = \frac{L_{\max}^2 \ln(S_a) \ln(S_b)}{L_{ab}^2 P_a P_b} \tag{9}$$

In the equation, G denotes the magnitude of the gravitational force; N is the weight value; D represents the standardized value of the potential corridor resistance between patches; P is the patch resistance value; S denotes the patch area; L represents the cumulative resistance value; and L_{max} is the maximum cumulative resistance of the regional corridors.

3. Results

3.1. Land Use Analysis

According to the remote sensing interpretation results for Chengmai County presented in Figure 3, significant changes in land use structure were observed between 2000 and 2020 (Table 6). The landscape was dominated by forest land, predominantly distributed in the western and southern regions, while cultivated land was primarily concentrated in the eastern areas. Collectively, these two land cover types accounted for nearly 90% of the total area. Driven by policies such as the Hainan International Tourism Island initiative, construction land underwent rapid expansion with an area increase of 289.81%. This urban growth was primarily achieved through the conversion of cultivated land and other ecological spaces, resulting in significantly negative impacts on regional ecological connectivity. Concurrently, influenced by ecological restoration policies including the Grain for Green Program, cultivated land area was reduced by 114.18 km^2, with substantial portions being converted to forest land and water bodies. Although this land use transformation affected food production capacity, it contributed to the overall enhancement of ecological regulation and service functions. The land use change pattern in Chengmai County reveals a distinct spatial divergence between ecological land and urban development land. A stable ecological barrier function is maintained in the central-western region, while significant urban expansion is observed in the northeast. This spatial pattern poses a potential threat to regional ecological security and requires coordination through ecological spatial optimization.

3.2. Ecological Sensitivity

The spatial patterns of individual ecological sensitivity factors are displayed in Figure 4a–k, revealing pronounced variations in the spatial distribution across different sensitivity types. As summarized in Table 7, both Elevation and Slope sensitivity were closely correlated with mountain distribution, with Extreme Sensitivity areas predominantly located in higher elevation zones. Specific Aspect orientations and Land Use Types dominated areas of Extreme Sensitivity. The Water Body Buffer Zone exhibited a distinct ribbon-like distribution along watersheds, while High Sensitivity and Extreme Sensitivity categories predominated in the NDVI and Rainfall Erosivity factors, indicating the substantial influence of vegetation cover and rainfall erosion on the ecological environment. Extreme Sensitivity areas demonstrated overwhelming dominance in both Population Density and Road Buffer Zone distributions, reflecting the pervasive pressure exerted by human activities on ecosystems. In summary, topographic factors, vegetation coverage extent, and human activities were all identified as exerting considerable influence on ecological sensitivity.

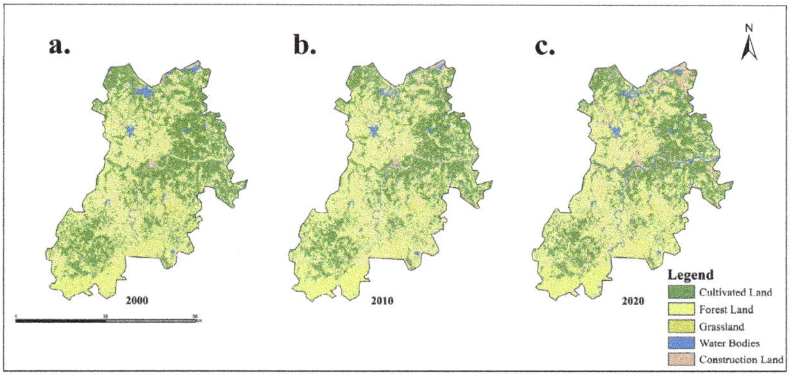

Figure 3. Land use changes in Chengmai County from 2000 to 2020: (**a**) 2000; (**b**) 2010; (**c**) 2020.

Table 6. Land use changes in Chengmai County from 2000 to 2020.

Land Use Type		Cultivated Land	Forest Land	Grassland	Water Bodies	Construction Land	Unutilized Land
2000	Area (km^2)	710.06	1252.50	24.39	45.14	29.45	0.00
	Proportion/%	34.44	60.76	1.18	2.19	1.43	0.00
2010	Area (km^2)	618.05	1356.40	12.15	40.07	34.85	0.00
	Proportion/%	29.98	65.80	0.59	1.94	1.69	0.00
2020	Area (km^2)	595.88	1276.52	12.07	62.32	114.81	0.00
	Proportion/%	28.90	61.92	0.59	3.02	5.57	0.00
Area Change (2000–2020)/km^2		−114.18	24.02	−12.32	17.18	85.36	0.00
Area Change Rate (2000–2020)/%		−16.08	1.92	−50.51	38.05	289.81	0.00

The comprehensive ecological sensitivity distribution presented in Figure 4l was generated through an overlay analysis of individual sensitivity factors and classified using the natural breaks method into four categories: Low Sensitivity, Moderate Sensitivity, High Sensitivity, and Extreme Sensitivity, as detailed in Table 8. This integrated assessment revealed a clear "high in the south, low in the north" spatial distribution pattern, with High Sensitivity and Extreme Sensitivity areas collectively accounting for 63.60% of the total study area. The predominance of higher sensitivity levels suggests that Chengmai County's ecological environment is more vulnerable to external environmental changes. Driven by complex topography and extensive vegetation coverage, the central-southern region displayed a distinct high ecological sensitivity pattern. This pattern signifies that the ecosystem in this region is particularly vulnerable and susceptible to degradation under external disturbances, which warrants its designation as a core protection area with decisive importance for maintaining regional ecological security. This spatial distribution reveals that the southern region of Chengmai County is characterized by steeper topography, denser vegetation coverage, and reduced human disturbance. These characteristics contribute to higher ecological sensitivity, identifying this area as a core zone for Soil Conservation and Biodiversity maintenance. In contrast, the northern low-sensitivity areas show substantial overlap with urban construction and agricultural reclamation zones. This spatial correspondence reflects the suppressive effect of intensive human activities on the self-regulatory capacity of ecosystems.

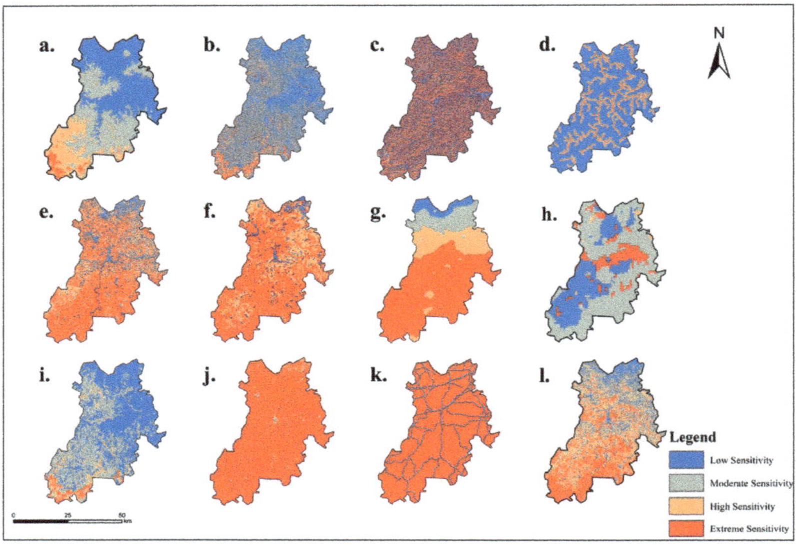

Figure 4. Ecological Sensitivity Assessment Results for Chengmai County. (**a**) Elevation; (**b**) Slope; (**c**) Aspect; (**d**) Water Body Buffer Zone; (**e**) NDVI; (**f**) Land Use Type; (**g**) Rainfall Erosivity; (**h**) Soil Erodibility; (**i**) Topographic Roughness; (**j**) Population Density; (**k**) Road Buffer Zone; (**l**) Comprehensive Sensitivity.

Table 7. Single-factor Ecological Sensitivity Areas in Chengmai County.

Evaluation Factors		Area\km^2			
		Low Sensitivity	Moderate Sensitivity	High Sensitivity	Extreme Sensitivity
Habitat Quality	Elevation	871.15	763.54	366.25	60.65
	Aspect	647.74	468.68	233.31	703.79
	Slope	1002.08	685.56	284.55	81.34
	Land Use Type	114.81	0.00	607.95	1338.83
	Water Body Buffer Zone	1488.02	216.41	232.35	124.75
Human Disturbance	Population Density	0.35	32.35	71.75	1956.66
	Road Buffer Zone	125.62	59.67	57.83	1818.42
Soil Erosion	Soil Erodibility	579.83	1191.25	19.23	264.99
	Rainfall Erosivity	114.77	259.97	384.70	1302.10
	NDVI	214.60	288.04	638.19	920.55
	Topographic Roughness	1091.90	829.13	160.38	59.48

Table 8. Comprehensive Ecological Sensitivity Areas in Chengmai County.

	Comprehensive Evaluation Index	Area\km^2	Proportion/%
Low Sensitivity	1.12–2.18	206.08	10.06
Moderate Sensitivity	2.18–2.61	539.87	26.34
High Sensitivity	2.61–2.96	791.69	38.63
Extreme Sensitivity	2.96–3.73	511.81	24.97

3.3. Spatiotemporal Distribution of Ecosystem Service Value

3.3.1. Ecosystem Service Value Changes

From 2000 to 2020, the total ecosystem services value in Chengmai County demonstrated a consistent upward trend, increasing from 10.078 billion yuan to 10.502 billion yuan, with a net gain of 424 million yuan as documented in Table 9. Among different land use types, water bodies exhibited the most substantial ESV increase of 470 million yuan, representing a growth rate of 38.05%. Forest land also showed an ESV increment of 151 million yuan, while grassland and cultivated land experienced substantial ESV declines, indicating degradation in their ecological functions. No ESV changes were recorded for Construction Land and unutilized land throughout the study period. Regarding ESV composition, forest land consistently constituted the dominant component, maintaining a share exceeding 76%. The proportion contributed by water bodies increased from 12.25% to 16.22%, with this expansion driven by ecological restoration projects directly resulting in significantly enhanced ESV. This transformation influenced the regional ecological baseline, specifically strengthening key service functions including Hydrological Regulation and Biodiversity maintenance, while creating new spatial clusters of exceptionally high ESV. Analysis of phased changes revealed pronounced ESV growth from 2000 to 2010, primarily attributable to the increase in forest land ESV. Between 2010 and 2020, the growth rate moderated, although substantial ESV gains from water bodies compensated for declines in forest land and cultivated land. In summary, the expansion of water bodies was identified as the principal driver behind the increased total ecosystem services value in Chengmai County. At the same time, reductions in cultivated land and grassland harmed ESV. A distinct spatial pattern of Ecosystem Services Value was observed in Chengmai County, characterized by higher values in the central and southern regions and lower values in the northeast (Figure 5). This distribution pattern was found to closely align with the spatial extent of forest land and water bodies. The consistent spatial correspondence highlights the crucial role of these ecological substrates in sustaining regional ecosystem service functions.

Table 9. Changes in Ecosystem Service Values by Land Use Type in Chengmai County, 2000–2020 (Unit: 10^8 Yuan).

Land Use Type		Cultivated Land	Forest Land	Grassland	Water Bodies	Construction Land	Unutilized Land	Total
2000	Area (km^2)	8.63	78.67	1.13	12.34	0.00	0.00	100.78
	Proportion/%	8.56	78.06	1.13	12.25	0.00	0.00	100.00
2010	Area (km^2)	7.51	85.20	0.56	10.96	0.00	0.00	104.23
	Proportion/%	7.21	81.74	0.54	10.51	0.00	0.00	100.00
2020	Area (km^2)	7.24	80.18	0.56	17.04	0.00	0.00	105.02
	Proportion/%	6.90	76.34	0.53	16.22	0.00	0.00	100.00
ESV Changes (2000–2020)		−1.39	1.51	−0.57	4.70	0.00	0.00	4.24
ESV Change Rate (2000–2020)/%		−16.08	1.92	−50.51	38.05	0.00	0.00	4.21

3.3.2. Spatial Distribution of Ecosystem Service Change Index

An analysis of the ESCI on a 1 km × 1 km grid system was conducted, revealing significant alterations in the spatial pattern of ESV in Chengmai County from 2000 to 2020, as illustrated (Figure 6). A continuous expansion of high-value zones was observed alongside a noticeable contraction of low-value zones. Newly established water bodies formed multiple punctate high-value centers, collectively creating a distribution pattern characterized as "high in the central-south, low in the northeast," which closely corresponded with topographic features and dominant land cover types. Analysis of the ESV hierarchical structure, as documented in Table 10, revealed a reduction of 87.14 km^2 in low-value areas, a steady decrease in medium-value zones, and significant increases in both high-value and extremely high-value categories. These changes reflect the positive impacts of ecological restoration and water body expansion on regional ecological quality. The spatial distribution of ESCI further revealed that significant ESV loss areas were concentrated in urban built-up zones, including Jinjiang Town and Laocheng Town, as well as along major transportation corridors, primarily resulting from the conversion of forest land to Construction Land. Conversely, significant ESV gain areas were distributed in zones of water body expansion and ecological restoration, such as the Lisha River and Nandu River, demonstrating the effectiveness of ecological conservation measures.

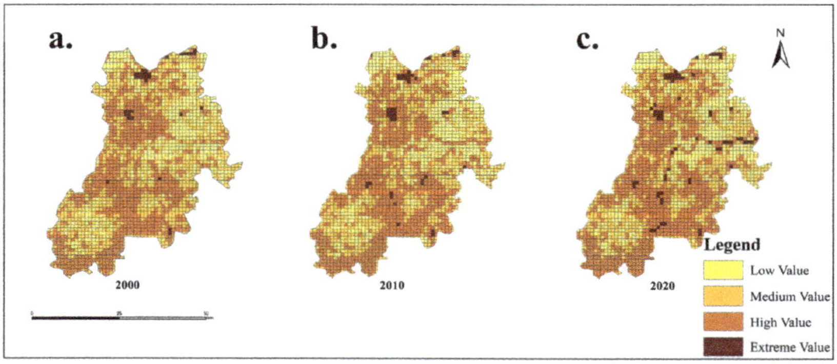

Figure 5. Spatial distribution of ecosystem service value in Chengmai County from 2000 to 2020: (**a**) 2000; (**b**) 2010; (**c**) 2020.

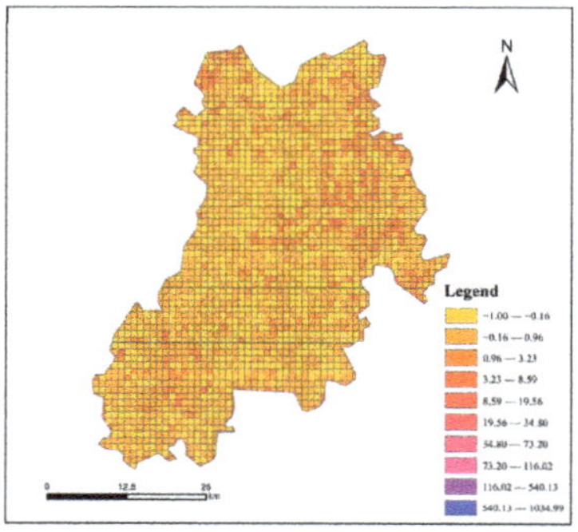

Figure 6. Ecosystem Service Value Index (ESCI) for Chengmai County.

Table 10. Changes in the distribution of ecosystem service value levels in Chengmai County from 2000 to 2020.

Value Level		Low Value	Medium Value	High Value	Extreme Value	Total
2000	Area (km^2)	535.88	684.52	808.52	32.67	2061.59
	Proportion/%	25.99	33.20	39.22	1.58	100.00
2010	Area (km^2)	469.90	634.37	913.30	44.00	2061.57
	Proportion/%	22.79	30.77	44.30	2.13	100.00
2020	Area (km^2)	448.74	623.21	919.97	69.66	2061.58
	Proportion/%	21.77	30.23	44.62	3.38	100.00

3.4. Landscape Connectivity Assessment

3.4.1. MSPA Analysis

The landscape pattern analysis employing MSPA methods, as detailed in Figure 7 and Table 11, demonstrated that the ecological landscape foreground in Chengmai County covered 1346.41 km^2, accounting for 65.31% of the total area. Core areas constituted 55.80% of the total landscape, representing 85.44% of the total ecological landscape area, which indicates well-connected natural habitats and a robust ecosystem foundation throughout the county. Edge areas accounted for 7.72% of the foreground area, providing essential buffering and protection for the core zones. However, bridges and loops were minimally represented, reflecting insufficient structural connectivity among core patches. Islets accounted for merely 0.28%, further indicating a certain degree of ecosystem fragmentation risk. The overall landscape was characterized by core dominance with weak connectivity, rendering core areas vulnerable to external disturbances. Comprehensive analysis indicated that the landscape pattern in Chengmai County was predominantly shaped by natural topography and vegetation coverage, forming a core-dominated configuration that provides complete habitat environments for organisms and exerts positive influences on maintaining regional Biodiversity. However, this configuration is constrained by a severe deficiency in landscape connectivity elements due to human-induced habitat fragmentation, which impedes ecological flows between patches and, without intervention, will generate long-term negative consequences for species migration and gene flow.

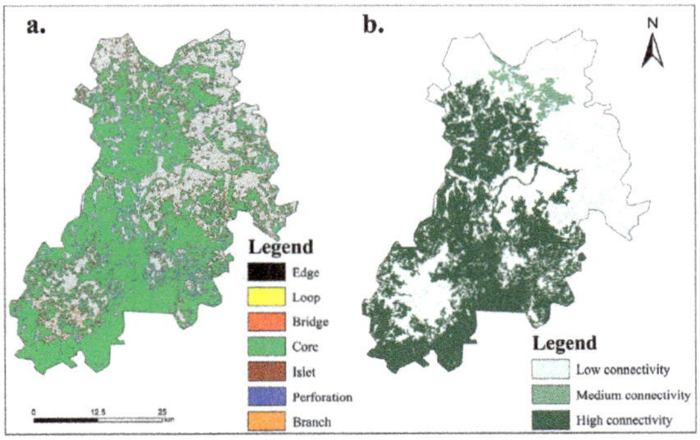

Figure 7. Landscape connectivity assessment results for Chengmai County. (**a**): MSPA; (**b**): Landscape connectivity.

Table 11. Area of mspa Landscape Types in Chengmai County.

Type	Edge	Branch	Islet	Core	Loop	Perforation	Bridge	Total
Area/km^2	103.89	17.17	3.77	1150.33	5.74	59.51	6.01	1346.41
Area of foreground/%	7.72	1.28	0.28	85.44	0.43	4.42	0.45	100
Proportion of total area/%	5.04	0.83	0.18	55.8	0.28	2.89	0.29	65.31

3.4.2. Landscape Connectivity Analysis

A connectivity analysis conducted using Conefor 2.6 software, as presented in Table 12 and Figure 7, demonstrated high overall connectivity within core areas, with a maximum dPC value of 98.66, indicating the hub functionality of specific patches. High-connectivity zones covered 84.13% of the core area and were predominantly distributed in the southern and western regions of Chengmai County. These patches serve as critical habitats for various species, providing secure, low-resistance migration corridors for birds, reptiles, and other wildlife. They facilitate gene flow among populations, maintain genetic diversity, ensure the continuity of ecological processes, and enhance ecosystem stability and resilience, thereby enabling the entire system to buffer disturbances and achieve recovery. These ecologically favorable areas provide a solid foundation for the subsequent selection of Ecological Sources. Low-connectivity areas were primarily located in the northern and eastern regions, dominated by cultivated land and Construction Land, where high landscape resistance impedes species migration and material flows. Natural ecological conditions and the intensity of human activity jointly drove the formation of the landscape connectivity pattern. In the southern and western regions, the continuous distribution of forest land and minimal human disturbance facilitated the development of highly connected ecological networks. Conversely, in the northern and eastern areas, rapid urbanization and agricultural expansion accelerated landscape fragmentation, constraining ecological flow processes and ultimately leading to significant degradation of ecological service functions. The high connectivity coverage of 84.13% within core areas signifies the establishment of a structurally intact and functionally stable ecological matrix in central-southern Chengmai County. This configuration provides crucial support for species migration, gene flow, and the circulation of matter and energy within ecosystems.

Table 12. Area of Landscape Connectivity in the Core Zone of Chengmai County.

Type	High Connectivity	Medium Connectivity	Low Connectivity	Total
Area/km^2	917.32	53.74	119.26	1090.32
Core area coverage/%	84.13	4.93	10.94	100

3.5. Ecological Security Pattern Construction

3.5.1. Resistance Surface

The comprehensive resistance surface presented in Figure 8a was constructed, revealing significant spatial heterogeneity in ecological resistance across Chengmai County, with resistance values ranging from 1.24 to 8.33. Areas of high resistance were concentrated in the northeastern and central regions, dominated primarily by Construction Land, where severe landscape fragmentation impedes ecological flows. In contrast, low resistance areas were distributed throughout the western and southeastern regions, characterized by forest land and grassland, which exhibit favorable connectivity and are identified as potential distribution zones for Ecological Sources. These results elucidate the interaction between human activities and ecological processes. The high resistance zones correspond precisely to areas that have experienced the most rapid expansion of Construction Land over the past two decades, as well as the highest concentrations of population and economic activity, as detailed in Section 3.1 and Figure 3. This spatial coupling demonstrates that rapid urbanization and infrastructure development serve as the primary drivers of ecological space fragmentation and connectivity impairment, representing key conflicts that require prioritized coordination in future planning. The resistance pattern revealed a clear directional preference for ecological flows in Chengmai County. Natural pathways for ecological functional flows were identified primarily in forest and grassland areas of the western and southeastern regions. In contrast, the concentration of Construction Land in the northeast was identified as the primary barrier to ecological connectivity.

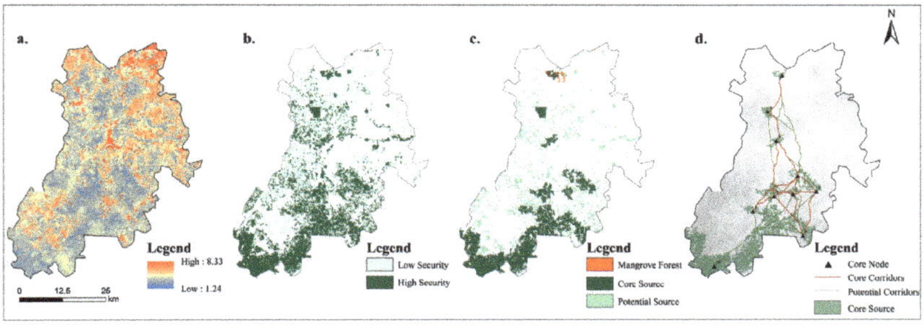

Figure 8. Distribution of ecological corridors and nodes. (**a**): Resistance surface; (**b**): Spatial distribution of ecological importance; (**c**): Source distribution; (**d**): Ecological corridors.

3.5.2. Source Distribution

The integration of ecosystem service importance, ecological sensitivity, and landscape connectivity assessments revealed favorable overall ecological security conditions. A High-Security Zone covering 437.27 km^2 was identified, representing 21.21% of the study area and primarily located in the central-southern forest land regions, as illustrated in Figure 8b. Through the final screening of core nodes, ten Core Ecological Sources were identified based on the actual ecological conditions of the study area, as shown in Figure 8c. The remaining areas with lower security were designated as Potential Ecological Sources. The spatial distribution of these Ecological Sources exhibited three distinct characteristics. Firstly, a patchy distribution pattern was observed, predominantly characterized by forest land as the primary Land Use Type, with additional distributions in grassland and water bodies. Secondly, Ecological Sources were primarily concentrated in forested areas with favorable ecological conditions, including Lingao Ridge, Jinshi Ridge, and agricultural regions, as well as surrounding water bodies such as Fushan Reservoir, Huachang Bay Mangrove Reserve, Nandu River Basin, and Nanfang Reservoir. Thirdly, overlay analysis between the identified Potential Ecological Sources and Chengmai County's ecological protection redlines and nature reserves demonstrated that most nature reserves, water sources, and ecological protection boundaries fall within the area of the Potential Ecological Sources, as depicted in Figure 8d. This spatial consistency validates the scientific basis of the methodology and supports the reasonableness of the evaluation results, establishing a foundation for subsequent ecological security pattern analysis. The distribution of Ecological Sources was found to be highly concentrated around key Ecological Nodes. This spatial pattern validates the scientific rationale underlying the existing nature reserve system in Chengmai County. Simultaneously, it offers precise spatial guidance for optimizing the ecological protection network.

3.5.3. Corridor Extraction

Forty-five Potential Corridors with a total length of 1186.91 km were extracted using the MCR model and the gravity model. Based on an evaluation of interaction intensity among sources as presented in Table 13, sixteen corridors with a total length of 193.12 km were identified as Core Corridors, primarily connecting ecologically functional and low-resistance sources. At the same time, the remaining twenty-nine were classified as Potential Corridors as shown in Figure 8d. These sixteen Core Corridors form an elongated network configuration that facilitates ecological connectivity across the landscape. The corridor system effectively interconnects various Ecological Sources, forming an ecological network with complementary functions. This integrated network serves as a critical spatial safeguard for mitigating habitat fragmentation and maintaining regional Biodiversity and ecological security.

Table 13. Importance of Ecological Corridors in Chengmai County (* indicates interaction > 430).

Node	1	2	3	4	5	6	7	8	9	10
1	0	3373.98 *	67.16	104.78	345.54	2913.64 *	3514.74 *	593.30 *	471.46 *	122.87
2		0	100.34	163.70	646.55 *	3019.94 *	1490.31 *	435.35 *	310.72	103.28
3			0	479.44 *	275.74	89.31	103.87	63.53	36.78	31.47
4				0	901.69 *	131.41	179.06	99.23	51.25	40.30
5					0	403.74	694.78 *	303.91	124.63	84.56
6						0	1065.36 *	344.60	522.79 *	120.23
7							0	2558.81 *	526.64 *	230.75
8								0	352.69	276.32
9									0	125.39
10										0

4. Discussion

4.1. The Necessity of Research

Through systematic analysis of the spatiotemporal evolution of land use and ESV between 2000 and 2020, rapid construction land expansion and widespread distribution of highly sensitive areas were revealed, demonstrating that Chengmai County is experiencing concurrent ecological preservation and development tensions. The ecological environment was shaped by inherent topographic and ecosystem distributions, which established a baseline of high sensitivity. In contrast, rapid urbanization introduced intensive human disturbance as an external stressor. The superposition of these factors substantially diminished the system's buffering capacity. Within this context, the constructed ESP essentially represents a spatial planning intervention strategy designed to enhance ecological resilience and preemptively mitigate the negative impacts of development. By identifying and protecting key ecological hubs and corridors, the capacity of ecosystems to reorganize their core functions following disturbances can be maintained or restored. This approach aligns closely with the principles of landscape resilience theory, emphasizing the preservation of diversity, connectivity, and modularity. Consequently, the value of this research extends beyond ecological pattern construction to provide an actionable framework integrating theoretical and practical approaches for enhancing ecosystem sustainability in vulnerable regions through spatial means, while also establishing early warning mechanisms for small-scale areas under development pressure.

4.2. Ecological Security Pattern and Optimization Recommendations

An ESP characterized by one axis, two belts, and three zones was constructed for Chengmai County based on ecological sources and resistance surfaces, through the extraction of ecological corridors and integration of ecological functional zones, as illustrated in Figure 9. This framework comprises 10 ecological nodes, 16 core corridors, and 29 potential corridors [57].

The one axis represents a core ecological corridor for water conservation and biodiversity protection, extending north–south across the county. It connects key ecological nodes, including the Huachang Bay Mangrove Reserve, Fushan Reservoir, and the Nandu River Basin, serving as the primary pathway that links core ecological sources and facilitates the flow of energy and materials. This axis is critical for ensuring regional water resource security, promoting natural forest restoration, and enhancing ecosystem stability. The two belts consist of a central water conservation belt and a southern mountain forest conservation belt. The water conservation belt encompasses aquatic ecological patches centered on reservoirs such as Jiatan, Jiaju, and Shilang, functioning as vital zones for water supply and conservation [58]. The mountain forest conservation belt encompasses northern mountainous regions, including forested areas such as Wuda Ridge and Fuzhong

Ridge, as well as contiguous farmlands, providing critical areas for soil and water conservation and serving as habitats for Biodiversity. Supported by the ecological axis, these two belts traverse east–west ecological sources, forming key passages that effectively connect these sources.

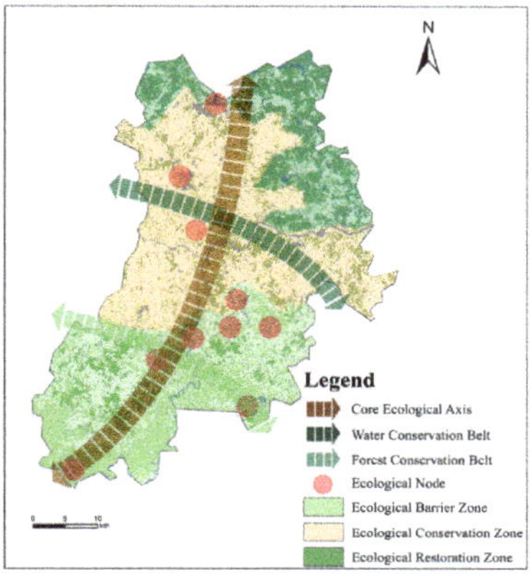

Figure 9. Ecological Security Pattern of Chengmai County.

The three zones correspond to protection and restoration divisions within the ESP, including ecological barrier zones, ecological restoration zones, and ecological conservation zones. Tailored optimization strategies were developed for each zone based on its specific ecological characteristics, ensuring a coordinated approach to conservation, restoration, and sustainable land use.

The ecological barrier zone comprises potential ecological sources within the region, primarily distributed in key areas such as the mangrove nature reserve, forest protection zones, and Fushan Reservoir. These areas exhibit high ecological flow intensity and function as core nodes and strategic pivots for maintaining regional ecological processes. The region is characterized by abundant forest resources and favorable ecological conditions, primarily attributable to the integrity of existing ecological resources and the level of Biodiversity. Consequently, the sustained functioning of ecological processes is maintained as the foundation of regional ecological security. At the same time, human disturbance is recognized as the principal threat to ecosystem stability and the degradation of ecological service capacity. Minimizing human disturbance and promoting biodiversity recovery are essential to consolidate its role as a regional ecological barrier and a hub for ecological radiation. The ecological restoration zone primarily encompasses areas with high ecological resistance, predominantly located in regions with limited potential corridors, such as Laocheng Town and Renxing Town. Here, cultivated land and construction land dominate, human activities are intensive, vegetation cover is low, and ecological disturbances are significant. The regular expression of ecological functions is constrained by high ecological resistance in this region. In contrast, the combined effects of ecological degradation and agricultural activities further exacerbate the obstruction of ecological processes. The ongoing expansion of Construction Land potentially intensifies this trend. Therefore, the key to ecological improvement in this area is considered to be the regulation of land use patterns, with the ecosystem's self-recovery capacity enhanced through the optimization of relationships between human activities, natural resources, and land use arrangements. The ecological conservation zone primarily consists of areas with low to moderate ecological resistance, located mainly in the central part of the study area. It serves as a peripheral buffer zone for core ecological sources [59]. These areas experience relatively light ecological disturbance and maintain generally good environmental quality. Conservation efforts should focus on building ecological infrastructure to enhance the flow and connectivity of natural elements, thereby promoting the health and resilience of ecosystems. These low-resistance areas provide favorable conditions for the flow of natural elements and the continuity of ecological processes. In contrast, locally distributed, ecologically degraded patches may compromise the integrity and functional stability of the buffer zone. Moderate intervention in degraded areas is recommended to facilitate the gradual restoration of ecosystems, thereby strengthening the supportive role of the entire buffer zone in maintaining regional ecological security.

4.3. Innovativeness and Rationality

This study employed an integrated approach to address common limitations in current ESP construction, such as excessive subjectivity and insufficient comprehensiveness. Ecological sources were identified through a synthesis of ecological sensitivity, ESV, and landscape connectivity. The MCR and gravity models were then applied in combination, overcoming biases inherent in traditional single-method approaches and enhancing the systematic and objective identification of ecological sources. During corridor extraction, the continuity of key ecological connections was ensured, and a graded assessment of their functional importance was achieved, strengthening the hierarchical structure and operational applicability of the ESP network. Moreover, many existing studies on ESP construction are limited to a single time point and do not investigate spatiotemporal changes in land use and ESV. However, significant land use changes can alter ESV trajectories and reshape ecosystem patterns [60], an aspect that has been overlooked mainly in previous research. By systematically analyzing the spatiotemporal evolution of land use and ESV from 2000 to 2020 [60], this study clarifies the causes and trends of ecological changes in Chengmai County, providing insight into the intensity and spatial distribution of human disturbance on ecological processes [61]. This approach supports the quantitative calibration of the ecological resistance surface, offering a dynamic perspective for early warning of ecosystem degradation and the rapid expansion of cultivated land.

In Chengmai County, ecological sources covering 437.27 km^2 were identified, showing a high degree of overlap with existing ecological protection redlines and nature reserves. This overlap validates the rationality of the current protected area system and underscores the scientific rigor of the study. Additionally, 45 ecological corridors and 10 key ecological nodes were extracted, collectively forming an ESP characterized as "one axis, two belts, and three zones." This pattern delineates clear boundaries between conservation and development, helps safeguard the ecological baseline under high-intensity development pressure, and provides a replicable framework for similar small-scale coastal regions in China. Particularly in the context of Free Trade Port construction, the Chengmai case demonstrates a potential pathway for less economically developed areas to achieve ecological primacy alongside green development.

4.4. Limitations and Prospects

While this study addresses several important issues, certain limitations should be acknowledged. First, due to data accessibility constraints, the influence of areas beyond Chengmai County's administrative boundaries was not incorporated in the ESP construction. Future research should be conducted across different regional ecological contexts. Second, the spatial data utilized in this study exhibited varying resolutions. Although resampling techniques were applied to standardize all data to a standard analytical grid, ensuring computational feasibility, this processing could not fundamentally alter the inherent precision of the original datasets [62]. Additionally, while the constructed ecological security pattern demonstrates high spatial consistency with officially protected areas, a significant limitation remains the absence of direct empirical validation based on field surveys of species distribution, biodiversity hotspots, or ecological process monitoring [63]. Third, the model complexity prevented the incorporation of influences from investment plans and development intensity on the modeling outcomes. Fourth, the analysis primarily relied on spatiotemporal distributions and current environmental conditions, without addressing the driving factors of ecological change or simulating ESP evolution under dynamic land use scenarios. Fifth, the ESP construction involved a complex multi-method framework. Although combined weighting methods were employed to enhance parameter objectivity, systematic sensitivity analysis and uncertainty quantification of key parameters were not performed. Finally, regardless of the methodology used for ESP construction, its effectiveness requires evaluation and subsequent optimization based on feedback.

Future research will be expanded in the following aspects: First, extensive connectivity between ESPs at different scales will be considered, extending beyond the current study area. Second, the introduction of higher-resolution remote sensing data integrated with species distribution surveys and ecological monitoring data will be prioritized for empirical validation, thereby verifying and refining the pattern construction results at finer scales [64]. Third, subsequent studies will incorporate socio-economic data, including territorial spatial planning and key project construction, directly integrating specific investment plans and development intensity into future policy scenarios to enhance the foresight and policy relevance of ESPs [65]. Fourth, building on ESV and land use spatiotemporal distributions, driving factor analysis and multi-scenario simulations will be combined to assess the underlying causes of ecological pattern changes and evaluate ESP stability and vulnerability under different development policies, enabling predictive adjustment of ecological networks through dynamic conservation concepts [66,67]. Fifth, systematic uncertainty quantification and sensitivity analysis will be implemented using methods such as Monte Carlo simulations to comprehensively assess model sensitivity, advancing the research from static pattern identification to dynamic reliability assessment. Sixth, methods for evaluating ESP effectiveness will be developed. These directions represent the primary focus for subsequent research efforts.

5. Conclusions

An integrated ecological security assessment framework, combining ecosystem service importance, ecological sensitivity, and landscape connectivity, was applied to evaluate the ecological security of Chengmai County. Based on this framework, areas with high ecological security were identified as ecological sources, and ecological corridors and nodes were extracted using the MCR and gravity models. The ESP of Chengmai County was subsequently constructed through overlay analysis. The key findings are summarized as follows:

(1) Rapid expansion of construction land was observed, particularly after 2010, with an overall increase of 289.81%. Ecological sensitivity exhibited a spatial pattern of "high in the south, low in the north," indicating that southern regions are more vulnerable to environmental changes. ESV displayed significant spatial differentiation, with higher values in the central–southern region and lower values in the northeast, showing an overall increase of 424 million Chinese Yuan.

(2) The overall ecological security status was relatively favorable. Potential ecological sources covering 437.27 km^2 were identified, showing a high degree of overlap with ecological protection red lines and nature reserves. A total of 10 ecological nodes and 45 ecological corridors were extracted, with a cumulative length of 1186.91 km.

Among these, 16 core corridors, totaling 193.12 km, were prioritized for conservation due to their ecological significance.

(3) An integrated ESP described as "one axis, two belts, and three zones" was established. This pattern provides a practical framework for guiding ecological conservation and regional development planning. Furthermore, the ecological security assessment framework developed in this study offers a novel and transferable approach for ESP research in small-scale coastal areas undergoing development.

Author Contributions: Conceptualization, Y.Z. and P.Z.; methodology, Y.Z. and P.Z.; software, Y.Z. and Q.L.; investigation, Y.Z., Y.F. and Q.L.; resources, Y.F. and Y.M.; data curation, Y.Z., Y.F. and S.Z.; writing—original draft preparation, Y.Z.; writing—review and editing, Y.Z. and P.Z. All authors have read and agreed to the published version of the manuscript.

Funding: This research received no external funding.

Institutional Review Board Statement: Not applicable.

Informed Consent Statement: Not applicable.

Data Availability Statement: The data supporting the findings of this study were obtained from publicly available datasets, and the websites providing access to these datasets are included in the article.

Acknowledgments: The authors would like to acknowledge all experts' contributions in building the model and formulating the strategies in this study. All individuals included in this section have consented to the acknowledgement.

Conflicts of Interest: The authors declare no conflicts of interest.

Abbreviations

The following abbreviations are used in this manuscript:

ESV	Ecosystem Service Value
ESP	Directory of open access journals
MCR	Minimum Cumulative Resistance
MSPA	Morphological Spatial Pattern Analysis
DEM	Digital Elevation Model
NDVI	Normalized Difference Vegetation Index
HWSD	Harmonised World Soil Database
AHP	Analytic Hierarchy Process

References

1. Bunting, P.; Hilarides, L.; Rosenqvist, A.; Lucas, R.M.; Kuto, E.; Gueye, Y.; Ndiaye, L. Global mangrove watch: Monthly alerts o mangrove loss for africa. *Remote Sens.* **2023**, *15*, 2050. [CrossRef]
2. Goto, G.M.; Goñi, C.S.; Braun, R.; Cifuentes-Jara, M.; Friess, D.A.; Howard, J.; Klinger, D.H.; Teav, S.; Worthington, T.A.; Busch, J Implementation costs of restoring global mangrove forests. *One Earth* **2025**, *8*, 101342. [CrossRef]
3. Mohan, M.; Selvam, P.P.; Ewane, E.B.; Moussa, L.G.; Asbridge, E.F.; Trevathan-Tackett, S.M.; Macreadie, P.I.; Watt, M.S.; Gillis L.G.; Cabada-Blanco, F.; et al. Eco-friendly structures for sustainable mangrove restoration. *Sci. Total Environ.* **2025**, *978*, 179393 [CrossRef] [PubMed]
4. Obonaga, L.D.; Ortiz, A.; Wilke, T.; Riascos, J.M. Plastic litter is rapidly bioeroded in mangrove forests. *Mar. Environ. Res.* **2025** *207*, 107027. [CrossRef]
5. Ram, M.; Sheaves, M.; Waltham, N.J. Restoring mangrove biodiversity: Can restored mangroves support fish assemblages comparable to natural mangroves over time? *Restor. Ecol.* **2025**, *33*, e70012. [CrossRef]
6. Frosi, L.; Arcioni, M.; Macamo, C.; Attorre, F.; Nicosia, E.; Americo, M.; Timba, I.L.; Ramoni-Perazzi, P. Temporal trends geographic scope, and research gaps in Mozambican mangrove studies. *Reg. Stud. Mar. Sci.* **2025**, *90*, 104422. [CrossRef]
7. Beltrao, M.G.; Goncalves, C.F.; Brancalion, P.H.S.; Carmignotto, A.P.; Silveira, L.F.; Galetti, P.M., Jr.; Galetti, M. Priority areas and implementation of ecological corridor through forest restoration to safeguard biodiversity. *Sci. Rep.* **2024**, *14*, 30837. [CrossRef]

Dewa, D.D.; Buchori, I.; Sejati, A.W.; Liu, Y. Integrating Google Earth Engine and regional ecological corridor modeling for remote sensing-based urban heat island mitigation in Java, Indonesia. *Remote Sens. Appl. Soc. Environ.* **2025**, *38*, 101573. [CrossRef]

Mardeni, V.D.N.; Dias, H.M.; Santos, A.R.D.; Santos, D.M.C.; Moreira, T.R.; Carvalho, R.D.C.F.; Santos, E.C.D.; Pautz, C.; Zandonadi, C.U. Delimitation of Ecological Corridor Using Technological Tools. *Sustainability* **2023**, *15*, 13696. [CrossRef]

Feng, Y.; Ping, L.; Wu, H.; Yao, J. Evaluation of the impact of climate change on the ecological resistance and ecological corridors based on set pair analysis theory. *Ecol. Indic.* **2024**, *163*, 112050. [CrossRef]

Saviano, I.R.; Gardon, F.R.; dos Santos, R.F. Ecological corridors and landscape planning: A model to select priority areas for connectivity maintenance. *Landsc. Ecol.* **2021**, *36*, 3311–3328. [CrossRef]

Pelorosso, R.; Noce, S.; De Notaris, C.; Gobattoni, F.; Apollonio, C.; Petroselli, A.; Recanatesi, F.; Ripa, M.N. The nexus between connectivity, climate, and land use: A scenario assessment of bio-energy landscape connectivity in central Italy. *J. Environ. Manag.* **2025**, *376*, 124521. [CrossRef]

Naidoo, R.; Aylward, C.; Elliott, W.; Keeley, A.; Kinnaird, M.; Knight, M.; Papp, C.-R.; Thapa, K.; Antelo, R. From science to impact: Conserving ecological connectivity in large conservation landscapes. *Proc. Natl. Acad. Sci. USA* **2025**, *122*, e2410937122. [CrossRef] [PubMed]

Montano, L.; Harrison, A.; Henderson, B.; Coleman, M.A.; Raoult, V.; Taylor, M.D.; Gaston, T.; Dwyer, P.G.; Costanza, R.; Dansie, A.; et al. The value of Australian aquatic ecosystem services. *Ecosyst. Serv.* **2025**, *76*, 101786. [CrossRef]

Abulaiti, H.; Liu, Y. Ecological sensitivity assessment and driving force analysis of the Tarim river basin. *Sci. Rep.* **2025**, *15*, 34630. [CrossRef]

Yang, X.; Liu, W.; Li, S.; Ma, Z.; Chen, C.; Gu, W.; Qu, M.; Zhang, C.; Tao, J.; Ding, Z.; et al. Restoration of urban waterbird diversity: A case study of the construction of a waterbird ecological corridor in the Guangdong-Hong Kong-Macao Greater Bay Area, Southern China. *Glob. Ecol. Conserv.* **2022**, *39*, e02277. [CrossRef]

Justeau-Allaire, D.; Ibanez, T.; Vieilledent, G.; Lorca, X.; Birnbaum, P. Refining intra-patch connectivity measures in landscape fragmentation and connectivity indices. *Landsc. Ecol.* **2024**, *39*, 24. [CrossRef]

Morin, E.; Razafimbelo, N.T.; Yengué, J.-L.; Guinard, Y.; Grandjean, F.; Bech, N. Are human-induced changes good or bad to dynamic landscape connectivity? *J. Environ. Manag.* **2024**, *352*, 120009. [CrossRef] [PubMed]

Nilander, C.; Lundmark, R. A review of forest ecosystem services and their spatial value characteristics. *Forests* **2024**, *15*, 919. [CrossRef]

Bishane, E.; Negash, E.; Getachew, T.; Gebrewahed, H.; Gidey, E.; Gebremedhin, M.A.; Mhangara, P. Changes in total and per-capital ecosystem service value in response to land-use land-cover dynamics in north-central Ethiopia. *Sci. Rep.* **2024**, *14*, 6540. [CrossRef]

Brander, L.M.; de Groot, R.; Schägner, J.P.; Guisado-Goñi, V.; van't Hoff, V.; Solomonides, S.; McVittie, A.; Eppink, F.; Sposato, M.; Do, L.; et al. Economic values for ecosystem services: A global synthesis and way forward. *Ecosyst. Serv.* **2024**, *66*, 101606. [CrossRef]

Rai, A.; Sharma, L.K. Spatial E-PSR modelling for ecological sensitivity assessment for arid rangeland resilience and management. *Ecol. Model.* **2023**, *478*, 110283. [CrossRef]

Khan, R.; Saxena, A.; Shukla, S.; Goel, P.; Bhattacharya, P.; Li, P.; Ali, E.F.; Shaheen, S.M. Appraisal of water quality and ecological sensitivity with reference to riverfront development along the River Gomti, India. *Appl. Water Sci.* **2022**, *12*, 13. [CrossRef]

de Morais, M.; Abdo, M.S.A.; dos Santos, C.; Sander, N.L.; da Silva Nunes, J.R.; Lázaro, W.L.; da Silva, C.J. Long-term analysis of aquatic macrophyte diversity and structure in the Paraguay river ecological corridor, Brazilian Pantanal wetland. *Aquat. Bot.* **2022**, *178*, 103500. [CrossRef]

De Feudis, C.; Torretta, E.; Orioli, V.; Tirozzi, P.; Bani, L.; Meriggi, A.; Dondina, O. Dispersal and settlement dynamics of wolves in a lowland ecological corridor in northern Italy: Effects of resource availability and human disturbance. *Biol. Conserv.* **2025**, *302*, 110936. [CrossRef]

Gao, J.; Du, F.; Zuo, L.; Jiang, Y. Integrating ecosystem services and rocky desertification into identification of karst ecological security pattern. *Landsc. Ecol.* **2021**, *36*, 2113–2133. [CrossRef]

Lin, H.; Wang, Z.; Zhang, L.; Tang, F.; Wang, G.; Li, M. Construction of an ecological security network in the Fenhe River Basin and its temporal and spatial evolution characteristics. *J. Clean. Prod.* **2023**, *417*, 137961. [CrossRef]

Xu, D.; Guo, X.; Watanabe, T.; Liang, K.; Kou, J.; Jiang, X. Ecological security pattern construction in rural settlements based on importance and vulnerability of ecosystem services: A case study of the southeast region of Chongqing, China. *Sustainability* **2023**, *15*, 7477. [CrossRef]

Wei, W.; Zhang, Y.; Wei, X.; Xie, B.; Ma, Z.; Liu, C.; Yu, L.; Zhou, J.; Shi, W.; Liu, T.; et al. Construction and optimization of ecological security patterns based on ecosystem service function and ecosystem sensitivity in the important ecological functional area—A case study in the Yellow River Basin. *Ecol. Eng.* **2025**, *215*, 107609. [CrossRef]

30. Xu, H.; Zhang, Z.; Yu, X.; Li, T.; Chen, Z. The construction of an ecological security pattern based on the comprehensive evaluatio of the importance of ecosystem service and ecological sensitivity: A case of Yangxin County, Hubei province. *Front. Environ. Sc* **2023**, *11*, 1154166. [CrossRef]
31. Nie, H.; Zhao, Y.; Zhu, J.; Ning, A.; Zheng, W. Ecological security pattern construction in typical oasis area based on ant colon optimization: A case study in Yili River Valley, China. *Ecol. Indic.* **2024**, *169*, 112770. [CrossRef]
32. Chen, J.; Wang, S.; Zou, Y. Construction of an ecological security pattern based on ecosystem sensitivity and the importance c ecological services: A case study of the Guanzhong Plain urban agglomeration, China. *Ecol. Indic.* **2022**, *136*, 108688. [CrossRef]
33. Meng, H.; Gong, Z.; Qian, C.; Zhao, X.; Liu, Q.; Bu, X.; Shen, C. Deciphering the spatial code: Identification and optimization c ecological security pattern—A case study of Jiangsu province, China. *Land* **2025**, *14*, 1928. [CrossRef]
34. Ran, Y.; Lei, D.; Li, J.; Gao, L.; Mo, J.; Liu, X. Identification of crucial areas of territorial ecological restoration based on ecologica security pattern: A case study of the central Yunnan urban agglomeration, China. *Ecol. Indic.* **2022**, *143*, 109318. [CrossRef]
35. Tong, H.-l.; Shi, P.-j. Using ecosystem service supply and ecosystem sensitivity to identify landscape ecology security patterns i the Lanzhou-Xining urban agglomeration, China. *J. Mt. Sci.* **2020**, *17*, 2758–2773. [CrossRef]
36. Horváth, G.F.; Mánfai, K.; Horváth, A. Relationship between landscape structure and the diet of common barn-owl (tyto alba) a different distances from the Drava River ecological corridor. *Ornis Hung.* **2023**, *31*, 88–110. [CrossRef]
37. Zheng, Y.; Tang, P.; Dong, L.; Yao, Z.; Guo, J. Evaluation of ecological carrying capacity and construction of ecological securit pattern in West Liaohe River Basin of China. *Front. Ecol. Evol.* **2024**, *12*, 1335671. [CrossRef]
38. Niu, X.; Zhang, J.; Wang, S.; Zong, L.; Zhou, M.; Zhang, M. Identification of priority areas for ecological restoration based o ecological security patterns and ecological risks: A case study of the Hefei Metropolitan Area. *Ecol. Indic.* **2025**, *175*, 11359 [CrossRef]
39. Wang, X.; Qiu, X.; Wang, Z.; Feng, C.; Fan, Z.; He, W. Construction of a social ecological security pattern based on the supply an demand of ecosystem services. *J. Nat. Conserv.* **2025**, *86*, 126962. [CrossRef]
40. Li, W.; Liu, Y.; Lin, Q.; Wu, X.; Hao, J.; Zhou, Z.; Zhang, X. Identification of ecological security pattern in the Qinghai-Tibe Plateau. *Ecol. Indic.* **2025**, *170*, 113057. [CrossRef]
41. Li, L.; Tang, H.; Lei, J.; Song, X. Spatial autocorrelation in land use type and ecosystem service value in Hainan Tropical Rai Forest National Park. *Ecol. Indic.* **2022**, *137*, 108727. [CrossRef]
42. García-Ontiyuelo, M.; Acuña-Alonso, C.; Vasilakos, C.; Álvarez, X. Strategies for detecting land-use change on the River Tea SC ecological corridor via satellite images. *Sci. Total Environ.* **2024**, *957*, 177507. [CrossRef]
43. Lei, J.; Chen, Z.; Chen, X.; Li, Y.; Wu, T. Spatio-temporal changes of land use and ecosystem services value in Hainan Island fror 1980 to 2018. *Acta Ecol. Sin.* **2020**, *40*, 4760–4773.
44. Manolaki, P.; Zotos, S.; Vogiatzakis, I.N. An integrated ecological and cultural framework for landscape sensitivity assessment i Cyprus. *Land Use Policy* **2020**, *92*, 104336. [CrossRef]
45. Duan, Y.; Zhang, L.; Fan, X.; Hou, Q.; Hou, X. Smart city oriented ecological sensitivity assessment and service value computin based on intelligent sensing data processing. *Comput. Commun.* **2020**, *160*, 263–273. [CrossRef]
46. Jin, X.; Wei, L.; Wang, Y.; Lu, Y. Construction of ecological security pattern based on the importance of ecosystem service function and ecological sensitivity assessment: A case study in Fengxian county of Jiangsu province, China. *Environ. Dev. Sustain.* **202** *23*, 563–590. [CrossRef]
47. Ying, B.; Liu, T.; Ke, L.; Xiong, K.; Li, S.; Sun, R.; Zhu, F. Identifying the landscape security pattern in karst rocky desertificatio area based on ecosystem services and ecological sensitivity: A case study of Guanling county, Guizhou province. *Forests* **202** *14*, 613. [CrossRef]
48. Liang, Y.-X.; Wu, D.-F.; Wu, Z.-J.; Xu, Y.; Zhu, Z.-W.; Zhang, Y.-C.; Zhu, H. Construction of ecological corridors in karst area based on ecological sensitivity and ecological service value. *Land* **2023**, *12*, 1177. [CrossRef]
49. Wei, L.; Li, M.; Ma, Y.; Wang, Y.; Wu, G.; Liu, T.; Gong, W.; Mao, M.; Zhao, Y.; Wei, Y.; et al. Construction of an ecological securit pattern for the national park of Hainan tropical rainforest on the basis of the importance of the function and sensitivity of it ecosystem services. *Land* **2024**, *13*, 1618. [CrossRef]
50. Xie, G.; Zhang, C.; Zhang, L.; Chen, W.; Li, S. Improvement of the evaluation method for ecosystem service value based on pe unit area. *J. Nat. Resour.* **2015**, *30*, 1243–1254. (In Chinese) [CrossRef]
51. Qin, J.-Z.; Dai, J.-P.; Li, S.-H.; Zhang, J.-Z.; Peng, J.-S. Construction of ecological network in Qujing city based on MSPA and MCR models. *Sci. Rep.* **2024**, *14*, 9800. [CrossRef]
52. Wang, F.; Yuan, X.; Zhou, L.; Zhang, M. Integrating ecosystem services and landscape connectivity to construct and optimiz ecological security patterns: A case study in the central urban area Chongqing municipality, China. *Environ. Sci. Pollut. Res.* **202** *29*, 43138–43154. [CrossRef] [PubMed]
53. Wei, Q.; Halike, A.; Yao, K.; Chen, L.; Balati, M. Construction and optimization of ecological security pattern in Ebinur Lake Basi based on MSPA-MCR models. *Ecol. Indic.* **2022**, *138*, 108857. [CrossRef]

- Qian, W.; Zhao, Y.; Li, X. Construction of ecological security pattern in coastal urban areas: A case study in Qingdao, China. *Ecol. Indic.* **2023**, *154*, 110754. [CrossRef]
- Zhang, F.; Jia, Y.; Liu, X.; Li, T.; Gao, Q. Application of MSPA-MCR models to construct ecological security pattern in the basin: A case study of Dawen River basin. *Ecol. Indic.* **2024**, *160*, 111887. [CrossRef]
- Ye, H.; Yang, Z.; Xu, X. Ecological corridors analysis based on mspa and mcr model—A case study of the tomur world natural heritage region. *Sustainability* **2020**, *12*, 959. [CrossRef]
- Zhang, S.; Zhang, Y.; Xiong, K.; Yu, Y.; He, C.; Zhang, S.; Wang, Z. Construction of forest ecological security patterns based on MSPA model and circuit theory in the Desertification Control forests in South China Karst. *npj Herit. Sci.* **2025**, *13*, 432. [CrossRef]
- Yuan, M.; Xian, Q.; Huang, Q.; Yang, C.; Shu, J.; Yao, C.; Pan, H. Research on ecological security pattern based on the paradigm of "portray-assessment-construction-validation"—Minjiang River Basin as an example. *J. Environ. Manag.* **2025**, *394*, 127553. [CrossRef]
- Kong, F.; Duan, S.; Xu, C. The construction of ecological security pattern based on ecosystem services and ecological sensitivity: A case study of the Qiantang River Basin. *Acta Ecol. Sin.* **2024**, *44*, 11359–11374.
- Li, Y.; Liu, W.; Feng, Q.; Zhu, M.; Yang, L.; Zhang, J.; Yin, X. The role of land use change in affecting ecosystem services and the ecological security pattern of the Hexi Regions, Northwest China. *Sci. Total Environ.* **2023**, *855*, 158940. [CrossRef]
- Berhanu, Y.; Dalle, G.; Sintayehu, D.W.; Kelboro, G.; Nigussie, A. Land use/land cover dynamics driven changes in woody species diversity and ecosystem services value in tropical rainforest frontier: A 20-year history. *Heliyon* **2023**, *9*, e13711. [CrossRef] [PubMed]
- Heddam, S.; Zhang, Z.; Ge, H.; Li, X.; Huang, X.; Ma, S.; Bai, Q. Spatiotemporal patterns and prediction of landscape ecological security in Xishuangbanna from 1996–2030. *PLoS ONE* **2023**, *18*, e0292875. [CrossRef]
- Zhou, G.; Huan, Y.; Wang, L.; Zhang, R.; Liang, T.; Han, X.; Feng, Z. Constructing a multi-leveled ecological security pattern for improving ecosystem connectivity in the Asian water Tower region. *Ecol. Indic.* **2023**, *154*, 110597. [CrossRef]
- Jin, B.; Geng, J.; Ding, Z.; Guo, L.; Rui, Q.; Wu, J.; Peng, S.; Jin, R.; Fu, X.; Pan, H.; et al. Construction and optimization of ecological corridors in coastal cities based on the perspective of "structure-function". *Sci. Rep.* **2024**, *14*, 27945. [CrossRef]
- Cai, L.; Liu, Y.; Luo, X. Integrating the MCR and DOI models to construct an ecological security network for the urban agglomeration around Poyang Lake, China. *Sci. Total Environ.* **2021**, *754*, 141868. [CrossRef]
- Dong, Z.; Liu, H.; Liu, H.; Chen, Y.; Fu, X.; Zhang, Y.; Xia, J.; Zhang, Z.; Chen, Q. Analysis of habitat quality changes in mountainous areas using the plus model and construction of a dynamic restoration framework for ecological security patterns: A case study of golog tibetan autonomous prefecture, Qinghai province, China. *Land* **2025**, *14*, 1509. [CrossRef]
- Huang, M.; Gong, D.; Zhang, L.; Lin, H.; Chen, Y.; Zhu, D.; Xiao, C.; Altan, O. Spatiotemporal dynamics and forecasting of ecological security pattern under the consideration of protecting habitat: A case study of the Poyang Lake ecoregion. *Int. J. Digit. Earth* **2024**, *17*, 2376277. [CrossRef]

sustainability

MDPI

Article

The Spatial–Temporal Evolution Analysis of Urban Green Space Exposure Equity: A Case Study of Hangzhou, China

Yuling Tang [1], Xiaohua Guo [1,*], Chang Liu [2], Yichen Wang [1] and Chan Li [1]

1 College of Horticulture and Gardening, Yangtze University, Jingzhou 434025, China; 2023710880@yangtzeu.edu.cn (Y.T.); 2024721010@yangtzeu.edu.cn (Y.W.); 2024720994@yangtzeu.edu.cn (C.L.)
2 School of Architecture and Urban Planning, Huazhong University of Science and Technology, Wuhan 430074, China; liuchang710865@163.com
* Correspondence: 500634@yangtzeu.edu.cn

Abstract

With the continuous expansion of high-density urban forms, residents' opportunities for daily contact with natural environments have been increasingly reduced, making the equity of urban green space allocation a critical challenge for sustainable urban development. Existing studies have largely focused on green space quantity or accessibility at single time points, lacking systematic investigations into the spatiotemporal evolution of green space exposure (GSE) and its equity from the perspective of residents' actual environmental experiences. GSE refers to the integrated level of residents' contact with urban green spaces during daily activities across multiple dimensions, including visual exposure, physical accessibility, and spatial distribution, emphasizing the relationship between green space provision and lived environmental experience. Based on this framework, this study takes the central urban area of Hangzhou as the study area and integrates multi-temporal remote sensing imagery with large-scale street view data. A deep learning–based approach is developed to identify green space exposure, combined with spatial statistical methods and equity measurement models to systematically analyze the spatiotemporal patterns and evolution of GSE and its equity from 2013 to 2023. The results show that (1) GSE in Hangzhou increased significantly over the study period, with accessibility exhibiting the most pronounced improvement. However, these improvements were mainly concentrated in peripheral areas, while changes in the urban core remained relatively limited, revealing clear spatial heterogeneity. (2) Although overall GSE equity showed a gradual improvement, pronounced mismatches between low exposure and high demand persisted in densely populated areas, particularly in older urban districts and parts of newly developed residential areas. (3) The spatial patterns and evolutionary trajectories of equity varied significantly across different GSE dimensions. Composite inequity characterized by "low visibility–low accessibility" formed stable clusters within the urban core. This study further explores the mechanisms underlying green space exposure inequity from the perspectives of urban renewal patterns, land-use intensity, and population concentration. By constructing a multi-dimensional and temporally explicit analytical framework for assessing GSE equity, this research provides empirical evidence and decision-making references for refined green space management and inclusive, sustainable urban planning in high-density cities.

Keywords: urban green space; green space exposure; equity; sustainable urban development; spatial–temporal evolution

https://doi.org/10.3390/su18021131

1. Introduction

Urbanization constitutes a central component of national modernization; however, the rapid expansion of high-density urban forms has profoundly reshaped residents' living environments and the configuration of public resources [1]. Currently, more than half of the global population resides in urban areas that occupy less than 3% of the world's land surface, intensifying spatial competition and exacerbating inequities in public resource allocation [2,3]. Among these challenges, inequitable access to urban green spaces has emerged as a critical concern not only in urban governance and public health, but also in the pursuit of sustainable and socially inclusive urban development.

As essential public service infrastructure, urban green spaces deliver multiple ecological, social, and health benefits. They contribute to mitigating urban heat island effects, improving air quality, reducing noise pollution, and alleviating urban flooding, while also providing spaces for recreation, social interaction, and physical activity [4–6]. A growing body of public health research indicates that both access to green spaces and the level of daily exposure are significantly associated with cardiovascular health, psychological stress, and overall well-being [7–9]. Consequently, green space equity has become a key issue linking environmental sustainability, social sustainability, and health equity, extending beyond conventional spatial planning concerns. In the context of urban governance in China, green space equity has also been incorporated into policy frameworks such as "Ecological Civilization" and the "Park City" initiative. These frameworks emphasize the institutionalized planning and fine-grained design of green public resources to ensure inclusive access and equitable sharing under conditions of high-density urban development, thereby enhancing urban sustainability and social inclusiveness.

In recent years, research on urban green space equity has expanded substantially. Nevertheless, due to differences in analytical perspectives and indicator systems, existing studies have often produced inconsistent or even contradictory conclusions. Most assessments rely on single indicators—such as green space coverage [10,11], accessibility [12,13], or the green view index (GVI) [12]—to evaluate distributional equity. While these metrics capture specific aspects of green space provision, they fall short of addressing the core question of whether residents actually encounter and experience green spaces in their daily lives. From a sustainability perspective, such single-indicator approaches risk overlooking experiential inequities that accumulate over time and undermine long-term urban livability and social sustainability.

Against this backdrop, GSE has gained increasing attention as a more experience-oriented and integrative concept. GSE emphasizes residents' actual contact with green spaces during everyday activities across multiple dimensions, including visual exposure, physical accessibility, and spatial availability [3,14,15]. Unlike traditional green space indicators, GSE is not a simple aggregation of metrics; rather, it adopts a human–environment interaction perspective to reveal potential structural mismatches between nominal green space provision and lived environmental experience. This conceptual shift enhances its explanatory power and policy relevance for sustainable urban governance, particularly in high-density cities where spatial resources are severely constrained.

Existing empirical evidence demonstrates substantial inter-city disparities in GSE levels and equity. For instance, Wu et al. (2023) [16], analyzing 1028 cities worldwide, found that improvements in per capita GSE in cities of the Global South occurred at nearly four times the rate observed in cities of the Global North. Han et al. (2022) [17], based on a comparative analysis of 1013 cities, further identified systematic inequalities in GSE between northern and southern cities globally. At the national scale, Lu et al. (2023) [18] examined 263 Chinese cities and revealed pronounced inequities across dimensions of green space quantity, quality, and spatial distribution. However, these studies predominantly

https://doi.org/10.3390/su18021131

adopt cross-sectional designs and focus on inter-city comparisons, limiting insights into the temporal evolution of intra-urban GSE inequity and its implications for long-term sustainable urban development.

Moreover, constraints related to data availability and methodological approaches have resulted in a predominant reliance on remotely sensed green space coverage at relatively coarse spatial resolutions. Such approaches are insufficient for capturing eye-level green space perception and often fail to incorporate micro-scale environmental exposure along streets and daily travel routes. These temporal and spatial limitations hinder a systematic understanding of the dynamic nature of GSE equity, weakening its applicability to evidence-based sustainability-oriented planning and policy evaluation.

Addressing these gaps, this study takes Hangzhou as a case study and integrates multi-source, multi-temporal data to examine the spatiotemporal evolution of urban GSE and its equity. By combining remote sensing imagery with large-scale street view data, this research adopts a multi-dimensional perspective to assess GSE across visual, accessibility, and availability dimensions. The specific objectives are to:

(1) Evaluate the overall level and spatiotemporal dynamics of GSE in Hangzhou from 2013 to 2023;

(2) Characterize the temporal evolution and spatial differentiation of GSE equity during the same period;

(3) Compare patterns across different dimensions and equity indicators, explore their underlying mechanisms, and propose planning and policy recommendations to enhance equitable and sustainable urban green space provision.

2. Materials and Methods

This study proposes a multi-dimensional framework to assess GSE and its spatiotemporal equity by integrating multi-source spatial data. Three dimensions of GSE—availability, accessibility, and visibility—are quantified using remote sensing imagery, vector-based green space and population data, and street view images with semantic segmentation techniques, respectively. Based on these indicators, GSE equity is evaluated using inequality metrics to reveal spatial and temporal patterns. Figure 1 illustrates the overall workflow of data acquisition, processing, GSE assessment, and equity evaluation.

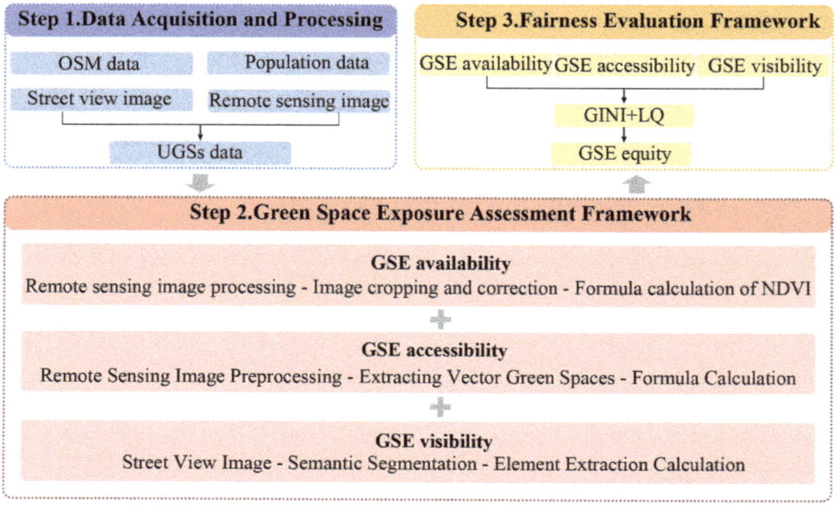

Figure 1. Workflow of data acquisition, processing, and GSE equity assessment.

https://doi.org/10.3390/su18021131

2.1. Case Study Area

This study selected the central urban area of Hangzhou, China, as the study area (Figure 2), primarily due to its typicality and representativeness in the co-evolution of rapid urbanization and green space development in Chinese cities. As one of the core cities in the Yangtze River Delta region, Hangzhou has experienced substantial population growth, intensified land development, and significant restructuring of its urban spatial pattern over the past decade. By the end of 2023, the city's permanent resident population had reached 12.376 million, while the built-up area continued to expand, exhibiting a distinctive coexistence of high-density urban development and high-intensity green space construction.

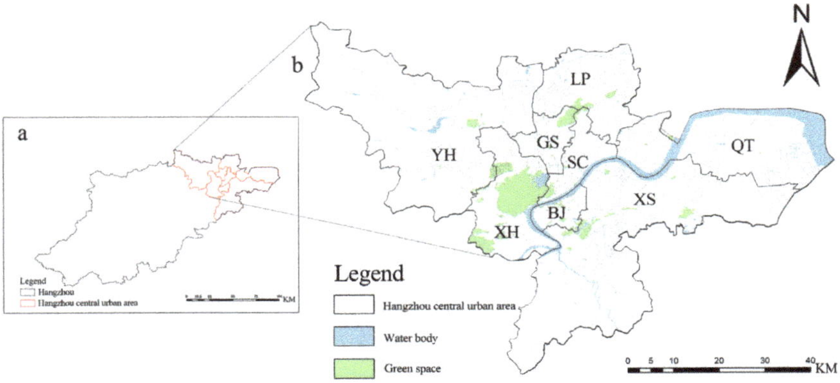

Figure 2. Location of Hangzhou (**a**) and Hangzhou central urban area (**b**).

From the perspective of internal urban structure, pronounced differences exist among subareas of Hangzhou's central urban area in terms of land-use composition, development stage, and residential population characteristics. For example, the old urban districts (e.g., Shangcheng and Xihu) are dominated by high-density residential land and relatively mature public green space systems, whereas the peripheral new districts (e.g., Yuhang, Qiantang, and Linping) are characterized by newly developed construction land and rapidly advancing green space planning. This marked spatial heterogeneity provides an ideal context for examining intra-urban disparities in green space exposure and associated equity issues.

Accordingly, selecting Hangzhou as the case city not only facilitates an in-depth investigation of the spatiotemporal evolution of green space exposure under rapid urbanization, but also enables a comparative exploration of equity differences in green space availability, accessibility, and visibility across areas at different development stages and with distinct spatial characteristics.

This study focuses on the most recent decade of accelerated urbanization and green space development (2013–2023) and selects 2013, 2018, and 2023 as representative time points. The analysis covers the central urban area of Hangzhou, including Gongshu (GS), Shangcheng (SC), Xihu (XH), Yuhang (YH), Binjiang (BJ), Qiantang (QT), Xiaoshan (XS), and Linping (LP) districts (Figure 2b), with a total area of 3334.17 km^2, to examine the spatiotemporal evolution of GSE equity.

https://doi.org/10.3390/su18021131

2.2. Data Sources and Processing

The data used in this study primarily include administrative boundary data, road network data, residential community data, remote sensing imagery, and street view images (Table 1).

Table 1. Data types and sources used in this study.

Data Type	Data Source
Administrative boundary data	MAPWORLD (National Platform for Common Geospatial Information Services of China)
Road network data	OpenStreetMap (OSM)
Remote sensing imagery	Geospatial Data Cloud, Computer Network Information Center, Chinese Academy of Sciences
Street view imagery	Baidu Maps Open Platform

2.2.1. Remote Sensing Imagery

Remote sensing imagery was obtained from the Geospatial Data Cloud of the Computer Network Information Center, Chinese Academy of Sciences (http://www.gscloud.cn/search, accessed on 10 May 2024). Landsat 8 satellite imagery was selected as the primary data source. Images acquired during the growing season (April–August) of 2013, 2018, and 2023, with cloud coverage less than 10%, were selected to minimize cloud-related interference in vegetation extraction and to ensure comparability of vegetation conditions across different years. All remote sensing images have a spatial resolution of 30 m.

2.2.2. Street View Imagery

Street view imagery was acquired through the Baidu Maps API (Application Programming Interface). Based on the urban road network data provided by OpenStreetMap (https://www.openstreetmap.org/, accessed on 10 May 2024), sampling points were randomly generated at 100 m intervals along the road network (Figure 3). For each point, panoramic static images were retrieved, covering a vertical field of view from 0° to 90° and a horizontal field of view from 0° to 360° [19].

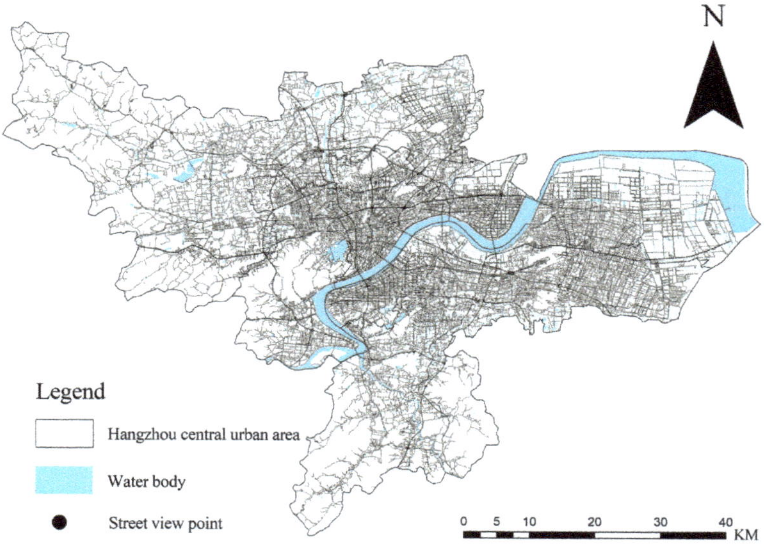

Figure 3. Street view sampling points.

https://doi.org/10.3390/su18021131

Using the temporal metadata associated with street view images, images captured in 2013 (from January 1 to 31 December 2013) were first identified. A manual visual inspection was then conducted to exclude images taken during non-green seasons (e.g., winter or periods of vegetation senescence). For sampling locations lacking street view images from the target year, images from the closest adjacent years (2012 or 2014) were used as substitutes to ensure spatial continuity and sample completeness.

The same screening and cleaning procedures were applied to construct the street view datasets for 2018 and 2023. Subsequently, all images underwent unified quality control to further remove images with severe occlusion, poor image quality, or those not meeting the definition of street view scenes. After preprocessing, a total of 42,410 valid street view images across the three time points were retained for the calculation and analysis of the GVI.

2.2.3. Population Data

Population data were sourced from two major Chinese real estate platforms, Anjuke (https://www.anjuke.com/, accessed on 13 May 2024) and Lianjia (https://wh.lianjia.com/xiaoqu/, accessed on 13 May 2024). Using Python-based web scraping techniques (Python version 3.9), data on residential communities in Hangzhou were collected as of the end of 2023. The extracted information includes the number of households, community name, year of completion, and administrative district. For records with missing household numbers or construction years, supplementary information was obtained through manual inspection of street view maps and consultation with real estate agents. After data cleaning, a total of 5600 valid residential community records were retained. Based on the year of completion, 4026 communities completed by or before 2013 were identified, and 5037 communities completed by or before 2018 were obtained using the same procedure.

Compared with conventional rectangular grid systems, hexagonal grids have equal distances from the centroid to all six directions, which helps reduce sample bias caused by boundary effects and improves the stability of spatial analysis results. Therefore, this study employed a regular hexagonal aggregation approach, constructing hexagonal grids with a side length of 250 m. Residential community points within each hexagon were aggregated to the grid centroid. This process yielded 1505 aggregated residential points for 2013, 1875 for 2018, and 2049 for 2023, facilitating subsequent quantitative analysis.

Population size at each aggregated residential point was calculated based on the total number of households within each hexagon, as shown in Equation (1):

$$P_K = \sum\nolimits_{i \epsilon A_k} R_i \times \overline{M}, \tag{1}$$

In Equation (1), P_K is the estimated population of hexagon k; A_k denotes the set of residential communities within hexagon k; R_i is the number of households in community A_i; M is the average number of people per household in the corresponding year.

To provide a basic empirical check of the population estimation approach, a small-scale validation was conducted at the residential community level. For example, in Hemuxincun, Hemu Subdistrict, Gongshu District, the officially reported population is 9757, while the estimated population derived from housing data is 9735.75, corresponding to a relative error of 0.22%. This comparison suggests that the housing-based method provides a close approximation of actual population counts at the residential community scale. This localized comparison serves as a consistency check rather than a comprehensive assessment of population estimation accuracy.

https://doi.org/10.3390/su18021131

2.3. Measurement of GSE

GSE in this study is measured through a multi-dimensional composite framework that integrates availability, accessibility and visibility. These dimensions reflect different aspects of residents' interaction with urban green space. The measurement approach is as follows.

2.3.1. Measurement of GSE Availability

The Normalized Difference Vegetation Index (*NDVI*) is a widely used remote sensing indicator that accurately reflects surface vegetation coverage. It is commonly employed to characterize the spatial distribution of and temporal variation in green spaces.

In this study, Landsat 8 satellite imagery was preprocessed using ENVI 5.6.2 software. First, radiometric calibration was performed to convert digital numbers into surface reflectance values. Atmospheric correction was then applied to reduce atmospheric effects on surface reflectance. Subsequently, image mosaicking and clipping were conducted to obtain imagery for the study area. *NDVI* was calculated using the Band Math function according to Equation (2):

$$NDVI = \frac{NIR - RED}{NIR + RED}, \tag{2}$$

In Equation (2), *NIR* represents the near-infrared band, and *RED* represents the red band of the electromagnetic spectrum. The resulting *NDVI* values were normalized using the Compute Statistics function, and mean *NDVI* values were obtained for subsequent analysis. *NDVI* serves as an important indicator of vegetation coverage and greenness, with higher values indicating denser vegetation. Following previous studies, *NDVI* values were classified into five levels using the thresholds ≤ 0.2, 0.2–0.4, 0.4–0.6, 0.6–0.8, and ≥ 0.8, corresponding to Level 1 through Level 5, which represent low, below-average, average, above-average, and high green space availability, respectively.

2.3.2. Measurement of GSE Accessibility

Accessibility refers to the extent to which residents can obtain GSE resources within a certain range. Due to its computational simplicity and the ease of data acquisition, it has been widely used to assess the supply-demand balance of urban green spaces. In this study, Landsat 8 imagery was preprocessed using ENVI 5.6.2 software, including image fusion and orthorectification. Urban green spaces were then extracted, and the resulting green space vector data were further processed for accessibility analysis.

This study adopts the Gaussian two-step floating catchment area (Gaussian 2SFCA) method, which incorporates a Gaussian decay function to simulate the gradual attenuation of service effectiveness with increasing distance. Compared with traditional methods using fixed boundaries, this approach better reflects the continuous variation in service intensity in real-world settings and effectively mitigates abrupt changes in accessibility outcomes.

Using ArcGIS 10.8, the service capacity of each park green space was calculated. For each green space j, a spatial distance threshold d_0 was defined to establish its service catchment. According to the *Guidelines for the Construction and Service of 15-Minute Quality Cultural Living Circles* [20] issued by the Hangzhou municipal government—which specify a 15 min walking distance and a maximum service radius of 1000 m for daily community public spaces—the distance threshold d_0 was set to 1000 m in this study.

Within the catchment area, the population of demand points i was weighted using a Gaussian distance decay function to estimate the total potential population served by green space j. The supply–demand ratio R_j of each green space was then calculated as follows:

$$R_j = \frac{S_j}{\sum_{k \in \{d_{kj} \leq d_0\}} G(d_{ij}) D_k}, \tag{3}$$

https://doi.org/10.3390/su18021131

where D_i denotes the population of demand unit i; d_{ij} represents the walking time cost between residential point i and green space j; S_j is the supply capacity (area) of green space j; and $G(d_{ij})$ is the Gaussian distance decay function accounting for spatial friction, defined as:

$$G(d_{ij}, d_0) = \begin{cases} \frac{e^{-\frac{1}{2}\left(\frac{d_{ij}}{d_0}\right)^2} - e^{-\frac{1}{2}}}{1-e^{-\frac{1}{2}}}, & d_{ij} \leq d_0 \ , \\ 0, & d_{ij} > d_0 \end{cases} \quad (4)$$

Finally, green space accessibility at each residential location i was calculated by summing the weighted supply–demand ratios of all green spaces within the threshold distance, as shown in Equation (5):

$$A_i^D = \sum_{j \in \{d_j \leq d_0\}} G(d_{ij}) R_j \ , \quad (5)$$

Following previous studies, the resulting accessibility indices were classified into five levels using the thresholds 0–1, 1–1.5, 1.5–2.5, 2.5–3.8, and ≥3.8, corresponding to Level 1 through Level 5, representing low, below-average, average, above-average, and high accessibility, respectively.

2.3.3. Measurement of GSE Visibility

The GVI is an important indicator for measuring urban three-dimensional greenery, and has been widely applied in recent years to assess urban greening levels. In this study, street view images were processed using the fully convolutional neural network DeepLabv3+ for semantic segmentation to extract green space–related pixel information. DeepLabv3+ has demonstrated stable and robust performance on mainstream benchmark datasets such as PASCAL VOC 2012 and Cityscapes, achieving pixel-level accuracies of 89.0% and 82.1% on their respective test sets, and has been widely applied in complex urban scene analysis [21].

In practice, pretrained model parameters based on the ADE20K dataset were adopted. The ADE20K dataset contains more than 20,000 high-resolution images with 150 categories of pixel-level annotations, covering typical urban scene elements such as vegetation, roads, and buildings, and is characterized by high scene diversity and annotation accuracy [22]. Benchmark evaluations indicate that leading models supported by this dataset exhibit strong performance in semantic segmentation tasks under complex scene conditions [23].

After semantic segmentation, different object types in the images were assigned distinct labels. Green elements visible to residents—including trees, grass, vegetation clusters, palm trees, and flowers—were identified and aggregated. The proportion of these green elements relative to all image elements was calculated to derive the GVI (Figure 4).

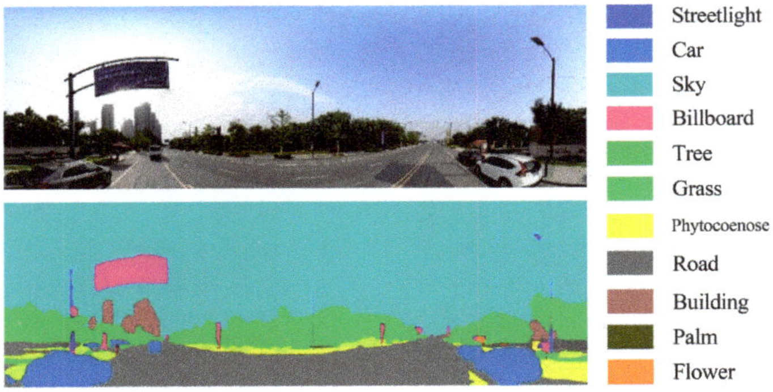

Figure 4. Example of the semantic segmentation result.

https://doi.org/10.3390/su18021131

It should be noted that this study did not conduct additional manual annotation or model retraining for street view images in the study area. Instead, the pretrained model was directly applied, a practice commonly adopted in existing studies that estimate GVI using street view imagery. Potential uncertainties associated with this approach are discussed in the limitations section.

With reference to previous studies [20,21], the calculated visibility results were classified into five categories based on ≤5%, 5%~15%, 15%~25%, 25%~35% and ≥35%,, corresponding to Level 1 through Level 5, where the intervals denote low, below average, average, above average, and high, respectively.

2.4. Gini Coefficient and Lorenz Curve

The Gini coefficient was originally developed to measure the equity of income distribution [22], but in recent years, it has been widely extended to the field of environmental equity, where it is commonly used to assess the distributional fairness of public resources, particularly in evaluating the distributional fairness of GSE [23–25]. One of its key advantages lies in its ability to capture overall inequality without relying on specific spatial form assumptions, making it applicable across different spatial scales.

In this study, the Lorenz curve is used to illustrate the cumulative distribution of GSE across the residential population. The X-axis represents the cumulative percentage of the population, while the Y-axis represents the cumulative percentage of GSE enjoyed by the corresponding population.

The Gini coefficient is derived based on the curvature of the Lorenz curve. The greater the curve's deviation from the 45-degree line of perfect equality, the higher the Gini coefficient, indicating a more unequal distribution of GSE resources. This deviation visually reflects the extent of inequity, and the corresponding Gini coefficient provides a quantitative measure of the degree of inequality. The Gini coefficient is calculated using the following formula:

$$Gini = 1 - \sum_{k=1}^{n} (P_k - P_{k-1})(S_k + S_{k-1}), \quad (6)$$

In Equation (6), n represents the total population, ranked in ascending order based on their level of GSE; k ranges from 0 to n; P_k is the cumulative proportion of the population, with $P_0 = 0$ and $P_n = 1$; S_k is the cumulative proportion of GSE received by the corresponding population, with $S_0 = 0$ and $S_n = 1$. The area under the line of perfect equality and above the axes is 0.5, which serves as the normalizing constant for calculating the Gini coefficient.

According to international standards and conventions for the Gini coefficient [8], its value ranges from 0 to 1, where 0 indicates perfect equality and 1 represents extreme inequality. In general: A Gini coefficient below 0.2 is considered absolutely equal; Between 0.2 and 0.3, it is considered relatively equal; Between 0.3 and 0.4, it is regarded as relatively reasonable; Between 0.4 and 0.5, it reflects a significant disparity; A Gini coefficient above 0.5 suggests severe inequality. In this study, these five ranges are further classified as Level 1 through Level 5, respectively.

2.5. Spatial Clustering Analysis

To further explore the spatial differentiation characteristics of GSE equity across different dimensions, this study further conducts spatial clustering analysis. Given that GSE equity is inherently multi-dimensional, a single indicator is insufficient to comprehensively capture the integrated disparities between areas. Therefore, it is necessary to identify spatial differentiation patterns from the perspective of multi-dimensional equity combinations.

 https://doi.org/10.3390/su18021131

Specifically, the natural breaks (Jenks) classification method is first applied to classify equity levels of GSE availability, accessibility, and visibility into five categories for each dimension. These levels, ranked from low to high, represent a transition from more unequal to more equitable distributions of GSE resources within each dimension. The Jenks method is well suited to environmental equity data with pronounced spatial heterogeneity, as it maximizes inter-class variance while minimizing intra-class variance.

Based on this classification, pairwise combinations of different equity dimensions within the same time slice are conducted to identify spatial clustering types of multi-dimensional equity characteristics. Through this approach, twelve typical equity combination types are constructed, including:

(1) Availability–Accessibility combinations:

High availability equity–high accessibility equity; Low availability equity–low accessibility equity; High availability equity–low accessibility equity; Low availability equity–high accessibility equity.

(2) Availability–Visibility combinations:

High availability equity–high visibility equity; Low availability equity–low visibility equity; High availability equity–low visibility equity; Low availability equity–high visibility equity.

(3) Visibility–Accessibility combinations:

High visibility equity–high accessibility equity; Low visibility equity–low accessibility equity; High visibility equity–low accessibility equity; Low visibility equity–high accessibility equity.

Among these, the "high–high (H–H)" type indicates areas where GSE resources across different dimensions are relatively evenly distributed in spatial terms, whereas the "low–low (L–L)" type reflects areas characterized by compounded deficiencies in multi-dimensional GSE and represents potential clusters of GSE inequity. In contrast, the "high–low (H–L)" or "low–high (L–H)" types reveal structural imbalances between different dimensions, indicating that improvements in a single dimension do not necessarily lead to a synchronous enhancement of overall equity.

By analyzing the spatial distribution and temporal evolution of these equity combination types, this study further elucidates the spatial clustering characteristics and evolutionary trajectories of GSE equity in Hangzhou. The results provide a basis for identifying priority intervention areas and formulating differentiated green space optimization strategies.

Through this pairing and spatial analysis, the study aimed to capture the interrelationships and spatial heterogeneity among different dimensions of GSE equity, as well as their spatial clustering characteristics and evolutionary trajectories, offering a more comprehensive understanding of urban green space distribution justice and providing a basis for identifying priority intervention areas and formulating differentiated green space optimization strategies.

3. Results and Analysis

3.1. Spatiotemporal Evolution of GSE in Hangzhou

3.1.1. Spatiotemporal Evolution of GSE Availability

From 2013 to 2023, the availability of GSE in Hangzhou's central urban districts exhibited slight fluctuations but an overall upward trend. Despite this general improvement, pronounced spatial inequality persisted throughout the study period. The spatial pattern was characterized by relatively higher levels of exposure availability in the western and southern areas, contrasted with consistently lower levels in the eastern and northern parts of the city.

https://doi.org/10.3390/su18021131

As shown in Table 2, the NDVI values for Hangzhou's central urban districts were 0.58 in 2013, decreased to 0.50 in 2018, and then rose again to 0.59 in 2023, exhibiting a decline followed by an increase trend, with a net increase of 0.01 over the decade. This indicates a slight overall improvement in GSE availability.

Table 2. Statistics of NDVI in Hangzhou Central Urban Districts (2013–2023).

District	NDVI		
	2013	2018	2023
BJ	0.42	0.36	0.46
GS	0.45	0.41	0.53
LP	0.53	0.42	0.53
QT	0.42	0.34	0.42
SC	0.42	0.38	0.48
XH	0.62	0.54	0.66
XS	0.60	0.51	0.59
YH	0.70	0.61	0.71
Hangzhou	0.58	0.50	0.59

At the administrative district level, both the magnitude and pace of NDVI change varied markedly across districts. In 2013, YH District had the highest NDVI value, followed by XH and XS districts, while SC District had the lowest, highlighting the substantial advantage of peripheral areas in terms of baseline natural vegetation. By 2018, NDVI values declined across all districts, with GS and SC districts experiencing the smallest decreases, whereas LP District saw the largest drop (exceeding 0.1) suggesting that this area bore particularly strong ecological pressure during the peak period of urban construction. In 2023, NDVI values rebounded in all districts; GS, LP and XH districts show the most pronounced increases, whereas QT and XS districts exhibited relatively limited recovery, reflecting inter-district differences in ecological restoration capacity and land-use adjustment strategies.

From a spatial perspective (Figure 5), GSE availability in Hangzhou exhibits marked spatial heterogeneity, with an overall west-high and east-low pattern. In 2013, low NDVI values were mainly concentrated in high-intensity built-up zones, including GS District, SC District, and BJ District, as well as in the northern industrial zones of QT District and the area south of the airport expressway in central XS District. In contrast, high-value areas were primarily found in the northwest corner of YH District, the XH (West Lake) Scenic Area, and the southwestern and southeastern corners of XS District. By 2018, low-value areas had further dispersed in a scattered pattern, while high-value zones gradually contracted from the urban center toward the periphery, with notable reductions in YH and XH Districts. In 2023, NDVI values increased significantly across the study area. High-value areas expanded from the urban periphery toward the urban center, forming clustered patches, while low-value areas shrank considerably and were mainly limited to localized zones such as the riverfront area in QT District. Overall, these changes suggest that recent ecological restoration and urban greening initiatives have begun to exert observable spatial effects.

https://doi.org/10.3390/su1802113

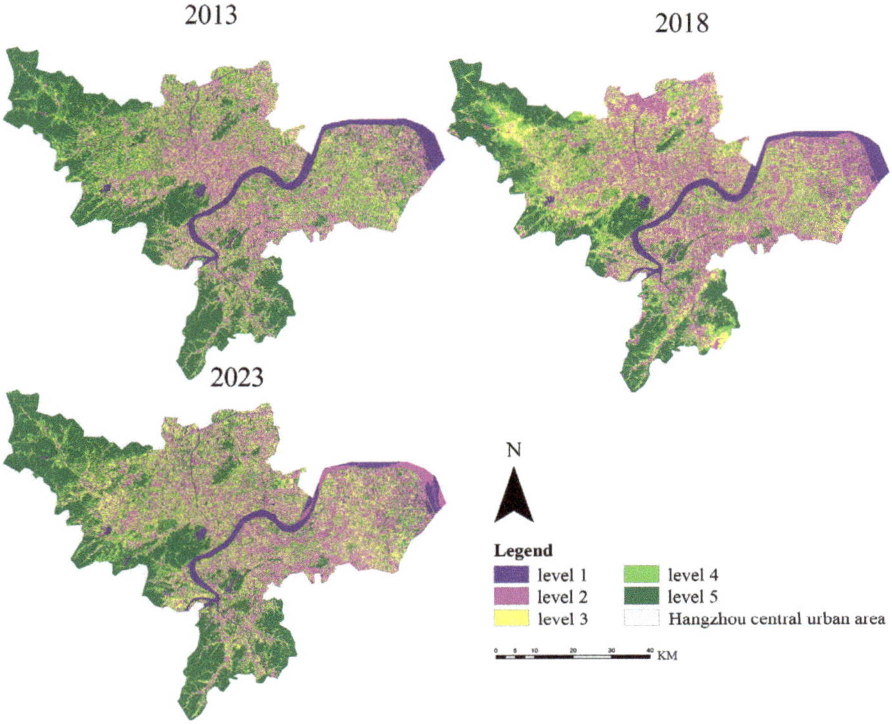

Figure 5. NDVI of various districts from 2013 to 2023.

3.1.2. Spatiotemporal Evolution of GSE Accessibility

From 2013 to 2023, the overall accessibility of GSE in Hangzhou's central urban area showed an upward trend (Table 3). The magnitude of change was substantially greater than that observed for GSE availability, making accessibility the most rapidly improving dimension among the three types of GSE.

Table 3. Statistics of accessibility in Hangzhou central urban districts (2013–2023).

District	Accessibility		
	2013	2018	2023
BJ	0.24	0.48	1.64
GS	0.07	1.02	1.71
LP	0.64	0.39	1.98
QT	0.45	0.47	1.70
SC	0.13	1.36	1.21
XH	0.25	2.16	1.48
XC	0.15	0.41	1.14
YH	0.59	0.47	1.74
Hangzhou	0.29	0.80	1.50

During the study period, the GSE accessibility values were 0.29 in 2013, 0.80 in 2018, and 1.50 in 2023—an overall increase of 1.21. This substantial improvement indicates that although urban development exerted pressure on natural green spaces, residents' practical access to green spaces continued to improve through the provision of new parks, greenway systems, and publicly accessible open spaces.

https://doi.org/10.3390/su18021131

At the district level, changes in accessibility exhibited clear stage-specific characteristics. In 2013, LP District had the highest accessibility, followed by YH District, while GS District had the lowest. By 2018, accessibility increased across most areas, except for slight decreases in LP and YH. XH District experienced the largest increase, with a rise of 1.91, reflecting a significant enhancement in the openness and accessibility of public green spaces surrounding scenic areas. In 2023, although XH and SC districts saw minor decreases, all other districts showed continued improvement, with all values exceeding 1.2, indicating a widespread spatial improvement in green space accessibility across the central urban area.

The spatial distribution patterns further illustrate the dynamic evolution of accessibility over time (Figure 6). In 2013, overall accessibility was low and spatially disordered, with extensive clusters of low-accessibility areas concentrated in XS District. By 2018, the spatial pattern had experienced its first major shift, evolving into a gradient distribution that declined from the urban center toward the periphery. High-accessibility areas were clustered around scenic zones such as the West Lake. In 2023, high-accessibility areas expanded further into the southern and central parts of the city, while low-accessibility areas became increasingly concentrated in border regions between BJ and XS Districts. This evolutionary trajectory demonstrates that green space accessibility is highly responsive to planning interventions and infrastructure development, with its spatial configuration adjusting rapidly in response to policy implementation.

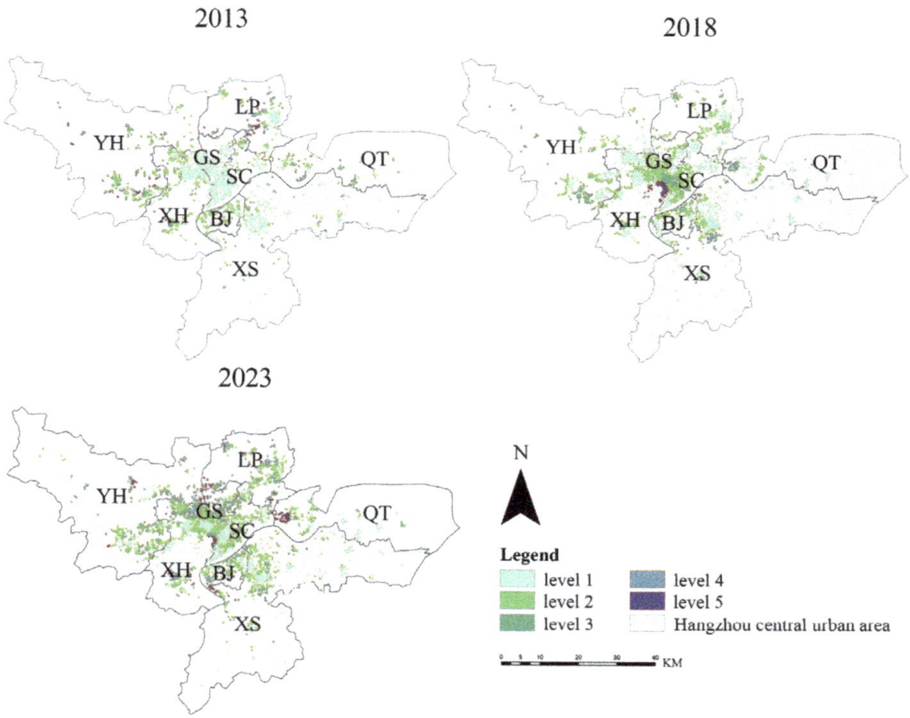

Figure 6. Accessibility of various districts from 2013 to 2023.

https://doi.org/10.3390/su18021131

3.1.3. Spatiotemporal Evolution of GSE Visibility

Unlike green space availability and accessibility, green space visibility exposure in Hangzhou exhibited an initial increase followed by a decline during the study period. Overall, the magnitude of change was limited, and the spatial pattern remained relatively stable.

From 2013 to 2023, the GSE based on the GVI in Hangzhou's central urban area increased from 13.68% in 2013 to 15.45% in 2018, before declining to 13.70% in 2023 (Table 4). The overall variation over the decade was modest. This pattern suggests that street-level green perception is more sensitive to factors such as road reconstruction, building renewal, and the expansion of transport infrastructure, resulting in lower temporal stability compared with green space availability and accessibility.

Table 4. Statistics of GVI in Hangzhou central urban districts (2013–2023).

District	GVI		
	2013	**2018**	**2023**
BJ	0.14	0.17	0.14
GS	0.13	0.15	0.13
LP	0.13	0.15	0.13
QT	0.14	0.15	0.14
SC	0.118	0.14	0.12
XH	0.20	0.22	0.21
XS	0.12	0.14	0.12
YH	0.14	0.16	0.15
Hangzhou	0.13	0.15	0.14

In 2013, the GVI in Hangzhou was 0.13, which increased to 0.15 by 2018, and then declined to 0.14 in 2023, resulting in a net increase of only 0.01 over the decade. This trend reflects the dynamic temporal fluctuation of GSE in the city.

From a spatial perspective (Figure 7), the GSE in Hangzhou exhibits a distinct west-high, east-low pattern. In 2013, areas with the highest GVI were mainly concentrated around the West Lake Scenic Area, China Water Expo Garden, and Liangzhu Cultural Village, while the lowest GVI values were found in the city center and along major expressways. By 2018, GVI expanded significantly in several newly developed residential areas along the Qiantang River, accompanied by a noticeable contraction of low-value zones in the urban center. However, in 2023, a significant decline in GVI was observed in the riverside areas of BJ, SC, and XS districts, with transport corridors and high-density built-up areas emerging as new low-value zones, whereas the core area of XH District remained highly stable.

https://doi.org/10.3390/su18021131

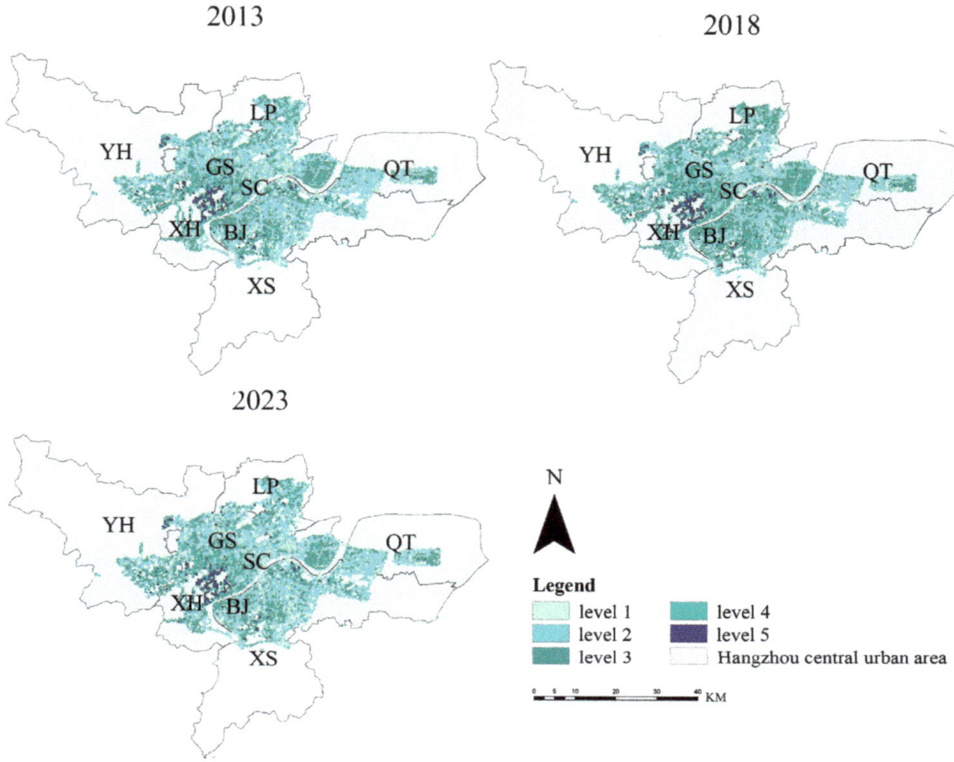

Figure 7. GVI of various districts from 2013 to 2023.

Overall, green space visibility exposure is highly sensitive to urban renewal approaches and street-space allocation. Changes in its spatial pattern primarily reflect planning orientations at the street scale rather than variations in the total amount of green space.

3.1.4. Temporal Trends of GSE Across Dimensions

To provide an overall synthesis of the temporal dynamics of GSE, this study summarizes the trends of availability, accessibility, and visibility from 2013 to 2023 (Table 5). Given the differences among these dimensions in data sources, measurement units, and value distributions, direct comparisons of absolute values across dimensions are not conducted. Instead, the analysis focuses on the direction and relative magnitude of change within each dimension over time.

Table 5. Summary of temporal trends in GSE across dimensions.

Dimension	2013–2018	2018–2023	Overall Trend
NDVI (Availability)	Increase	Slight increase	Moderate increase
Accessibility	Significant increase	Increase	Significant increase
GVI (Visibility)	Stable	Stable	Largely stable

The results show that all three dimensions of GSE exhibited varying degrees of improvement during the study period, although their trajectories differed markedly. Accessibility exposure demonstrated pronounced growth in both sub-periods. Availability

https://doi.org/10.3390/su18021131

exposure showed a generally gradual upward trend, whereas visibility exposure experienced relatively limited variation and remained largely stable over time.

These findings indicate clear inter-dimensional disparities in the improvement of GSE in Hangzhou, which may be associated with distinct driving mechanisms and spatial constraints underlying each dimension.

3.2. Spatiotemporal Evolution of GSE Equity in Hangzhou

3.2.1. Spatiotemporal Evolution of GSE Equity Based on Availability

Over the past decade, the equity of GSE based on availability in Hangzhou has improved significantly but remained at a moderately low level overall. Meanwhile, its spatial distribution exhibited relatively stable structural disparities.

From an overall perspective (Table 6; Figure 8a–d), the Gini coefficients of GSE based on NDVI in Hangzhou were 0.50, 0.47, and 0.46 for the years 2013, 2018, and 2023, representing a cumulative reduction of 0.03 over the decade. These values indicate a considerable disparity in the distribution of GSE resources, although equity improved over the period remained limited. The period from 2013 to 2018 constituted the primary phase of equity improvement, whereas progress slowed during 2018–2023, suggesting diminishing marginal gains in availability-based equity.

Table 6. Statistics of Gini coefficient based on NDVI in Hangzhou central urban districts (2013–2023).

District	Gini Coefficient Based on NDVI		
	2013	**2018**	**2023**
BJ	0.47	0.42	0.39
SC	0.63	0.57	0.55
LP	0.56	0.51	0.49
QT	0.62	0.57	0.58
GS	0.25	0.33	0.31
XH	0.55	0.54	0.50
XS	0.74	0.76	0.74
YH	0.63	0.71	0.72
Hangzhou	0.50	0.47	0.46

At the district level, equity trajectories differed substantially. The Gini coefficients in BJ, SC, LP, and XH districts showed a downward trend, with BJ district experiencing the most notable improvement, decreasing by 8 percentage points, indicating a notable improvement in the balance of green space distribution. Conversely, YH district's Gini coefficient increased continuously, reflecting a progressive deterioration in equity and a growing mismatch between green space distribution and population restructuring in rapidly expanding areas. QT and GS districts displayed fluctuating patterns, suggesting that green space allocation in these areas was more strongly influenced by changing development phases.

From the spatial distribution perspective (Figure 8e), the equity of GSE in Hangzhou exhibits a pattern of higher equity in the central areas and lower equity in the peripheral regions. In 2013, peripheral areas showed lower equity compared to the center, with particularly poor equity in the southeast. By 2018, the equity in the central area remained relatively stable, while peripheral areas experienced more pronounced changes. By 2023, although the overall pattern persisted, areas with relatively higher equity became increasingly concentrated toward the urban core, indicating a strong path dependence in the spatial distribution of natural vegetation–based green space resources.

https://doi.org/10.3390/su18021131

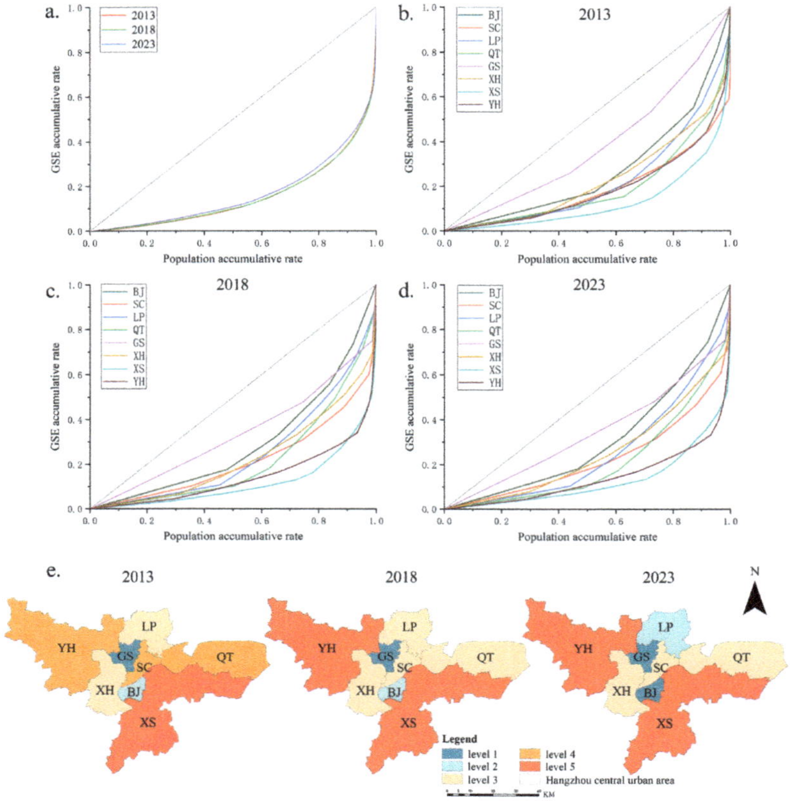

Figure 8. Lorenz curves of GSE availability in Hangzhou (**a**) and its districts (**b**–**d**), and Gini coefficients of district-level GSE availability (**e**) from 2013 to 2023.

3.2.2. Spatiotemporal Evolution of GSE Equity Based on Accessibility

Over the past decade, the equity of GSE based on accessibility in Hangzhou first experienced a slight decline, followed by a significant improvement with a considerable magnitude. Spatially, notable disparities persisted, exhibiting distinct evolutionary trends across different areas.

Overall (Table 7; Figure 9a–d), the Gini coefficient of accessibility-based GSE in the central urban area was 0.77 in 2013, slightly increased to 0.78 in 2018, and then declined markedly to 0.55 in 2023. Although all three values exceeded 0.50—indicating persistent inequality—the cumulative reduction of 0.22 over the study period demonstrates that accessibility-based equity experienced the most substantial improvement among the three dimensions.

At the district level, most areas followed a trajectory similar to the overall trend, with slight deterioration between 2013 and 2018 and pronounced improvement between 2018 and 2023. SC District recorded the largest decline in its Gini coefficient (−0.26), transitioning from one of the most inequitable areas to a relatively equitable one. LP and YH districts also exhibited continuous declines, indicating steady improvements in accessibility equity. In contrast, although BJ District showed a decrease after 2018, its 2023 Gini coefficient remained higher than in 2013, suggesting a comparatively limited degree of improvement.

Spatially (Figure 9e), accessibility-based equity in 2013 displayed a pattern of "higher equity in central areas and lower equity in peripheral areas", with northwestern areas performing better than southeastern ones. This pattern largely persisted in 2018, though local disparities intensified. By 2023, while the overall structure remained stable, the equity center of gravity in peripheral areas shifted, with notable improvements in the northeastern and southwestern directions, and relatively slower progress in the northwestern and southeastern zones. Overall, accessibility equity was highly responsive to planning interventions and adjusted more rapidly than availability- and visibility-based equity.

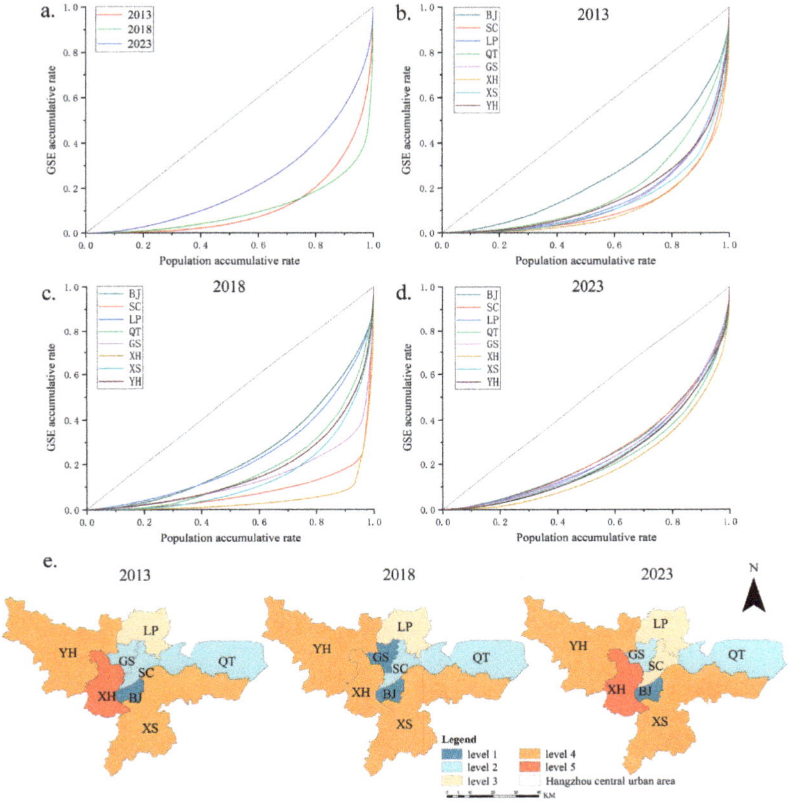

Figure 9. Lorenz curves of GSE accessibility in Hangzhou (**a**) and its districts (**b–d**), and Gini coefficients of district-level accessibility (**e**) from 2013 to 2023.

Table 7. Statistics of Gini coefficient based on accessibility in Hangzhou central urban districts (2013–2023).

District	Gini Coefficient Based on Accessibility		
	2013	2018	2023
BJ	0.49	0.51	0.50
SC	0.76	0.84	0.50
LP	0.69	0.53	0.51
QT	0.62	0.64	0.57
GS	0.70	0.73	0.52
XH	0.78	0.91	0.62
XS	0.73	0.72	0.54
YH	0.68	0.66	0.55
Hangzhou	0.77	0.78	0.55

https://doi.org/10.3390/su18021131

3.2.3. Spatiotemporal Evolution of GSE Equity Based on Visibility

During the study period, the equity of GSE in Hangzhou, based on GVI, showed slight improvement; however, the overall level of equity remained low. Spatially, GSE equity exhibited a distinct west-high, east-low pattern across the city.

From an overall perspective (Table 8; Figure 10a–d), the Gini coefficient based on GVI in the central urban area remained within the range of 0.50–0.60 from 2013 to 2023, showing a pattern of slight decline followed by rebound. Between 2013 and 2018, improvements in overall GVI were accompanied by modest gains in equity, whereas between 2018 and 2023, the decline in GVI coincided with an increase in the Gini coefficient, indicating the high sensitivity of visibility equity to changes in street-space environments.

Table 8. Statistics of Gini coefficient based on GVI in Hangzhou central urban districts (2013–2023).

District	Gini Coefficient Based on GVI		
	2013	**2018**	**2023**
BJ	0.42	0.44	0.45
SC	0.50	0.47	0.51
LP	0.54	0.52	0.51
QT	0.50	0.48	0.50
GS	0.47	0.45	0.46
XH	0.61	0.60	0.61
XS	0.59	0.57	0.56
YH	0.60	0.60	0.59
Hangzhou	0.56	0.54	0.55

At the district level, disparities were pronounced. BJ District consistently exhibited relatively low Gini coefficients, making it one of the most equitable areas in terms of visibility exposure. In contrast, XH District persistently showed the highest inequality and was the only district with Gini coefficients exceeding 0.60 in all three years, indicating a strong concentration of visual green resources in a limited number of advantaged areas. Although some districts (e.g., YH, LP, and XS) displayed slight improvements in later years, overall progress remained limited.

Spatially (Figure 10e), visibility-based equity exhibited a stable "high-in-the-west, low-in-the-east" pattern. While equity improved in most areas in 2018, Gini coefficients increased again in 2023 in most districts, suggesting that street-level greening improvements are temporally unstable and difficult to sustain in terms of equity.

https://doi.org/10.3390/su18021131

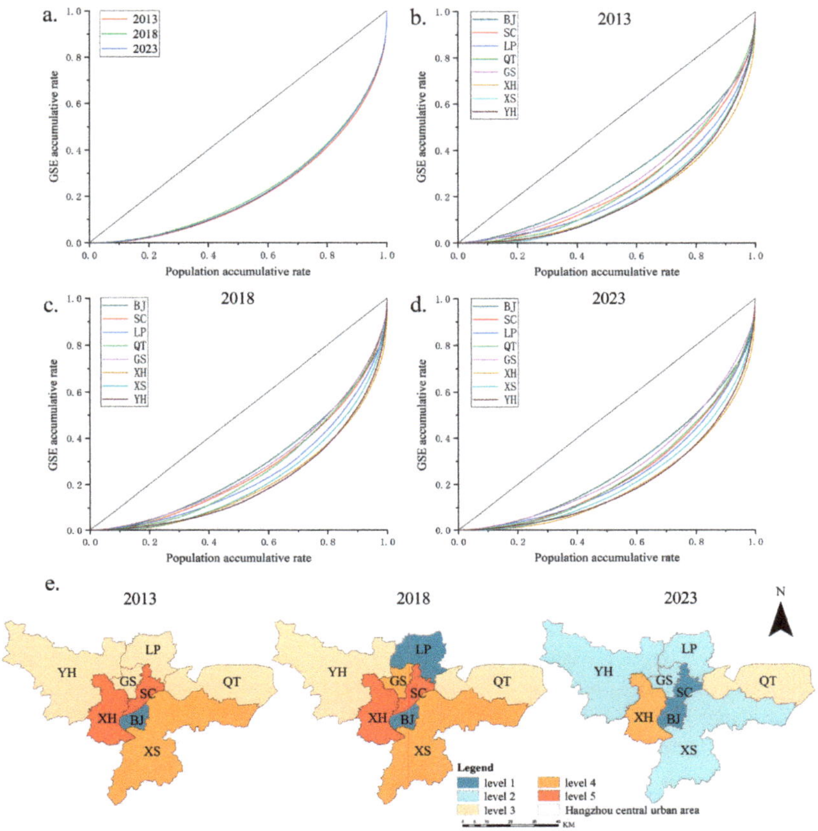

Figure 10. Lorenz curves of GSE visibility in Hangzhou (**a**) and its districts (**b–d**), and Gini coefficients of district-level visibility (**e**) from 2013 to 2023.

3.2.4. Comparative Trends of GSE Equity Across Dimensions

Table 9 summarizes changes in GSE equity across availability, accessibility, and visibility dimensions from 2013 to 2023. Overall, equity trajectories differed markedly across dimensions. Accessibility-based equity improved most substantially, with the Gini coefficient declining by 0.22, reflecting the role of park development and improvements in pedestrian and slow-mobility systems in reducing spatial disparities over the past decade. In contrast, availability-based equity showed a continuous but modest improvement, suggesting that inequalities in green space supply are structurally constrained by land availability and ecological baselines. Visibility-based equity exhibited minimal fluctuation and remained above 0.50 throughout the study period, indicating persistent inequity in street-level green exposure and relatively slow responsiveness to short-term policy interventions.

Table 9. Trends in GSE equity across dimensions (2013–2023).

Dimension	Gini Coefficient			Trend	Magnitude
	2013	**2018**	**2023**		
NDVI (Availability)	0.50	0.47	0.46	Continuous improvement	−0.04
Accessibility	0.77	0.78	0.55	Increase followed by sharp decline	−0.22
GVI (Visibility)	0.56	0.54	0.55	Fluctuating stability	−0.01

https://doi.org/10.3390/su18021131

3.3. *Spatial Cluster Matching Analysis of GSE Equity in Hangzhou*

During the study period, the spatial matching relationships among the three dimensions of GSE equity in Hangzhou exhibited marked differences. Overall, pairwise combinations of equity dimensions were most prominently characterized by stable and significant Low–Low clusters, followed by High–High clusters, whereas High–Low and Low–High clusters were largely insignificant in most years. This pattern indicates that GSE inequity is more likely to manifest as the spatial accumulation and persistence of multidimensional disadvantages, while the co-location of multidimensional advantages tends to be less stable.

3.3.1. Clustering Between Accessibility-Based and Visibility-Based GSE Equity

For the spatial matching between accessibility-based and visibility-based equity (Figure 11a–c), XH District exhibited a Low accessibility–Low visibility cluster in 2013, while BJ District showed a High accessibility–High visibility cluster. All other areas did not form significant spatial clusters. This spatial configuration remained largely unchanged in 2018. By 2023, XH District continued to display a Low–Low clustering pattern, whereas BJ District no longer exhibited a High–High cluster.

These results indicate that Low–Low clusters demonstrate greater temporal stability. Although XH District possesses relatively abundant natural green space resources, its high density of historical built-up areas, concentrated population, and limited flexibility for street-space retrofitting have placed residents at a persistent disadvantage in both green space accessibility and street-level visual exposure. In contrast, BJ District initially benefited from its riverside landscape resources and relatively well-developed slow-mobility systems, forming a dual advantage in accessibility and visibility. However, with rapid population growth and increasing functional intensification, this equity advantage gradually weakened, suggesting the vulnerability of High–High clusters under conditions of accelerated urban development.

3.3.2. Clustering Between Availability-Based and Accessibility-Based GSE Equity

The spatial matching between availability-based and accessibility-based equity was generally weak (Figure 11d–f). In both 2013 and 2018, no significant spatial clusters were identified across the study area, indicating that the level of green space supply and residents' actual ease of accessing green spaces did not exhibit synchronous spatial patterns. By 2023, Xiaoshan District emerged as a Low availability–Low accessibility cluster, while all other areas remained insignificant.

This result reflects differences in the planning logic and implementation pathways underlying availability and accessibility. Availability is primarily determined by the scale and structure of green space land use, whereas accessibility is influenced by the combined effects of road networks, slow-mobility infrastructure, and spatial access routes. As a rapidly expanding urban area, Xiaoshan District has experienced construction land growth that outpaced public green space provision, while transport and pedestrian systems have not yet fully adapted to this expansion. Consequently, insufficient green space supply and limited accessibility became spatially compounded, resulting in the formation of a Low–Low cluster.

3.3.3. Clustering Between Availability-Based and Visibility-Based GSE Equity

The spatial clustering between availability-based and visibility-based equity exhibited pronounced instability (Figure 11g–i). In 2013, no significant clusters were observed. In 2018, Yuhang and Xiaoshan Districts emerged as Low availability–Low visibility clusters.

 https://doi.org/10.3390/su18021131

By 2023, these clusters disappeared, and the overall spatial pattern returned to a non-significant state.

This temporal variability suggests that availability and visibility are unlikely to form long-term stable spatial coupling. Availability largely depends on the natural ecological base and the structural layout of green spaces, whereas visibility is highly sensitive to street-scale greening forms, vegetation configuration, and routine maintenance. Under conditions of continuous urban renewal and streetscape improvement, visibility exhibits greater dynamism and malleability, thereby weakening its spatial consistency with availability over time.

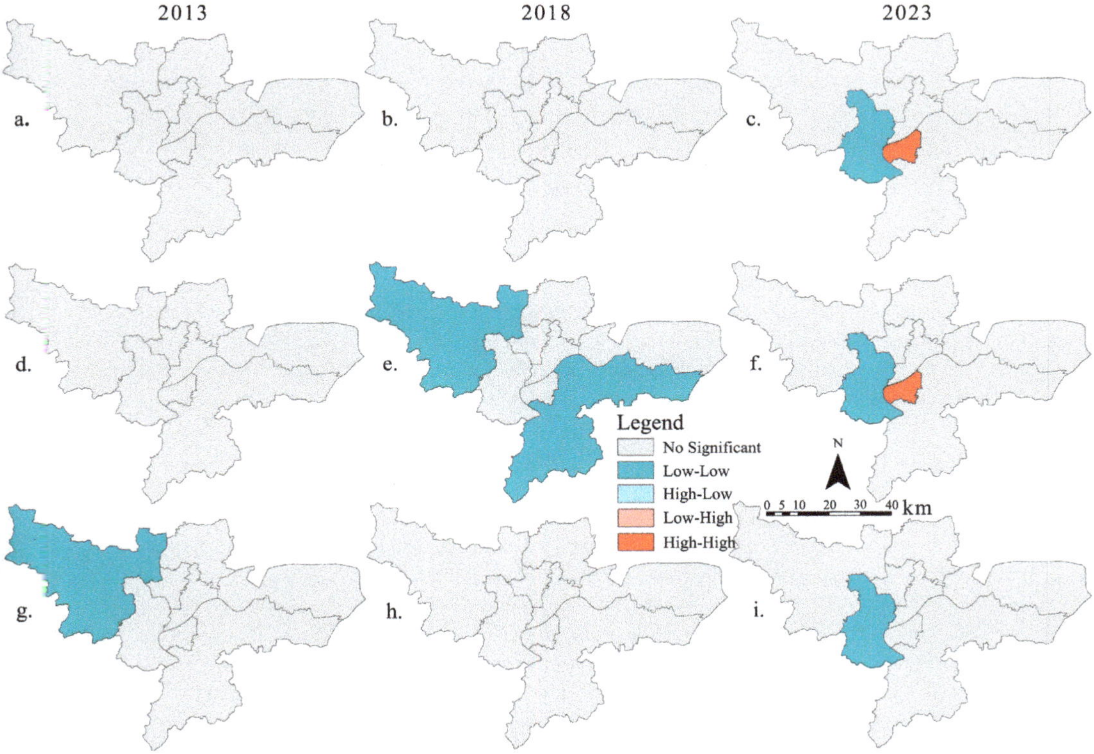

Figure 11. Spatial clustering of GSE equity (2013–2023): (**a–c**) accessibility-based vs. visibility-based; (**d–f**) availability-based vs. accessibility-based; (**g–i**) availability-based vs. visibility-based.

4. Discussion

4.1. Driving Mechanisms of Overall GSE Improvement and Causes of Spatial Inequality

This study finds that GSE in the central urban area of Hangzhou exhibited a sustained upward trend during the study period, with the most pronounced improvement observed in the accessibility dimension. However, this growth was not spatially uniform; instead, it displayed clear regional differentiation. This result indicates that improvements in GSE are not driven solely by the expansion of green space area, but are more fundamentally shaped by the combined effects of urban development strategies, land-use patterns, and the orientation of public investment.

At the national level, policies promoting new-type urbanization and Ecological Civilization have stimulated continuous increases in urban greening investment. At the local

 https://doi.org/10.3390/su18021131

level, Hangzhou has adopted the "Park City" vision and systematically advanced the development of park systems and greenway networks, significantly enhancing residents' opportunities to access green spaces on foot. This trend is consistent with findings reported by Wang et al. (2020) [26] and Cao et al. (2022) [27], who documented overall improvements in GSE across Chinese cities, further confirming the pivotal role of policy-driven interventions in enhancing GSE.

Nevertheless, the uneven spatial growth of GSE reflects substantial differences in land resource endowments and functional positioning across urban areas. Newly developed districts, characterized by relatively abundant land supply and greater planning flexibility, are better positioned to enhance GSE through the provision of new parks and green spaces. In contrast, older urban districts are constrained by high development intensity and rigid land-use structures, leaving limited room for green space enhancement and resulting in comparatively lagging improvements in GSE. While this "increment-oriented" greening pathway contributes to overall gains, it may also reinforce pre-existing spatial disparities. This spatial imbalance aligns with findings from a similar study conducted in Wuhan [25].

4.2. Spatial Mechanisms of GSE Inequity and Planning Implications

Although overall GSE equity in Hangzhou improved during the study period, the magnitude of improvement remained limited for most dimensions except accessibility. This pattern suggests that gains in GSE equity have not progressed in tandem with overall increases in GSE, with the primary constraint stemming from supply–demand mismatches under conditions of rapid population concentration.

Previous studies have highlighted a structural contradiction common to Chinese cities, wherein the pace of green space expansion struggles to keep up with population growth (Huang et al., 2017 [28]; Xie et al., 2023 [29]). Despite substantial annual increases in green space provision during the study period, Hangzhou experienced sustained population inflows into its central urban areas, significantly diluting improvements in per capita GSE. This effect is particularly pronounced in older urban districts, where high population density combined with land scarcity further constrains the flexibility of public green space provision, thereby limiting the scope for equity improvement.

At the same time, functional differentiation across urban areas has intensified spatial mismatches in GSE allocation. Districts with high concentrations of educational, commercial, and administrative functions attract large permanent and transient populations but face difficulties in synchronously providing green spaces of corresponding scale. Conversely, some newly developed areas exhibit relatively favorable green space conditions but lower population intensity and utilization levels, resulting in a structural imbalance characterized by "green space surplus–demand deficit". By shifting the analytical focus from single indicators to population–exposure relationships, this study extends previous approaches to green space equity assessment and underscores the necessity of evaluating equity from a resident-centered exposure perspective.

4.3. Formation Mechanisms of Different GSE Equity Types and Policy Implications

This study further reveals pronounced spatial differentiation among different dimensions of GSE equity, with Low–Low clusters emerging as the most prominent and spatially stable pattern across all three dimensions. This finding indicates that GSE inequity exhibits a distinctly structural character within urban space, rather than arising from random spatial variation.

Low–Low clusters are closely associated with historical land-use patterns and infrastructure configurations. In older urban areas, early planning stages placed limited emphasis on green space provision, while road spaces were predominantly designed to

https://doi.org/10.3390/su1802113

prioritize motorized traffic, restricting opportunities for street greening and improvements in pedestrian-accessible green spaces. In addition, some industry-oriented districts have deprioritized the provision of livability-oriented public spaces due to their functional positioning. Similar path-dependent effects have also been identified in Wu et al.'s (2022) study of Shenzhen [30].

The pronounced clustering of "low visibility–low accessibility" equity further highlights the critical role of street space in shaping GSE equity. Compared with centralized parks, street greenery and green exposure along daily travel routes exert a more direct influence on residents' everyday experiences. When street design prioritizes vehicular efficiency, GSE is constrained both visually and behaviorally, resulting in stable zones characterized by low exposure and low equity.

By contrast, spatial matching between availability-based equity and other dimensions is generally weaker and lacks long-term stable clustering patterns, reflecting fundamental differences in the underlying formation mechanisms of GSE dimensions. Availability is more strongly dependent on the city's natural ecological base and land-use structure, whereas accessibility and visibility are more directly influenced by transportation systems and street-space design. This lack of coordinated evolution across dimensions further demonstrates that single indicators are insufficient for capturing the complex realities of GSE equity in cities.

Accordingly, differentiated planning strategies are required for different types of inequitable areas. For stable Low–Low clusters, priority should be given to improving daily exposure through enhanced street greening, optimization of slow-mobility systems, and the embedded development of pocket parks. For newly developed areas, GSE equity indicators should be incorporated into land-use and transportation system design at early planning stages to prevent the emergence of structural inequities.

5. Conclusions

5.1. Main Findings and Research Contributions

By integrating multi-temporal remote sensing imagery, street-view big data, and population data, this study systematically examined the spatiotemporal evolution of GSE (GSE) and its equity in the central urban area of Hangzhou from 2013 to 2023. The main findings are summarized as follows:

(1) During the study period, overall GSE in Hangzhou exhibited a sustained upward trend, with accessibility showing the most pronounced improvement. This indicates that policy-oriented greening practices centered on park development and the expansion of green space networks have been effective in enhancing residents' opportunities for green space contact.

(2) The growth of GSE displayed significant spatial unevenness. Areas with higher growth were mainly concentrated in regions with relatively abundant ecological resources or lower development intensity, whereas densely built-up areas and industry-oriented districts experienced more limited improvement. This suggests that expansion models driven primarily by land supply and functional zoning are insufficient to simultaneously mitigate intra-urban disparities in GSE.

(3) Although overall equity in GSE improved to some extent, the magnitude of improvement remained limited, with only accessibility-based equity exhibiting a substantial enhancement. This reflects the persistence of structural mismatches between population concentration and green space provision under rapid urbanization.

(4) Different dimensions of GSE equity demonstrated markedly differentiated spatial patterns and evolutionary trajectories. In particular, Low–Low clusters characterized by low

 https://doi.org/10.3390/su18021131

availability, low accessibility, and low visibility showed strong spatial stability, indicating the long-term existence of structurally disadvantaged greening areas within the city.

(5) By integrating availability, accessibility, and visibility dimensions, this study developed a multidimensional and time-series-based framework for assessing GSE equity. This framework provides methodological support and empirical evidence for identifying urban greening blind spots, optimizing green space spatial allocation, and promoting inclusive urban planning.

5.2. Limitations and Future Directions

Despite integrating multi-source, time-series remote sensing and street-view data to systematically characterize the spatiotemporal evolution of GSE and its equity in Hangzhou from the perspectives of availability, accessibility, and visibility, this study has several limitations that warrant further investigation.

First, at the data level, population distribution was primarily represented using residential address point data obtained from real estate platforms. While this approach captures the general spatial pattern of residential population, it does not explicitly account for housing vacancy rates or variations in actual occupancy intensity. In newly developed or investment-oriented areas in particular, high vacancy rates may lead to overestimation of population size, thereby underestimating per capita GSE and affecting the accuracy of equity measurements. Future studies could integrate census data, mobile signaling data, or nighttime light data to dynamically adjust population distributions and assess the impacts of such systematic biases through scenario analysis or sensitivity testing.

Second, in the measurement of accessibility, this study adopted a Gaussian two-step floating catchment area (2SFCA) method with a fixed distance threshold and decay function parameters based on local planning standards. While this policy-consistent parameterization enhances interpretability and comparability, accessibility outcomes may still be sensitive to alternative choices of catchment size or decay settings. Future research could conduct systematic sensitivity analyses across multiple distance thresholds and decay parameters to further assess the robustness of accessibility and equity estimates.

Third, regarding street-view data and semantic segmentation methods, although the DeepLabv3+ model employed in this study was pre-trained on well-established datasets such as ADE20K and has been widely validated across various urban contexts, no additional manual annotation was conducted to verify segmentation accuracy for the study area. Moreover, due to constraints in the availability of historical street-view imagery, inconsistencies in acquisition timing and update frequency across different years may introduce uncertainty into GVI estimation. Future research could improve robustness by constructing small-scale manually labeled samples or applying multi-model comparison approaches to validate semantic segmentation results.

Fourth, in terms of analytical scope, this study primarily focused on the supply-side characteristics of GSE and its spatial equity, while individual-level differences and actual usage behaviors were not explicitly examined. Residents with different age structures, income levels, or travel modes may exhibit substantially different demands for and interactions with green spaces, yet such socioeconomic factors were not fully incorporated into the current framework. Future studies could integrate socioeconomic statistics or survey data to deepen the interpretation of GSE inequities from the perspectives of environmental justice and health equity.

Finally, regarding spatial scale and generalizability, this study focused on the central urban area of Hangzhou as a case study, and its findings are influenced to some extent by the city's development stage, planning institutions, and natural geographic conditions. Future research could extend this multidimensional GSE equity framework to other types

https://doi.org/10.3390/su18021131

of cities through cross-city comparisons or multi-case studies, in order to identify both common patterns and contextual differences in GSE inequities. In addition, alternative equity indicators and evaluation methods could be explored to develop a more robust and generalizable assessment system for urban GSE equity.

Author Contributions: Conceptualization, Y.T. and X.G.; methodology, C.L. (Chang Liu); software, Y.W.; validation, Y.T., Y.W. and C.L. (Chan Li); formal analysis, Y.T.; investigation, C.L. (Chan Li); resources, X.G.; data curation, Y.T.; writing—original draft preparation, Y.T.; writing—review and editing, Y.T.; visualization, Y.W. and C.L. (Chan Li); supervision, X.G.; project administration, X.G.; funding acquisition, X.G. All authors have read and agreed to the published version of the manuscript.

Funding: This research was funded by the National Natural Science Foundation of China (51908063), the Philosophy and Social Science Foundation of Hubei Province (21Q049), the Research Planning Project on Higher Education Science of the China Association of Higher Education (25JZ0301) and the 2023 Provincial Teaching Reform Research Project for Undergraduate Universities in Hubei Province (270).

Institutional Review Board Statement: Not applicable.

Informed Consent Statement: Not applicable.

Data Availability Statement: The original contributions presented in this study are included in the article. Further inquiries can be directed to the corresponding author.

Conflicts of Interest: The authors declare no conflicts of interest.

References

1. Yan, M.; Yao, B.; Cenci, J.; Liao, C.; Zhang, J. Visualisation of High-Density City Research Evolution, Trends, and Outlook in the 21st Century. *Land* **2023**, *12*, 485. [CrossRef]
2. Zhu, Z.; Li, J.; Chen, Z. Green Space Equity: Spatial Distribution of Urban Green Spaces and Correlation with Urbanization in Xiamen, China. *Environ. Dev. Sustain.* **2023**, *25*, 423–443. [CrossRef]
3. Chen, J.; Kinoshita, T.; Li, H.; Luo, S.; Su, D.; Yang, X.; Hu, Y. Toward Green Equity: An Extensive Study on Urban Form and Green Space Equity for Shrinking Cities. *Sustain. Cities Soc.* **2023**, *90*, 104395. [CrossRef]
4. Xu, H.; Zheng, G.; Lin, X.; Jin, Y. Study on the Microclimatic Effects of Plant-Enclosure Conditions and Water-Green Space Ratio on Urban Waterfront Spaces in Summer. *Sustainability* **2024**, *16*, 2957. [CrossRef]
5. Beele, E.; Aerts, R.; Reyniers, M.; Somers, B. Spatial Configuration of Green Space Matters: Associations between Urban Land Cover and Air Temperature. *Landsc. Urban Plan.* **2024**, *249*, 105121. [CrossRef]
6. Kim, H.W.; Kim, J.-H.; Li, W.; Yang, P.; Cao, Y. Exploring the Impact of Green Space Health on Runoff Reduction Using NDVI. *Urban For. Urban Green.* **2017**, *28*, 81–87. [CrossRef]
7. Zhang, Y.; Luo, F. Linkages among Socio-Economic Status, Green Space Accessibility, and Health Outcomes: An Environmental Justice Perspective in Australia. *Sustain. Cities Soc.* **2024**, *114*, 105784. [CrossRef]
8. Zhang, Y.; Lin, T.; Hamm, N.A.S.; Liu, J.; Zhou, T.; Geng, H.; Zhang, J.; Ye, H.; Zhang, G.; Wang, X.; et al. Quantitative Evaluation of Urban Green Exposure and Its Impact on Human Health: A Case Study on the 3-30-300 Green Space Rule. *Sci. Total Environ.* **2024**, *924*, 171461. [CrossRef]
9. Bell, S.L.; Phoenix, C.; Lovell, R.; Wheeler, B.W. Green Space, Health and Wellbeing: Making Space for Individual Agency. *Health Place* **2014**, *30*, 287–292. [CrossRef]
10. Ko, Y.J.; Cho, K.-H.; Kim, W.-C. Analysis of Environmental Equity of Green Space Services in Seoul—The Case of Jung-gu, Seongdong-gu and Dongdaemun-gu-. *J. Korean Inst. Landsc. Archit.* **2019**, *47*, 100–116. [CrossRef]
11. Zhang, Y.; Zhao, J.; Mavoa, S.; Smith, M. Inequalities in Urban Green Space Distribution across Priority Population Groups: Evidence from Tamaki Makaurau Auckland, Aotearoa New Zealand. *Cities* **2024**, *149*, 104972.
12. Chen, Y.; La Rosa, D.; Yue, W.; Xu, Z.; Zhuo, Y. Do Larger Cities Enjoy Better Green Space Accessibility? Evidence from China. *Environ. Impact Assess. Rev.* **2024**, *107*, 107544. [CrossRef]
13. Huang, B.-X.; Li, W.-Y.; Ma, W.-J.; Xiao, H. Space Accessibility and Equity of Urban Green Space. *Land* **2023**, *12*, 766. [CrossRef]
14. van Heezik, Y.; Freeman, C.; Falloon, A.; Buttery, Y.; Heyzer, A. Relationships between Childhood Experience of Nature and Green/Blue Space Use, Landscape Preferences, Connection with Nature and pro-Environmental Behavior. *Landsc. Urban Plan.* **2021**, *213*, 104135. [CrossRef]

 https://doi.org/10.3390/su18021131

15. Cheng, Y.; Browning, M.H.E.M.; Zhao, B.; Qiu, B.; Wang, H.; Zhang, J. How Can Urban Green Space Be Planned for a 'Happ City'? Evidence from Overhead- to Eye-Level Green Exposure Metrics. *Landsc. Urban Plan.* **2024**, *249*, 105131. [CrossRef]
16. Wu, S.; Chen, B.; Webster, C.; Xu, B.; Gong, P. Improved Human Greenspace Exposure Equality during 21st Century Urbanizatio *Nat. Commun.* **2023**, *14*, 6460. [CrossRef]
17. Han, Y.; He, J.; Liu, D.; Zhao, H.; Huang, J. Inequality in Urban Green Provision: A Comparative Study of Large Cities throughou the World. *Sustain. Cities Soc.* **2023**, *89*, 104229. [CrossRef]
18. Lu, Y.; Chen, R.; Chen, B.; Wu, J. Inclusive Green Environment for All? An Investigation of Spatial Access Equity of Urban Gree Space and Associated Socioeconomic Drivers in China. *Landsc. Urban Plan.* **2024**, *241*, 104926. [CrossRef]
19. Huang, Z.; Tang, L.; Qiao, P.; He, J.; Su, H. Socioecological Justice in Urban Street Greenery Based on Green View Index-A Cas Study within the Fuzhou Third Ring Road. *Urban For. Urban Green.* **2024**, *95*, 128313. [CrossRef]
20. Construction and Service Standards for the 15-Minute High-Quality Cultural Life Circle. Available onlin https://zlzx.zjamr.zj.gov.cn/bzzx/public/news/view/STD_PUBLIC/ac945905326449f088111fb9a757bdc2.html (accessed on January 2026).
21. Chen, L.-C.; Zhu, Y.; Papandreou, G.; Schroff, F.; Adam, H. Encoder-Decoder with Atrous Separable Convolution for Semanti Image Segmentation. In *Proceedings of the Computer Vision—ECCV 2018*; Ferrari, V., Hebert, M., Sminchisescu, C., Weiss, Y., Eds Springer International Publishing: Cham, Switzerland, 2018; pp. 833–851.
22. Wang, R.; Feng, Z.; Pearce, J.; Yao, Y.; Li, X.; Liu, Y. The Distribution of Greenspace Quantity and Quality and Their Associatio with Neighbourhood Socioeconomic Conditions in Guangzhou, China: A New Approach Using Deep Learning Method an Street View Images. *Sustain. Cities Soc.* **2021**, *66*, 102664. [CrossRef]
23. Zhou, B.; Zhao, H.; Puig, X.; Xiao, T.; Fidler, S.; Barriuso, A.; Torralba, A. Semantic Understanding of Scenes Through the ADE20 Dataset. *Int. J. Comput. Vis.* **2019**, *127*, 302–321. [CrossRef]
24. Lorenz, M.O. Methods of Measuring the Concentration of Wealth. *Publ. Am. Stat. Assoc.* **1905**, *9*, 209–219.
25. Guo, X.; Liu, C.; Bi, S.; Tang, Y. Identification of Inequities in Green Visibility and Ways to Increase Greenery in Neighborhoods: Case Study of Wuhan, China. *Appl. Sci.* **2025**, *15*, 742. [CrossRef]
26. Wang, L.; Cheng, R.; Wang, X.; Song, W.; Zhang, S.; Huang, S. A Dynamic Assessment for Greenness Exposure and Socioeconomi Drivers: Evidence from 314 Chinese Cities (2000–2020). *Urban For. Urban Green.* **2025**, *105*, 128717. [CrossRef]
27. Cao, Y.; Li, G.; Huang, Y. Spatiotemporal Evolution of Residential Exposure to Green Space in Beijing. *Remote Sens.* **2023**, *15*, 154 [CrossRef]
28. Huang, Y.; Lin, T.; Zhang, G.; Jones, L.; Xue, X.; Ye, H.; Liu, Y. Spatiotemporal Patterns and Inequity of Urban Green Spac Accessibility and Its Relationship with Urban Spatial Expansion in China during Rapid Urbanization Period. *Sci. Total Enviro* **2022**, *809*, 151123.
29. Xie, Y.; Shang, C.; Deng, X. Evolution of Urban Vitality Drivers from 2014 to 2022: A Case Study of Kunming, China. *Int. Environ. Sci. Technol.* **2025**, *22*, 11459–11472. [CrossRef]
30. Wu, C.; Yang, S.; Ma, Y.; Liu, P.; Ye, X. Urban Green Space Assessment: Spatial Clustering Method Based on Multisource Data t Facilitate Zoning Planning. *J. Urban Plan. Dev.* **2024**, *150*, 04024032. [CrossRef]

 https://doi.org/10.3390/su1802113

sustainability

MDPI

Article

Mapping Spatial Synergies and Trade-Offs: A Geographically Weighted Analysis of Ecosystem Services and Carbon Sequestration in Southern Italy

Federica Isola [1], Bilge Kobak [1,2], Sabrina Lai [1], Francesca Leccis [1], Federica Leone [1] and Corrado Zoppi [1,*]

[1] Department of Civil and Environmental Engineering, and Architecture, University of Cagliari, 09128 Cagliari, Italy; federica.isola@unica.it (F.I.); bilge.kobak@iusspavia.it (B.K.); sabrinalai@unica.it (S.L.); francesca.leccis@unica.it (F.L.); federicaleone@unica.it (F.L.)

[2] University Institute of Superior Studies of Pavia, 27100 Pavia, Italy

* Correspondence: zoppi@unica.it

Abstract

The transition towards climate neutrality requires the development of spatially explicit planning approaches that account for territorial differences and land-use dynamics. Within this conceptual framework, this study has the objective of identifying and discussing spatially explicit planning approaches that can support the transition to climate neutrality in different regional spatial contexts. With reference to this research question, a methodological framework is introduced and applied that is designed to support climate neutrality through spatial planning strategies. Carbon sequestration (CS) serves as a key metric to evaluate both the current state and the temporal evolution of this process, examined in connection with the provision of specific ecosystem services (ESs) within the relevant spatial setting. The work is structured as follows. An approach is developed to define the provision of ESs. Drawing on previous research and detailed assessments of environmental, landscape, and socio-cultural features, the study considers the following ESs: maintaining or improving habitat quality to sustain the life cycles of wild species valuable to humans; regulating climate by mitigating land surface temperature; agricultural and forestry production; and nature-based recreational opportunities. Moreover, spatial relationships between CS capacity and ES provision are examined through geographically weighted regressions, allowing comparisons across Basilicata, Campania, and Sardinia, three Regions in southern Italy forming the Italian Mezzogiorno. The multifunctional characteristics of ES supply contributes to optimizing CS capacity and advancing climate neutrality goals. In particular, in all three regional contexts, high values of CS capacity elasticity are recognized in relation to habitat quality and ground temperature mitigation, and very low elasticity conditions as regards the supply of recreational ESs and agricultural and forestry production.

Keywords: carbon sequestration; climate neutrality; ecosystem services; geographically weighted regression (GWR); InVEST model; spatial planning

1. Introduction

Climate change (CC) currently represents one of the most significant challenges to global stability, increasingly affecting both rural and urban environments by compromising human health, physical well-being, and overall quality of life [1]. The evident consequences of global warming necessitate urgent and decisive action to meet the targets established by the Paris Agreement, which seeks to limit the temperature increase to 1.5 °C above pre-industrial levels. Achieving this objective requires a long-term balance between greenhouse gas emissions and removals, a pursuit often referred to as climate neutrality. This task is made more arduous by the persistence of carbon dioxide (CO_2) in the atmosphere; even if all human-induced emissions were to cease immediately, existing concentrations are projected to cause an additional warming of approximately 0.6 °C over the next one hundred years [2]. Consequently, the European Union (EU) has formalized its commitment through the European Climate Act, the Green Deal, and the "Ready for 55" package, targeting a 55% reduction in emissions by 2030 and full climate neutrality by 2050.

Human activities, particularly the combustion of fossil fuels and extensive land clearing, have fundamentally disrupted the global carbon cycle. Currently, anthropogenic activities influence over 70% of the Earth's ice-free land surface, utilizing a third of its potential biological productivity. In the Italian context, this pressure is manifest in increasing land consumption and urban sprawl, which have led to biodiversity loss and heightened environmental risks such as floods and heat islands [3,4].

Within this framework, regulating ecosystem services (ESs), specifically Carbon Capture and Storage (CCAS), are vital for mitigation and adaptation. CCAS is a natural phenomenon, primarily driven by photosynthesis, through which terrestrial ecosystems remove CO_2 from the atmosphere and store it in biomass and soil [5]. Soil, in particular, constitutes the largest terrestrial carbon pool; the quantity of carbon locked within the soil is significantly greater than that in above-ground biomass [6]. Therefore, even minor fluctuations in soil organic carbon concentration can exert a substantial impact on atmospheric CO_2 levels [7].

Natural ecosystems provide a multifunctional supply of services that interact in complex ways. While the principle of multifunctionality suggests that a single area can provide multiple benefits, certain functions may be mutually conflicting, leading to trade-offs where the enhancement of one service (e.g., carbon storage) may negatively affect another (e.g., biodiversity or agricultural production). Conversely, synergies occur when multiple services improve simultaneously.

Multiple studies document trade-offs and synergies between CCAS and other services. Früh-Müller et al. [8] mapped carbon storage alongside timber supply, crop production, and outdoor recreation across two German counties, Wetterau and Vogelsberg, finding positive relationships among these services except where food production created trade-offs. Bateman et al. [9] quantified land-use conflicts between agriculture, carbon sequestration, recreation, and biodiversity across the UK, demonstrating that targeted planning can increase overall ecosystem service values. Viglizzo et al. [10] provides a comprehensive regression-based analysis, using both parametric and non-parametric regression to examine how carbon sequestration partitions with water-flow regulation, soil protection, and climate regulation across 1348 studies. On the other hand, although regression analysis is a method used by several studies in order to analyze relationships among multiple ESs [10,11], these relationships are analyzed in a stationary way. This study seeks to go beyond the limitations of stationary regression analysis by studying non-stationary relationships in Mediterranean landscapes using a geographically weighted regression (GWR).

From this perspective, this work aims to analyzes the relationship between CCAS and multiple ESs, such as habitat quality, land surface temperature mitigation, agricultural and forestry production, and nature-based outdoor activity potential using a GWR model, in order to determine and critically examine geographically delineated planning methodologies capable of facilitating the progression toward climate neutrality across heterogeneous regional territorial configurations. A GWR-based approach is preferred over ordinary regression models in ES studies for its ability to handle spatial non-stationarity, where

 https://doi.org/10.3390/su18042146

relationships vary across locations. ESs like CCAS exhibit spatial variability due to factors such as land use, topography, and climate. Moreover, the correlations are identified in heuristic terms and may form the basis for future studies aimed at identifying causal links between the CCAS supply and the supply of the various ecosystem services considered. This in-depth analysis should be considered an important future development of the research based on the results described and discussed in this study, which show, through the implementation of a GWR-based approach, that there is statistically significant evidence of the existence of these correlations.

The work presents and implements an integrated methodological framework aimed at promoting climate neutrality through spatially oriented planning policies. CCAS is employed as a central indicator to assess both the existing conditions and the dynamic progression of sequestration processes, analyzed in relation to the supply of selected ESs. The methodological approach is structured into two main steps. In the first step, the provision of multiple ESs is assessed. In particular, the study focuses on the following four ESs: maintaining or improving habitat quality to sustain the life cycles of wild species valuable to humans; regulating climate by mitigating land surface temperature; agricultural and forestry production; nature-based recreational opportunities. Each ecosystem service is assessed and mapped using different methods. For example, the use of the InVEST (Integrated Valuation of Ecosystem Services) Suite is, in this conceptual framework, exclusively aimed at identifying, in direct and clear terms, the extent of CCSA provision and habitat quality. In the second part, the spatial interdependencies between CS potential and ES distribution are investigated using a GWR model, enabling comparative analyses among Basilicata, Campania, and Sardinia, three regions located in southern Italy that collectively constitute the Italian Mezzogiorno.

The paper is structured as follows. Section 2 describes the contexts of the three regional case studies and the methodologies used to evaluate and map the selected ESs and for implementing the GWR. Section 3 presents the results obtained, which are then discussed in Section 4. Finally, Section 5 concludes with final remarks and future research directions.

2. Materials and Methods

2.1. Study Area

The study area comprises three regions of the Italian "Mezzogiorno": Basilicata, Campania, and Sardinia (Figure 1).

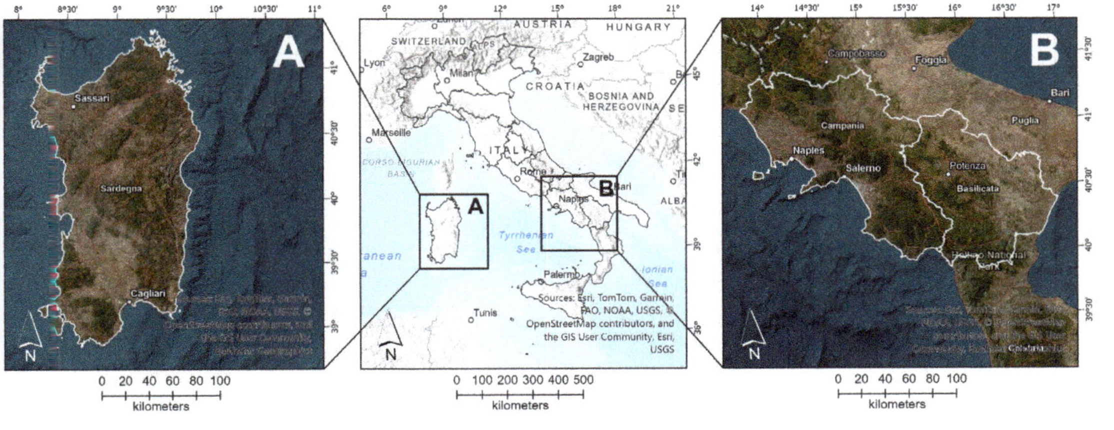

Figure 1. Study areas: Sardinia (panel **A**), Basilicata, and Campania (panel **B**).

https://doi.org/10.3390/su18042146

Basilicata covers an area of approximately 10,000 km^2 [12]. The region is predominantly hilly, with plains accounting for about 10 percent of the surface and mainly concentrated along the Ionian coastline, and its landscape is predominantly agricultural, with extensive farming practices and high biodiversity [13]. Mountain ranges are mainly located to the west and include Mount Pollino and the Vulture volcanic complex [14]. The main watercourses, characterized by a predominantly torrential regime, flow into the Ionian Sea and include Bradano, Basento, and part of the Ofanto River. Concerning land covers, according to the latest CORINE data [15], around 57 percent of the region is agricultural, around 41 percent is covered by forests and semi-natural vegetation, and only 1.6 percent is artificial. As of 1 January 2025, 530,005 people [16] lived in Basilicata's 131 municipalities, of which the main important are Potenza and Matera, the latter being a designated UNESCO site [17].

Campania extends over approximately 13,700 km^2 [12]. The regional territory is mainly hilly and mountainous, while alluvial plains account for about 14 percent of the surface. The main mountain systems include the Matese Mountains and the Cilento Massif. Volcanic landforms are a distinctive feature of the region, with two major active volcanoes: Vesuvius and the Campi Flegrei–Ischia volcanic district [18]. The main watercourses are the Volturno and Sele rivers, which exhibit a predominantly torrential regime. Concerning land covers, according to the latest CORINE data [15], around 55 percent of the region is agricultural, around 37 percent is covered by forests and semi-natural vegetation, and 7.5 percent is artificial. As of 1 January 2025, 5,582,337 people [16] lived in Campania's 553 municipalities, of which the most important are Avellino, Benevento, Caserta, Salerno, and Naples.

Sardinia is the second-largest island in the Mediterranean Sea and has a landmass of approximately 24,000 km^2. The regional territory is predominantly hilly and characterized by numerous isolated mountain groups, including Gennargentu, Sette Fratelli, and Limbara [19]. The island's geomorphology reflects a complex volcanic history and diverse lithological formations [20], and distinctive granite landscapes, particularly to the north of the island, shape unique landforms including inselbergs and tors [21]. Watercourses exhibit a marked torrential regime due to the characteristics of the terrain, dry summers, and limited rainfall [19], and coastal wetlands constitute a distinctive landscape feature. Concerning land covers, according to the latest CORINE data [15], around 46 percent of the region is agricultural, around 50 percent is covered by forests and semi-natural vegetation, and only 3 percent is classed artificial. As of 1 January 2025, 1,562,381 people [16] lived in Sardinia's 377 municipalities, of which the most important are Cagliari and Sassari.

2.2. Carbon Capture and Storage

We assess carbon capture and storage (CCAS) assessed using the InVEST Suite, which quantifies overall carbon sequestration by integrating four distinct carbon pools: living biomass located both above the surface (aboveground biomass) and in subsurface layers (belowground biomass, confined to the upper 30 cm of the soil profile), dead organic material, and soil organic matter. According to the InVEST User Guide, aboveground biomass (AGB) encompasses all living vegetation located above the soil surface, including structural and foliar components such as stems, bark, branches, and foliage. In contrast, belowground biomass (BGB) refers to the living root systems associated with aboveground vegetation. Dead organic material includes both surface litter and woody debris, whether standing or downed. Lastly, soil organic matter, constituting the largest pool of terrestrial carbon, comprises the organic constituents embedded within the soil matrix [22].

The InVEST Model requires only two inputs: a land cover map and a table in .CSV format showing carbon density values corresponding to the classes in the land cover map.

https://doi.org/10.3390/su1804214

The table should have the density values for at least one of the four pools. We utilize Corine Land Cover (CLC) 2018 as the land cover map. The information on the carbon density values for different carbon pools is sourced from various data sources such as NASA'S ORNL DAAC [23], used for aboveground biomass and belowground biomass, National Inventory of Forests and Carbon Pools in Italy [24,25], used for dead organic material and soil organic matter, ISPRA (Italian Institute for Environmental Protection and Research) [26], used for soil organic matter, and AGRIS Sardegna (Regional Agency for Agricultural Research of Sardinia) [27], used for soil organic matter.

The methodology is implemented in two stages. First, we upload to InVEST information on three pools except belowground biomass. Then, we overlay the result with the ORNL DAAC BGB map, as data for BGB was sourced from this single map; therefore, we integrated it as a separate step by summing cell by cell the values of the ORNL DAAC BGB map and the map related to the three pools using the "Raster calculator" tool in ArcGIS. The decision to use the sum reflects the methodological approach used by InVEST to calculate the value of the final map. In fact, InVEST adds up the values for the three different pools to obtain the final values.

2.3. Habitat Quality

We assessed habitat quality (HQ) using the InVEST Habitat Quality Model, which requires four key inputs: a land cover map, a directory containing threat maps, a sensitivity table, and a threats table. We adopted threats from the study by Sallustio et al. [28], which identifies the following:

- Roads 1: Motorways; trunks; primary roads;
- Roads 2: Secondary and tertiary roads;
- Roads 3: Residential and service roads;
- Roads 4: Tracks and bridleways;
- Railways;
- Intensive agricultural lands;
- Extensive agricultural lands;
- Buildings and other artificial areas or impervious soils.

Threat maps were generated using the Geofabrik Download Server [29] and CLC 2018 [15]. The land cover map was derived from the CLC 2018, reclassified into 12 habitat types following the classification by Sallustio et al. [28]:

- Beaches, dunes, and sands
- Water bodies
- Wetlands
- Grasslands
- Shrublands
- Broadleaves forests
- Conifers forests
- Inland unvegetated or sparsely vegetated areas
- Intensive agricultural lands
- Extensive agricultural lands
- Buildings and other artificial areas or impervious soils
- Open urban areas

In the case of Sardinia, we introduced the additional habitat type "mixed forest" to reflect local characteristics.

The sensitivity table details how each habitat responds to the defined threats. The average values of "broadleaves forests" and "conifers forests" were utilized for the "mixed forest" category.

https://doi.org/10.3390/su18042146

The threats table provides three parameters: MAX_DIST, the maximum distance at which each threat impacts HQ; WEIGHT, the relative weight of each threat compared to other threats; and DECAY, the spatial decay function of each threat. We assumed the decay parameter as linear in this study as there are no available data supporting alternative functions.

The model integrates these inputs to calculate the HQ scores using a degradation equation, which employs the half-saturation constant (K) to calibrate the spread and variation of scores along the 0–1 scale. K selection directly affects the absolute HQ scores, which serve as key inputs for the GWR analysis and therefore influence the scale of the resulting outputs. To maximize visual contrast, we performed a preliminary run with the default K = 0.05 to identify the maximum degradation value; we the adjusted K to half this value for the final run, following recommendations in the InVEST model's user guide [30], consistently with previous applications [31–33]. The final output is a continuous gradient map that illustrates the spatial distribution of HQ.

2.4. Regulating Local Climate: Land Surface Temperature Mitigation

We assessed the Land Surface Temperature (LST) using satellite images provided by USGS Earth Explorer [34]. First, the images from the specified date range were downloaded for selection, which was set from the last week of June to the first week of September to cover the summer season to capture the highest average temperatures in each of the three study areas. The maximum cloud cover was set to a 6%. All downloaded images were analyzed further to identify the one with the highest median LST. In all cases considered, i.e., five Landsat images selected for Sardinia, four selected for Campania, and one selected for Basilicata, the selection of images is such that those with the highest median coincide with those with the highest means. To do so, the digital number (DN) values in the raster image were converted into degree Celsius (°C) by using the formula (1), as per the USGS guide [35].

$$0.00341802 * DN + 149.0 - 273.15, \tag{1}$$

Depending on the research area, one or more satellite pictures may be necessary to encompass the entire region. Table 1 shows the full set of images utilized to create an LST map for the three regions. It is important to note that temporal variability in temperature measurement may lead to minor inconsistencies. This limitation is absent for the Basilicata Region, where LST is derived from a single image (Table 1), whereas for Campania and Sardinia full regional coverage is obtained by combining multiple LANDSAT images as indicated in Table 1.

Table 1. Selected Landsat images to create an LST map for Basilicata, Campania, and Sardinia Regions.

Region	Image Code
Basilicata	LC08_L2SP_188032_20230718_20230725_02_T1_ST_B10
Campania	LC09_L2SP_189031_20230717_20230719_02_T1_ST_B10
	LC09_L2SP_189032_20230717_20230719_02_T1_ST_B10
	LC09_L2SP_190031_20230825_20230827_02_T1_ST_B10
	LC09_L2SP_190032_20230825_20230827_02_T1_ST_B10
Sardinia	LC08_L2SP_192033_20230730_20230805_02_T1_ST_B10
	LC08_L2SP_192032_20230730_20230805_02_T1_ST_B10
	LC09_L2SP_193033_20230814_20230817_02_T1_ST_B10
	LC09_L2SP_193032_20230814_20230817_02_T1_ST_B10
	LC09_L2SP_193031_20230814_20230817_02_T1_ST_B10

https://doi.org/10.3390/su18042146

2.5. Agricultural and Forestry Production

We classified agricultural and forestry production among provisioning ESs, a category that encompasses material goods such as food, freshwater, timber, fiber, and energy resources. Recognized by the Millennium Ecosystem Assessment [36], provisioning ESs represent a fundamental component of sustainable land management, particularly in regions characterized by high landscape heterogeneity and extensive agro-forestry systems, as is the case in Italy. In recent years, also in response to the reforms of the European Common Agricultural Policy [37] and international frameworks on climate and sustainable development, the economic valuation of provisioning ESs has increasingly been incorporated into land-use planning and strategic environmental assessment.

In the three regions analyzed, the evaluation of the value of agricultural and forestry production (AFPL) relies on the market value on the production related to land uses, as a proxy indicator based on geographic and environmental variables such as location, altitude, morphology, and orography. This methodological approach, consolidated in recent studies, employs two primary datasets: (i) the national agricultural land value database produced by the Council for Agricultural Research and Economics (CREA), which provides average monetary values per unit area for agricultural lands; (ii) the dataset issued by the National Revenue Agency (NRA) reporting mean land values for forestry areas. Information on agrarian and forestry regions, organized by municipality, province, and elevation zone, enables detailed spatial analyses [38]. To integrate the datasets, correspondence was established between the CLC 2018 classes, and the crop typologies identified by CREA for agricultural land, as well as the NRA's taxonomy for forest land. The resulting data were combined through spatial overlay of the land-cover vector map with the agro-forestry value map, allowing the assignment of production values to each classified spatial unit.

2.6. Nature-Based Outdoor Activity Potential

We conducted the assessment of nature-based outdoor activity potential using the ESTIMAP (Ecosystem Service Mapping Tool) methodology, a GIS-based framework, developed by the Joint Research Centre (JRC) [39]. The study specifically implements the first part of the original model developed by Vallecillo et al. [40], Barton et al. [41], and Isola et al. [42] to quantify the recreational potential, defined as the capacity of ecosystems to support nature-based outdoor activities. We structured the methodological approach into three distinct phases. The first phase evaluates the availability of recreational areas based on their degree of naturalness. Central to this assessment is the hemeroby index, which measures the divergence of an ecosystem from its natural state due to human intervention on a scale from zero (no modification) to nine (maximum modification) [43]. For agricultural land, we determined the index by cross-referencing nitrogen nutrient inputs (utilizing ISTAT data from 2019 for Sardinia, and 2023 for Basilicata and Campania) and livestock density measured in Livestock Units (LSU) [44]. Forested areas are qualitatively assessed by comparing current land cover with potential natural vegetation maps to determine the level of disturbance [43]. The final values are inverted and normalized into a 0–1 range, where 1 represents maximum naturalness. The second phase integrates the presence of recreationally significant natural features, such as national and regional parks, Ramsar sites, and Natura 2000 sites. Scores are expert-based and attributed according to IUCN management categories, resulting in a raster map with discrete values [0; 0.8; 1] representing the legal and environmental significance of the land [39,45]. The final phase focuses on the attractiveness of coastal zones through three parameters: (i) distance from the coastline, modeled via an inverse logistic function; (ii) coastal geomorphology, using the Eurosion dataset [46]; and, (iii) bathing water quality, based on European Bathing Water Directive data [47]. We calculated the total potential supply of nature-based outdoor

https://doi.org/10.3390/su18042146

activity potential by summing the normalized outputs of all three phases, providing a comprehensive spatial representation of recreational potential for outdoor activities across the diverse topographies of Basilicata, Campania, and Sardinia.

2.7. Geographically-Weighted Regressions

We investigated the relationship between CCAS capacity and the supply of the four ESs selected for the study proposed here using a GWR model, according to the formalization proposed by [48], which is briefly described below.

The model provides information on how the dependent variable, representing CCAS, is associated with the explanatory variables, which express the supply of ESs, when the relationship is not spatially stationary and, therefore, presents a more or less pronounced variability across the territory.

We formalize the spatially stationary multiple linear regression according to the following formulation, in which the coefficients of the explanatory variables are considered constant in the spatial context of reference, therefore, in our case, in the three regional study areas: Basilicata, Campania, and Sardinia:

$$CCAS = a_0 + a_1\ LOHM + a_2\ QOHA + a_3\ AFPL + a_4\ NOAP, \quad (2)$$

where (units of measurement in parentheses):

- CCAS is the density of carbon capture and storage capacity (Mg/m^2);
- LOHM is the heat mitigation reference, i.e., the land surface temperature (LST), which serves as a measure of urban heat fluctuation and, consequently, to assess changes; if it were to decrease, it would indicate an improvement in the quality of life for users of local environments (°C);
- QOHA measures the level of habitat quality, and takes values ranging from 0 and 1, as described in Section 2.3;
- AFPL is the value of agricultural and timber production (€/ha);
- NOAP measures the potential for nature-based outdoor activities, and takes values ranging from 0 to 1, as described in Section 2.6.

The estimated coefficients of the explanatory variables show the marginal impacts of these covariates on the CCAS. The regression model is typically used when no prior assumptions can be made about the functional relationships between the variables [49–52].

The linear regression operationalizes the equation of a hyperplane, which is tangent to a hypersurface of an unknown shape within an n-dimensional hyperspace that includes the dependent variable CCAS, together with four explanatory variables. Near the point of tangency, the linear correlations between the five factors, which come from the estimation of model (2), effectively approximate the unknown equation of the hypersurface of unknown shape mentioned above [53,54].

Significance tests for the estimated coefficients are usually assessed using *p*-values, and the adjusted R-squared is the reference for evaluating the goodness of fit.

The GWRs model replaces the linear regression model when, as in the case of the three regions considered in this study, there is a large number of records, in our case tens of thousands, located in a vast and non-homogeneous territory, such as the regions of Basilicata, Campania, and Sardinia. In this case, the linear relationship between the variables, expressed by model (2), is not stationary in the spatial contexts of reference, but varies with location point, which corresponds to one of the thousands of records, in which it is estimated.

The GWRs model proposed by [48], therefore, estimates a cloud of multiple linear regressions represented by the following formulation:

https://doi.org/10.3390/su18042146

$$CCAS(x_i,y_i) = \alpha_0\ (x_i,y_i) + \alpha_1\ (x_i,y_i)\ LOHM + \alpha_2\ (x_i,y_i)\ QOHA + \alpha_3\ (x_i,y_i)\ AFPL + \alpha_4\ (x_i,y_i)\ NOAP, \quad (3)$$

where x_i and y_i represent the geographical coordinates, in the regional territories of Basilicata, Campania, and Sardinia, of the various records represented by quintuples of values of the dependent variable and explanatory variables.

For each of the points where the records of the three spatial contexts are located, we implement the regression estimate (3) by assuming a decay function of the influence of the explanatory variables on the dependent variable, according to which this influence decreases as the distance of the point (x_i,y_i), where the i-th record is located and where the i-th GWR is estimated, increases. We express the decay function, as a function of distance, according to the Gaussian weighting scheme [48], using a weight w_{ij}, according to the following formula:

$$w_{ij} = \exp[-(d_{ij}/b)^2], \quad (4)$$

where w_{ij} is the weight of the j-th record, located at the point with coordinates (x_j, y_j) at a Euclidean distance d_{ij} from the point with coordinates (x_i, y_i), where the i-th record is located, with respect to whose location the i-th GWR is being estimated. Formulation (4) is such that: (i) $w_{ij} = 1$ when the record (x_j, y_j) coincides with the one detected at the point with coordinates (x_i, y_i), with respect to which the i-th GWR is estimated; (ii) w_{ij} decreases exponentially as the distance of the point (x_j, y_j) from the point (x_i, y_i) increases; (iii) records located at a distance d_{ij} greater than some multiple of b have a progressively reduced influence in the estimation of the GWR estimated with reference to the point (x_i, y_i).

According to the detailed and analytical discussion of the model proposed by [48], the value of b, called bandwidth, must be chosen in such a way as to bear in mind that there is a trade-off in terms of unbiasedness and reliability of the GWR estimate. This trade-off must be considered when choosing the bandwidth, which defines the decay function, and therefore the weight w_{ij} to be assigned to the records based on their distance from the point at which the GWR is estimated.

On the one hand, the smaller the bandwidth, i.e., the faster the weight w_{ij} decreases with increasing distance d_{ij}, the more records become irrelevant and are therefore essentially excluded from the GWR estimate. Thus, the smaller the bandwidth, the higher the biasedness of the biased estimator of the variable $CCAS(x_i, y_i)$, expressed by model (3).

On the other hand, the larger the bandwidth, the more records are included in the GWR estimate. Therefore, in this case, although the bias decreases, the variance of the biased estimator of the variable $CCAS(x_i, y_i)$ of model (3) is greater.

It is therefore very important, for the purposes of applying the GWRs model, to identify the bandwidth in such a way as to find the best compromise between minimizing the variance and the bias of the estimate of the dependent variable $CCAS(x_i, y_i)$.

As suggested by [48], the choice of the bandwidth measure is defined by the solution to the following problem of minimizing the variance of the $CCAS(x_i, y_i)$ estimate.

The minimization problem is operationalized as follows, with reference to the GWR estimate at the coordinate point (x_i, y_i), where the i-th record is located:

$$\min_b \sum_{i=1,n} (Y_i - \hat{Y}_i)^2, \quad (5)$$

In our case, Y_i corresponds to the value of the dependent variable $CCAS(x_i,y_i)$ relative to the i-th record, and $\hat{Y}_i$ is the estimate of the variable based on the implementation of GWRs, determined using the estimates of the coefficients α_j, $\hat{a}_j$, $j = 0, \ldots 4$, of (2).

https://doi.org/10.3390/su18042146

The estimates $\hat{a}_j$'s are defined by the following matrix formula [48]:

$$\hat{\mathbf{a}}(x_i,y_i,b) = [\mathbf{X}^T\ \mathbf{W}(x_i,y_i,b)\ \mathbf{X}]^{-1}\ \mathbf{X}^T\ \mathbf{W}(x_i,y_i,b)\ \mathbf{Y}, \tag{6}$$

where

- $\hat{\mathbf{a}}(x_i,y_i)$ is the vector of the estimates of the α's coefficients of (3);
- $\mathbf{X}$ is the matrix ($n \times 5$) of the n records related to the constant and the four explanatory variables of (3);
- $\mathbf{W}$ is the diagonal matrix ($n \times n$) whose diagonal elements represent the weights of the n records representing the observations relating to the explanatory variables; the n weight values are defined by formula (4);
- $\mathbf{Y}$ is the vector of n observations relating to the dependent variable CCAS(x_i,y_i).

The estimates $\hat{a}_j$, $j = 0, \ldots 4$, derived from (6), allow us to express, in (5), $\hat{Y}_i$, i.e., the estimate of Y_i, value of the dependent variable of the record detected in the location (x_i,y_i), using (3) and, therefore, to solve, in b, the minimization problem (5). The search for the value of the bandwidth b that solves the minimization problem (5) can be carried out, as suggested by [48], by applying the golden section share algorithm described by [55], with reference to the field defined by the minimum and maximum distances of the locations of the records detected.

With regard to the goodness-of-fit of the GWRs estimate, the following statistics are important: (i) the adjusted R-squared or adjusted coefficient of determination, which measures the share of the overall variance of the phenomenon represented by the dependent variable, in our case the CCAS, explained by the GWRs estimate, purifying this measurement from the increase in the coefficient linked to the number of explanatory variables; (ii) the adjusted critical value of pseudo-t statistics, which is used to verify the statistical significance of the estimated coefficients using a two-tailed *t*-test, at a significance level of 5%; this test takes into account the degree of freedom of the estimate and checks in relation to the family-wise error rate (FWER).

The spatial distribution of the vector $\hat{\mathbf{a}}(x_i,y_i)$, which derives from the GWRs estimate, locally characterizes the correlations between the dependent variable CCAS(x_i,y_i) and the four explanatory variables that express the availability of the four ESs.

The methodology described above, based on the aforementioned seminal article by Fotheringham et al. [48], has been applied in the field of analysis relating to the spatial profiles of ES supply. The following are some examples of these studies. Shi and Xu [56] analyze the distribution of ESs in an urban green space system in the coastal metropolis of Suzhou, Jiangsu Province, China, through the integrated use of the InVEST suite, machine learning techniques, and the GWR approach, in relation to the spatial heterogeneity of ES supply. Another very interesting study is that of Neill et al. [57], which applies a spatial analysis based on GWRs to examine the relationship between environmental indicators and cultural ecosystem services in multiple sites in Europe, with the aim of bypassing the limitations of models that do not take into account the non-stationarity of ES supply.

3. Results

3.1. Carbon Capture and Storage

The CCAS taxonomy is intrinsically linked to the biophysical characteristics of the territory, with particular reference to land cover, especially in relation to forest types. As can be seen below, the spatial distribution of this ES supply shows the highest values in forest-covered soils, albeit with significant differences depending on forest types, while significantly lower values are found where vegetation is lower and sparser, and, above all, in more or less densely built-up areas.

https://doi.org/10.3390/su18042146

The CCAS assessment yields that in all three regions highest capacity is observed in forests (Figure 2). Campania exhibits the highest density at 23.82 kg/m^2, followed by Basilicata at 21.21 kg/m^2, both observed in broad-leaved forests, whilst Sardinia's maximum value is 17.75 kg/m^2, noted in the coniferous category. In all three regions, forest ecosystems (CLC 311, 312, 313) represent the primary carbon pool. However, average forest productivity varies significantly. Campania's forests range from 15.07 to 17.40 kg/m^2, followed by Basilicata (12.88–14.76 kg/m^2) and Sardinia (10.99–12.65 kg/m^2). A notable regional anomaly is identified in Sardinia, where broad-leaved forests consistently store less carbon (10.99 kg/m^2) than coniferous or mixed forests across all carbon pools. Spatially, high-density clusters are strongly correlated with mountainous topography and protected areas, such as the Cilento and Vesuvius parks in Campania, the Pollino and Appennino Lucano in Basilicata, and the eastern/southwestern massifs in Sardinia. Intermediate storage levels (7.77–10.54 kg/m^2) are found in shrublands and marshes, while artificial areas and water bodies consistently yield the lowest results. Moreover, these regions act as natural reservoirs. Campania serves as a high-capacity tank, Basilicata as a well-distributed network of medium basins, and Sardinia as a specialized reservoir where specific cluster, coniferous forests, are more efficient than others.

3.2. Habitat Quality

The analysis of HQ shows consistent contrasts between high values characterizing natural areas and low values found in urban and intensive agricultural areas. Human infrastructures are recognized as major sources of degradation, reducing HQ scores according to their proximity and their threat level.

Sardinia records the highest average HQ (0.54) and the greatest proportion of high-quality habitats (40.86%). Basilicata follows with an average score of 0.51, but with the highest percentage of low-quality habitats (44.45%). Campania reports the lowest average HQ (0.49) and the smallest share of high-quality habitats (31.43%). In all regions, the best habitats are forested areas; broadleaved forests exhibit the highest mean scores (ranging from 0.85 to 0.86), followed by grasslands and wetlands. Conversely, artificial surfaces and impervious areas score near zero (0.09) across all regions, with open urban fabric and intensive arable land also exhibiting very low values (0.21–0.25). Spatial patterns reflect this gradient: high values cluster in national and regional parks, such as Pollino and Appennino Lucano in Basilicata, Cilento and Vesuvius in Campania, and Gennargentu, Limbara, and Gutturu Mannu in Sardinia, while low values characterize major urban centers, like Matera, Potenza, Naples, Salerno, and Cagliari, and intensively cultivated plains. Differently, extensive agricultural systems and sparsely vegetated inland areas occupy an intermediate position (0.43–0.53).

The biophysical characteristics of the reference territory, in all three regional cases analyzed, are decisive in identifying the HQ level. First of all, we note that some of these characteristics are decisive in relation to the classification of threats, which are linked to agricultural land use, whether extensive or intensive, or to the degree of soil permeability. This is strongly influenced by the biophysical characteristics of the soil as modified by human activities and, in particular, by the level and type of infrastructure. In this regard, it is worth highlighting the radical difference between threats related to green infrastructure and those related to gray infrastructure, with particular reference to transport infrastructure.

Secondly, we note how the biophysical characteristics of soils come into play, in a structural and pervasive manner, with reference to the assessment of HQ in the InVEST Suite, where these characteristics identify the quality of soil cover through the Corine Land Cover mapping units and their sensitivity to threats in terms of loss of ESs supply capacity.

https://doi.org/10.3390/su18042146

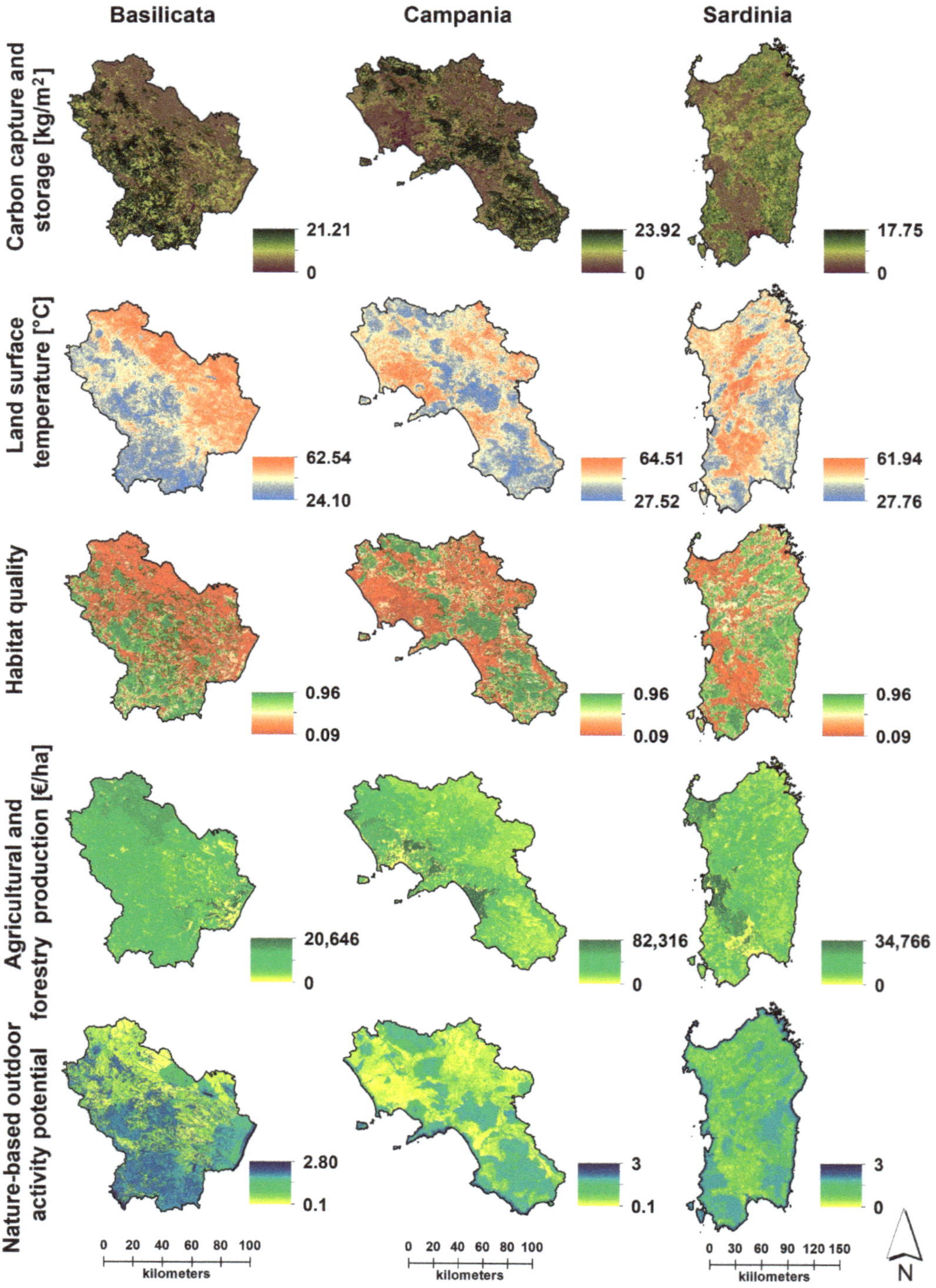

Figure 2. Spatial distribution of the five ecosystem services in the three study areas.

https://doi.org/10.3390/su18042146

3.3. Regulating Local Climate: Land Surface Temperature Mitigation

In Campania, LST ranges from about 27.5 to 64.5 °C (Figure 2), with the highest values concentrated in densely urbanized and artificial areas. Hotspots include the Metropolitan City of Naples, the Campanian and Phlegraean plains, the areas in the crown of the Vesuvius Volcano, and agricultural sectors in Avellino, Benevento, and the Sele Plain. Lower temperatures occur along the Apennine ridge, where mountainous and hilly landscapes dominate. CLC data show that artificial fabrics (CLC 111–112) exhibit the highest mean LSTs (47.6–53.3 °C), while water bodies, wetlands, forests, and herbaceous or shrub-covered areas (CLC 512, 521, 311–313, 323–324) display the lowest values (34.3–39.4 °C). Overall, thermal patterns align with land-use intensity, with cooler conditions in vegetated and high-elevation zones.

In Sardinia, LST ranges from around 27.8 to 61.9 °C (Figure 2). The highest temperatures occur mainly in agricultural zones, especially in the eastern sector, while cooler values characterize mountainous, forested, and protected areas such as Limbara, Sulcis-Iglesiente, Gennargentu, and Sette Fratelli. CLC data show that while the highest temperatures are observed in agricultural fields and discontinuous urban fabric (both 61.9 °C), the highest mean values are observed in artificial areas such as construction sites (54.0 °C), dump sites (51.2 °C), and airports (50.1 °C). Water bodies and forests remain the coolest (32–40 °C).

Basilicata's LST, derived from a single Landsat scene covering the entire region, ranges from 24.1 to 62.5 °C (Figure 2). Higher values occur in industrial, commercial, and agricultural areas of the northeast, including the Bradano basin, Vulture Hills, and the urban zone of Matera, as well as the Metapontino Plain and Ionian coast. Cooler temperatures characterize the Lucanian Apennines. CLC data indicates the highest mean LSTs in irrigated and mixed agriculture and artificial surfaces, and the lowest in water bodies, forests, and rocky areas.

The relationship between the territorial distribution of LST and spatial biophysical characteristics is evident from the above results. For example, the lowest temperatures in Basilicata and Campania are recorded in the Apennine Mountain range, while in Sardinia they are found in the mountainous areas of Sulcis-Iglesiente, Limbara, Sette Fratelli, and Gennargentu. All these areas are also characterized by their proximity to or inclusion of wetlands, waterways, lakes, woods, and forests.

In the three regions, the highest temperatures are consistently found in intensive agricultural areas, in urbanized areas with high building density, and in peri-urban areas characterized by scattered settlements with marked fragmentation of the spatial continuity of permeable soils and habitats. These areas are, in particular, as already partly highlighted: in Basilicata, the Bradano basin, Vulture Hills, and the urban zone of Matera, the Metapontino Plain, and Ionian coast; in Campania, the Phlegraean plains, the areas in the crown of the Vesuvius Volcano, and agricultural sectors in Avellino, Benevento, and the Sele Plain; in Sardinia, the Campidano Plain, the Upper Oristano area, and the south-eastern coast.

3.4. Agricultural and Forestry Production

The 2023 AFPL map for Campania reveals a highly uneven spatial configuration, with values ranging from zero, where no agricultural or forestry classification is present, to a maximum of EUR 82,316/ha (Figure 2). About 22.08% of the region records a null value, while the mean reaches EUR 12,552/ha. Over half of the territory (52.11%) shows values below EUR 10,000/ha, mainly in the eastern mountain sectors of the Campanian Apennines and the Cilento area. The highest AFPL values are concentrated in the coastal hills of Naples and the Sorrento Peninsula, characterized by high-value arable and horticultural production.

A similar distribution emerges in Sardinia, where the 2023 AFPL values range from zero to EUR 34,766/ha (Figure 2). Around 17.41% of the region falls below EUR 1000/ha,

https://doi.org/10.3390/su18042146

with an overall mean of EUR 5091/ha. Values above EUR 10,000/ha are concentrated in fertile plains such as Campidano di Cagliari, Campidano di Oristano, and Nurra, areas typically dedicated to vineyards, orchards, citrus groves, and horticultural crops. Peaks exceeding EUR 30,000/ha occur in the coastal hills of Sarrabus, in the Lower Tirso plain, and in the Campidano of Serrenti.

In Basilicata, AFPL values range from zero to EUR 20,646/ha (Figure 2). Approximately 9.86% of the Region shows a null value, while 11.44% falls below EUR 1000/ha. The average value is EUR 5689/ha. Higher AFPL values, above EUR 20,000/ha, are found in the Metapontino Plain near the river mouths of the Sinni, Agri, Cavone, Basento, and Bradano, where agricultural production is dominated by orchards and vineyards.

Even in the case of agricultural and forestry production, it is clear that supply is closely linked to the biophysical characteristics of the territory in all three regions considered. In all three cases, the highest values are found where there are flat or hilly, arable soils characterized by vegetable crops, olive groves, vineyards, and citrus groves, located in the Metapontino area in Basilicata, in the plain and coastal hills of Naples in Campania, and in Campidano and Nurra in Sardinia.

More than with reference to the characterization of other ESs, in this case the relationship between ES supply and biophysical characterization is almost deterministic. Finally, it should be noted that the values for Campania are significantly higher than those for the other two regions.

3.5. Nature-Based Outdoor Activity Potential

The assessment of nature-based outdoor activity potential in Basilicata, Campania, and Sardinia reveals significant spatial heterogeneity driven by naturalness and coastal proximity. According to the sources, potential values range from 0.1 to 2.8 in Basilicata and 0.1 to 3.0 in Campania, while Sardinia covers the full 0 to 3 scale with a mean of 1.21 (Figure 2). A common geographical trend emerges, represented by high-value clusters predominantly located along coastal strips, such as the Tyrrhenian and Ionian coasts in Basilicata and the Campania coastline, and within mountainous protected areas. Notable inland clusters include the Pollino and Lucanian Apennines National Parks in Basilicata, the Cilento and Vesuvius National Parks in Campania, and the Gennargentu and Supramonte massifs in Sardinia. Statistical analysis of CLC classes confirms that natural and semi-natural areas provide the highest service levels. Specifically, salt marshes (CLC 421), beaches (CLC 331), and water bodies (CLC 512/522) consistently show the highest mean values across all regions. Conversely, the lowest values are systematically associated with anthropized environments, including industrial units, urban fabrics, and intensive agricultural plains like the Volturno in Campania and the Campidano in Sardinia. These results emphasize that nature-based outdoor activity potential is intrinsically linked to environmental conservation and the presence of significant landscape features.

In the case of outdoor recreational potential, the biophysical characteristics of the territory relate, for all three regions, to the spatial taxonomy of the ES offer in a homogeneous and easily recognizable manner.

In all three cases, in fact, there are two profiles that play a fundamental positive role in identifying areas with high recreational potential: coastal areas and protected natural areas. With regard to the latter, the ES offer is very high: in the Pollino and Lucanian Apennines National Parks in Basilicata; in the Cilento and Vesuvius National Parks in Campania; and in Gennargentu and Supramonte in Sardinia. In terms of land cover, the highest recreational potential values are found on sandy beaches, in coastal lagoons, and in water bodies. The lowest values, on the other hand, are generally found on artificial land and, in particular, in urban areas with high building density.

 https://doi.org/10.3390/su18042146

3.6. Geographically Weighted Regressions

To run the GWR, we created a vector layer composed of one-kilometer square polygons arranged in a regular grid. For each polygon, we calculated the average values of CCAS, LOHM, QOHA, AFPL, and NOAP from the ES maps described in Sections 3.1–3.6 and shown in Figure 2, using zonal statistics.

Prior to the GWR, we performed an ordinary least squares regression to check for global multicollinearity. For all explanatory variables, the Variance Inflation Factor was always lower than four, and therefore well below ten, a commonly used threshold for identifying potentially problematic multicollinearity [58–60].

Subsequently, we implemented the GWR model presented in Equation (3), in which the dependent variable CCAS is locally regressed against the four explanatory variables—LOHM, QOHA, AFPL, and NOAP—by using the tool "Geographically Weighted Regression (GWR) (Spatial Statistics)" in ESRI ArcGIS Pro 3.6.0 separately for each of the three study areas. This tool requires, as input, a shapefile where for each feature the attribute table provides the value of the dependent and of the explanatory variables; users need to set a few parameters, among which include the model type, the neighborhood type, the neighborhood selection method and, optionally, the local weighting scheme.

As for the model type, we selected the "Continuous (Gaussian)" one, since the dependent variable CCAS is continuous. We constructed the neighborhood type as a fixed distance and, consistent with the theoretical explanation provided in Section 2.7, in the "Neighborhood selection method" dropdown menu we chose the golden search selection option to define the neighborhood size, without predefining any lower or upper search distance. This means that the tool itself identifies the optimal distance, by testing the Akaike Information Criterion at various distances between a minimum and maximum distance. As per the ArcGIS Pro documentation [61], the lower bound corresponds to the distance at which each feature has at least 5 percent of the dataset features as neighbors, while the upper bound represents the distance at which each feature includes half of the total input features as neighbors. Finally, as for the local weighting scheme, we adopted a Gaussian-based weighting approach to implement the decay function of the influence of the explanatory variables on the dependent variable. With this option, every feature is assigned a weight; however, the magnitude of these weights decreases exponentially as the distance from the target feature increases.

For each GWR, Table 2 reports the main details, namely the number of features and the bandwidth, together with the model diagnostics, i.e., adjusted R^2 and adjusted critical value of pseudo-t statistics already introduced in Section 2.7. Good levels of model goodness-of-fit are indicated by the adjusted R^2 values, very high in Basilicata and Campania and high in Sardinia, and by the extremely low p-values corresponding to the adjusted critical values of pseudo-t statistics reported in Table 2. Moreover, Tables 3–5 provide the descriptive statistics of the coefficients α_1, α_2, α_3, α_4, as well as of the local R^2, for Basilicata, Campania, and Sardinia, respectively.

Table 2. GWR: details and main model diagnostics for the three regional case studies.

Region	No. of Records	Bandwidth [Meters]	Adjusted R^2	Adjusted Critical Value of Pseudo-t Statistics
Basilicata	10,429	6708.38	0.931	3.203
Campania	14,079	6708.51	0.935	3.285
Sardinia	24,565	8062.39	0.774	3.351

https://doi.org/10.3390/su18042146

Table 3. GWR: descriptive statistics of the coefficients of the models presented in Equation (3) and of the local coefficient of determination (R^2) for Basilicata.

Region	Explanatory Variable	Descriptive Statistics	Coefficient	t-Statistic	% Significant Records at 0.95 Level
Basilicata	LOHM	*minimum*	−0.3873	−19.5713	87.35
		maximum	0.2263	12.1438	
		mean	−0.1712	−8.4322	
	QOHA	*minimum*	1.3038	1.2611	99.88
		maximum	16.7384	50.0964	
		mean	11.1508	24.1314	
	AFPL	*minimum*	−0.0002	−4.6978	79.83
		maximum	0.0013	28.4700	
		mean	0.0003	10.1066	
	NOAP	*minimum*	−3.6491	−13.1826	62.43
		maximum	6.5122	16.7003	
		mean	1.0311	3.6148	

No. of records: 10,429; Local R^2: min 0.4921; max 0.9812; mean 0.8764.

Table 4. GWR: descriptive statistics of the coefficients of the models presented in Equation (3) and of the local coefficient of determination (R^2) for Campania.

Region	Explanatory Variable	Descriptive Statistics	Coefficient	t-Statistic	% Significant Records at 0.95 Level
Campania	LOHM	*minimum*	−0.8388	−30.1517	96.78
		maximum	0.1788	5.6609	
		mean	−0.3097	−11.9467	
	QOHA	*minimum*	5.8714	7.3816	100.00
		maximum	20.3862	35.9835	
		mean	13.6702	24.0552	
	AFPL	*minimum*	−0.00018	−5.7950	62.01
		maximum	0.00042	14.9714	
		mean	0.00004	4.2351	
	NOAP	*minimum*	−4.8632	−12.6429	38.14
		maximum	3.5577	8.5280	
		mean	−0.0973	−0.9962	

No. of records: 14,079; Local R^2: min 0.4389; max 0.9644; mean 0.896.

Table 5. GWR: descriptive statistics of the coefficients of the models presented in Equation (3) and of the local coefficient of determination (R^2) for Sardinia.

Region	Explanatory Variable	Descriptive Statistics	Coefficient	t-Statistic	% Significant Records at 0.95 Level
Sardinia	LOHM	*minimum*	−0.2817	−15.1436	70.84
		maximum	0.6353	19.8301	
		mean	−0.0419	−3.0587	
	QOHA	*minimum*	2.4342	4.7891	100.00
		maximum	15.5463	54.8433	
		mean	8.8116	21.8081	
	AFPL	*minimum*	−0.0003	−7.7093	38.75
		maximum	0.0003	15.4141	
		mean	0.0001	2.5219	
	NOAP	*minimum*	−4.7603	−18.8702	45.54
		maximum	2.2537	9.2818	
		mean	−0.3690	−2.3808	

No. of records: 24,565; Local R^2: min 0.1703; max 0.9352; mean 0.7173.

https://doi.org/10.3390/su1804214

The spatial distribution of the local R^2 values, which represent for each element in the vector grid the goodness of fit of the local correlation between CCAS and the four explanatory variables at that location, is shown in Figure 3. Among the three regions, Sardinia exhibits not only the lowest adjusted R^2 in Table 2, but also the lowest local R^2 statistics when looking at Tables 3–5; however, as shown in the map, values below 0.5 are identified only as below the first decile, highlighting the general strength of the local correlation even on the island. The highest values tend to cluster in all three regions, with two notable agglomerations in Basilicata and three in Campania, all located in the mountainous Appennine area, and four in Sardinia, two of which are in the island's main alluvial plain, with the remaining two in the mountainous regions of Marghine, to the north, and Linas, to the southwest.

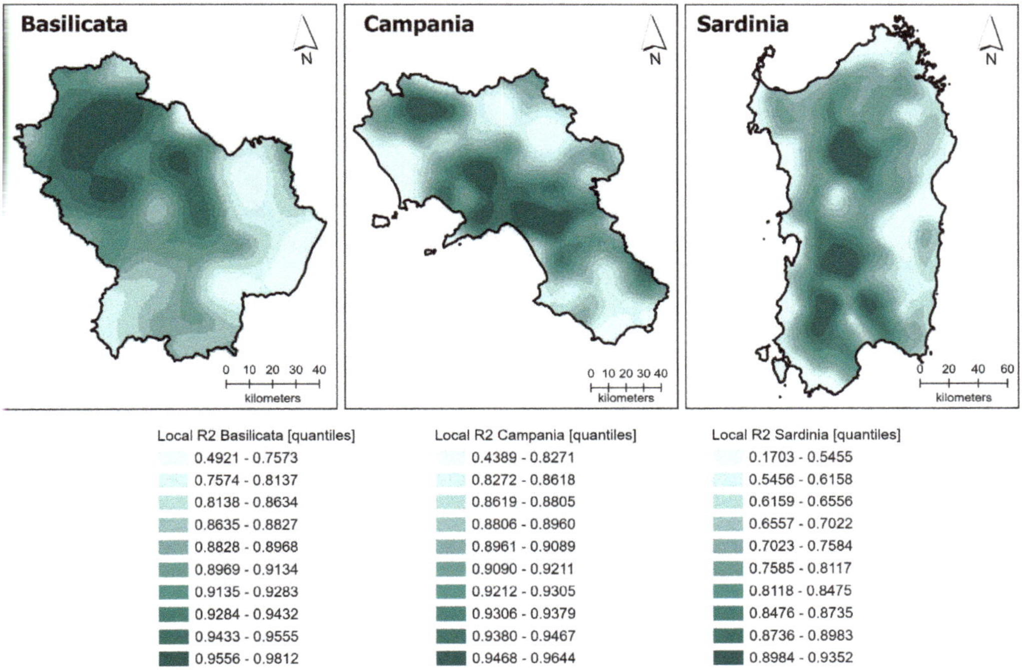

Figure 3. Spatial distribution of the local R^2's of the GWR model in the three study areas.

As shown in Tables 3–5, the most important variable in terms of significance levels is QOHA, whose estimates are nearly always significant at the 0.95 level. LOHM ranks second, with significance at the 0.95 level ranging from approximately 71% to approximately 97% of the records in Sardinia and Campania, respectively. As for the remaining two variables, the percentage of records for which the estimates are significant is high in Basilicata (AFPL $\simeq$ 80% and NOAP $\simeq$ 62%) and low in Sardinia (AFPL $\simeq$ 39% and NOAP $\simeq$ 46%), with Campania showing intermediate values (AFPL $\simeq$ 62% and NOAP $\simeq$ 38%).

Figures 4–6 show the spatial distribution of the coefficients of the GWR model presented in Equation (3) for the three case studies, overlaid on a basemap that shows the location of the mail urban centers and provides some qualitative information on the local geomorphology. Shades of red indicate negative values, shades of blue positive values, and grey null values.

https://doi.org/10.3390/su18042146

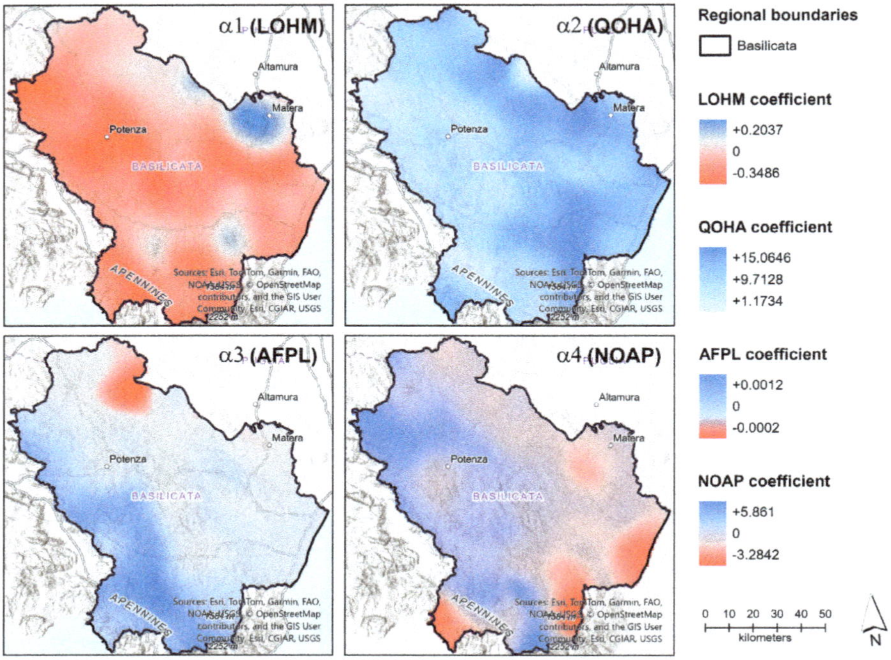

Figure 4. GWR coefficients in Basilicata.

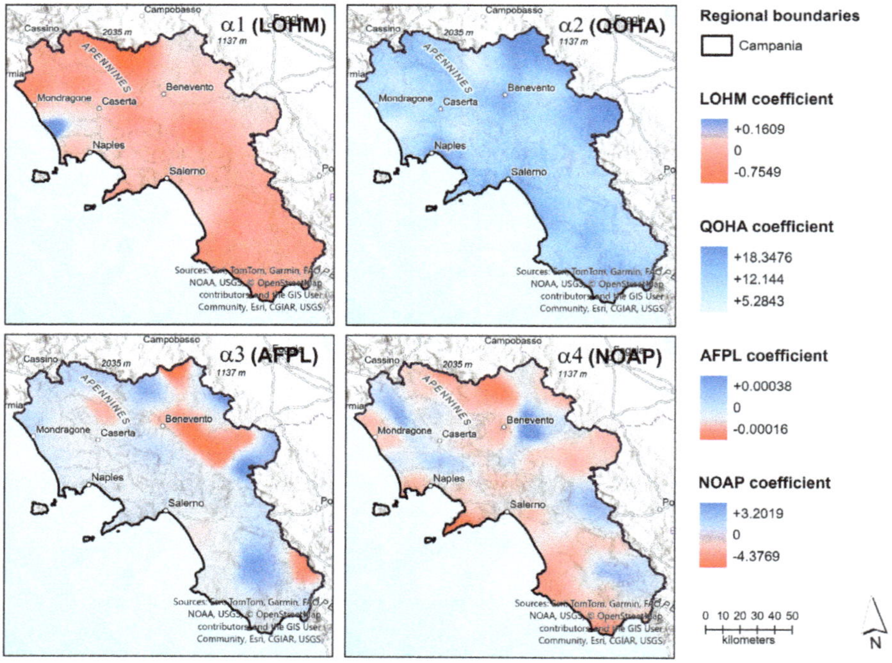

Figure 5. GWR coefficients in Campania.

https://doi.org/10.3390/su18042146

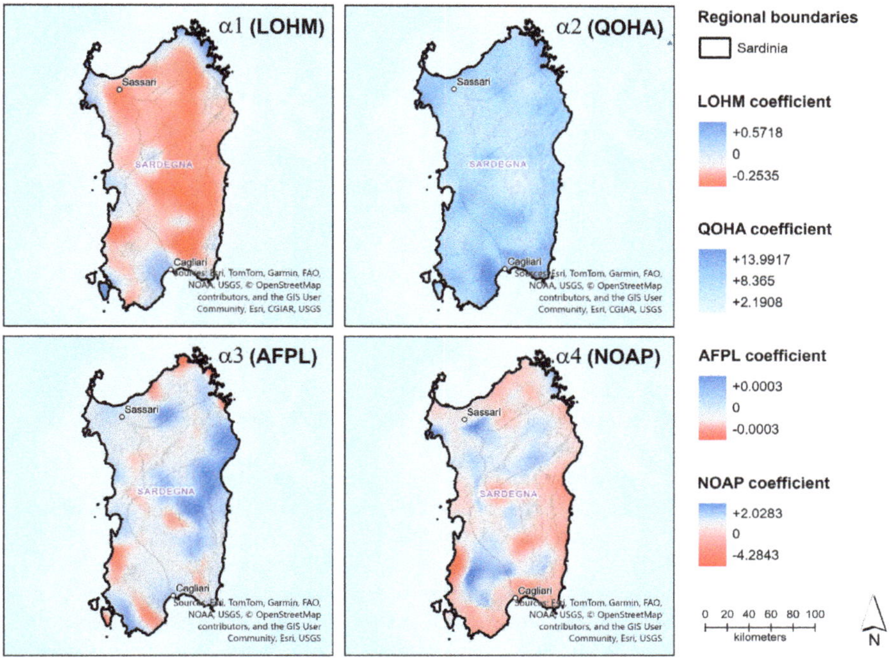

Figure 6. GWR coefficients in Sardinia.

For LOHM, the mean value reported in Tables 3–5 is always negative, and, accordingly, red shades prevail in Figures 4–6, as expected, indicating that an increase in LOHM is generally associated with a decrease in CCAS and vice versa; the impact of an increase in LST of one degree Celsius, on average, results in a decrease of CCAS by 0.17 kg/m^2 in Basilicata, 0.30 in Campania, and 0.05 in Sardinia. This association is more pronounced in Basilicata and Campania, whereas in Sardinia positive (blue) values are mainly distributed along the coastline and in some inland clusters, with very low absolute magnitudes. As shown in Figures 4–6, the main settlements in the three regions are mostly characterized by red shades, indicating negative coefficients. The main exceptions are Matera (Figure 4), where, however, the significance of the correlation is low (see Figure 3), and the Cagliari area (Figure 6), where the coefficients are positive and therefore shown in blue, most likely due to the city's proximity to the sea and its location between two large wetlands. Elevated areas characterized by natural land cover, such as the Appennines in Campania and Basilicata, and the Gennargentu-Supramonte massifs in Sardinia tend to display darker shades of red in Figures 4–6. This indicates that in these areas an increase in CCAS is associated with a decrease in temperature, which is fairly intuitive from the maps in Figure 2.

QOHA is the only explanatory variable whose coefficients in Tables 3–5, in addition to being relatively high, are always positive across the three regions. This pattern is clearly illustrated in Figures 4–6, which display only varying shades of blue and no red tones, signaling that an increase in habitat quality is consistently associated with an increase in CCAS. An increase in QOHA of ten percent, on average, results in an increase of CCAS by 1.12 kg/m^2 in Basilicata, 1.37 in Campania, and 0.88 in Sardinia, which correspond to percent changes close to 10 percent in all three cases. Although the coefficient is always positive, lower values tend to correspond to areas where the main settlements and transport

https://doi.org/10.3390/su18042146

infrastructures are located, as indicated by the light shades of blue in Figures 4 and 5. In contrast, the darker shades are generally associated with hilly landscapes in Basilicata and Campania, namely, the area bordering the river Agri (Figure 4), and two sparsely populated areas to the east of Benevento (Figure 5). Conversely, in Sardinia (Figure 6) the darker shades are found in a flat, low-elevation area in the southeastern part of the island, corresponding to the Santa Gilla wetland and Natura 2000 site west of Cagliari.

AFPL takes an average positive, but very small value in the three regions, and even the minimums and the maximum values are in the order of $|10^{-3}|$ to $|10^{-4}|$, highlighting the negligible impact of this variable on CCAS. The dominance of blue shades in Figures 4–6 indicates that the coefficients are mostly positive, which is consistent with the fact that the mean values in in Tables 3–5 are always positive, with negative (red) values clustering in a distinct area in northeast Basilicata bordering the Alta Murgia National Park in Apulia and in a few isolated clusters in Campania, most notably the one including the town of Benevento, whereas in Sardinia several negative clusters are spread across the island. An increase in AFPL of one euro, on average, does not produce any changes in CCAS in Campania, and an increase of only 0.1 g/m^2 in Sardinia and 0.3 in Basilicata. Because of the extremely low coefficient values and the resulting narrow range between the minimum and maximum, no clear pattern can be identified in relation to either geomorphology or land cover, including the artificial–natural dichotomy.

Finally, as for NOAP, the coefficients are the highest after those of QOHA, and their spatial distribution varies considerably across the three regions. An increase in NOAP of one unit, on average, is associated with an increase of CCAS by 1.03 kg/m^2 in Basilicata and a decrease by 0.1 kg/m^2 in Campania, and 0.37 in Sardinia. In Basilicata, the mean value reported in Table 3 is positive; accordingly, the prevalence of blue shades in Figure 4 indicates that coefficients are mostly positive except along the coastal strips on both the Ionian and Tyrrhenian coasts where red tones are visible. Conversely, in Campania the mean value in Table 4 is negative, as signaled by the prevailing red shades in Figure 5, with four notable positive (blue) agglomerations in inland areas, largely coinciding with natural protected areas: the Roccamonfina and Foce Garigliano Regional Park to the northwest, and the Monti Picentini Regional Park together with the Cilento and Vallo di Diano National Park to the southeast. These blue clusters are separated by red wedges that essentially connect the coastal strip with hilly and mountain areas, where diffuse small settlements are located within a predominantly rural landscape featuring both agricultural and natural land covers. In Sardinia the mean value reported in Table 5 is also negative. Moreover, values are less clustered than in the other two regions, and are predominantly negative (red) in Figure 6, with positive (blue) clusters mostly concentrated inland, similarly to what is observed in Campania. In addition, and still similar to what happens in Campania, negative coefficient values also characterize the island's most important mountain areas with natural land covers, namely the Gennargentu-Supramonte massifs. As shown in Figures 4–6, the areas surrounding all the main settlements in the three regions' light shades of red, indicating negative but low coefficients, prevail.

4. Discussion

4.1. CCAS and Habitat Quality

Based on the figures reported in Tables 3–5 and the discussion in the preceding section, the GWR results for all three regions consistently show that increases in QOHA correspond to increases in CCAS at its mean level. Specifically, the estimated supply elasticity of CCAS is about 72% in Basilicata, 78% in Campania, and 68% in Sardinia.

The positive correlation between CCAS and the habitat quality level is consistent with several studies available in the current literature concerning this issue, some of

https://doi.org/10.3390/su1804214

which are mentioned below, among many. The findings of Spohn et al. [62], related to grasslands of 84 sites located on five continents, suggest that species richness in plant communities affects the accumulation of carbon in soils, primarily by modifying the characteristics of organic inputs rather than their overall amount. Their results further indicate that management practices aimed at recovering plant diversity can promote greater soil organic carbon (SOC) sequestration, with especially pronounced benefits in hot and dry environmental conditions.

Using 146 soil samples from urban plots across Beijing, Xie et al. [63] find that SOC is strongly linked to the structural attributes of the habitat quality of urban forests. Their results suggest that increasing tree cover and density, within the limits of urban planning, can effectively enhance SOC storage in cities by targeting the spatial framework and species complexity of urban forests.

Schittko et al. [64] show, as regards the metropolitan area of Berlin, that the quality of habitats has a consistently positive impact on both soil multifunctionality and soil organic carbon stocks. Structural equation modeling indicates that these effects are largely mediated by an increase in the diversity of soil-dwelling organisms. Importantly, the positive influence of plant diversity on soil functions and fauna is observed not only for native plant species but also, although to a somewhat lesser extent, for non-native species. Similar findings are reported, among many, by Reiff et al. [65] and Lange et al. [66].

Furthermore, a comparison of the spatial distribution of the GWR coefficients of QOHA and LOHM shows a significant association between growth in carbon capture and storage capacity, improvement in habitat quality, and a decrease in LST, and therefore a positive correlation between the latter two variables. In other words, in all three regional cases examined, where carbon storage capacity is higher, habitat quality is also comparatively higher and LST is lower. This result is confirmed in the literature. For example, a study by Hu et al. [67], referring to the urban context of Zhengzhou, a city of over four million inhabitants in Central China, shows that higher levels of LST are associated with lower habitat quality values, thus with opposite spatial patterns. Similarly, in a study by Parvar and Salmanmahiny [68], referring to the provinces of Gilan, Golestan, and Mazandaran in Northern Iran, we note that areas with degraded habitats show comparatively higher average LST values than areas with good habitat quality, thus highlighting a significant negative correlation. Finally, a study by Müllerová and Šiffel [69], conducted with reference to the Bohemian Switzerland National Park in the Czech Republic, shows that areas with better vegetation conditions and greater water availability tend to have lower LSTs, while areas with more or less pronounced vegetation cover degradation show higher LSTs, thus revealing a negative correlation.

4.2. CCAS and Local Climate Mitigation

According to the results shown in Tables 3–5 and described in the previous section, in all three regions the GWR estimates indicate, with fairly small differences, that a decrease in LOHM is associated with an increase in CCAS, at its average value, more or less proportional in the case of Basilicata and Campania, and lower in the case of Sardinia. The elasticity of CCAS supply for the three regions is, in fact, approximately 89%, 150%, and 2%, respectively.

This result of the GWR estimation reflects a spatial characterization of LOHM coefficients that is consistent for Basilicata and Campania, while it is significantly different for Sardinia. As can be seen from a comparison of Figures 3 and 4 (Basilicata and Campania) with Figure 5 (Sardinia), the LOHM coefficients are practically always negative for Basilicata and Campania, while for Sardinia they take on positive values in a significant part of the coastal areas, namely in the south, west, northeast, and southeast.

https://doi.org/10.3390/su18042146

This different pattern, which is clearly reflected in the different elasticity values found in Basilicata and Campania on the one hand, and in Sardinia on the other, can be explained by the significant presence of wetlands in the Sardinian coastal areas, which are much more widespread than in Basilicata and Campania and distributed over much longer stretches of coastline: the coastline of Sardinia is almost 1900 km long, that of Campania about 500 km, and that of Basilicata about 70 km. As Zhang et al. [70] point out with regard to Bohai Bay, China, the spatial distribution of SOC is associated with environmental variables that include various climatic factors, with areas of higher SOC corresponding to areas with higher temperatures, which favor carbon input into wetland soils. The study uses satellite images, climate data, and machine learning models. Similar results are also found in the study by Osland et al. [71], referring to US coastal wetlands in the Gulf of Mexico, which highlights positive relationships between average temperature values and CCAS, linked to higher plant productivity and greater biomass in warmer sites. The inland areas of Sardinia, on the other hand, offer a spatial pattern very similar to those of Basilicata and Campania, characterized by a generalized positive correlation between CCAS and decrease in LOHM. This result finds numerous and consistent confirmation in the current literature, some of which are reported here below. The findings of Tan et al. [72], based on 400 mm isohyet sampling in the Chinese Country, indicate that higher temperatures are associated with a reduction in soil organic carbon reserves, with an estimated sensitivity coefficient of 0.24. According to this study, thermal conditions alone accounted for over 50% of the observed variation in SOC stocks, exceeding the joint influence of moisture conditions, soil characteristics, vegetation cover, and soil classification. Moreover, the inverse association between temperature and SOC levels was consistent across different vegetation and soil types, implying that this relationship represents a broadly applicable trend. Such positive correlation is supported by the results proposed by Hartley et al. [73], who, by analyzing over 9000 soil profiles located in the five continents, show that stored organic carbon decreases sharply with an increase in average annual temperature, thus finding that lower soil temperatures are associated with higher amounts of organic carbon.

Zeng at al. [74], with reference to two distinct Chinese sites, such as the Shennongjia Mountain and the Tibetan Plateau, show that site-specific warming is associated with declines in both microbial residue carbon and soil organic carbon. Moreover, they show through additional examination that temperature increases led to lower SOC levels primarily through a reduction in microbial residue carbon, providing a mechanistic explanation for the expected loss of soil carbon under future warming scenarios. Overall, the results suggest that ongoing climate warming may impair soil carbon sequestration by accelerating the breakdown of microbial-derived carbon, thereby intensifying the reinforcing feedback between elevated temperatures and carbon dioxide emissions. Analogous outcomes can also be retrieved, among many, from Huang et al. [75] and Kirschbaum [76].

4.3. CCAS and Agricultural and Forestry Production

Based on Tables 3–5, the consistently positive but very small AFPL coefficients indicate that increases in AFPL are associated with only modest increases in CCAS at mean values. Overall, the impact of AFPL on CCAS is quite weak, and less significant than that of QOHA and LOHM, particularly in Sardinia, where significance occurs for only about 38% of observations. This limited effect may reflect the heterogeneous and sometimes contradictory evidence from the literature.

Aryal et al. [77] identify the trade-off between crop production and carbon or climate services as the most frequently studied relationship between 2005 and 2020. Such trade-offs have been observed in Quebec, Canada [78], and the Beijing area, China [79], and often emerge from scenario analyses [80–82]. Conversely, some studies report synergistic rela-

 https://doi.org/10.3390/su18042146

tionships between food production and carbon sequestration or storage, such as in a region along the Yellow River in China [83]. These differing outcomes depend on site-specific characteristics, including altitude and shade [84], and especially on management practices. Crop residue retention, crop rotation, and reduced tillage are generally considered effective for conserving soil organic carbon [85], although this view has been questioned [86,87].

For forested areas, most studies report a positive association with carbon sequestration; however, as with agriculture, effective management is essential [88] to ensure the synergistic effect: timber harvesting can either increase or decrease carbon storage depending on management strategies [89], while species composition, biodiversity, and forest age also play key roles [90,91].

Moreover, the inner hilly and mountain landscapes in all three regions are characterized by traditional Mediterranean grazing practices, with livestock living on natural pastures and wooded grassland [92] and, in the case of sheep and goats, even on mediterranean maquis and garrigue or forest undergrowth [93]. While these practices are not only culturally and socially rooted but also generally sustainable due to their low intensity, long-term grazing is associated with reductions in carbon stocks, as shown in numerous studies (among many: [94–96]). This might explain the absence of a strong and consistent regional signal even in forested areas, with only a few exceptions, i.e., the blue spots emerging to the west in Basilicata (Figure 4) and to the south in Campania (Figure 5), where two major national parks (Cilento-Vallo di Diano and Pollino, respectively) are located, and in a central eastern belt in Sardinia (Figure 6), hosting some large Natura 2000 sites.

Future research could therefore test whether GWR results improve in significance by distinguishing agricultural from forested areas and, especially, by incorporating management practices; although, to the best of our knowledge, such data are not currently available for the study areas.

4.4. CCAS and Nature-Based Recreational Potential

The figures reported in Tables 3–5 indicate that the GWR results differ across the three regions. An increase in NOAP is associated with higher CCAS values at the mean level in Basilicata, whereas it corresponds to modest decreases in Campania and Sardinia. The estimated supply elasticity of CCAS at the mean level is approximately 13% in Basilicata, −1% in Campania, and −5% in Sardinia; in the latter two regions, fewer than 50% of the records are statistically significant.

A clear pattern emerges from the spatial distribution of the negative coefficients, which cluster along coastal areas in all three study regions, meaning that in these areas an increase in recreation is associated with a decrease in carbon sequestration and storage. The recreation potential in coastal areas, as modeled in ESTIMAP, is high due to the combined presence of natural ecosystems such as sandy beaches, proximity to water, and protected areas [97]. This pattern is intuitive and consistent with empirical evidence worldwide, which confirms the popularity of coastal areas for daily recreation due to the abundance of recreational opportunities and high environmental quality [98,99], including presence of ecosystem that are intrinsically perceived as significant [100], as well as concentration of urban areas, and thus beneficiaries of this ES, near coastlines [101]. On the other hand, carbon storage tends to be lower in arid regions such as Mediterranean coastlines, driven by the relatively limited forest cover and by calcareous and acidic soils [102], and in spite of the presence of coastal wetlands, which function as important carbon sinks [103]. The key takeaway, therefore, is that in Mediterranean coastal areas, a balance must be struck between promoting recreation and optimizing carbon storage, as enhancing one may come at the expense of the other.

https://doi.org/10.3390/su18042146

By contrast, positive coefficients characterize most of Basilicata and large inland areas of both Campania and Sardinia, possibly reflecting the presence of forests and other vegetated areas that simultaneously support carbon storage and outdoor recreation. Proximity to urban settlements is a key determinant of daily recreation, as walking and cycling trips typically occur within 8 km of the place of residence [40]. From a policy perspective, therefore, increasing recreational opportunities through afforestation in inland areas near human settlements could also lead to enhanced carbon sequestration and storage.

5. Conclusions

This study establishes a comprehensive methodological framework for analyzing the functional relationships between CCAS, assumed as the reference ES for evaluating the role of regional spatial contexts in achieving global climate neutrality, and the provision of four additional ESs. These include two regulating ESs, namely HQ and LST; one provisioning service, represented by AFPL; and one cultural service, nature-based outdoor activity potential. The empirical application of this framework is conducted within the spatial contexts of the Italian regions of Basilicata, Campania, and Sardinia. The core of the analysis involves correlating the spatial distribution of each of these four ESs with the taxonomy of CCAS in urban, peri-urban, rural, and natural areas.

Since these relationships are not stationary in the spatial contexts of reference, they are analyzed by employing a GWR, run on 1 km grid cells, adopting a Gaussian-based weighting approach, and choosing bandwidths that are optimized through the golden search selection method. HQ consistently showed a strong positive relationship with CCAS, indicating that areas with higher ecological integrity tend to support greater carbon stocks. Conversely, LST shows negative associations, confirming that cooler microclimatic conditions are associated with higher carbon stocks. AFPL shows a negligible association with CCAS, which can be positive or negative, depending on management strategies. Similarly, nature-based outdoor activity potential displays spatially variable results, negative correlations in coastal areas, and synergistic relationships in forested inland regions.

Maps obtained through GWR visualize spatially varying correlations among ESs that enable the identification of leverage areas, where local coefficients are highest and targeted interventions (e.g., habitat restoration) would yield the greatest synergistic co-benefits (e.g., CCAS), as well as the delineation of zones of inevitable trade-offs, where conflicting ecosystem service priorities, such as tensions between coastal tourism and CCAS, must be negotiated. By making these spatial asymmetries explicit, GWR provides a direct, evidence-based foundation for differentiated regional planning, supporting targeted measures, priority-setting, and context-sensitive policy design that cannot be derived from non-spatial models. Thanks to GWR outputs, it is possible to focus restoration or protection efforts in high-leverage areas to maximize multifunctionality and ecosystem service synergies, while implementing mitigation measures or zoning restrictions in trade-off areas to minimize losses and balance competing demands.

Despite its innovative contribution and rigorous approach, the proposed methodology presents some critical limitations. The first concerns the nature of the data employed and is caused by the reliance on proxy-based estimations drawn from secondary sources, which gives rise to uncertainties regarding their accuracy and spatial representativeness. The second limitation relates to the methods applied for assessing and mapping the ESs. For example, the habitat quality model relies on expert evaluations, which may introduce subjectivity and potential biases, and it only considers a limited number of threats under static provision. The third limitation concerns the adoption of a linear decay function for threats to habitat quality because no available data support alternative functions. The fourth limitation pertains to the exclusion of climate change mitigation actions, which often

 https://doi.org/10.3390/su1804214

involve land use changes having implications for land functionality, from the analytical framework. The fifth limitation relates to the lack of an explicit assessment of how settlements and human infrastructure influence trade-offs among ESs. Promising directions for further research involve the following: i. exporting and applying the methodology to other geographical regions to identify similarities and differences in outcomes, and in the consequent related territorial policy implications; ii. broadening the set of ESs included in integrated assessments of their interactions with climate neutrality, beyond the four considered here, including, for example, hydraulic and hydrogeological risk mitigation; iii. implementing alternative statistical techniques within the same data infrastructure; iv. exploring alternative, non-linear forms of decay functions concerning threats to habitat quality; v. extending the framework to explicitly incorporate climate change mitigation scenarios, dynamic land-use modeling, or mitigation-specific interventions, in order to investigate long-term synergies and trade-offs in ecosystem multifunctionality; vi. exploring the specific role of settlements and human infrastructure in modulating ES trade-offs.

Author Contributions: Conceptualization, Methodology, Investigation, Data Curation, Writing—Original Draft Preparation, and Writing—Review & Editing: F.I., B.K., S.L., F.L. (Francesca Leccis), F.L. (Federica Leone), and C.Z. Specific task-level contributions were distributed as follows: F.L. (Federica Leone) contributed to Sections 1, 2.6 and 3.5; S.L. to Sections 2.1, 3.6, 4.3 and 4.4; B.K. and F.L. (Federica Leone) to Sections 2.2 and 3.1; B.K. and F.L. (Francesca Leccis) to Sections 2.3 and 3.2; B.K. and F.I. to Sections 2.4 and 3.3; F.I. to Sections 2.5 and 3.4; C.Z. to Sections 2.7, 4.1 and 4.2; and F.L. (Francesca Leccis) to Section 5. All authors have read and agreed to the published version of the manuscript.

Funding: This study was carried out as follows: i. within the RETURN Extended Partnership and received funding from the European Union Next-GenerationEU (National Recovery and Resilience Plan—NRRP, Mission 4, Component 2, Investment 1.3—D.D. 1243 2/8/2022, PE0000005); ii. with the financial support under the National Recovery and Resilience Plan (NRRP), Mission 4, Component 2, In-vestment 1.1, Call for tender No. 1409 published on 14.9.2022 by the Italian Ministry of University and Research (MUR), funded by the European Union—NextGenerationEU—Project Title "Definition of a guidelines handbook to implement climate neutrality by improving ecosystem service effectiveness in rural and urban areas"—CUP F53D23010760001- Grant Assignment Decree No. 1378 adopted on 1 September 2023, by the Italian Ministry of University and Research (MUR); iii. the research grant CUP F73C23001680007 for the project "Geodesign for climate change mitigation and adaptation in the Mediterranean region" funded in 2022 by Fondazione di Sardegna.

Institutional Review Board Statement: Not applicable.

Informed Consent Statement: Not applicable.

Data Availability Statement: The original contributions presented in the study are included in the article, further inquiries can be directed to the corresponding author.

Conflicts of Interest: The authors declare no conflicts of interest.

References

Driga, A.M.; Drigas, A.S. Climate change 101: How everyday activities contribute to the ever-growing issue. *iJES* **2019**, *7*, 22. [CrossRef]

Riebeek, H. The Carbon Cycle. NASA Earth Observatory. 2011. Available online: https://earthobservatory.nasa.gov/features/CarbonCycle (accessed on 2 February 2026).

Mussinelli, E.; Tartaglia, A.; Castaldo, G.; Riva, R.; Cerati, D.; Sereni, A. The mitigation potential of environmental compensations: A challenge for the carbon neutrality transition. *IOP Conf. Ser. Earth Environ. Sci* **2024**, *1402*, 012064. [CrossRef]

Munafò, M. (Ed.) *Consumo di Suolo, Dinamiche Territoriali e Servizi Ecosistemici [Land Consumption, Territorial Dynamics and Ecosystem Services]; Edizione 2023 [2023 Edition]*; Report Sistema Nazionale per la Protezione dell'Ambiente (SNPA) 37/23; Sistema Nazionale per la Protezione dell'Ambiente: Rome, Italy, 2023; p. 443.

Lal, R. Carbon sequestration. *Philos. T. Roy. Soc. B* **2008**, *363*, 815–830. [CrossRef]

https://doi.org/10.3390/su18042146

6. European Environment Agency. Climate Change Impacts and Vulnerability in Europe 2012. An Indicator-Based Report. 201 Available online: https://www.eea.europa.eu/publications/climate-impacts-and-vulnerability-2012 (accessed on 2 Februar 2026).
7. Munoz-Rojas, M.; Jordán, A.; Zavala, L.M.; González-Peñaloza, F.A.; De la Rosa, D.; Pino-Mejias, R.; Anaya-Romero, M. Modellin soil organic carbon stocks in global change scenarios: A CarboSOIL application. *Biogeosciences* **2013**, *10*, 8253–8268. [CrossRef]
8. Früh-Müller, A.; Hotes, S.; Breuer, L.; Wolters, V.; Koellner, T. Regional patterns of ecosystem services in cultural landscapes. *Lan* **2016**, *5*, 17. [CrossRef]
9. Bateman, I.J.; Harwood, A.R.; Mace, G.M.; Watson, R.T.; Abson, D.J.; Andrews, B.; Binner, A.; Crowe, A.; Day, B.H.; Dugdale, S et al. Bringing ecosystem services into economic decision-making: Land use in the United Kingdom. *Science* **2013**, *341*, 45–5 [CrossRef] [PubMed]
10. Viglizzo, E.F.; Jobbágy, E.G.; Ricard, M.F.; Paruelo, J.M. Partition of some key regulating services in terrestrial ecosystem Meta-analysis and review. *Sci. Total Environ.* **2016**, *562*, 47–60. [CrossRef] [PubMed]
11. Tian, S.; Wu, W.; Chen, S.; Li, Z.; Li, K. Global mismatch between ecosystem service supply and demand driven by climate chang and human activity. *Environ. Sci. Ecotechnol* **2025**, *26*, 100573. [CrossRef]
12. ISTAT (Istituto Nazionale di Statistica) [National Institute of Statistics]. Superficie di Comuni, Province e Regioni Italian al 9 Ottobre 2011 [Area of Italian Local Municipalities, Provinces and Regions as of 9 October 2011]. Available onlin https://www.istat.it/classificazione/principali-statistiche-geografiche-sui-comuni/ (accessed on 2 February 2026).
13. Fiorentino, C.; D'Antonio, P.; Toscano, F.; Donvito, A.; Modugno, F. New Technique for Monitoring High Nature Value Farmlan (HNVF) in Basilicata. *Sustainability* **2023**, *15*, 8377. [CrossRef]
14. Gizzi, F.T.; Proto, M.; Potenza, M.R. The Basilicata Region (Southern Italy): A Natural and 'Human-Built' Open-Air Laborator for Manifold Studies. Research trends over the last 24 years (1994–2017). *Geomat. Nat. Haz. Risk* **2019**, *10*, 433–464. [CrossRef]
15. Copernicus Land Monitoring Service. CORINE Land Cover. Available online: https://land.copernicus.eu/en/products/corin land-cover (accessed on 2 February 2026).
16. ISTAT (Istituto Nazionale di Statistica) [National Institute of Statistics. Popolazione Residente al 1° Gennaio [Resident Population a of 1st January]. Available online: https://esploradati.istat.it/databrowser/#/it/dw/categories/IT1,POP,1.0/POP_POPULATION DCIS_POPRES1/IT1,22_289_DF_DCIS_POPRES1_1,1.0 (accessed on 2 February 2026).
17. Torrieri, F.; Crisopulli, A.; Rossitti, M. Assessing the Feasibility of PPPs for Cultural Heritage Enhancement in UNESCO Site The Case of Matera (Italy). *Land* **2025**, *14*, 898. [CrossRef]
18. Scandone, R.; Bellucci, F.; Lirer, L.; Rolandi, G. The Structure of the Campanian Plain and the Activity of the Neapolitan Volcano (Italy). *J. Volcanol. Geoth. Res.* **1991**, *48*, 1–31. [CrossRef]
19. Pungetti, G.; Marini, A.; Vogiatzakis, I. Sardinia. In *Mediterranean Island Landscapes*; Vogiatzakis, I., Pungetti, G., Mannion, A.M Eds.; Landscape Series; Springer: Dordrecht, The Netherlands, 2008; Volume 9, pp. 143–169. [CrossRef]
20. Ginesu, S. Aspetti geomorfologici delle montagne sarde [Geomorphological aspects of the Sardinian mountains]. In *Montagn di Sardegna [Sardinian Mountains]*; Camarda, I., Ed.; Carlo Delfino Editore: Sassari, Italy, 1993; pp. 27–56. Available onlin https://www.sardegnaforeste.it/notizia/montagne-di-sardegna (accessed on 2 February 2026).
21. Melis, R.T.; Di Gregorio, F.; Panizza, V. Granite landscapes of Sardinia: Long-term evolution of scenic landforms. In *Landscape and Landforms of Italy*; Soldati, M., Marchetti, M., Eds.; World Geomorphological Landscapes; Springer: Cham, Switzerland, 201 [CrossRef]
22. Natural Capital Project. InVEST User Guide. n.d. Available online: http://releases.naturalcapitalproject.org/invest userguide/latest/en/carbonstorage.html (accessed on 2 February 2026).
23. Spawn, S.A.; Gibbs, H.K. Global Aboveground and Belowground Biomass Carbon Density Maps for the Year 2010 (Version 1 ORNL Distributed Active Archive Center. 2020. Available online: https://www.earthdata.nasa.gov/data/catalog/ornl-cloud global-maps-c-density-2010-1763-1 (accessed on 2 February 2026). [CrossRef]
24. INFC (Inventario Nazionale delle Foreste e dei Serbatoi Forestali di Carbonio) [National Inventory of Forests and Carbon Pools] Inventario Forestale Nazionale Italiano [National Inventory of Forests and Forest Carbon Reservoirs]. 2005. Available online https://www.inventarioforestale.org/it/statistiche-infc/ (accessed on 2 February 2026).
25. INFC (Inventario Nazionale delle Foreste e dei serbatoi forestali di Carbonio) [National Inventory of Forests and Carbon Pools] Inventario Forestale Nazionale Italiano [National Inventory of Forests and Forest Carbon Reservoirs]. 2015. Available online https://www.inventarioforestale.org/it/statistiche_infc/ (accessed on 2 February 2026).
26. ISPRA (Istituto Superiore per la Protezione e la Ricerca Ambientale) [Institute for Environmental Protection and Research] Linee Guida per la Redazione dei Piani di Monitoraggio o di Gestione Dell'impatto Sulla Qualità del Suolo e Sul Carbonic nel Suolo [Guidelines for Preparing Monitoring or Management Plans Concerning Impacts on Soil Quality and Soil Carbon] 2022. Available online: https://www.mase.gov.it/portale/documents/d/guest/linee_guida_ispra_implementazione_red_ii-pd (accessed on 2 February 2026).

https://doi.org/10.3390/su1804214

AGRIS. [Regional Agricultural Research Agency of Sardinia]. Carta della Distribuzione del Carbonio Organico del Progetto CUT alla Scala 1:50.000 [Organic Carbon Distribution Map, Scale 1:50,000]. 2016. Available online: https://www.sardegnaportalesuolo.it/cartografia/ (accessed on 2 February 2026).

Sallustio, L.; De Toni, A.; Strollo, A.; Di Febbraro, M.; Gissi, E.; Casella, L.; Geneletti, D.; Munafò, M.; Vizzarri, M.; Marchetti, M. Assessing habitat quality in relation to the spatial distribution of protected areas in Italy. *J. Environ. Manag.* **2017**, *201*, 129–137. [CrossRef] [PubMed]

Geofabrik. OpenStreetMap Data for Italy. Geofabrik GmbH. 2024. Available online: https://download.geofabrik.de/europe/italy.html (accessed on 2 February 2026).

Sharp, R.; Douglass, J.; Wolny, S.; Arkema, K.; Bernhardt, J.; Bierbower, W.; Chaumont, N.; Denu, D.; Fisher, D.; Glowinski, K.; et al. *InVEST 3.9.0 User's Guide*. The Natural Capital Project, Stanford University, University of Minnesota, The Nature Conservancy, and World Wildlife Fund. Available online: https://storage.googleapis.com/releases.naturalcapitalproject.org/invest/3.9.0/userguide/index.html (accessed on 2 February 2026).

Zhang, Y.; Zhang, C.; Zhang, X.; Wang, X.; Liu, T.; Li, Z.; Lin, Q.; Jing, Z.; Wang, X.; Huang, Q.; et al. Habitat Quality Assessment and Ecological Risks Prediction: An Analysis in the Beijing-Hangzhou Grand Canal (Suzhou Section). *Water* **2022**, *14*, 2602. [CrossRef]

Hack, J.; Molewijk, D.; Beißler, M.R. A Conceptual Approach to Modeling the Geospatial Impact of Typical Urban Threats on the Habitat Quality of River Corridors. *Remote Sens.* **2020**, *12*, 1345. [CrossRef]

Yang, Q.; Zhang, P.; Qiu, X.; Xu, G.; Chi, J. Spatial-Temporal Variations and Trade-Offs of Ecosystem Services in Anhui Province, China. *Int. J. Environ. Res. Public Health* **2023**, *20*, 855. [CrossRef]

U.S. Geological Survey. EarthExplorer. Landsat 8–9 OLI/TIRS Collection 2 Level-2 Science Product (Data Set). Available online: https://earthexplorer.usgs.gov/ (accessed on 2 February 2026).

U.S. Geological Survey (USGS). Landsat 8-9 Collection 2 (C2) Level 2 Science Product (L2SP) Guide. 2024. Available online: https://www.usgs.gov/media/files/landsat-8-9-collection-2-level-2-science-product-guide (accessed on 2 February 2026).

Millennium Ecosystem Assessment. Ecosystems and Human Well-Being: Synthesis; Island Press: Washington, DC, 2005. Available online: https://www.millenniumassessment.org/documents/document.356.aspx.pdf (accessed on 2 February 2026).

Common Agricultural Policy. Available online: https://agriculture.ec.europa.eu/common-agricultural-policy_en (accessed on 2 February 2026).

Agrarian Regions Survey by MF INEA. Available online: www.crea.gov.it/documents/68457/0/Regioni+agrarie+indagine+MF+INEA.xlsx/8019a0cb-f3d4-dcd9-6639-d178e9f2e89e?t=1561366035978 (accessed on 2 February 2026).

Zulian, G.; Paracchini, M.L.; Maes, J.; Liquete, C. *ESTIMAP: Ecosystem Services Mapping at European Scale. JRC (European Commission—Joint Research Centre—Institute for Environment and Sustainability) Technical Report EUR 26474 ENG*; Publications Office of the European Union: Luxembourg, 2013. [CrossRef]

Vallecillo, S.; La Notte, A.; Zulian, G.; Ferrini, S.; Maes, J. Ecosystem services accounts: Valuing the actual flow of nature-based recreation from ecosystems to people. *Ecol. Model.* **2019**, *392*, 196–211. [CrossRef]

Barton, D.N.; Obst, C.; Day, B.; Caparrós, A.; Dadvand, P.; Fenichel, E.; Havinga, I.; Hein, L.; McPhearson, T.; Randrup, T.; et al. Discussion Paper 10: Recreation Services from Ecosystems. 2019. Available online: https://www.researchgate.net/publication/333263149_Recreation_services_from_ecosystems (accessed on 2 February 2026).

Isola, F.; Lai, S.; Leone, F.; Zoppi, C. *Green Infrastructure and Regional Planning—An Operational Framework*; FrancoAngeli: Milan, Italy, 2022.

Paracchini, M.L.; Capitani, C. *Implementation of a EU Wide Indicator for the Rural-Agrarian Landscape*; Publications Office of the European Union: Luxembourg, 2011.

Eurostat. Glossary: Livestock Unit (LSU). n.d. Available online: https://ec.europa.eu/eurostat/statistics-explained/index.php?title=Glossary:Livestock_unit_(LSU) (accessed on 2 February 2026).

La Notte, A.; Vallecillo, S.; Polce, C.; Zulian, G.; Maes, J. *Implementing an EU System of Accounting for Ecosystems and Their Services. Initial Proposals for the Implementation of Ecosystem Services Accounts. JRC Technical Report EUR 28681 EN*; Publications Office of the European Union: Luxembourg, 2017. [CrossRef]

European Environment Agency—Eurosion. Costal Erosion and Defense. Available online: https://www.eea.europa.eu/en/datahub/datahubitem-view/ba6d7fe6-c79f-48c7-b738-f78260730538?activeAccordion=1090218 (accessed on 2 February 2026).

European Environment Agency—Bathing Water Directive. Status of Bathing Water. Available online: https://www.eea.europa.eu/en/datahub/datahubitem-view/c3858959-90da-4c1b-b9ca-492db0e514df (accessed on 2 February 2026).

Fotheringham, A.S.; Charlton, M.E.; Brunsdon, C. Geographically Weighted Regression: A natural evolution of the Expansion method for spatial data analysis. *Environ. Plann. A* **1998**, *30*, 1905–1927. [CrossRef]

Cheshire, P.; Sheppard, S. On the price of land and the value of amenities. *Econ. New Ser.* **1995**, *62*, 247–267. [CrossRef]

https://doi.org/10.3390/su18042146

50. Sklenicka, P.; Molnarova, K.; Pixova, K.C.; Salek, M.E. Factors affecting farmlands in the Czech Republic. *Land Use Policy* **2013**, *3*
130–136. [CrossRef]
51. Stewart, P.A.; Libby, L.W. Determinants of farmland value: The case of DeKalb County, Illinois. *Rev. Agric. Econ.* **1998**, *20*, 80–9
[CrossRef]
52. Zoppi, C.; Argiolas, M.; Lai, S. Factors influencing the value of houses: Estimates for the City of Cagliari, Italy. *Land Use Polic*
2015, *42*, 367–380. [CrossRef]
53. Byron, R.P.; Bera, A.K. Linearised estimation of nonlinear single equation functions. *Int. Econ. Rev.* **1983**, *24*, 237–248. [CrossRef
54. Wolman, A.L.; Couper, E.A. Potential consequences of linear approximation in economics. *Fed. Reserve Bank Econ. Q.* **2003**, *1*
51–67. Available online: https://www.richmondfed.org/-/media/richmondfedorg/publications/research/economic_quarterly
2003/winter/pdf/wolman.pdf (accessed on 2 February 2026).
55. Greig, D.M. *Optimisation*; Longman: Harlow, UK, 1980.
56. Shi, T.; Xu, H. Study on ecosystem service values of urban green space systems in Suzhou City based on the extreme gradien
boosting geographically weighted regression method: Spatiotemporal changes, driving factors, and influencing mechanism
Land **2025**, *14*, 564. [CrossRef]
57. Neill, A.M.; O'Donoghue, C.; Stout, J.C. Spatial analysis of cultural ecosystem services using data from social media: A guide t
model selection for research and practice. *One Ecosyst.* **2023**, *8*, e95685. [CrossRef]
58. Brunsdon, C.; Charlton, M.; Harris, P. Living with Collinearity in Local Regression Models. Accuracy 2012—Proceedings o
the 10th International Symposium on Spatial Accuracy Assessment in Natural Resources and Environmental Sciences. 201
pp. 67–72. Available online: https://mural.maynoothuniversity.ie/id/eprint/5755/ (accessed on 2 February 2026).
59. O'Brien, R.M. A caution regarding rules of thumb for variance inflation factors. *Qual. Quant.* **2007**, *41*, 673–690. [CrossRef]
60. Wheeler, D.C. Diagnostic tools and a remedial method for collinearity in Geographically Weighted Regression. *Environ. Plann.*
Econ. Space **2007**, *39*, 2464–2481. [CrossRef]
61. Geographically Weighted Regression (GWR) (Spatial Statistics). Available online: https://pro.arcgis.com/en/pr
app/latest/tool-reference/spatial-statistics/geographically-weighted-regression.htm (accessed on 2 February 2026).
62. Spohn, M.; Bagchi, S.; Biederman, L.A.; Borer, E.T.; Bråthen, K.A.; Bugalho, M.N.; Caldeira, M.C.; Catford, J.A.; Collins, S.L
Eisenhauer, N.; et al. The positive effect of plant diversity on soil carbon depends on climate. *Nat. Commun.* **2023**, *14*, 662
[CrossRef]
63. Xie, R.; Shah, S.M.H.; Xu, C.; Li, X.; Li, S.; Ma, B. Positive relationships between soil organic carbon and tree physical structur
highlights significant carbon co-benefits of Beijing's urban forests. *Forests* **2025**, *16*, 1206. [CrossRef]
64. Schittko, C.; Onandia, G.; Verdier, M.B.; Heger, T.; Jeschke, J.M.; Kowarik, I.; Maass, S.; Joshi, J. Biodiversity maintains so
multifunctionality and soil organic carbon in novel urban ecosystems. *J. Ecol.* **2022**, *110*, 916–934. [CrossRef]
65. Reiff, J.; Jungkunst, H.F.; Mauser, K.M.; Kampel, S.; Regending, S.; Rösch, V.; Zaller, J.G.; Entling, M.H. Permaculture enhance
carbon stocks, soil quality and biodiversity in Central Europe. *Commun. Earth Environ.* **2024**, *5*, 305. [CrossRef]
66. Lange, M.; Eisenhauer, N.; Sierra, C.A.; Bessler, H.; Engels, C.; Griffiths, R.I.; Mellado-Vázquez, P.G.; Malik, A.A.; Roy, J.; Sche
S.; et al. Plant diversity increases soil microbial activity and soil carbon storage. *Nat. Commun.* **2015**, *6*, 6707. [CrossRef]
67. Hu, Y.; Xu, E.; Kim, G.; Liu, C.; Tian, G. Response of spatio-temporal differentiation characteristics of habitat quality to lan
surface temperature in a fast urbanized city. *Forests* **2021**, *12*, 1668. [CrossRef]
68. Parvar, Z.; Salmanmahiny, A. Assessing and predicting habitat quality under climate change including changes in land surfac
temperature. *J. Nat. Conserv.* **2025**, *86*, 126903. [CrossRef]
69. Müllerová, J.; Šiffel, E. Cooling the land surface: Ecosystem health and water availability drive the landscape capacity to mitigat
climate change. *Ecol. Indic.* **2025**, *172*, 113265. [CrossRef]
70. Zhang, Y.; Kou, C.; Liu, M.; Man, W.; Li, F.; Lu, C.; Song, J.; Song, T.; Zhang, Q.; Li, X.; et al. Estimation of coastal wetland so
organic carbon content in Western Bohai Bay using remote sensing, climate, and topographic data. *Remote Sens.* **2023**, *15*, 4241
[CrossRef]
71. Osland, M.J.; Gabler, C.A.; Grace, J.B.; Day, R.H.; McCoy, M.L.; McLeod, J.L.; From, A.A.S.; Enwright, N.M.; Feher, L.C.; Stagg
C.L.; et al. Climate and plant controls on soil organic matter in coastal wetlands. *Glob. Change Biol.* **2018**, *24*, 5361–5379. [CrossRef
72. Tan, Q.; Han, W.; Li, X.; Wang, G. Clarifying the response of soil organic carbon storage to increasing temperature throug
minimizing the precipitation effect. *Geoderma* **2020**, *374*, 114398. [CrossRef]
73. Hartley, I.P.; Hill, T.C.; Chadburn, S.E.; Hugelius, G. Temperature effects on carbon storage are controlled by soil stabilisatior
capacities. *Nat. Commun.* **2021**, *12*, 6713. [CrossRef]
74. Zeng, X.-M.; Feng, J.; Yu, D.-L.; Wen, S.-H.; Zhang, Q.; Huang, Q.; Delgado-Baquerizo, M.; Liu, Y.-M. Local temperature increase
reduce soil microbial residues and carbon stocks. *Glob. Change Biol.* **2022**, *28*, 6433–6445. [CrossRef]
75. Huang, J.; Minasny, B.; McBratney, A.B.; Padariana, J.; Triantafilis, J. The location- and scale- specific correlation betweer
temperature and soil carbon sequestration across the globe. *Sci. Total Environ.* **2018**, *615*, 540–548. [CrossRef]

 https://doi.org/10.3390/su1804214

Kirschbaum, M.U.F. The temperature dependence of soil organic matter decomposition, and the effect of global warming on soil organic C storage. *Soil Biol. Biochem.* **1995**, *27*, 753–760. [CrossRef]

Agyal, K.; Maraseni, T.; Apan, A. How much do we know about trade-offs in ecosystem services? A systematic review of empirical research observations. *Sci. Total Environ.* **2022**, *806*, 151229. [CrossRef]

Raudsepp-Hearne, C.; Peterson, G.D.; Bennett, E.M. Ecosystem service bundles for analyzing tradeoffs in diverse landscapes. *Proc. Natl. Acad. Sci. USA* **2010**, *107*, 5242–5247. [CrossRef] [PubMed]

Chen, T.; Feng, Z.; Zhao, H.; Wu, K. Identification of ecosystem service bundles and driving factors in Beijing and its surrounding areas. *Sci. Total Environ.* **2020**, *711*, 134687. [CrossRef]

Tang, L.; Ke, X.; Zhou, T.; Zheng, W.; Wang, L. Impacts of cropland expansion on carbon storage: A case study in Hubei, China. *J. Environ. Manag.* **2020**, *265*, 110515. [CrossRef]

O'Connell, C.S.; Carlson, K.M.; Cuadra, S.; Feeley, K.J.; Gerber, J.; West, P.C.; Polasky, S. Balancing tradeoffs: Reconciling multiple environmental goals when ecosystem services vary regionally. *Environ. Res. Lett.* **2018**, *13*, 064008. [CrossRef]

Briner, S.; Huber, R.; Bebi, P.; Elkin, C.; Schmatz, D.; Grêt-Regamey, A. Trade-offs between ecosystem services in a mountain region. *Ecol. Soc.* **2013**, *18*, 35. [CrossRef]

Liu, R.; Clarke, K.C.; Zhang, J.; Feng, J.; Jia, X.; Li, J. Spatial correlations among ecosystem services and their socio-ecological driving factors: A case study in the City Belt along the Yellow River in Ningxia, China. *Appl. Geogr.* **2019**, *108*, 64–73. [CrossRef]

Cerda, R.; Allinne, C.; Gary, C.; Tixier, P.; Harvey, C.A.; Krolczyk, L.; Mathiot, C.; Clément, E.; Aubertot, J.-N.; Avelino, J. Effects of shade, altitude and management on multiple ecosystem services in coffee agroecosystems. *Eur. J. Agron.* **2017**, *82*, 308–319. [CrossRef]

Power, A.G. Ecosystem Services and Agriculture: Tradeoffs and Synergies. *Philos. T. Roy. Soc. B* **2010**, *365*, 2959–2971. [CrossRef] [PubMed]

Dimassi, B.; Cohan, J.-P.; Labreuche, J.; Mary, B. Changes in soil carbon and nitrogen following tillage conversion in a long-term experiment in northern France. *Agric. Ecosyst. Environ.* **2013**, *169*, 12–20. [CrossRef]

Govaerts, B.; Verhulst, N.; Castellanos-Navarrete, A.; Sayre, K.D.; Dixon, J.; Dendooven, L. conservation agriculture and soil carbon sequestration: Between myth and farmer reality. *Crit. Rev. Plant Sci.* **2009**, *28*, 97–122. [CrossRef]

Schwaiger, F.; Poschenrieder, W.; Biber, P.; Pretzsch, H. Ecosystem service trade-offs for adaptive forest management. *Ecosyst. Serv.* **2019**, *39*, 100993. [CrossRef]

Ciesielski, M.; Gołos, P.; Wysocka-Fijorek, E.; Kaliszewski, A. Relationships between forest ecosystem services—Current state of knowledge. *Folia For. Pol.* **2024**, *66*, 228–248. [CrossRef]

Brockerhoff, E.G.; Barbaro, L.; Castagneyrol, B.; Forrester, D.I.; Gardiner, B.; González-Olabarria, J.R.; Lyver, P.O.; Meurisse, N.; Oxbrough, A.; Taki, H.; et al. Forest biodiversity, ecosystem functioning and the provision of ecosystem services. *Biodivers. Conserv.* **2017**, *26*, 3005–3035. [CrossRef]

Bryant, R.L.; Kothari, S.; Cavender-Bares, J.; Curran, S.J.; Grossman, J.J.; Hobbie, S.E.; Nash, C.; Neumiller, G.C.; See, C.R. Independent effects of tree diversity on aboveground and soil carbon pools after six years of experimental afforestation. *Ecol. Appl.* **2024**, *34*, e3042. [CrossRef]

Seddaiu, G.; Bagella, S.; Pulina, A.; Cappai, C.; Salis, L.; Rossetti, I.; Lai, R.; Roggero, P.P. Mediterranean cork oak wooded grasslands: Synergies and trade-offs between plant diversity, pasture production and soil carbon. *Agroforest Syst.* **2018**, *92*, 893–908. [CrossRef]

Scotti, R.; Cadoni, M. A historical analysis of traditional common forest planning and management in Seneghe, Sardinia—Lessons for sustainable development. *For. Ecol. Manag.* **2007**, *249*, 116–124. [CrossRef]

Conant, R.T.; Paustian, K. Potential soil carbon sequestration in overgrazed grassland ecosystems. *Glob. Biogeochem. Cycles* **2002**, *16*, 1143. [CrossRef]

Derner, J.D.; Schuman, G.E. Carbon sequestration and rangelands: A synthesis of land management and precipitation effects. *J. Soil Water Conserv.* **2007**, *62*, 77–85. [CrossRef]

Zhou, G.; Luo, Q.; Chen, Y.; He, M.; Zhoud, L.; Frank, D.; He, Y.; Fu, Y.; Zhang, B.; Zhou, X. Effects of livestock grazing on grassland carbon storage and release override impacts associated with global climate change. *Glob. Change Biol.* **2019**, *25*, 1119–1132. [CrossRef] [PubMed]

Paracchini, M.L.; Zulian, G.; Kopperoinen, L.; Maes, J.; Schägner, J.P.; Termansen, M.; Zandersen, M.; Perez-Soba, M.; Scholefield, P.A.; Bidoglio, G. Mapping cultural ecosystem services: A framework to assess the potential for outdoor recreation across the EU. *Ecol. Indic.* **2014**, *45*, 371–385. [CrossRef]

Kline, J.D.; Swallow, S.K. The demand for local access to coastal recreation in southern New England. *Coast. Manag.* **1998**, *26*, 177–190. [CrossRef]

Paudel, K.P.; Caffey, R.H.; Devkota, N. An Evaluation of factors affecting the choice of coastal recreational activities. *J. Agric. Appl. Econ.* **2011**, *43*, 167–179. [CrossRef]

https://doi.org/10.3390/su18042146

100. Ghermandi, A.; Nunes, P.A.L.D. A global map of coastal recreation values: Results from a spatially explicit meta-analysis. *Eco Econ.* **2013**, *86*, 1–15. [CrossRef]
101. Littles, C.J.; Jackson, C.A.; DeWitt, T.H.; Harwell, M.C. Linking people to coastal habitats: A meta-analysis of final ecosystem goods and services on the coast. *Ocean Coast. Manag.* **2018**, *165*, 356–369. [CrossRef]
102. Romanyà, J.; Rovira, P. An appraisal of soil organic C content in Mediterranean agricultural soils. soil use and management. *Soil Use Manag.* **2011**, *27*, 321–332. [CrossRef]
103. Romo, S.; Soria, J.; Olmo, C.; Flor, J.; Calvo, S.; Ortells, R.; Armengol, X. Nutrients and carbon in some Mediterranean dune ponds. *Hydrobiologia* **2016**, *782*, 97–109. [CrossRef]

https://doi.org/10.3390/su18042146

sustainability

MDPI

ticle

ulti-Dimensional Coupling Perspective on the Compatibility f Ecosystem Service Supply and Demand in Megacities and uture Scenario Simulation: The Case of Shanghai

afang Huang [1,†], Shaofeng Chen [1,†], Chenxi Su [1], Miaomiao Yan [1], Han Chen [2] and Zheng Ding [1,*]

1 College of Landscape Architecture and Art, Fujian Agriculture and Forestry University, Fuzhou 350002, China
2 School of Arts, Lancaster University, Lancaster LA1 4YW, UK
* Correspondence: 000q151039@fafu.edu.cn
† These authors contributed equally to this work.

Abstract

Amid global climate change and rapid urbanization, megacities such as Shanghai confront prominent ecological challenges. A critical issue is the growing mismatch between the supply of and demand for urban green space (UGS) ecosystem services. This study aims to explore the supply–demand compatibility of Shanghai's UGS ecosystem services and simulate future scenarios. Guided by the SSP1-2.6 scenario, it integrates the PLUS model, InVEST model, and nSFCA method to conduct dynamic analysis, quantifying supply–demand alignment and identifying imbalance areas. Results show a significant spatial mismatch: high demand but low supply in Shanghai's inner ring and low demand but high supply in the outer ring. UGS attractiveness presents a core-concentrated and peripheral-diffused pattern by level. By 2030, a coordinated supply framework of "city-level dominance, community-level support, and neighborhood-level supplementation" will form, improving supply–demand alignment, though accessibility gaps persist. The study reveals that urbanization, planning policies, and population–spatial expansion asynchrony drive these patterns, providing scientific decision-making support for optimizing Shanghai's green space planning and building an ecologically livable city.

Keywords: urban green space; ecosystem service supply–demand matching; scenario simulation; accessibility; multi-model coupling

1. Introduction

Driven by both global climate change and rapid urbanization, ecological and environmental issues in megacities are becoming increasingly prominent. Urban Green Space (UGS), as the core vehicle for maintaining urban ecological balance and safeguarding residents' health and well-being, exhibits a dual-emphasized value: On the one hand, UGS delivers irreplaceable ecological benefits by regulating, supplying, cultivating, and supporting multiple ecosystem services. These include curbing soil erosion, safeguarding water resources, preserving biodiversity, and mitigating climate change [1,2]. On the other hand, as residents increasingly demand a higher quality of life, the importance of the leisure, entertainment, wellness, and recreational services provided by UGS has become increasingly prominent, serving as a key pillar for enhancing urban livability and residents' well-being. Urban green spaces also serve as nature-based solutions (NBS) to mitigate risks and vulnerabilities associated with climate change [3]. Achieving the United Nations

 https://doi.org/10.3390/su18052195

Sustainable Development Goals (such as Goal 11) is crucial for providing universal access to safe, inclusive, green, and public spaces [4].

Existing research on environmental justice has primarily focused on the disparate circumstances experienced by different groups, encompassing residents of various urban areas [5] and individuals of differing socioeconomic statuses [6], as well as populations of diverse racial [7], age [8], and gender backgrounds. At the methodological level, some scholars have employed statistical data to evaluate urban green space accessibility (UGSA) patterns [9], with assessment indicators including green space quantity, species community size, and distribution density. Other studies have employed spatial syntax, network analysis, kernel density estimation, and other technical methods to measure spatial proximity [8,10,11]. The core characteristic of such research is that it defines and measures the geographic accessibility of green spaces based solely on supply-side factors, failing to adequately incorporate demand-side dimensions. Meanwhile, some studies have also examined demand-side factors influencing residents' willingness to use green spaces, such as the distance between green spaces and residential areas, as well as the size and quality of the green spaces themselves. These studies predominantly employ gravity models for analysis [6,12,13]. As an important derivative of the gravity model, the Two-Step Floating Catchment Area Method (2SFCA) and its optimized variants—such as the Gaussian Two-Step Floating Catchment Area Method (G2SFCA) [14], the Kernel Density Two-Step Floating Catchment Area Method [15], and the Huff–Gaussian Two-Step Floating Catchment Area Method (H-G2SFCA) [16]—can simultaneously account for both supply and demand factors in estimation. These factors include spatial distance, population distribution, and green space catchment area coverage. The aforementioned method has been widely applied in assessing the accessibility of urban green spaces and has been proven to demonstrate superior capability in fitting real-world scenarios [17]. Nevertheless, existing research still demonstrates notable multidimensional limitations, as key factors that significantly influence research outcomes have not been systematically incorporated. Specifically, comprehensive demand factors have not been effectively integrated into analytical frameworks; the spatial characteristics and magnitude of accessibility supply–demand imbalances remain difficult to accurately characterize; the differentiated impacts of various travel modes on green space accessibility are often overlooked; and there is a lack of prediction and analysis of green space accessibility under future development scenarios.

However, current traditional static analysis frameworks often fall short in supporting the adaptive planning of future UGS [18,19]. Currently, one-size-fits-all spatial standards, such as per capita green space area, often fail to effectively address challenges posed by spatial and temporal population dynamics and evolving local needs. While it is widely assumed that larger urban settlements should be allocated more green space to accommodate their resident populations [20], the expansion of urban green space frequently lags behind the pace of population growth and urban density increases. This mismatch, in turn, exacerbates the existing disparities in the accessibility of urban green space [21]. Despite their limitations, spatial standards remain essential within top-down spatial planning systems. They provide normative benchmarks and regulatory guidance, safeguarding the fundamental rights of urban residents to a livable daily existence. The release of the latest Coupled Model Intercomparison Project Phase 6 (CMIP6) provides robust global development trajectories that support projections of future ecosystems. The approach proposed by the Intergovernmental Panel on Climate Change (IPCC) predicts the future state of human and natural systems under climate change by coupling Shared Socioeconomic Pathways (SSPs) with Representative Concentration Pathways (RCPs) [22,23]. An increasing number of studies integrate SSP-RCP scenarios with land-use models to predict the spatial and temporal patterns of ES and their drivers [24,25]. Quantitative

https://doi.org/10.3390/su1805219

estimates of key ES are widely adopted through tools such as the InVEST model [26–29]. Among these, SSP1-2.6—as a representative pathway balancing sustainable development with low greenhouse gas emissions [30]—has been confirmed by multiple studies to effectively mitigate heat-related health risks at both global and regional scales [31,32]. This scenario provides scientifically grounded planning guidance for promoting ecological conservation and enhancing climate resilience in high-density cities like Shanghai [33]. As a mainstream land-use simulation tool, the PLUS model can accurately capture land-use conversion patterns based on historical data, thereby supporting predictions of future green space distribution.

In recent years, with the advancement of global carbon neutrality goals and the increasing frequency of extreme climate events, the synergistic mechanisms between urban green spaces and carbon neutrality, as well as the service effectiveness of green spaces under heatwave conditions, have become frontier research hotspots. In the field of carbon neutrality, studies have shown that urban green spaces play a key role in mitigating climate change through their carbon storage function. Sun et al. [34] evaluated urban green spaces in Beijing and found that their aboveground carbon storage makes a significant contribution to the urban carbon balance. More recent studies further indicate that the carbon sink benefits of park green spaces at multiple scales need to be systematically assessed by integrating vegetation structure, spatial configuration, and land-use types [35,36]. In terms of heatwave adaptation, green spaces have attracted widespread attention as nature-based cooling solutions. Some studies have shown that increasing urban tree canopy coverage to 30% can reduce air temperature by 1.5 °C [37], highlighting the core regulatory role of green spaces during extreme high-temperature events. Meanwhile, Matias et al. [38] emphasized the trade-off between heatwave adaptation and water resource use in Mediterranean urban vegetation. By simulating the evolution of green space ecosystem services in Shanghai under the SSP1-2.6 scenario, this study essentially responds to the core concerns regarding the synergy of green space carbon sinks and heatwave mitigation effectiveness, providing a theoretical background and comparative benchmark for subsequent discussions.

Currently, as a densely populated core node within a vast metropolitan area with extensive construction land, Shanghai faces multiple pressures, including further population concentration, spatial expansion, and intensifying climate risks. Issues such as the encroachment and uneven distribution of UGS have become increasingly prominent [39]. The Shanghai Ecological Space Special Plan (2021–2035) lists the goal of achieving full coverage of a 500 m service radius for parks and green spaces larger than 3000 m^2 within the city's development boundary as a core planning indicator for 2035. The policy intention is highly aligned with the development concepts of a "People's City" and a "Park City." On the one hand, through a multi-level green space layout including community parks and micro (pocket) parks, it aims to address the shortage of ecological space in high-density urban areas, enabling residents to "see greenery from their windows and access parks upon leaving their homes," thereby meeting daily recreation and health needs. On the other hand, this target supports the ecological spatial structure of the city characterized by the "double rings, nine corridors, and ten zones," strengthening the inclusiveness and equity of green space ecosystem services and supporting the construction of a resilient ecological city. However, this goal faces multiple implementation challenges due to the characteristics of Shanghai's high-density living environment and practical planning constraints. First, construction land in core urban areas is already saturated, and land resources are scarce, making it difficult to locate green spaces larger than 3000 m^2, while high demolition costs further limit implementation; consequently, micro (pocket) parks are often the only feasible supplement, but they struggle to meet comprehensive service demands. Second, Shanghai is promoting population inflow to new towns and population decentralization from old

https://doi.org/10.3390/su18052195

urban areas, leading to dynamic population redistribution, yet green space layouts lack dynamic adjustment mechanisms aligned with population changes, resulting in lagging green space supply in new towns and declining supply–demand matching in old urban areas. Third, although the plan proposes the construction of a greenway network system, some areas still suffer from insufficient greenway connectivity and inadequate public transit station coverage, meaning that green spaces may be spatially reachable but lack practical accessibility. Fourth, while the plan explicitly proposes strategies for improving ecological space quality, it also indirectly reflects significant regional disparities in current green space quality in Shanghai; in some older urban areas, green spaces have outdated facilities and single ecological functions, and even when the 500 m spatial accessibility standard is met, they still fail to satisfy residents' demand for high-quality ecosystem services. This in return raises a series of core research questions that demand urgent attention: What characteristics does the comprehensive service capacity of UGS exhibit under the current spatial pattern? Under current conditions and the future SSP1-2.6 scenario, how will the supply level and spatial characteristics of UGS ecosystem services change? Can the accessibility and comprehensive capacity of existing UGS services effectively match current and future population demands? Under different development scenarios, what are the adaptive characteristics between the evolution of population demand and changes in ecosystem service supply of UGS, and what are the key areas and underlying causes of supply–demand imbalance? Resolving these issues is crucial for precisely enhancing the effectiveness of UGS planning and safeguarding residents' well-being. Therefore, studying UGS accessibility, the future evolution of ecosystem services, and shifts in population demand holds significant practical and scientific value. Land-use change (LUCC) plays a major role in shaping the ecosystem service supply of urban green space. The advanced n-step Floating Catchment Area (nSFCA) method enables comprehensive quantification of urban green space service supply based on accessibility, providing robust support for analyzing supply–demand matching relationships. Against this backdrop, integrating land-use simulation, ecosystem service assessment, and nSFCA accessibility analysis emerges as an effective pathway to address these questions.

Compared with existing analytical frameworks for supply–demand matching and coupling coordination of urban green space ecosystem services, the multidimensional coupling perspective constructed in this study achieves three core theoretical advancements. First, it breaks through the limitations of traditional two-dimensional coupling analyses centered on ecological supply and population demand by incorporating transportation accessibility as an independent core dimension, thereby establishing a three-dimensional coupling analysis framework integrating ecological supply, population demand, and transportation accessibility, which compensates for the theoretical deficiency in existing studies that insufficiently consider intermediary dimensions in the transmission of supply–demand relationships. Second, it overcomes the static coupling logic of traditional research by integrating the dynamic driving mechanisms of land-use evolution and future development scenario simulations, enabling full temporal coupling analysis that characterizes current conditions, predicts future trends, and evaluates dynamic supply–demand adaptation, thus addressing the theoretical limitations of static analytical frameworks in supporting adaptive planning of urban green spaces. Third, it surpasses the limitations of single-method or single-model applications by achieving deep integration of multiple models, including land-use simulation, ecosystem service assessment, and accessibility measurement, rather than simple methodological or model aggregation, thereby constructing an integrated coupling analysis framework that unifies elements, methods, and scales and improves the fragmented theoretical system of existing methodological applications. This multidimensional coupling perspective provides a new theoretical analytical approach for

 https://doi.org/10.3390/su1805219

studying supply–demand adaptation of green space ecosystem services in megacities and offers valuable practical exploration for expanding research dimensions and promoting methodological innovation in ecosystem service coupling studies.

In summary, this study, grounded in the high-density urban characteristics of Shanghai and guided by the SSP1-2.6 scenario, integrates the PLUS model, InVEST model, and nSFCA methodology. It addresses the core research questions by conducting a dynamic analysis of urban green space ecosystem service supply characteristics and accessibility under future scenarios. The study quantifies the alignment between ecosystem service supply and population demand, identifies areas of supply–demand imbalance, and provides more concrete and effective scientific decision-making support for optimizing Shanghai's green space planning and building an ecologically livable city. The specific related work is summarized as follows.

2. Research Area and Data

2.1. Research Area

This study focuses on Shanghai, a megacity in eastern China. Against the backdrop of rapid urbanization, Shanghai's urban expansion and shifts in land-use patterns have led to an imbalance between the supply and demand of ecosystem services [40]. This has exacerbated the urban heat island effect and increased exposure to extreme weather events, posing threats to public health [41]. In recent years, Shanghai has established a multi-tiered, integrated urban green space system through initiatives such as pocket park development, wetland conservation and restoration, and the interconnection of greenway networks. The Shanghai Ecological Space Special Plan (2021–2035) explicitly sets the goal of achieving full coverage of a 500 m service radius for parks and green spaces larger than 3000 m^2 within the city's development boundary. Shanghai's green space development has continuously improved in both quality and quantity. From 2015 to 2021, the per capita recreational green space increased from 7.6 m^2 to 8.8 m^2, with the total number of green spaces reaching 532. However, recreational green spaces remain limited, necessitating improvements in service efficiency and scale [42]. This study examined all recreational green spaces (ranging from 0.03 to 434.50 km^2 in size) across Shanghai's 16 districts (Figure 1). These green spaces were rated by the Shanghai Municipal Administration of Greening and City Appearance using a five-star rating system (from five stars to one star), based on multiple factors including park classification, area, facilities, safety, and services. The rating results are published on the official website (http://lhsr.sh.gov.cn/, accessed on 16 December 2025) [43].

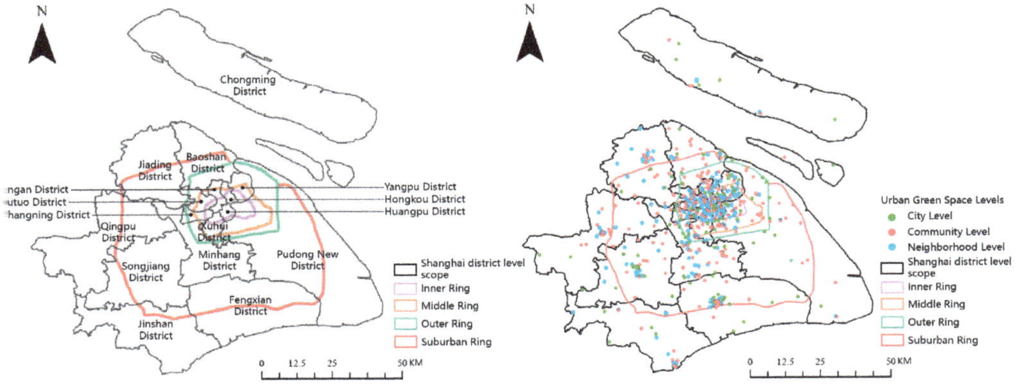

Figure 1. The research area.

https://doi.org/10.3390/su18052195

2.2. Data Collection and Preprocessing

Based on the scale characteristics and service functions of urban green spaces (UGS), this study categorizes recreational urban green spaces into three tiers: neighborhood-level, community-level, and city-level. Considering that walking and public transportation are the primary modes of travel for residents in large cities to access green spaces for leisure, this study focuses its analysis on these two travel patterns. Using 2020 accessibility data for green spaces across all residential areas in Shanghai, this study systematically examines the supply characteristics, service efficiency, and development status of recreational urban green spaces.

To ensure the accuracy and reliability of research data, this study implemented a rigorous data validation process. The core data required for the study—including the distribution of urban green spaces, residential area locations, road networks, transportation infrastructure, and urban service points (covering roads, public transit routes and stations, urban green spaces, and residential area entrances/exits)—were collected through multiple sources and subjected to cross-validation. Specific data sources include official government databases and authoritative geospatial platforms (such as the Bigemap platform, http://www.bigemap.com/, accessed on 16 December 2025). Discrepancies identified during data verification were resolved through on-site inspections and consultations with local urban planning authorities. This multi-tiered data validation approach significantly enhanced the credibility of the research findings. Following comprehensive data cleansing, a road network optimized for walking and public transit was constructed based on the verified data. This network was further refined into transportation models tailored to different travel modes by integrating actual traffic characteristics and resident travel scenarios.

The shortest travel times for each mode of transportation were calculated using origin–destination (OD) analysis. Specifically, ArcGIS 10.7 software was employed to determine the minimum time cost between residential area entrances/exits and urban green space entrances/exits [13]. This data provides foundational support for subsequent calculations of accessibility threshold times (i.e., maximum acceptable travel times) for urban green spaces at different hierarchical levels. Specifically, the 500 m accessibility target outlined in the Shanghai Ecological Space Special Plan (2021–2035) applies exclusively to neighborhood-level urban green spaces. Travel time thresholds for community-level and city-level green spaces are set based on differences in travel modes and green space hierarchy. The travel threshold time and per capita green space area standards for urban green spaces at various levels comprehensively reference prior research findings on Shanghai's urban green space service radius [44], as well as relevant domestic and international standards and technical guidelines for urban green space or park system planning. These include the Urban Landscape and Greening Evaluation Standard (GB/T 50563-2010) (China, 2010) [45], Shanghai Urban Park Planning and Construction Guidelines (https://lhsr.sh.gov.cn/ghjhl/, accessed on 16 December 2025), and the Accessible Natural Greenspace Standard (ANGSt) [46]. The population size of each residential area was estimated by calculating the ratio of the total residential floor area in each Shanghai community to the per capita residential floor space [43].

3. Methods

This study established a multidimensional coupled analytical framework that integrates land-use simulation, ecosystem service assessment, and accessibility measurement to investigate the compatibility between supply and demand of ecosystem services in Shanghai's urban green spaces and their future scenario characteristics. The specific methodology is as follows:

- Based on land-use/cover change data from 2010 to 2020, the PLUS model was employed to simulate Shanghai's land use spatial patterns under the SSP1-2.6 scenario for

https://doi.org/10.3390/su1805219

the year 2030. Land-use types were reclassified into six categories: forest, grassland, water bodies, wetland, cropland, and non-vegetation. Model accuracy was validated using the Kappa coefficient, and a 30 m resolution land use map was generated to analyze the evolution characteristics of green spaces.

- Using the InVEST model, the supply capacity of urban green spaces in 2023 and 2030 was quantitatively assessed for five core ecosystem services: soil retention, water purification, habitat quality, carbon storage, and water yield. Each service module was quantified using corresponding methodologies and formulas, including carbon density calculations, water balance principles, erosion factor measurements, and human activity threat assessments.
- Building upon this foundation, the integrated urban green space accessibility index (UgsA) was constructed using the n-step Floating Catchment Area method (nSFCA). This index encompasses both walking and public transit modes by integrating three key factors: comprehensive carrying capacity, population demand, and transportation supply. The model incorporates the Huff model to calculate residents' green space selection probability, employs a Gaussian function to correct for distance impedance effects, and assigns differentiated mode weights based on green space hierarchy (neighborhood, community, and city levels). Simultaneously, the study calculates complementary indices: the Comprehensive Attractiveness Index for Urban Green Spaces, Population Demand Index, and Transportation Supply Index.

By integrating land use simulation, ecosystem service assessment, and accessibility measurement results, this study systematically analyzes the supply–demand matching relationship of ecosystem services provided by urban green spaces in Shanghai (Figure 2).

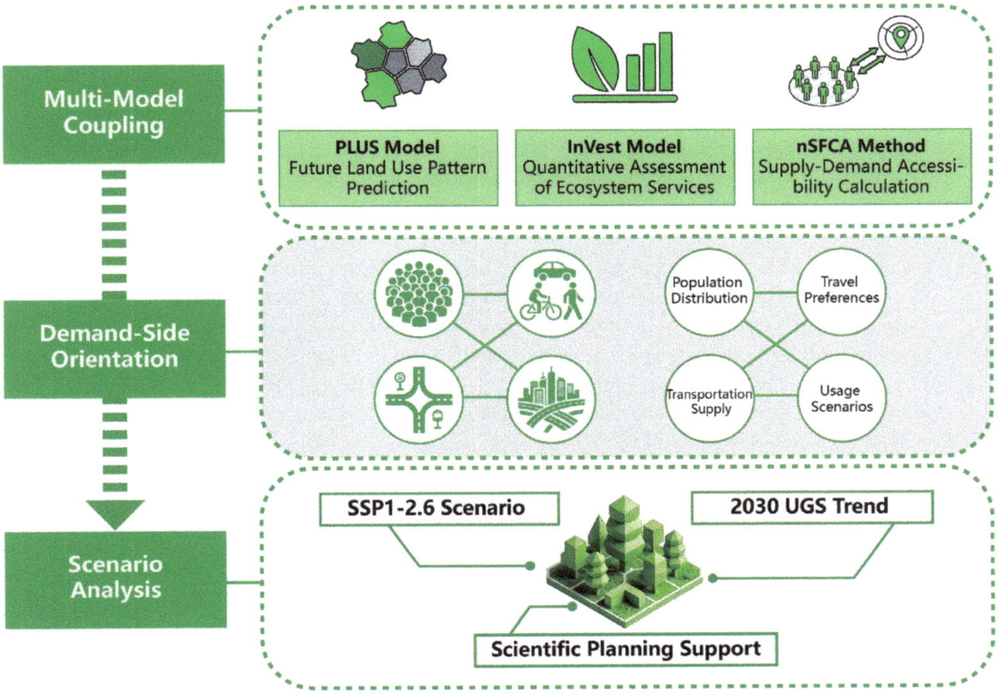

Figure 2. Technology Roadmap.

https://doi.org/10.3390/su18052195

3.1. Land Use Simulation

To forecast future development, this study employs the PLUS model to simulate land distribution in 2030 based on land use/land cover change (LUCC) data from 2010 to 2020. This study adopts a spatial resolution of 30 m, which not only matches the raster-based computational requirements of the InVEST model for ecosystem service assessment but also aligns with the commonly used resolution in studies of green space spatial patterns at the municipal scale in Shanghai. This resolution balances computational efficiency while accurately capturing the spatial boundaries and distribution characteristics of small-scale green spaces such as neighborhood- and community-level green areas, thereby avoiding the loss or blurring of green space spatial information caused by overly coarse resolution. This study, based on data from 19 CMIP6 model projections by the Yuan Jiacan research group at Fudan University, investigated the disease burden associated with hot and humid weather in Shanghai under different emission scenarios. Results indicate that as a humid megacity, Shanghai experiences significantly increased risks of all-cause, cardiovascular, and respiratory outpatient visits during hot and humid weather, with nighttime risks potentially exceeding daytime levels. Compared to the high-emission scenario SSP5-8.5, maintaining the low greenhouse gas emission pathway SSP1-2.6 by 2100 could effectively prevent 13.05% of heat-related illnesses. This study provides core evidence demonstrating how the SSP1-2.6 pathway safeguards the health of Shanghai residents and aligns with the requirements for building a livable city [47]. To ensure consistency across datasets, land use types were reclassified into six categories: forest, grassland, water bodies, wetlands, cropland, and non-vegetation (including impervious surfaces and unutilized land). Model accuracy was assessed using the Kappa coefficient, and the resulting 30 m land use map served as the foundation for subsequent analyses. Detailed information on scenario settings, model parameters, and modeling factors can be found in Supplementary Material Table S1.

3.2. Assessment of Urban Green Space Ecosystem Service Supply

Based on existing empirical research findings [48,49], this study identified five core provisioning types of urban green space ecosystem services (UGS-ES): soil retention (SR), water purification (WP), habitat quality (HQ), carbon storage (CS), and water yield (WY). This indicator system encompasses the multidimensional ecological benefits of UGS in resource supply assurance, environmental process regulation, and ecological integrity maintenance [50,51]. Moreover, its role in enhancing human well-being and sustaining habitat quality has been extensively validated by the academic community [52–54].

Specifically, SR effectively curbs soil erosion and ecosystem degradation, providing support for the biophysical foundation of human settlements while safeguarding habitat ecological integrity [55]; WY and WP play crucial roles in ensuring human water security [56]; HQ reflects the core capacity of ecosystems to support biodiversity and human habitability [57]; and CS, as a critical ecological regulation service, contributes to climate change mitigation through carbon sequestration and storage functions, thereby supporting long-term ecosystem stability [57,58]. To objectively reflect the relative importance of each service and enhance the academic rigor and scientific reliability of the assessment results, this study applies the Analytic Hierarchy Process (AHP) to determine the weights of the five ecosystem services, thereby avoiding biases caused by subjective weighting. The weight calculation process refers to existing related studies [59]. Ultimately, this study employed the InVEST model to conduct quantitative assessments of the aforementioned five ecosystem services under the SSP1-2.6 scenario. The evaluation results lay the foundation for subsequent spatial analyses of ecosystem service (ES) dynamics [60,61]. The rationale for selecting the indicators is detailed in Supplementary Material Table S2.

 https://doi.org/10.3390/su18052195

3.2.1. Carbon Storage

This module assesses regional ecosystem carbon storage based on the spatial distribution patterns, cover types, and corresponding carbon densities of each land use category and automatically generates spatial distribution maps of carbon storage. The final values used in this study are provided in Supplementary Material, Table S3. The formula is as follows:

$$CS = \sum_{k=1}^{n} A_k \cdot (C_{above} + C_{below} + C_{soil} + C_{dead})$$

In the formula: CS represents total carbon storage, t; A_k denotes the area of land category k in the study area, hm^2; n is the number of land categories; C_{above} represents above-ground carbon density; C_{below} represents below-ground carbon density; C_{soil} represents soil organic carbon density; C_{dead} represents dead wood and litter carbon density, measured in t/hm^2.

3.2.2. Water Yield

In the InVEST model, water yield is defined as the runoff volume across all land use categories. Based on fundamental water balance principles, the annual water yield for each grid cell is precisely quantified by calculating the difference between the annual average precipitation and the actual annual evapotranspiration. The biophysical input tables used in this module are derived from existing studies [23,62] and are included in Supplementary Material, Table S4. The Z-value parameter was determined through iterative calibration and empirical validation, ultimately set to 11.5 [23]. The formula is as follows:

$$\mathrm{WY}_{ij} = (1 - \frac{\mathrm{AET}_{ij}}{P_i}) \cdot P_i$$

In the formula: WY_{ij} represents the annual average runoff from land use type j in grid cell i, in mm; AET_{ij} represents the actual annual evapotranspiration from land use type j in grid cell i, in mm; P_i represents the annual average precipitation in grid cell i, in mm.

3.2.3. Soil Retention

The module consists of two components: reduced soil erosion, expressed as the difference between potential erosion and actual erosion; and sediment retention, expressed as the product of sediment volume and sediment retention rate, reflecting the reduction in potential erosion at each site [63]. Specific parameters are provided in Supplementary Material, Table S5. The formula is as follows:

$$\mathrm{SC} = R \cdot K \cdot \mathrm{LS} \cdot (1 - C \cdot P)$$

In the formula: SC represents soil retention, $t/(hm^2 \cdot a)$; R denotes the rainfall erosion index calculated based on precipitation; K is the soil erodibility factor; LS is the terrain factor derived from the slope length factor (L) and slope gradient factor (S); C represents the vegetation cover factor; P denotes the soil conservation measure factor.

3.2.4. Habitat Quality

The module reflects the impact of human activities on the environment. The higher the intensity of human activities, the greater the threat to habitats, and the lower the quality and level of biodiversity. Habitat quality depends on the relative impact of threats, the sensitivity of habitats to threats, and the distance between habitats and threat sources [64].

https://doi.org/10.3390/su18052195

The corresponding weights and sensitivity parameters are based on existing literature [65], as detailed in Supplementary Material Tables S6 and S7. The formula is as follows:

$$D_{xj} = \sum_{r=1}^{R} \sum_{y=1}^{Y_r} (W_r / \sum_{r=1}^{R} W_r) r_y i_{rxy} \beta_x S_{jr}$$

In the formula: D_{xj} denotes the habitat degradation index (habitat quality) of grid cell x under land use/cover type j; R represents the total number of ecological threat factors; W_r denotes the weight of threat factor type r; Y_r denotes the quantity of threat factor type r; r_y denotes the quantity or intensity of the rth threat factor at the yth specific threat point; i_{rxy} represents the distance decay function value between grid cell x and threat source point y; β_x indicates the legal protection level of grid cell x; S_{jr} denotes the sensitivity coefficient of land use/cover type j to the rth threat factor.

$$HQ_{xj} = H_j[1 - (\frac{D_{xj}^z}{D_{xj}^z + k^z})]$$

In the formula: HQ_{xj} denotes the habitat quality of grid cell x under land use/cover type j, with values ranging from 0 to 1; H_j represents the habitat suitability index for land use/cover type j; D_{xj} indicates the degree of habitat degradation; k is the semi-saturation constant; z is the exponent parameter, with the model specifying $z = 2.5$. This formula indicates that habitat quality varies with changes in the habitat suitability index, and when $H_j = 0$, $HQ_{xj} = 0$.

3.3. Measuring Urban Green Space Supply Levels from an Accessibility Perspective and Its Three Key Elements

To analyze the supply characteristics of urban green spaces from an accessibility perspective, this study integrates three key factors—comprehensive carrying capacity, population demand, and transportation supply—to estimate the accessibility supply levels of green space areas across cities. Subsequently, the n-step Floating Catchment Area method (nSFCA) is employed. Through calculations, the urban green space comprehensive accessibility index (UgsA) is derived to characterize the supply level of urban green spaces. The three key factors—comprehensive carrying capacity, population demand, and transportation supply—were quantified through the Urban Green Space Comprehensive Carrying Capacity Index (UgsCC), the Urban Green Space Population Demand Index (UgsPD), and the Urban Green Space Transportation Supply Index (UgsTS), respectively.

3.3.1. Calculation Method for the Urban Green Space Accessibility Index (UgsA)

The comprehensive accessibility index ($UgsA_j$) of urban green space j characterizes the accessibility supply level of that green space. It is calculated by aggregating the accessibility values from all residential points k within the time threshold t_0 range of green space j to that green space. The comprehensive accessibility index (A_{ij}) between residential point i and green space j characterizes the accessibility level from each residential point to its corresponding green space. This index encompasses both walking and public transportation modes to reflect travel diversity, and it references relevant Shanghai-based studies to determine the weights of walking and public transportation [66], thereby enhancing the rationality and scientific robustness of the analysis. The calculation formula is as follows:

https://doi.org/10.3390/su18052195

$$UgsA_j = \sum_{k\in\{t_{ij}\le t_0\}} A_{ij},$$

$$A_{ij} = \begin{cases} 75\%Probb_{ij}^{w}G\left(t_{ij}^{w}, t_0^{w}\right)\frac{s_j}{\sum_{k\in\{t_{ij}^{w}\le t_0^{w}\}} Prob_{b_{ij}^{w}} P_i} + 25\%Probb_{ij}^{p}G\left(t_{ij}^{p}, t_0^{p}\right)\frac{s_j}{\sum_{k\in\{t_{ij}^{p}\le t_0^{p}\}} Prob_{ij}^{p}P_i}, L = 1 \\ 50\%Probb_{ij}^{w}G\left(t_{ij}^{w}, t_0^{w}\right)\frac{s_j}{\sum_{k\epsilon\{t_{ij}^{w}\le t_0^{w}\}} Prob_{i_{ij}^{w}} P_i} + 50\%Probb_{ij}^{p}G\left(t_{ij}^{p}, t_0^{p}\right)\frac{s_j}{\sum_{k\in\{t_{ij}^{p}\le t_0^{p}\}}^{Prob} b_{ij}^{p}P_i}, L = 2 \\ 25\%Probb_{ij}^{w}G\left(t_{ij}^{w}, t_0^{w}\right)\frac{s_j}{\frac{Prob_{ij}^{w}P_i}{\sum_{k\epsilon\{t_{ij}^{w}\le t_0^{w}\}}}} + 75\%Probb_{ij}^{p}G\left(t_{ij}^{p}, t_0^{p}\right)\frac{s_j}{\sum_{k\in\{t_{ij}^{p}\le t_0^{p}\}} Prob_{ij}^{p}P_i}, L = 3 \end{cases}$$

In the formula: k represents all residential points i within the service area of green space j (i.e., satisfying $t_{ij} \le t_0$); t_{ij} denotes the travel time between residential point i and green space j; t_{ij}^{w} and t_{ij}^{p} represent the travel times for walking and public transportation, respectively. $Prob_{ij}$ denotes the probability of residents choosing to travel from settlement i to green space j. This indicator, based on the Huff model, quantifies the probability of residents selecting a green space service point by integrating two factors: green space attractiveness and travel distance [67]. Here, $Prob_{ij}^{w}$ and $Prob_{ij}^{p}$ respectively represent the probabilities of residents choosing to travel from i to j via walking and public transportation. This study introduces the Gaussian function G as a modified distance impedance coefficient, where $G(t_{ij}^{w}, t_0^{w})$ and $G(t_{ij}^{p}, t_0^{p})$ correspond to walking and public transportation modes, respectively. t_0 represents the travel time threshold for urban green spaces, determined based on the service radius and travel mode for different tiers of green spaces. Under this definition, t_0^{w} and t_0^{p} respectively denote the travel time thresholds for distance decay effects under walking and public transportation modes; simultaneously, t_0^{w} also serves as the critical time threshold at which residents abandon walking in favor of public transportation. For trip combinations ij where the walking travel time is below this critical threshold, residents typically disregard travel costs. At this point, the distance impedance coefficient G exhibits no decay effect, and residents will not opt for public transportation. P_i represents the population size of settlement i, and S_j denotes the area of green space j. The weights for walking and public transportation modes corresponding to different levels of green space (L) were determined based on the importance of various travel modes in urban green space commuting in Shanghai, referencing questionnaire survey results from existing research [68].

The formula for calculating $Prob_{ij}^{p}$ and the coefficient G is as follows:

$$Prob_{ij} = \frac{M_jG\left(t_{ij}, t_0\right)}{\sum_{h\in\{t_{ij}\le t_0\}} M_jG(t_{ij}, t_0)},$$

$$G(t_{ij}, t_0) = \{ \begin{matrix} 1, if\ t_{ij}^{w} \le t_0^{w} \\ \frac{e^{-(\frac{1}{2})\times(\frac{t_{ij}}{t_0})^2} - e^{-(\frac{1}{2})}}{1-e^{-(\frac{1}{2})}}\}, if\ t_{ij}^{w} > t_0^{w}\ and\ t_{ij} \le t_0 \\ 0, if\ t_{ij} > t_0 \end{matrix}$$

In the formula: M_j represents the attractiveness of urban green space j, quantified based on the star rating of the green space. In this study, the time threshold t_0^{w} for walking trips is set at 7.5 min (corresponding to a walking distance of 500 m at a walking speed of 4 km/h). Research indicates that 500 m is the distance threshold at which residents are more likely to choose walking over other modes of transportation [11]. This distance also aligns with Shanghai's target for enhancing the accessibility of urban green spaces.

3.3.2. Definition and Calculation of the Urban Green Space Comprehensive Carrying Capacity Index (UgsCCj)

The comprehensive carrying capacity index ($UgsCC_j$) of urban green space j characterizes its overall carrying capacity, calculated based on two key factors: the green space's

https://doi.org/10.3390/su18052195

attractiveness M_j and its visitor capacity. This index is derived from the ratio of green space j's area (S_j) to the per capita suitable area for L level green space ($Pcgsa^L$), as expressed by the following formula:

$$UgsCC_j = M_j \frac{S_j}{Pcgsa^L}$$

The standards for $Pcgsa$ in this study are shown in Tables 1 and 2.

Table 1. Per Capita Green Space Area (PCGSA), Service Radius, and Standards for Urban Green Space (UGS) Across Different City Levels.

Level (L)	Urban Green Space Level	Urban Green Space Area	Per Capita Green Space Area (PCGSA) Standard	Transportation Mode	t_0 (Threshold Travel Time)
1	Neighborhood Level	Less than 2 hectares	15 m^2	Walking	7.5 min
2	Community Level	2–10 hectares	30 m^2	Walking	15 min
				Driving or public transportation	20 min
3	City Level	10 hectares or more	60 m^2	Walking	30 min
				Driving or public transportation	40 min

Table 2. Transportation Network Based on Actual Travel Behavior Across Different Modes of Transportation.

Transportation Mode	Transportation Network	Transportation Behavior	Road or Route Type	Speed
Walking	Walking Network	Walking	Urban Roads—Major Arterials, Secondary Arterials, Collector Roads, Sidewalks	4 km per h
Public Transportation	Integrated Walking and Public Transit Network	Walking	Urban Roads—Major Arterials, Secondary Arterials, Collector Roads, Sidewalks	4 km per h
		Public Transportation—Bus	Downtown Bus Routes	20 km per h
			Suburban Bus Routes	30 km per h
			Downtown–Suburban Commuter Bus Routes (Highway)	70 km per h
			Downtown–Suburban Commuter Bus Routes (Non-Highway)	40 km per h
		Public Transportation—Ferry	Ferry Routes	20 km per h
		Public Transportation—Rail Transit	Maglev Rail	430 km per h
			Subway Lines 6, 10, 15	30 km per h
			Subway Lines 1, 4, 7, 8, 9, 11, 12, 14	35 km per h

3.3.3. Calculation Method for the Urban Green Space Population Demand Index (UgsPD)

UgsPD represents the population demand for each Urban Green Space (UGS) location. The comprehensive population demand index ($UgsPD_j$) for UGS location j is measured by the sum of population demands from all residential locations (k) within threshold time t_0 to j that can access UGS supply at location i. Its calculation formula is

$$UgsPD_j = \sum_{k \in \{t_{ij} \leq t_0\}} P_i$$

3.3.4. Calculation Method for the Urban Green Space Transportation Supply Index (UgsTS)

The Comprehensive Transportation Supply Index of Urban Green Space j ($UgsTS_j$) is used to characterize the traffic supply level of the green space. Its calculation method is as follows: for all residential areas k within the time threshold t_0 of green space j, superimpose the accessibility traffic supply volume from each residential area i to this green space. The formula is

https://doi.org/10.3390/su18052195

$$UgsTS_j = \begin{cases} \sum_{k\in\{t_{ij}\le t_0\}} \left[75\%G\left(t_{ij}^{w}, t_0^{w}\right) + 25\%G\left(t_{ij}^{p}, t_0^{p}\right)\right], L = 1 \\ \sum_{k\in\{t_{ij}\le t_0\}} \left[50\%G\left(t_{ij}^{w}, t_0^{w}\right) + 50\%G\left(t_{ij}^{p}, t_0^{p}\right)\right], L = 2 \\ \sum_{k\in\{t_{ij}\le t_0\}} \left[25\%G\left(t_{ij}^{w}, t_0^{w}\right) + 75\%G\left(t_{ij}^{p}, t_0^{p}\right)\right], L = 3 \end{cases}$$

4. Results

4.1. Analysis of Supply and Demand Characteristics of Urban Green Space Resources in Shanghai

As shown in Figure 3, the diagram illustrates various dimensions of Shanghai's urban green space services, population demand, and green space functions. The overall image reveals a mismatch in green space resources: high demand but low supply within the inner ring area and low demand but high supply in the outer ring area. Particularly concerning the spatial alignment between population demand and ecosystem service functions, the supply–demand imbalance is most pronounced in the central urban districts. Specifically: Figure 3a shows that core districts within the inner ring, such as Huangpu and Jing'an, possess strong service capacity, while outer ring areas are relatively weaker, exhibiting a decreasing trend from the core to the periphery. Figure 3b indicates that green space service capacity and transportation supply within the inner ring are largely balanced, whereas transportation supply in the outer and suburban rings exhibits pronounced deficiencies. Figure 3c reveals that population demand is high in the inner ring core areas, and there is strong demand for green spaces. However, in some areas, transportation supply capacity fails to keep pace with demand growth, resulting in a clear mismatch of high demand yet low accessibility. Figure 3d reveals that while inner ring areas exhibit high population demand, ecosystem service supply remains inadequate, whereas suburban ring areas like Songjiang and Qingpu, benefiting from high-quality natural endowments, demonstrate significant advantages in ecosystem service supply, thereby creating a marked contrast of abundant supply but low demand.

This study selected four representative urban parks in Shanghai to analyze their supply and demand characteristics, revealing significant spatial differentiation and functional specialization (Figure 4), as detailed below.

Wanli Park in Putuo District is located in the central urban area between the inner ring and middle ring. Despite the overall high population density in this area, the population demand index shown in Figure 3 exhibits a light shade, indicating low to medium values. This reflects that the park's actual recreational demand has not been fully activated, or that existing green spaces in the surrounding area have diverted potential visitors. On the supply side, both the park's comprehensive service capacity index and ecosystem service supply index register at low to moderate levels. This phenomenon directly correlates with the central urban area's constrained land resources, the park's relatively small scale, and its monotonous vegetation structure. These factors collectively result in weak ecological and recreational service supply capabilities. Overall, it exhibits a weak coupling state of low supply and low demand, representing a mismatch of supply and demand characterized by insufficient service efficiency. This occurs because limited supply capacity fails to attract more demand, while unmet demand struggles to drive supply upgrades.

Songjiang Central Park is situated in the core area of Songjiang New Town, between the outer ring and the suburban ring. With the influx of population into the new town, its population demand index remains in the medium-to-high range. The park maintains relatively stable supply capacity and delivers outstanding ecosystem services. Leveraging ample space reserved in the new town's master plan, its vegetation layout and landscape design excel in both ecological functionality and aesthetic value. While the park currently achieves a fundamental equilibrium between supply and demand, ongoing attention is

https://doi.org/10.3390/su18052195

required to dynamically align the pace of population growth with the green space's service carrying capacity.

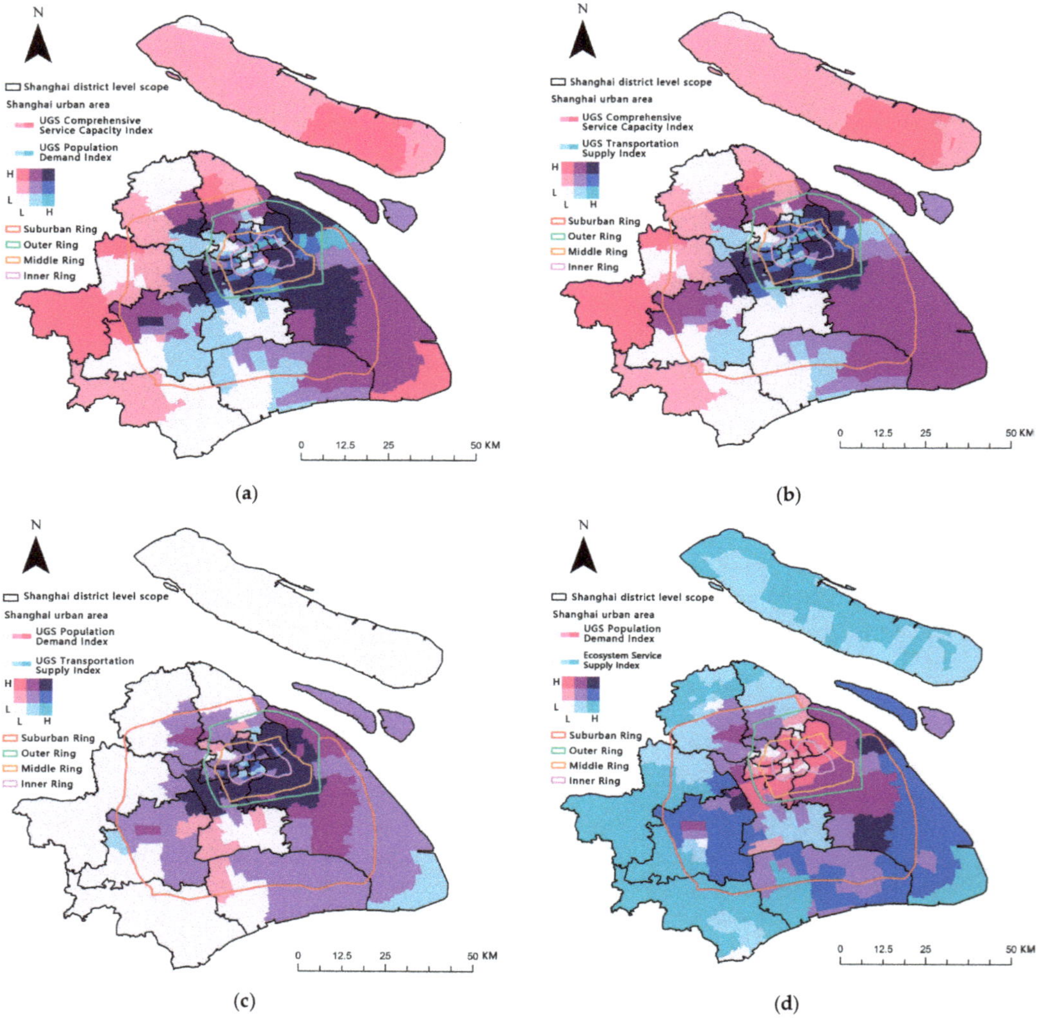

Figure 3. Shanghai Urban Green Space (UGS) Service Spatial Differentiation of Multi-dimensional Supply and Demand: (**a**) Matching distribution diagram of the UGS Comprehensive Service Capacity Index and the UGS Population Demand Index; (**b**) Matching distribution diagram of the UGS Comprehensive Service Capacity Index and the UGS transportation supply index; (**c**) Matching distribution diagram of the UGS population demand index and the UGS transportation supply index; (**d**) Matching distribution diagram of the UGS population demand index and the ecosystem service supply index.

Nanxiang Water Ecological Park is located in Nanxiang Town, Jiading District, situated between the outer ring and the suburban ring. The surrounding area primarily features mixed residential and industrial functions, with a medium-to-high population demand index. The park's comprehensive service capacity is moderate to high, with its water ecological features enabling good synergy between ecosystem services and recreational offerings. However, inadequate transportation connections significantly limit its service reach, resulting in a structural imbalance where ecological supply is sufficient but service accessibility is insufficient—a case of accessibility-constrained supply–demand mismatch.

Luchaogang Park is situated in the Lingang New Area along the outer suburban ring. Currently exhibiting low population density and a low population demand index, it demonstrates significant potential for demand growth alongside the ongoing development of the Lingang New Area. The park's ecological supply ranks high, with prominent cultural and aesthetic advantages stemming from its coastal setting. Its comprehensive service capacity is moderate, presenting an overall profile of advanced ecological provision and reserved service capacity. This configuration reserves ample ecological buffer space for future population influx.

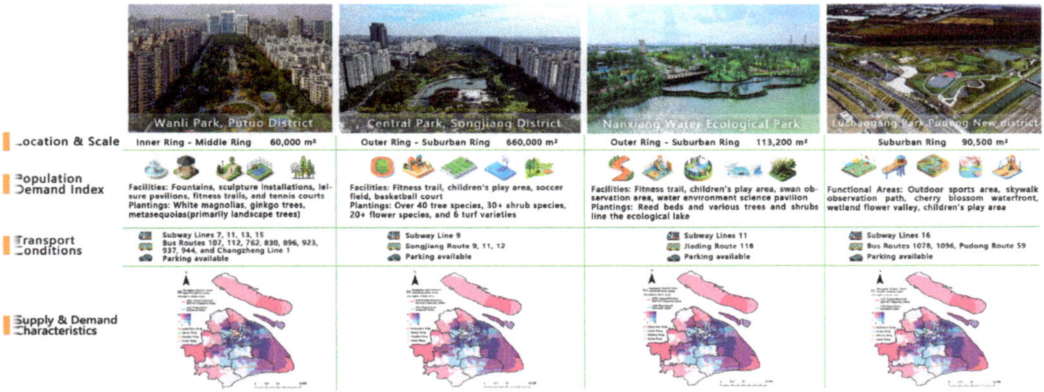

Figure 4. Typical Case Analysis Image.

In summary, from an overall perspective, the supply–demand match for urban parks in Shanghai exhibits distinct gradient characteristics: The low supply–low demand pattern observed at Wanli Park in Putuo District, a central urban area, reveals that some existing green spaces in central districts suffer from inadequate service efficiency. In the near-suburban new towns, Songjiang Central Park and Nanxiang Water Ecological Park demonstrate a medium demand–ecosystem service synergy transition pattern, with their supply–demand matching status mainly influenced by two key factors: the pace of population inflow and transportation accessibility. In the far-suburban Lingang area, Luchaogang Park demonstrates a low demand–ecologically pre-allocated reserve characteristic. The abundant ecosystem service supply provides critical ecological support for the future development of the Lingang New Area, reflecting the forward-looking nature of strategic urban spatial planning.

4.2. Analysis of Attractiveness Characteristics of Urban Green Spaces at Different Levels in Shanghai

From the perspective of spatial distribution characteristics (Figure 5), the comprehensive attractiveness of urban green spaces across different tiers in Shanghai exhibits a pattern of gradual diffusion from the core to the periphery. Neighborhood-level urban green spaces demonstrate highly concentrated attractiveness within the inner ring, particularly in core urban districts such as Huangpu and Xuhui, while areas within the outer ring and suburban ring generally maintain low to moderate attractiveness levels. This overall configuration

https://doi.org/10.3390/su18052195

reveals a pronounced single-core concentration pattern. At the community-level urban green space scale, high-attraction areas began extending toward the middle ring zone. Scattered points of extremely high attraction emerged in areas near the middle ring, such as Putuo District and Minhang District. However, most areas within the outer ring and beyond still showed no significant increase in attraction. When green space classification is elevated to the citywide level, the coverage of attractiveness further expands to the outer ring and even suburban ring areas. Multiple highly attractive points also emerge in peripheral districts like Baoshan and Fengxian, collectively forming a spatially dispersed multi-node pattern.

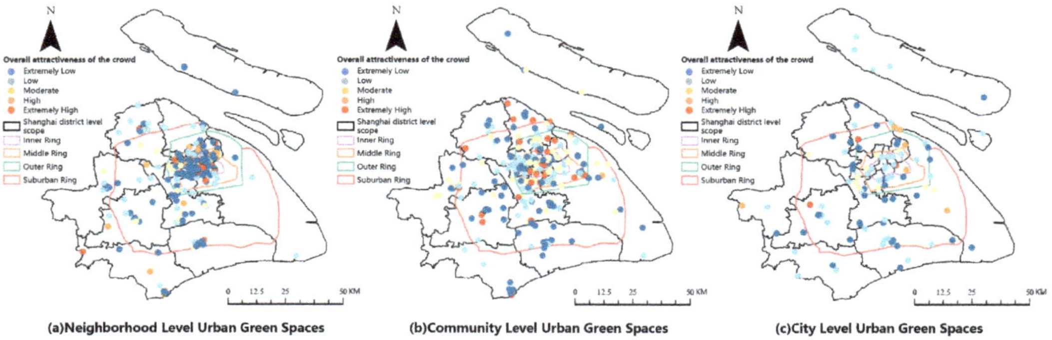

Figure 5. Overall Attractiveness of the Crowd at Different Urban Green Space Levels in Shanghai.

From the perspective of differences in ring zone scope, the attractiveness of green spaces across various tiers exhibits significant divergence across each ring zone. The inner ring zone maintains high or extremely high attractiveness levels for all three tiers of green spaces, serving as the core advantage area for all types. Attractiveness in the middle ring zone increases with green space tier: neighborhood-level spaces predominantly offer medium–high attractiveness, while community-level and city-level spaces reach high or extremely high levels. The outer ring exhibits more pronounced changes: neighborhood-level green spaces show only low-to-moderate appeal, rising to moderate-to-high at the community level, and reaching high or extremely high at the city level. The suburban ring's appeal deficit requires higher-grade green spaces to compensate, with only city-level green spaces elevating its appeal to moderate-to-high levels. At the neighborhood and community levels, its appeal generally remains low or low-to-moderate.

From the perspective of the correlation between green space tiers and attractiveness, as green spaces upgrade from neighborhood-level to city-level, the spatial coverage of their attractiveness expands progressively from the inner ring to the suburban ring, while the distribution of high-attractiveness areas gradually shifts from a concentrated pattern in the urban core to a more dispersed pattern toward peripheral areas. Meanwhile, the higher the green space level, the more significant its role in enhancing attractiveness in urban peripheral areas. City-level green spaces effectively compensate for the insufficient attractiveness of lower-level green spaces in peripheral zones, serving as key carriers for attracting populations to these areas. In contrast, lower-level green spaces primarily focus on strengthening localized attractiveness within the urban core, thereby meeting residents' demand for nearby recreational opportunities.

4.3. Assessing the Distribution Characteristics of Different Levels of UGS by 2030

The current comprehensive capacity of Shanghai's UGS exhibits a tiered differentiation: city-level green spaces possess higher comprehensive capacity but are limited in number;

https://doi.org/10.3390/su18052195

community-level green spaces maintain moderate comprehensive capacity and serve as the primary coverage for the population; neighborhood-level green spaces demonstrate relatively low comprehensive capacity and are predominantly characterized by small-scale, dispersed layouts. These patterns align closely with the common development characteristics observed in green spaces across international megacities [69].

As shown in Figure 6, between 2023 and 2030, green spaces of different levels in Shanghai exhibit significant evolutionary characteristics in terms of spatial distribution, ecosystem service capacity, supply–demand matching, and functional structure. During this development process, green spaces at different levels demonstrate differentiated development trends as well as distinct existing issues.

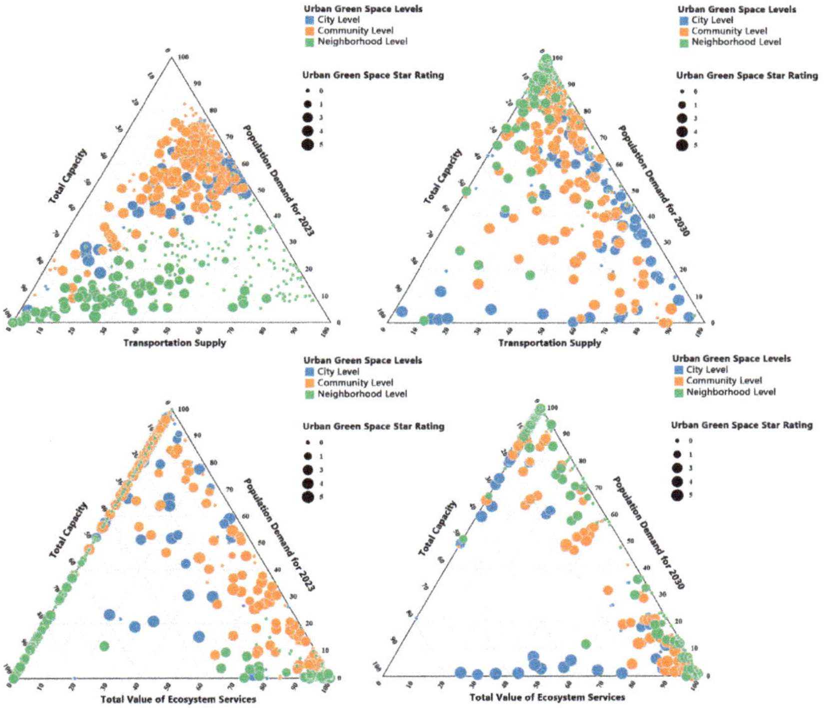

Figure 6. Changes in the Distribution Characteristics of Different Levels of Green Space in Shanghai in 2023 and 2030.

For neighborhood-level green spaces, the distribution pattern in 2023 was mainly characterized by high comprehensive capacity (60–100%) and moderate transportation supply levels (20–60%), with overall service capacity and coverage remaining relatively limited. With the continuous advancement of urban renewal and residential development, the layout of neighborhood-level green spaces has gradually extended toward areas with high population density and strong ecological demand, with service supply becoming more aligned with residents' daily needs for recreation, leisure, fitness, and nearby use. By 2030, the corresponding population demand level of neighborhood-level green spaces is projected to cover the full range of 0–100%, and service coverage is expected to expand significantly compared with previous conditions, showing notable progress in meeting residents' demand for nearby green spaces. However, in terms of overall construction scale, per capita service level, and comprehensive service efficiency, neighborhood-level green

https://doi.org/10.3390/su18052195

spaces still face challenges such as insufficient total supply, uneven spatial distribution, and the need for improved service quality, making it difficult to fully match the continuously growing population demand and residents' increasing expectations for high-quality living environments.

By 2030, community-level green spaces are expected to exhibit clear characteristics of functional upgrading and quality improvement, with the ecosystem service range further expanding to 50–90% compared with previous levels. Service functions become more diversified, while ecological benefits, recreational value, and social benefits are simultaneously enhanced. Correspondingly, the population demand range increases from 50–80% to 60–95%, indicating that community-level green spaces achieve effective expansion in service coverage and population reach, with simultaneous improvements in service level and overall performance, thereby playing an increasingly important supporting role in building community living circles, improving residential environmental quality, and enhancing residents' well-being. However, in some areas, transportation supply levels remain relatively low at only 10–40%. Issues such as insufficient accessibility to green spaces, weak transportation connections, and incomplete slow-mobility systems remain prominent, which to some extent constrain the full realization of green space service efficiency and become key bottlenecks limiting functional performance.

By 2030, city-level green spaces are expected to achieve a significant increase in total ecosystem service value, with a more optimized ecological structure and enhanced comprehensive ecological benefits, landscape value, and social service functions, thereby occupying an important position in the urban ecological security pattern and green space system. However, the corresponding population demand proportion remains mainly concentrated within the 0–40% range, indicating a relatively low spatial coupling with areas of high population density and high service demand and reflecting a certain mismatch between green space services and population distribution. The overall spatial layout still shows a certain degree of dispersion, and substantial room for improvement remains in terms of service balance, spatial equity, and supply–demand matching. Therefore, future urban green space system planning needs to further strengthen spatial layout regulation and service enhancement.

5. Discussion

5.1. An In-Depth Analysis of the Underlying Causes of Supply–Demand Matching Characteristics in Shanghai's Green Space Resources

The mismatched pattern within Shanghai's inner ring—characterized by high demand and low supply—and the outer ring's low demand and high supply stems from the long-term interplay of multiple factors. These include core elements such as land use prioritization during urbanization, historical planning legacies, and the asynchrony between population growth and spatial expansion, alongside natural endowments and planning directives.

(1) The inherent constraints of the evolutionary logic of urbanization. The development of Shanghai's central urban area originated within the inner ring. During the early industrialization phase, industrial layout and population concentration were the core priorities for land use, squeezing out ecological space and forming a pattern that prioritized production over ecology. As urbanization deepens, population density within the inner ring increases and urban functions become concentrated. With limited scope for renovating existing green spaces and constraints on new green areas due to high land costs and difficulties in relocation, the growth of ecosystem service supply lags behind the expansion of demand [70]. In highly market-oriented urban core areas, the reallocation of land resources often faces extremely high opportunity

costs and transaction costs [71]. Due to the extremely high land value in the inner ring area of Shanghai, the addition or expansion of green spaces typically involves complex processes such as land acquisition, functional replacement, and property rights negotiations, which means that even when planning objectives are clearly defined, implementation is often hindered by excessively high costs. In addition, the existing property rights structure further reduces the financial attractiveness of high-quality redevelopment projects for district-level governments and market entities [72]; property ownership is fragmented among multiple stakeholders, and the institutional costs associated with coordinating property rights further constrain the contiguous provision of green spaces. The outer ring and suburban ring areas were developed later, featuring low land use intensity and preserving a sound natural ecological foundation. Designated as ecological reserve zones, they offer ample green space supply. However, their distance from core functional zones and low population density have led to relative oversupply—essentially reflecting a mismatch between population growth and spatial expansion.

(2) The phased orientation of planning policies and historical legacy issues has further entrenched this mismatch. Early green space planning focused solely on ecological conservation, disconnecting from population distribution patterns—a problem difficult to swiftly reverse. Although the Shanghai Ecological Space Special Plan (2021–2035) explicitly sets a "500 m accessibility" target, spatial constraints in central urban areas [73] and issues such as sparse distribution and small scale of neighborhood-level green spaces [74] persist. In addition, city-level green spaces are predominantly located in suburban areas, further widening disparities in green space accessibility across different regions. Meanwhile, the planning shortcoming of emphasizing construction over operation has led to problems in some outer ring green spaces, such as inadequate facility maintenance and insufficient transportation connections. As a result, even with sufficient ecological supply capacity, their service effectiveness is difficult to fully realize.

(3) Spatial differentiation in population and socioeconomic development serves as the core driving factor. The inner and middle rings concentrate high-quality public services and employment opportunities, attracting dense populations. Young families and working professionals exhibit strong demand for premium green space services, yet the area predominantly features small pocket parks or outdated green spaces whose scale, functionality, and facilities fall short of meeting these needs. The outer and suburban rings exhibit lower population density and weaker demand intensity, resulting in a more pronounced supply–demand disparity. Songjiang Central Park, with its 660,000-square-meter footprint, composite spatial structure, convenient transportation, and round-the-clock operation, serves the needs of 200,000 permanent residents. This demonstrates that scale, functionality, accessibility, and management are key factors in adapting large-scale parks. In contrast, Putuo District's Wanli Park, with its mere 6-hectare space, single-function design, and operational shortcomings, highlights the core requirements for small-scale green spaces: spatial efficiency, functional diversity, and robust operational support.

(4) The spatial heterogeneity of ecological foundations further intensifies the pattern of supply–demand differentiation. The suburban ring boasts premium resources like the Sheshan Ecological Belt and Dianshan Lake, endowed with significant ecosystem service advantages. In contrast, the central urban area has suffered from long-term human intervention, resulting in inherently deficient ecosystem service capacity. Even with newly added green spaces, their effectiveness struggles to match that of the suburban ring. This disparity, compounded by the scale and functional constraints of

https://doi.org/10.3390/su18052195

different green space types, ultimately creates a dual mismatch: "high demand but low supply efficiency in the inner ring, and ample supply but weak demand intensity in the outer ring." The four major parks selected above provide micro-level empirical evidence for interpreting this mismatched pattern.

5.2. Mechanistic Explanation of Attractiveness and Accessibility Characteristics of UGS at Different Levels

The spatial patterns of green space attraction—characterized by core concentration at lower levels and peripheral diffusion at higher levels—stem fundamentally from the combined effects of green space attributes, transportation accessibility, and resident travel preferences [75,76]. The hierarchical structure—manifesting as single-core concentration at the neighborhood level, middle-ring extension at the community level, and multi-node dispersion at the city level—is closely linked to differences in service radius, functional complexity, and travel costs across these tiers:

(1) Neighborhood-level green spaces operate within a 7.5 min walking radius, focusing on daily, nearby recreation with infrastructure as their primary function. The inner ring core area features dense populations and a well-developed pedestrian network, resulting in highly efficient green space coverage, comprehensive facilities, and concentrated attractiveness. In contrast, the outer and suburban rings exhibit sparse populations and insufficient road density, leading to scattered green space distribution, rudimentary facilities, limited walkability, and generally low attractiveness, forming a single-core pattern.
(2) Community-level green space service radius expanded to a 15 min walk or 20 min public transit threshold, integrating multifunctional services such as sports fitness and cultural activities. The middle ring, serving as a transitional zone between the central city and outer areas, features moderate population density and a well-developed transportation network, keeping residents' travel costs manageable. Some green spaces have been upgraded into regional leisure hubs, extending the city's appeal into the middle ring. The outer ring, however, still maintains a low-to-medium level of appeal due to insufficient public transportation coverage and higher travel costs.
(3) City-level green spaces are accessible within a 30 min walk or 40 min public transit threshold, combining scarce ecological resources with diverse service functions to generate significant cross-regional appeal. As rail transit extends into suburban areas, city-level green spaces in the outer ring reduce travel costs through transit connections. Coupled with the uniqueness of premium natural landscapes, this creates a dispersed multi-node pattern.

Current accessibility and alignment with population demand exhibit significant shortcomings: Low-level green spaces are characterized by proximity but insufficient capacity. While neighborhood-level and community-level green spaces offer nearby accessibility, their limited overall capacity means their facility provision and ecosystem service levels struggle to match the robust population demand in core areas. Higher-level green spaces, however, are abundant but distant. While city-level green spaces have strong supply capacity, their scattered distribution and inadequate transportation connections make it difficult to fully meet residents' cross-regional usage needs. In the 2030 scenario, the core drivers for improved adaptability stem from increased green space capacity and optimized layout. Community-level green spaces will upgrade toward higher capacity and enhanced ecosystem service levels, neighborhood-level green spaces will expand their coverage, and city-level green spaces will focus on improving transportation accessibility. This will ultimately form a spatially complementary hierarchical structure. Furthermore, residents' travel preferences will intensify spatial differentiation. Older adults and low-income

https://doi.org/10.3390/su18052195

groups will increasingly favor walkable neighborhood-level green spaces, reinforcing the concentration of core areas [77]. Meanwhile, younger adults and high-income groups will be willing to incur travel costs for high-quality green services, extending the appeal of community-level and city-level green spaces outward. This will further solidify the pattern of central concentration and peripheral expansion.

From the perspective of green space hierarchy in global megacities, the green space system of London is centered on the Green Belt, forming an ecological barrier through peri-urban open spaces. However, the Green Belt policy has long focused primarily on controlling urban sprawl, resulting in fragmented layouts of community- and neighborhood-level green spaces and weak connectivity between the Green Belt and internal urban green spaces, showing a pattern of concentrated large green spaces in the periphery and scattered small green spaces within the city [78]. In contrast, New York is centered on city-level landmark green spaces and a network of community parks. While the ecological and recreational functions of city-level green spaces are well integrated, the distribution of green spaces shows significant inequality [79]. Both London and New York share a common issue of emphasizing core landmark green spaces while underdeveloping grassroots green space networks.

In comparison, the uniqueness of Shanghai lies in two aspects. First, it explicitly constructs a three-tier linkage system of city-level, community-level and neighborhood-level green spaces, rather than relying solely on single-point control or landmark-based models, and incorporates 500 m accessibility into statutory planning, thereby strengthening the service function of grassroots green spaces. Second, under the SSP1-2.6 scenario, it forms a supply framework characterized by city-level dominance, community-level support, and neighborhood-level supplementation, breaking through the limitations of London's protection-oriented but service-limited model and New York's core-focused but grassroots-weak pattern, thus achieving an upgrade from spatial distribution to functional coordination.

The Shanghai model provides valuable reference for global megacities: green space planning should balance ecological protection and residents' needs rather than following a single development orientation; it should strengthen the linkage and connectivity among green spaces of different hierarchical levels; and statutory planning should explicitly define grassroots green space supply standards to alleviate supply–demand imbalances in core urban areas and insufficient accessibility in suburban areas. This also aligns with the emerging trend of transforming London's Green Belt toward multifunctionality.

5.3. The Significance of Green Space Ecosystem Service Supply Evolution Under the SSP1-2.6 Scenario by 2030

Under the SSP1-2.6 scenario for 2030, the evolution of Shanghai's green space ecosystem service supply reflects synergistic effects achieved through land use optimization guided by sustainable development principles, alongside coordinated advancement of ecological conservation and urban development. This holds significant importance for building ecologically livable cities, achieving carbon peak targets, and enhancing residents' well-being. A tiered ecosystem service supply framework—dominated at the city level, supported at the community level, and supplemented at the neighborhood level—is gradually taking shape. Its rationality and core values are particularly evident in the practice of sustainable development goals.

The SSP1-2.6 scenario represents an ideal pathway balancing sustainable development and low-carbon emissions. Based on this scenario, this study conducted simulation analyses of green space ecosystem services in Shanghai. However, in real-world development, various uncertainties—such as lower-than-expected policy implementation effectiveness and population growth exceeding planning projections—may affect the practical applica-

https://doi.org/10.3390/su18052195

bility of the findings. Therefore, building on existing climate scenario research, this study further supplements the analysis by examining changes in land use and ecosystem services under the SSP2-4.5 medium-development scenario [80]. The results indicate that under SSP2-4.5, increased urban land development intensity and expansion of construction land will further encroach on green space, significantly slowing the growth of ecosystem service supply. Meanwhile, continued population concentration will further increase urban green space demand, intensifying the contradiction of high demand and low supply within the inner ring and exacerbating supply–demand spatial mismatches in outer ring areas.

In terms of spatial patterns, the overall attractiveness pattern of green spaces of different hierarchical levels does not fundamentally change, but disparities in attractiveness among different urban rings continue to widen. Service pressure on community- and neighborhood-level green spaces increases significantly, while city-level green spaces exhibit more pronounced accessibility limitations due to dispersed layouts and insufficient transportation connectivity. Nevertheless, the green space ecosystem service supply framework proposed in this study—characterized by city-level dominance, community-level support, and neighborhood-level supplementation—remains valid under SSP2-4.5, requiring only strengthened implementation measures, such as accelerating the provision of neighborhood-level green spaces in core urban areas and enhancing the functions and facilities of community-level green spaces. Overall, the core conclusions of this study demonstrate strong robustness: even when the development scenario shifts from the ideal SSP1-2.6 to the moderate SSP2-4.5, the fundamental understanding of green space supply–demand characteristics and spatial patterns remains unchanged, with only flexible adjustments needed in planning implementation strategies across scenarios.

From the perspective of spatial adaptability, this model overcomes the fragmentation of traditional green space ecosystem services, achieving a transition toward structured and networked provision. It enhances ecosystem stability and service efficiency, precisely aligning with core demands for carbon peaking and ecological livability, thereby demonstrating significant dual ecological and social value. Ecologically, city-level green spaces reinforce regional ecological barriers and carbon sink functions, community-level green spaces bridge ecological gaps and transmit ecological benefits, while neighborhood-level green spaces ensure universal service access. Together, they form a comprehensive system of ecosystem services covering the entire area. Socially, differentiated functional configurations create a service network aligned with residents' daily life radii, enhancing living satisfaction and well-being.

Enhancing the alignment between green space ecosystem service supply and population demand offers an effective pathway to alleviate urban ills and support the optimization of the spatial structure encompassing the main urban area and five new towns. Neighborhood-level green spaces in the core area should be optimized to cover high-density population zones, consolidating the accessibility of recreational green spaces and alleviating the imbalance between ecosystem service supply and demand. Community-level green spaces in the five new cities should be densely distributed to accommodate population influx, enhancing livability appeal while reducing pressure on city-level green spaces. This approach will foster a spatial development pattern characterized by the revitalization of the main urban area and the rise of new cities.

The synergistic evolution of green spaces across different tiers provides quantitative support and spatial pathways for achieving carbon peak and carbon neutrality goals. Under the SSP1-2.6 scenario, this framework establishes a multi-tiered, comprehensive carbon sink spatial system: city-level green spaces enhance carbon storage functions, becoming regional carbon sink cores; community- and neighborhood-level green spaces fill carbon sink gaps in built-up areas, boosting overall carbon sink efficiency. Concurrently, upgraded ecosystem

https://doi.org/10.3390/su18052195

services indirectly drive energy structure optimization and foster low-carbon lifestyles. Through dual-engine propulsion—enhanced carbon sinks and low-carbon guidance—this approach provides comprehensive support for achieving carbon peak targets.

6. Conclusions

Guided by the SSP1-2.6 scenario, this study systematically investigates the supply levels, accessibility characteristics, and supply–demand matching relationships of urban green space ecosystem services in Shanghai by integrating the PLUS model, InVEST model, and nSFCA method. This research provides scientific basis for optimizing urban green space planning. Findings reveal a significant mismatch in Shanghai's green space resources: high demand but low supply within the inner ring and low demand but high supply in the outer ring. The contradiction between population concentration in central urban areas and insufficient ecosystem service supply is particularly pronounced. Meanwhile, the suburban ring area possesses superior ecological endowments but exhibits weaker demand intensity. The attractiveness of green spaces at different hierarchical levels follows a spatial pattern of core concentration and peripheral diffusion. Neighborhood-level green spaces cluster around the inner ring core, forming a single-core pattern. Community-level green spaces extend toward the middle ring, while city-level green spaces are distributed across the entire city in a multi-node, dispersed pattern. These differences are closely related to hierarchical variations in service radius, functional complexity, and travel costs.

The 2030 scenario simulation indicates that Shanghai's green space ecosystem services will form a coordinated framework characterized by city-level dominance, community-level support, and neighborhood-level supplementation. Neighborhood-level green space coverage will continue to expand, community-level green spaces will undergo functional upgrades and capacity enhancements, and the total value of city-level green space ecosystem services will significantly increase. Together, these three tiers will progressively improve the alignment between ecosystem service supply and population demand. However, it should be noted that accessibility gaps persist: low-grade green spaces are often nearby but insufficient, while high-grade green spaces are sufficient but not easily accessible. Transportation connections and facility configurations in some areas still require optimization.

From the perspective of driving mechanisms, the mismatch between green space supply and demand, along with divergent attractiveness, stems from core factors such as urbanization processes, planning policy orientations, and the asynchrony between population growth and spatial expansion. These are the result of long-term interplay among multiple factors, including natural endowments and planning directives. Comparative case studies of Songjiang Central Park and Putuo District's Wanli Park further validate the critical impact of green space scale, functional configuration, transportation accessibility, and operational management on supply–demand alignment. The multidimensional, multi-tiered green space supply assessment framework developed in this study effectively integrates ecosystem service supply and accessibility factors, offering a novel technical pathway for green space planning in megacities. Moving forward, Shanghai must continuously optimize green space layouts, enhance functional coordination and transportation connectivity among green spaces at different tiers, prioritize addressing neighborhood-level green space supply gaps in central urban areas, and simultaneously boost the demand–absorption capacity of green spaces in suburban ring zones. This integrated approach will ultimately advance the synergistic development of an ecologically livable city and carbon peak goals.

Although this study has achieved certain results in terms of methodological coupling, dimensional coverage, and scenario simulation, shortcomings remain. First, the InVEST model has inherent limitations in fine-scale assessment, including coarse land-use represen-

https://doi.org/10.3390/su18052195

tation and parameter generalization, and it has limited capability in characterizing cultural and recreational services. Second, demand-side factors have not adequately addressed population-specific needs or temporal fluctuations in demand [81], and uncertainties in population projections may influence future supply–demand matching results; overestimation of population size may lead to misjudgment of insufficient supply in certain areas, whereas underestimation may conceal actual supply gaps. Third, transportation network simulations did not sufficiently account for variations in travel efficiency or emerging mobility modes, such as shared bicycles and electric scooters [82,83]. These modes may improve green space accessibility in peripheral areas but may also increase congestion and safety risks in urban core areas, potentially affecting the accuracy of overall supply–demand matching assessments. In addition, refined characteristics of public transportation are not fully captured. Future research could integrate mobile signaling and travel platform data for dynamic assessment to better reflect real-time travel behavior and green space use patterns [84]. Furthermore, future scenario simulations rely on rule-based and parameter-driven settings in the PLUS and InVEST models, limiting their ability to capture resident behavioral heterogeneity and the dynamic impacts of unexpected events (e.g., policy changes or extreme weather) on land use and ecosystem services. Future studies could incorporate agent-based models (ABM) to simulate individual decision-making, including interactions among residents' travel preferences, developers' land-use choices, and government planning interventions, thereby improving micro-level realism and robustness. Previous studies have identified related single-factor mechanisms through empirical or case studies, such as demand heterogeneity, the influence of emerging transport modes on spatial accessibility, and the potential of mobile data-based dynamic assessment. By developing a three-dimensional coupling framework integrating ecological supply, population demand, and transportation accessibility, and coupling the PLUS, InVEST, and nSFCA models, this study enables dynamic scenario simulation of green space supply–demand adaptation in megacities and enhances multi-factor integrated analysis and scenario prediction capacity.

Supplementary Materials: The following supporting information can be downloaded at: https://www.mdpi.com/article/10.3390/su18052195/s1, Table S1: Participation in the MaxEnt modeling factor; Table S2: Selection basis for supply services in urban green ecosystem; Table S3: Carbon density of carbon pools in Shanghai (t/ha); Table S4: Biophysical table using water yield (Shanghai); Table S5: P value and C value of different land use types in the USLE models (Shanghai); Table S6: Habitat suitability and habitat sensitivity to threats (Shanghai); Table S7: Weights and effective distances of degradation sources (Shanghai) [85–92].

Author Contributions: Conceptualization, J.H.; methodology, J.H. and S.C.; software, S.C. and C.S.; writing—original draft preparation, J.H. and C.S.; writing—review and editing, S.C. and M.Y.; visualization, H.C.; supervision, Z.D.; funding acquisition, Z.D. All authors have read and agreed to the published version of the manuscript.

Funding: This research work was supported by the Special Funding Project of the China Agricultural and Forestry University Design Art Alliance under Grant No. 111900050.

Institutional Review Board Statement: Not applicable.

Informed Consent Statement: Not applicable.

Data Availability Statement: The original contributions presented in this study are included in the article. Further inquiries can be directed to the corresponding author.

Conflicts of Interest: The authors declare no conflicts of interest.

https://doi.org/10.3390/su18052195

References

Gao, K. Research on Green Space Ecosystem Services under the Guidance of Healthy City Orientation. *Int. J. Soc. Sci. Public Adm.* **2024**, *2*, 445–455. [CrossRef]

Zhao, L. Environmental ecosystem services assessment based on urban green infrastructure structure indicators with case study in Shanghai. In *Proceedings of the E3S Web of Conferences*; EDP Sciences: Les Ulis, France, 2019; p. 03003.

Goodwin, S.; Olazabal, M.; Castro, A.J.; Pascual, U. Global mapping of urban nature-based solutions for climate change adaptation. *Nat. Sustain.* **2023**, *6*, 458–469. [CrossRef]

Guterres, A. *The Sustainable Development Goals Report 2020*; United Nations Department of Economic and Social Affairs: New York, NY, USA, 2020. [CrossRef]

Lan, M.; Yuan, L.; Guo, R.; Wu, Y.; Liu, X. Sustainable development: Investigating the correlations between park equality and mortality by multilevel model in Shenzhen, China. *Sustain. Cities Soc.* **2021**, *75*, 103385. [CrossRef]

Sharifi, F.; Nygaard, A.; Stone, W.M.; Levin, I. Accessing green space in Melbourne: Measuring inequity and household mobility. *Landsc. Urban Plan.* **2021**, *207*, 104004. [CrossRef]

Zhang, R.; Zhang, C.-Q.; Cheng, W.; Lai, P.C.; Schüz, B. The neighborhood socioeconomic inequalities in urban parks in a High-density City: An environmental justice perspective. *Landsc. Urban Plan.* **2021**, *211*, 104099. [CrossRef]

Wang, S.; Yung, E.H.K.; Sun, Y. Effects of open space accessibility and quality on older adults' visit: Planning towards equal right to the city. *Cities* **2022**, *125*, 103611. [CrossRef]

Zepp, H.; Groß, L.; Inostroza, L. And the winner is? Comparing urban green space provision and accessibility in eight European metropolitan areas using a spatially explicit approach. *Urban For. Urban Green.* **2020**, *49*, 126603. [CrossRef]

Seifu, S.; Stellmacher, T. Accessibility of public recreational parks in Addis Ababa, Ethiopia: A GIS based analysis at sub-city level. *Urban For. Urban Green.* **2021**, *57*, 126916. [CrossRef]

Wolff, M. Taking one step further–Advancing the measurement of green and blue area accessibility using spatial network analysis. *Ecol. Indic.* **2021**, *126*, 107665. [CrossRef]

Park, Y.; Guldmann, J.-M. Understanding disparities in community green accessibility under alternative green measures: A metropolitan-wide analysis of Columbus, Ohio, and Atlanta, Georgia. *Landsc. Urban Plan.* **2020**, *200*, 103806. [CrossRef]

Liang, H.; Zhang, Q. Assessing the public transport service to urban parks on the basis of spatial accessibility for citizens in the compact megacity of Shanghai, China. *Urban Stud.* **2017**, *55*, 1983–1999. [CrossRef]

Guo, S.; Song, C.; Pei, T.; Liu, Y.; Ma, T.; Du, Y.; Chen, J.; Fan, Z.; Tang, X.; Peng, Y. Accessibility to urban parks for elderly residents: Perspectives from mobile phone data. *Landsc. Urban Plan.* **2019**, *191*, 103642. [CrossRef]

Zheng, Z.; Shen, W.; Li, Y.; Qin, Y.; Wang, L. Spatial equity of park green space using KD2SFCA and web map API: A case study of zhengzhou, China. *Appl. Geogr.* **2020**, *123*, 102310. [CrossRef]

Zhang, J.; Cheng, Y.; Zhao, B. Assessing the inequities in access to peri-urban parks at the regional level: A case study in China's largest urban agglomeration. *Urban For. Urban Green.* **2021**, *65*, 127334. [CrossRef]

Zhang, W.; Li, S.; Gao, Y.; Liu, W.; Jiao, Y.; Zeng, C.; Gao, L.; Wang, T. Travel changes and equitable access to urban parks in the post COVID-19 pandemic period: Evidence from Wuhan, China. *J. Environ. Manag.* **2022**, *304*, 114217. [CrossRef]

Cai, G.; Xiong, J.; Wen, L.; Weng, A.; Lin, Y.; Li, B. Predicting the ecosystem service values and constructing ecological security patterns in future changing land use patterns. *Ecol. Indic.* **2023**, *154*, 110787. [CrossRef]

Gou, M.; Li, L.; Ouyang, S.; Wang, N.; La, L.; Liu, C.; Xiao, W. Identifying and analyzing ecosystem service bundles and their socioecological drivers in the Three Gorges Reservoir Area. *J. Clean. Prod.* **2021**, *307*, 127208. [CrossRef]

Boulton, C.; Dedekorkut-Howes, A.; Byrne, J. Factors shaping urban greenspace provision: A systematic review of the literature. *Landsc. Urban Plan.* **2018**, *178*, 82–101. [CrossRef]

Sun, C.; Song, X.; Cheng, J.; Peng, M.; Li, X. An equitable spatial strategy for urban green space based on reconciling growing and shrinking neighbourhoods at the parcel level. *Land Use Policy* **2025**, *158*, 107713. [CrossRef]

Ai, X.; Zheng, X.; Zhang, Y.; Liu, Y.; Ou, X.; Xia, C.; Liu, L. Climate and land use changes impact the trajectories of ecosystem service bundles in an urban agglomeration: Intricate interaction trends and driver identification under SSP-RCP scenarios. *Sci. Total Environ.* **2024**, *944*, 173828. [CrossRef]

Lin, G.; Lan, S.; Chi, B.; Wu, K.; Liu, Y.; Li, X. Synergistic optimization of ecological supply–demand: Integrating urban green space ecosystem services and dual ecological demands across multiple scenarios. *Ecol. Indic.* **2025**, *178*, 114133. [CrossRef]

Chen, Q.; Ning, Y. Projecting LUCC dynamics and ecosystem services in an emerging urban agglomeration under SSP-RCP scenarios and their management implications. *Sci. Total Environ.* **2024**, *949*, 175100. [CrossRef] [PubMed]

Huang, J.; Zheng, F.; Dong, X.; Wang, X.-C. Exploring the complex trade-offs and synergies among ecosystem services in the Tibet autonomous region. *J. Clean. Prod.* **2023**, *384*, 135483. [CrossRef]

Zhang, J.; Liu, Z.; Guan, Z.; Wang, L.; Zhang, J.; Han, Z. Balancing future urban development and carbon sequestration: A multi-scenario InVEST model analysis of China's urban clusters. *J. Environ. Manag.* **2025**, *380*, 125003. [CrossRef]

https://doi.org/10.3390/su18052195

27. Huang, Z.; Qian, L.; Cao, W. Developing a novel approach integrating ecosystem services and biodiversity for identifying priorit ecological reserves. *Resour. Conserv. Recycl.* **2022**, *179*, 106128. [CrossRef]
28. Lyu, K.; Li, Z. Scenario based assessment of carbon storage and habitat quality under land use change in Shandong Provinc China. *Sci. Rep.* **2025**, *15*, 38098. [CrossRef]
29. Zhang, Q.; Huang, R.; Zhu, C.; Huang, L.; Yang, D. Integrating land use simulation and carbon assessment for sustainable urba planning in Fuzhou metropolitan area using PLUS and InVEST models. *Sci. Rep.* **2025**, *15*, 30382. [CrossRef]
30. O'Neill, B.C.; Tebaldi, C.; van Vuuren, D.P.; Eyring, V.; Friedlingstein, P.; Hurtt, G.; Knutti, R.; Kriegler, E.; Lamarque, J.-F.; Lowe, The Scenario Model Intercomparison Project (ScenarioMIP) for CMIP6. *Geosci. Model Dev.* **2016**, *9*, 3461–3482. [CrossRef]
31. Cui, M.; Hong, C.; Liu, W.; He, K.; Zhang, Q. Future warming exacerbates heatwave-ozone compound extremes in China. *np Clim. Atmos. Sci.* **2025**, *8*, 341. [CrossRef]
32. Usta, D.F.B.; Teymouri, M.; Chatterjee, U. Assessment of temperature changes over Iran during the twenty-first century usin CMIP6 models under SSP1-26, SSP2-4.5, and SSP5-8.5 scenarios. *Arab. J. Geosci.* **2022**, *15*, 416. [CrossRef]
33. Zhang, Y.; Cui, L.; Xie, D.; Jiang, J. Simulation and prediction of sea level rise impact on the distribution of mangrove and Spartin alterniflora in coastal China. *Forests* **2023**, *14*, 831. [CrossRef]
34. Sun, Y.; Xie, S.; Zhao, S. Valuing urban green spaces in mitigating climate change: A city-wide estimate of aboveground carbo stored in urban green spaces of China's Capital. *Glob. Change Biol.* **2019**, *25*, 1717–1732. [CrossRef] [PubMed]
35. Li, Y.-Y.; Ren, B.-T.; Chen, Y.-S.; Huang, L.-C.; Sun, C.-G. Multiscale spatiotemporal dynamics analysis of urban green spac Implications for green space planning in the rapid urbanizing Hefei City, China. *Front. Ecol. Evol.* **2022**, *10*, 998111. [CrossRef]
36. Zhao, W.; Li, X.; Zhang, L.; Peng, L.; Liu, Y.; Wang, Z.; Jiao, L.; Wang, H. The multi-scale spatial heterogeneity of ecosysten services' supply–demand matching and its influencing factors on urban green space in China. *Forests* **2023**, *14*, 2091. [CrossRef
37. Zaerpour, M.; Papalexiou, S.M.; Pietroniro, A. Increasing tree canopy lowers urban air temperature by up to 1.5 °C in heat-pron areas. *npj Urban Sustain.* **2025**, *5*, 92. [CrossRef]
38. Matias, P.; Moreira da Silva, M.; Teigão, J.; Duarte, A. Urban vegetation benefits in mediterranean cities for climate chang adaptation and water usage efficiency–a case study in Algarve, Portugal. *Front. Environ. Sci.* **2025**, *13*, 1520934. [CrossRef]
39. Liu, S.; Shen, P.; Huang, Y.; Jiang, L.; Feng, Y. Spatial distribution changes in nature-based recreation service supply from 2008 t 2018 in Shanghai, China. *Land* **2022**, *11*, 1862. [CrossRef]
40. Baró, F.; Haase, D.; Gómez-Baggethun, E.; Frantzeskaki, N. Mismatches between ecosystem services supply and demand in urba areas: A quantitative assessment in five European cities. *Ecol. Indic.* **2015**, *55*, 146–158. [CrossRef]
41. Yang, Y.; Guangrong, S.; Chen, Z.; Hao, S.; Zhouyiling, Z.; Shan, Y. Quantitative analysis and prediction of urban heat islan intensity on urban-rural gradient: A case study of Shanghai. *Sci. Total Environ.* **2022**, *829*, 154264. [CrossRef]
42. Li, J.; Fu, J.; Gao, J.; Zhou, R.; Wang, K.; Zhou, K. Effects of the spatial patterns of urban parks on public satisfaction: Evidenc from Shanghai, China. *Landsc. Ecol.* **2023**, *38*, 1265–1277. [CrossRef]
43. Liang, H.; Yan, Q.; Yan, Y.; Zhang, Q. Using an improved 3SFCA method to assess inequities associated with multimoda accessibility to green spaces based on mismatches between supply and demand in the metropolitan of Shanghai, China. *Sustair Cities Soc.* **2023**, *91*, 104456. [CrossRef]
44. Liang, H.; Zhang, Q. Temporal and spatial assessment of urban park visits from multiple social media data sets: A case study o Shanghai, China. *J. Clean. Prod.* **2021**, *297*, 126682. [CrossRef]
45. *GB/T 50563-2010*; Evaluation Standard for Urban Landscaping and Greening. Ministry of Housing and Urban-Rural Developmen Beijing, China, 2010.
46. Comber, A.; Brunsdon, C.; Green, E. Using a GIS-based network analysis to determine urban greenspace accessibility for differen ethnic and religious groups. *Landsc. Urban Plan.* **2008**, *86*, 103–114. [CrossRef]
47. Liang, C.; Yuan, J.; Tang, X.; Kan, H.; Cai, W.; Chen, J. The influence of humid heat on morbidity of megacity Shanghai in China *Environ. Int.* **2024**, *183*, 108424. [CrossRef]
48. Li, Z.; Chang, J.; Wang, Z.; Chen, Y.; Li, C. Stability of Regional Ecological Supply–Demand Is Enhanced by Complex Networl Modeling: Evidence From the Xuzhou Metropolitan Area, China. *IEEE J. Sel. Top. Appl. Earth Obs. Remote Sens.* **2024**, *17* 1857–1873. [CrossRef]
49. Zhang, Y.; Hu, W.; Min, M.; Zhao, K.; Zhang, S.; Liu, T. Optimization of ecological connectivity and construction of supply-demanc network in Wuhan Metropolitan Area, China. *Ecol. Indic.* **2023**, *146*, 109799. [CrossRef]
50. Pinto, L.V.; Inácio, M.; Ferreira, C.S.S.; Ferreira, A.D.; Pereira, P. Ecosystem services and well-being dimensions related to urbar green spaces—A systematic review. *Sustain. Cities Soc.* **2022**, *85*, 104072. [CrossRef]
51. Xu, L.; You, H.; Li, D.; Yu, K. Urban green spaces, their spatial pattern, and ecosystem service value: The case of Beijing. *Habita Int.* **2016**, *56*, 84–95. [CrossRef]
52. Hu, Y.n.; Zhang, J.; Yang, Y.; Hu, T.; Wang, S. Integrating supply and beneficiaries into assessing the benefit of water purificatior service: A spatial flow perspective. *Landsc. Ecol.* **2024**, *39*, 200. [CrossRef]

https://doi.org/10.3390/su1805219

Liu, X.; Zhao, Y.; Zeng, D.; Yang, Y.; Li, W.; Kang, Y.; Wei, G.; Yuan, X.; Bo, S.; Si, X. Characterizing bird species for achieving the win-wins of conserving biodiversity and enhancing regulating ecosystem services in urban green spaces. *Urban For. Urban Green.* **2023**, *87*, 128064. [CrossRef]

Liu, Y.; Lü, Y.; Zhao, M.; Fu, B. Integrative analysis of biodiversity, ecosystem services, and ecological vulnerability can facilitate improved spatial representation of nature reserves. *Sci. Total Environ.* **2023**, *879*, 163096. [CrossRef] [PubMed]

Wang, C.; Zhao, M.; Xu, Y.; Zhao, Y.; Zhang, X. Ecosystem Service Synergies Promote Ecological Tea Gardens: A Case Study in Fuzhou, China. *Remote Sens.* **2023**, *15*, 540. [CrossRef]

He, L.; Xie, Z.; Wu, H.; Liu, Z.; Zheng, B.; Wan, W. Exploring the interrelations and driving factors among typical ecosystem services in the Yangtze river economic Belt, China. *J. Environ. Manag.* **2024**, *351*, 119794. [CrossRef] [PubMed]

Liu, Q.; Qiao, J.; Li, M.; Huang, M. Spatiotemporal heterogeneity of ecosystem service interactions and their drivers at different spatial scales in the Yellow River Basin. *Sci. Total Environ.* **2024**, *908*, 168486. [CrossRef]

Dou, H.; Li, X.; Li, S.; Dang, D.; Li, X.; Lyu, X.; Li, M.; Liu, S. Mapping ecosystem services bundles for analyzing spatial trade-offs in inner Mongolia, China. *J. Clean. Prod.* **2020**, *256*, 120444. [CrossRef]

Fan, G.; Wang, Y.; Zhao, S.; Qiao, X.; Yang, G.; Lin, Z. Optimization of social values for ecosystem services assessment based on the analytic hierarchy process: A case study of the core district parks in Zhengzhou. *Environ. Sustain. Indic.* **2025**, *27*, 100786. [CrossRef]

Jia, Q.; Jiao, L.; Lian, X.; Wang, W. Linking supply-demand balance of ecosystem services to identify ecological security patterns in urban agglomerations. *Sustain. Cities Soc.* **2023**, *92*, 104497. [CrossRef]

Zhang, Z.; Wang, Q.; Yan, F.; Sun, Y.; Yan, S. Revealing spatio-temporal differentiations of ecological supply-demand mismatch among cities using ecological Network: A case study of typical cities in the "Upstream-Midstream-Downstream" of the Yellow River Basin. *Ecol. Indic.* **2024**, *166*, 112468. [CrossRef]

Huang, W.; Wang, P.; He, L.; Liu, B. Improvement of water yield and net primary productivity ecosystem services in the Loess Plateau of China since the "Grain for Green" project. *Ecol. Indic.* **2023**, *154*, 110707. [CrossRef]

Xiang, H.; Zhang, J.; Mao, D.; Wang, Z.; Qiu, Z.; Yan, H. Identifying spatial similarities and mismatches between supply and demand of ecosystem services for sustainable Northeast China. *Ecol. Indic.* **2022**, *134*, 108501. [CrossRef]

Terrado, M.; Sabater, S.; Chaplin-Kramer, B.; Mandle, L.; Ziv, G.; Acuña, V. Model development for the assessment of terrestrial and aquatic habitat quality in conservation planning. *Sci. Total Environ.* **2016**, *540*, 63–70. [CrossRef] [PubMed]

Sun, C.; König, H.J.; Uthes, S.; Chen, C.; Li, P.; Hemminger, K. Protection effect of overwintering water bird habitat and defining the conservation priority area in Poyang Lake wetland, China. *Environ. Res. Lett.* **2020**, *15*, 125013. [CrossRef]

Liang, H.; Yan, Q.; Yan, Y. Evaluating green space provision development in Shanghai (2012–2021): A focus on accessibility and service efficiency. *Sustain. Cities Soc.* **2024**, *103*, 105269. [CrossRef]

Luo, J. Integrating the huff model and floating catchment area methods to analyze spatial access to healthcare services. *Trans. GIS* **2014**, *18*, 436–448. [CrossRef]

Zhang, J.; Yue, W.; Fan, P.; Gao, J. Measuring the accessibility of public green spaces in urban areas using web map services. *Appl. Geogr.* **2021**, *126*, 102381. [CrossRef]

Collins, R.M.; Spake, R.; Brown, K.A.; Ogutu, B.O.; Smith, D.; Eigenbrod, F. A systematic map of research exploring the effect of greenspace on mental health. *Landsc. Urban Plan.* **2020**, *201*, 103823. [CrossRef]

Liu, S.; Zhang, X.; Feng, Y.; Xie, H.; Jiang, L.; Lei, Z. Spatiotemporal Dynamics of Urban Green Space Influenced by Rapid Urbanization and Land Use Policies in Shanghai. *Forests* **2021**, *12*, 476. [CrossRef]

Liang, Z. Strategic land allocation in China's urban renewal: The role of non-market mechanisms. *SN Bus. Econ.* **2025**, *5*, 47. [CrossRef]

Li, J. Development Control in Transitional Urban China from an Institutional Perspective: Shanghai in the "Era of Redevelopment" as Case Study. Ph.D. Thesis, University of Cambridge, Cambridge, UK, 2023.

Liu, B.; Zheng, S.; Zhang, L.; Liu, J.; Fu, T.; Hao, R.; Yin, M. Identification and Analysis of Potential Open-Sharing Subjects of Unit-Affiliated Green Spaces in Shanghai Based on POI Data. *Land* **2023**, *12*, 2162. [CrossRef]

Qiu, T.; Zhou, D.; Li, W. Fitness culture and green space equity: Accessibility evaluation of shanghai communities. *Front. Environ. Sci.* **2022**, *10*, 958649. [CrossRef]

Jin, Y.; He, R.; Hong, J.; Luo, D.; Xiong, G. Assessing the accessibility and equity of urban green spaces from supply and demand perspectives: A case study of a mountainous city in China. *Land* **2023**, *12*, 1793. [CrossRef]

Xie, H.; Wang, X.; Hu, X.; Shi, Z.; Lin, H.; Xie, X.; Chen, L.; Dai, H.; Zhang, J.; Xu, M. Exploring urban green space optimization of the urban walking life circle in Fuzhou, China. *Int. J. Environ. Res. Public Health* **2023**, *20*, 1180. [CrossRef] [PubMed]

Jiaxin, L.; Bing, C.; Yinke, Z.; Chuanchun, H.; Yu, W.; Feng, S. Accessibility and social equity of urban parks. *J. Zhejiang AF Univ.* **2024**, *41*, 820–829.

Kirby, M.G.; Scott, A.J.; Walsh, C.L. A greener Green Belt? Co-developing exploratory scenarios for contentious peri-urban landscapes. *Landsc. Urban Plan.* **2025**, *255*, 105268. [CrossRef]

79. Tian, Y.; Van Leeuwen, E.; Tsendbazar, N.-e.; Jing, C.; Herold, M. Urban green inequality and its mismatches with human deman across neighborhoods in New York, Amsterdam, and Beijing. *Landsc. Ecol.* **2024**, *39*, 60. [CrossRef]
80. Li, Y.; Wang, C.; Sun, M.; Zhang, H. Dynamic Simulation of Land Use Change and Assessment of Ecosystem Services Unde Climate Change Scenarios: A Case Study of Shanghai, China. *Land* **2025**, *14*, 1791. [CrossRef]
81. Zhang, T.; Zeng, Y.; Zhang, Y.; Song, Y.; Li, H. Dynamic and heterogeneous demand for urban green space by urban resident Evidence from the cities in China. *Sustainability* **2020**, *12*, 9384. [CrossRef]
82. Abouelela, M.; Durán-Rodas, D.; Antoniou, C. Do we all need shared E-scooters? An accessibility-centered spatial equit evaluation approach. *Transp. Res. Part A Policy Pract.* **2024**, *181*, 103985. [CrossRef]
83. McKenzie, G. Spatiotemporal comparative analysis of scooter-share and bike-share usage patterns in Washington, DC. *J. Transp Geogr.* **2019**, *78*, 19–28. [CrossRef]
84. Mears, M.; Brindley, P.; Barrows, P.; Richardson, M.; Maheswaran, R. Mapping urban greenspace use from mobile phone GP data. *PLoS ONE* **2021**, *16*, e0248622. [CrossRef]
85. Guo, W.; Teng, Y.; Li, J.; Yan, Y.; Zhao, C.; Li, Y.; Li, X. A new assessment framework to forecast land use and carbon storage unde different SSP-RCP scenarios in China. *Sci. Total. Environ.* **2023**, *912*, 169088. [CrossRef] [PubMed]
86. Tang, J.; Song, P.; Hu, X.; Chen, C.; Wei, B.; Zhao, S. Coupled effects of land use and climate change on water supply in SSP–RC scenarios: A case study of the Ganjiang River Basin, China. *Ecol. Indic.* **2023**, *154*, 110745. [CrossRef]
87. He, N.; Guo, W.; Wang, H.; Yu, L.; Cheng, S.; Huang, L.; Jiao, X.; Chen, W.; Zhou, H. Temporal and Spatial Variations in Landscap Habitat Quality under Multiple Land-Use/Land-Cover Scenarios Based on the PLUS-InVEST Model in the Yangtze River Basir China. *Land* **2023**, *12*, 1338. [CrossRef]
88. Li, X.; Fu, J.; Jiang, D.; Lin, G.; Cao, C. Land use optimization in Ningbo City with a coupled GA and PLUS model. *J. Clean. Pro* **2022**, *375*, 134004. [CrossRef]
89. Gao, L.; Tao, F.; Liu, R.; Wang, Z.; Leng, H.; Zhou, T. Multi-scenario simulation and ecological risk analysis of land use based o the PLUS model: A case study of Nanjing. *Sustain. Cities Soc.* **2022**, *85*, 104055. [CrossRef]
90. Liao, W.; Liu, X.; Xu, X.; Chen, G.; Liang, X.; Zhang, H.; Li, X. Projections of land use changes under the plant functional typ classification in different SSP-RCP scenarios in China. *Sci. Bull.* **2020**, *65*, 1935–1947. [CrossRef]
91. Liu, X.; Liu, Y.; Wang, Y.; Liu, Z. Evaluating potential impacts of land use changes on water supply–demand under multipl development scenarios in dryland region. *J. Hydrol.* **2022**, *610*, 127811. [CrossRef]
92. Popp, A.; Calvin, K.; Fujimori, S.; Havlik, P.; Humpenöder, F.; Stehfest, E.; Bodirsky, B.L.; Dietrich, J.P.; Doelmann, J.C.; Gusti, M et al. Land-use futures in the shared socio-economic pathways. *Glob. Environ. Change* **2017**, *42*, 331–345. [CrossRef]

https://doi.org/10.3390/su1805219

sustainability

MDPI

Article

What Drives Residents' Divergent Perceptions of Cultural Ecosystem Services in Urban Park Green Spaces? A Dual-Source Analysis Synergizing Social Media and Survey Data

Xiaokang Li [1], Zhuofan Ye [1,*], Lin Lei [2], Yiwu Wen [1] and Junwen Huang [1]

1 School of Civil Engineering, Architecture and Environment, Hubei University of Technology, Wuhan 430068, China; lixiaokang@hbut.edu.cn (X.L.); 102411041@hbut.edu.cn (Y.W.); 102411038@hbut.edu.cn (J.H.)
2 Changjiang Ecology (Hubei) Technology Development Co., Ltd., Wuhan 430011, China; happy_aliya@163.com
* Correspondence: 102310949@hbut.edu.cn

Abstract

In the context of rapid urbanization and the pursuit of the United Nations Sustainable Development Goals (SDGs), particularly SDG 11 (Sustainable Cities and Communities), cities face multifaceted challenges such as high population density, limited green space, ecosystem degradation, and an insufficient supply of ecological products, all of which undermine urban sustainability. As crucial ecological units, urban park green spaces (UPGS) play a vital role in alleviating environmental pressures and providing cultural ecosystem services (CES) that are essential for human well-being and social sustainability. However, systematic insight into how residents perceive and value CES, along with the underlying drivers, remains underdeveloped, impeding the advancement of refined park management practices. Based on 12,083 social media texts, this study employed BERTopic topic modeling to identify five core dimensions of CES perception: recreational services (RS), aesthetic experiences (AE), health-promoting activities (HA), social interactions (SI), and educational services (ES). Additionally, four underlying drivers with corresponding measurable indicators were also identified: residents' socioeconomic backgrounds (RSB), external built environment of parks (EBE), internal landscape composition (ILC), and quality of services management (QSM). Subsequently, using 313 valid questionnaires and geographic park data, a Partial Least Squares Structural Equation Modeling (PLS-SEM) framework was constructed to analyze the influence mechanisms of EBE, ILC, and QSM on CES perception differences, with residents' satisfaction with CES serving as the measure of their perceived CES levels. Hierarchical regression analysis was further employed to examine the moderating effects of RSB on these driving pathways. The findings reveal the following: (1) Significant synergies and heterogeneities existed among CES dimensions, with notable synergistic effects observed between AE and SI, as well as between HA and RS. (2) EBE, ILC, and QSM significantly influenced CES perception differences ($p < 0.05$). EBE affected these differences through pathways such as EBE → ILC → QSM → CES and EBE → QSM → CES. Notably, QSM was identified as the most critical mediating factor affecting CES perception differences. (3) Age exerted a significant positive moderating effect on the QSM → CES pathway, while monthly income showed a marginally significant negative moderating trend on the ILC → QSM pathway. This study elucidates the multi-level driving mechanisms underlying differences in residents' perceptions of CES in UPGS. A key innovation lies in the integration of large-scale social media text data with questionnaire surveys, combined with the application of the BERTopic model and PLS-SEM to analyze these perceptual differences. The findings offer both theoretical foundations and practical insights for landscape optimization and service enhancement in park planning and management, contributing to the development of more equitable, resilient, and sustainable urban environments.

Keywords: urban park green spaces; cultural ecosystem services; residents' divergent perceptions; social media-survey analysis; driving mechanisms; PLS-SEM; socioeconomic background moderating effects

1. Introduction

The United Nations Sustainable Development Goals (SDGs), particularly SDG 11 (Sustainable Cities and Communities) and SDG 15 (Life on Land), highlight the critical need to foster inclusive, safe, resilient, and sustainable urban environments [1]. Central to this global agenda is the provision of urban green spaces, which deliver essential ecosystem services that support both human well-being and ecological integrity [2]. However, the rapid urbanization observed worldwide, and especially in China, has generated significant socio-environmental pressures that directly challenge these sustainability targets. These pressures, including population overcrowding, resource depletion, ecosystem degradation, the loss of historic urban character, and a shortfall in ecological service provision [3–7], represent fundamental obstacles to realizing the SDGs. Consequently, a critical challenge confronting governments at all levels in China today is how to reconstruct new orientations, approaches, and pathways for urban development in the new era, facilitating urban transformation and sustainable development while meeting the people's growing aspirations for a better quality of life and enhancing their sense of fulfillment and well-being [8]. In response, the Chinese government has advanced the innovative concept of "Park City" development as a new urban paradigm, aiming to implement the vision of ecological civilization and the people-centered development philosophy in the new era.

Urban park green spaces (UPGS) are defined as "publicly accessible green areas primarily designated for recreation, while also incorporating ecological, aesthetic, cultural, and educational functions, and emergency shelter capabilities, equipped with certain recreational and service facilities" [9]. Serving as a pivotal spatial medium for Park City development and ecological infrastructure construction, UPGS form essential structural components of urban ecological and landscape systems. Beyond their demonstrated efficacy in alleviating urban environmental stressors [10–12], UPGS are fundamental to delivering cultural ecosystem services (CES), namely the non-material benefits that are increasingly recognized as crucial determinants of urban livability and social sustainability [13]. However, the prevailing top-down, government-led development model has resulted in significant homogenization of urban parks. This manifests in features such as morphological monotony, functional oversimplification, and imbalanced integration of greening and cultural elements [14–16]. Park homogenization not only leads to inefficient utilization of scarce urban land resources but also fails to satisfy the public's diverse CES demands. This deficiency directly undermines the social dimension of sustainable urban development. A primary underlying cause is the insufficient attention paid to the needs and preferences of residents with varying backgrounds and behavioral patterns. This study focuses on residents' subjective perceptions of CES. This concept refers to individuals' cognitive and emotional interpretations and experiences of the non-material benefits provided by UPGS. As a psychological construct, it is distinct from the objective provision of ecosystem services or actual service flows. To quantify the degree of subjective perception, this

 https://doi.org/10.3390/su1805257

study adopts satisfaction with CES as the core reflective indicator, serving as a quantifiable proxy for residents' subjective evaluations and the extent to which CES meets their needs. How to gain an in-depth understanding of the differences in residents' perceptions of CES in UPGS across diverse social groups, and how to internalize such insights into the planning, construction, renovation, and quality enhancement of urban parks, remains a critical area of research for realizing the vision of "a city built by and for the people," which lies at the core of socially sustainable urban development.

Cultural ecosystem services refer to the non-material benefits humans obtain from ecosystems through spiritual fulfillment, cognitive development, reflection, recreation, and aesthetic experiences. CES constitute one of the four main categories of ecosystem services, alongside provisioning, supporting, and regulating services [2]. CES encompasses diverse non-material benefits, including aesthetic appreciation, physical recreation, social interaction, mental health enhancement, spiritual inspiration, place identity, religious fulfillment, educational value, cultural heritage, and ecotourism opportunities [17–20]. These services are intrinsically linked to human well-being—a central pillar of sustainability that extends beyond environmental and economic considerations. As human perceptions of CES are inherently subjective and not easily quantifiable, and given the differences in cognition, cultural background, and other contextual factors among individuals, perceptions of CES may vary significantly across different population groups, even when engaging with the same ecosystem [21–23]. The intangible nature of CES, coupled with the difficulty of direct tangible valuation, often leads to their oversight in urban ecological space utilization and governance processes [24,25]. This oversight poses a significant risk to equitable sustainability, as it may systematically disadvantage certain social groups whose needs and perceptions are not adequately captured by conventional planning approaches.

Existing research on residents' perceptions of CES in UPGS has primarily focused on three key areas. First, research on heterogeneous public perceptions of CES in UPGS. Residents, as primary end-users of CES in UPGS, exhibit perceptual patterns that constitute critical metrics for evaluating the efficacy and performance of park-delivered CES. Scholars have long employed the resident perception lens to research CES in UPGS. Key themes include: analyzing cognitive differences in CES types among demographic groups [23,26], assessing perceived importance and satisfaction [27], and estimating CES value [28]. Methodologies encompass questionnaire surveys [29], interviews [30], willingness-to-pay analysis [31], public participation GIS (PPGIS) [32,33], the SolVES model [26,28], and social media analytics [34].

Second, research on the influence of demographic attributes on CES perceptual differences. Heterogeneous population composition within a region necessitates UPGS to fulfill diverse and heterogeneous demands. Variations in CES preferences among resident groups highlight this heterogeneity. Differences in residents' demographic, sociological, and behavioral backgrounds significantly influence their UPGS usage and CES perceptions [18,27]. Researchers often stratify respondents by demographics, social attributes, or personal characteristics to investigate group-specific CES preference influences. For instance, Zhang et al. examined the relational dynamics between generational CES demand patterns (youth vs. elderly cohorts) and landscape spatial configurations within urban park ecosystems [26]. Ghermandi et al. compared CES perceptions among locals, domestic, and international tourists, revealing distinct beneficiary patterns [35]. Zhou et al. examined the influence of gender on perceived CES preferences, revealing that male and female subgroups exhibited similar perceptions regarding cultural, educational, and recreational services, whereas preferences diverged for aesthetic appreciation and biodiversity perception [28]. Riechers et al. delimited their respondents to experts and laypersons, revealing through comparative

https://doi.org/10.3390/su18052578

analysis that while professionals exhibited stronger orientations toward environmental stewardship, non-specialists primarily valued nature for recreational enjoyment [36].

Third, research examining the underlying drivers shaping heterogeneous resident perceptions of CES. Research on the formative drivers of these differences remains limited. Existing studies primarily analyze the impacts of socioeconomic status, landscape characteristics, and spatial distribution on perceptual variations [36,37]. Landscape influence analyses often leverage social media data and remote sensing imagery to assess how environmental features shape CES preferences [38]. Accessibility exerts significant positive effects on CES perceptions [39]. Perceptions of health-promoting services strongly influence park usage intentions among younger residents [40]. Diverse UPGS characteristics, including location, size, and amenities, also significantly impact usage patterns and CES provision [19,37,41].

Despite growing attention to CES perceptions, the field remains constrained by a reliance on single-method approaches, which compromises both the scope and validity of its findings. First, existing knowledge is divided between two primary data sources, each with inherent limitations. Traditional surveys, while demographically representative, are epistemologically anchored in a priori constructs. Their retrospective design captures post hoc rationalizations rather than immediate lived experience, which makes them susceptible to recall bias and fundamentally unable to capture in-the-moment affective responses [42]. Conversely, social media big data, while offering real-time, context-specific expressions, is subject to intrinsic self-selection bias [43]. This bias systematically over-represents digitally active demographics while leaving digitally disadvantaged groups unrepresented [44]. Moreover, the unstructured nature of such data presents a methodological challenge, as reliably distinguishing ephemeral sentiments from enduring attitudes remains difficult. Hence, the field finds itself caught between data that is demographically rich but experientially thin, and data that is experientially rich but demographically opaque. Second, this methodological schism has fragmented theoretical development regarding the drivers of perceptual heterogeneity, with studies isolating either socio-demographic factors or environmental attributes rather than exploring their interactive pathways [45,46]. To transcend these dual limitations, this study employs a sequential mixed-methods approach that methodologically triangulates online social media data with offline survey and spatial data. This design leverages their complementary strengths to correct for each other's inherent weaknesses: large-scale online data is used to surface emergent perceptions and generate hypotheses, while offline data ensures demographic representativeness and enables rigorous causal testing. Ultimately, this integrated approach reveals both the mediating and moderating processes underlying the heterogeneity of CES perceptions, thereby providing a socially inclusive evidence base for optimizing the structural composition of urban green spaces, refining the functional zoning of parks, and promoting the equitable allocation of public green space resources. This represents a core tenet of environmental justice and social sustainability. Integrating CES into planning thus helps bridge perceptual disparities between urban cores and peripheries, fostering more inclusive and demand-responsive urban greening that contributes directly to the achievement of SDGs.

Operationalizing this framework, the research followed a two-phase sequential design to systematically analyze the driving mechanisms of residents' perceptions of CES in UPGS. In the first (online) phase, data were collected from user comments on the social media platform Weibo through web crawling. The BERTopic topic modeling was then employed to identify five key CES perception dimensions: Recreational Services (RS), Aesthetic Experiences (AE), Health-promoting Activities (HA), Social Interactions (SI), and Educational Services (ES). Corresponding potential drivers and their observable indicators were also identified, including residents' socioeconomic backgrounds (RSB), external built

https://doi.org/10.3390/su18052578

environment (EBE), internal landscape composition (ILC), and quality of services management (QSM). In the subsequent (offline) phase, informed by these inductive insights, a structured questionnaire was developed and administered through field surveys in urban parks, with responses integrated alongside park-related geospatial datasets. A Partial Least Squares Structural Equation Modeling (PLS-SEM) framework was applied to construct and test a comprehensive model of the driving pathways among EBE, ILC, QSM, and CES perception variations. Finally, hierarchical regression analysis was conducted to examine the moderating effects of RSB within these causal pathways.

This study aims to address the following core research question: In the context of UPGS, how do the EBE, ILC, QSM, and RSB synergistically drive variations in residents' perceptions of CES?

Based on preliminary exploratory findings and a synthesis of relevant theoretical frameworks, the following specific hypotheses are proposed for subsequent quantitative testing:

H1. *The EBE has a positive effect on residents' perception variations of CES in UPGS.*

H2. *The EBE has a positive effect on the ILC.*

H3. *The EBE has a positive effect on the QSM.*

H4. *The ILC has a positive effect on residents' perception variations of CES in UPGS.*

H5. *The ILC has a positive effect on the QSM.*

H6. *The QSM has a positive effect on residents' perception variations of CES in UPGS.*

Furthermore, hierarchical regression analysis will be employed to examine the moderating effects of RSB on the hypothesized causal pathways. This analytical step aims to achieve a more comprehensive understanding of the heterogeneity in CES perceptions across different social groups.

By constructing and empirically testing this integrated model, this study seeks to theoretically elucidate the complex pathways that shape CES perceptions in UPGS. From a practical and sustainability-oriented perspective, the findings are intended to provide targeted empirical evidence to inform park planning, design, and fine-scale management, thereby contributing to the maximization of the socio-ecological benefits provided by UPGS and advancing the broader transition toward socially equitable and ecologically resilient urban development.

2. Materials and Methods

2.1. Study Area

Wuhan (29°58′–31°22′ N, 113°41′–115°05′ E) is situated in eastern Jianghan Plain along the middle reaches of the Yangtze River, covering a total area of 8569.15 km^2. As a central Chinese metropolis, Wuhan has undergone rapid urbanization. By the end of 2024, Wuhan had achieved an urbanization rate of 85% among its permanent resident population [47]. This substantial population concentration has not only intensified urban ecological pressures but also challenged the capacity to meet the diverse and growing spiritual and cultural demands of its large resident base. To support high-quality urban development and enhance resident well-being, Wuhan municipal authorities have actively promoted the expansion and quality enhancement of urban green spaces in recent years. As of the end of 2024, Wuhan had cumulatively established 1024 parks of various categories,

 https://doi.org/10.3390/su18052578

achieving a per capita urban park green space area of 15.03 m^2 and an urban built-up area green coverage rate of 43.14% [47]. Wuhan thus presents a representative case study context characterized by rapid urban park expansion, significant population agglomeration, and rising resident demand for CES. This empirical focus offers valuable insights for optimizing UPGS allocation and informing planning, construction, and renewal strategies in comparable cities facing land, ecological, and cultural resource constraints.

Regarding sample selection, this study implemented a stratified sampling strategy for both online and offline research subjects, guided by the principles of representativeness, feasibility, and data quality. For the online component, priority was given to data accessibility and the effectiveness of topic identification. Twenty parks in Wuhan's central urban area were selected, covering a range of sizes and types, all of which exhibited active discussion on social media platforms to ensure diverse sources of textual reviews. For the offline component, emphasis was placed on geographical coverage, typological representativeness, and survey practicality. Thirteen parks were chosen from various administrative districts across Wuhan, representing different locations, scales, functions, and visitor profiles to strengthen the sample's explanatory power. The study area and the locations of the online and offline parks are presented in Figure 1, with detailed park characteristics provided in Table 1.

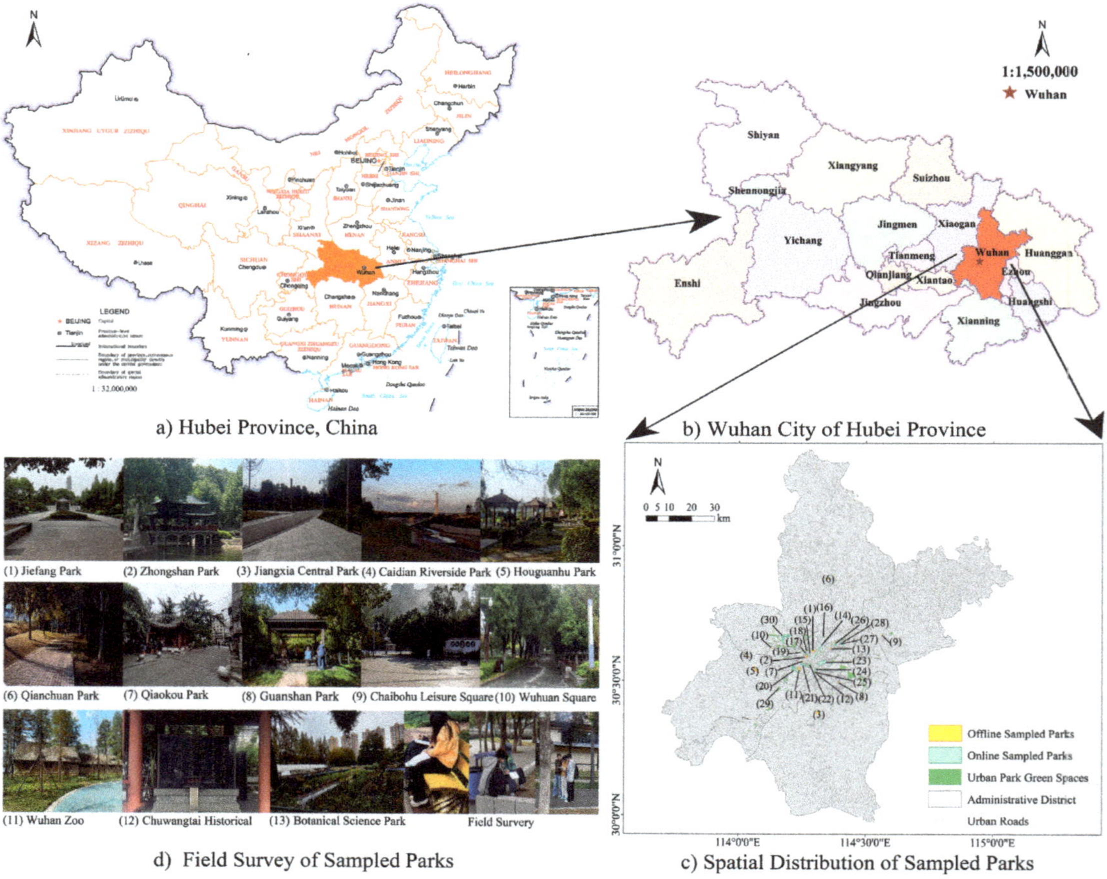

Figure 1. Study area and locations of the sampled parks.

https://doi.org/10.3390/su18052578

Table 1. Profile of Sample Parks.

No.	Parks	Year	Area (Hectares)	Location	Type	Application
(1)	Jiefang Park	1952	46	Jiang'an District	Comprehensive Parks	Survey/Online
(2)	Zhongshan Park	1910	32.80	Jianghan District	Comprehensive Parks	Survey/Online
(3)	Jiangxia Central Park	2017	95.60	Jiangxia District	Comprehensive Parks	Survey
(4)	Caidian Riverside Park	2004	26.17	Caidian District	Ecological Parks	Survey
(5)	Houguanhu Park	2011	28.55	Hannan District	Ecological Parks	Survey/Online
(6)	Qianchuan Park	2019	21.07	Huangpi District	Ecological Parks	Survey
(7)	Qiaokou Park	1957	3.27	Qiaokou District	Neighborhood Parks	Survey
(8)	Guanshan Park	2008	8.23	Hongshan District	Neighborhood Parks	Survey
(9)	Chaibohu Leisure Square	2010	6.95	Xinzhou District	Ecological Parks	Survey
(10)	Wuhuan Square	2008	3.70	Dongxihu District	Neighborhood Parks	Survey
(11)	Wuhan Zoo	1985	64.66	Hanyang District	Thematic Parks	Survey
(12)	Chuwangtai Historical Park	2008	3.10	Wuchang District	Thematic Parks	Survey
(13)	Botanical Science Park	2007	11.2	Qingshan District	Thematic Parks	Survey
(14)	Hankou Riverside Park	2002	153.71	Jiang'an District	Comprehensive Parks	Online
(15)	Baodao Park	1985	13.46	Jiang'an District	Comprehensive Parks	Online
(16)	Dijiao Park	2004	19.89	Jiang'an District	Comprehensive Parks	Online
(17)	Xibeihu Green Square	1999	24.15	Jianghan District	Ecological Parks	Online
(18)	Lingjiaohu Park	2008	14.12	Jianghan District	Comprehensive Parks	Online
(19)	Wangjiadun Park	2017	12.03	Jianghan District	Comprehensive Parks	Online
(20)	Moshuihu Park	2016	78.22	Hanyang District	Ecological Parks	Online
(21)	Yuehu Park	2006	106	Hanyang District	Comprehensive Parks	Online
(22)	Guishan Park	1983	35.33	Hanyang District	Comprehensive Parks	Online
(23)	Shahu Park	1917	377	Wuchang District	Comprehensive Parks	Online
(24)	Shouyi Park	1923	16.6	Wuchang District	Comprehensive Parks	Online
(25)	Ziyang Park	1951	29.61	Wuchang District	Comprehensive Parks	Online
(26)	Heping Park	1998	49.4	Qingshan District	Comprehensive Parks	Online
(27)	Qingshan Park	1962	35.43	Qingshan District	Comprehensive Parks	Online
(28)	Daijiahu Park	2015	70.28	Qingshan District	Ecological Parks	Online
(29)	Tanghu Park	2004	35.44	Hannan District	Comprehensive Parks	Online
(30)	Jinyinhu Park	2001	77	Dongxihu District	Ecological Parks	Online

As shown in Table 1, differences in composition exist between the online and offline park samples, which can be attributed to the underlying data generation mechanisms. The Weibo data are generated from users' voluntary posts on a public platform. Consequently, the online sample naturally emphasizes parks with greater visibility and discussion frequency on social media, such as large comprehensive parks and iconic urban parks. These types of parks are more likely to stimulate public willingness to share, whereas smaller, community-based parks intended for daily use tend to have limited online presence on Weibo.

Online data may amplify the visibility of certain high-profile services, whereas offline data provide a more balanced representation of diverse service types. To ensure systematic representation of urban parks in terms of geographic location, scale, and typology, this study employed a stratified systematic sampling method for the offline questionnaire survey. One representative park was selected from each administrative district, covering four park types: comprehensive parks, neighborhood parks, ecological parks, and thematic parks. This approach aimed to capture the perceptions of park users under diverse functional, spatial, and locational conditions.

In summary, online data capture spontaneous and expansive public discourse generated by parks with heightened public appeal and online discussion, while offline data enables systematic measurement of park user experiences. Together, these complementary data sources contribute to a more multidimensional and comprehensive understanding of CES perceptions in UPGS.

https://doi.org/10.3390/su18052578

2.2. Data Sources and Processing

The data utilized in this study encompass social media data, RSB, EBE, ILC, QSM, and residents' perceptions of CES. Detailed information regarding each data source is provided in Table 2. This study collected review texts from Sina Weibo 15.5.0 as the primary data source. Following initial data cleaning of the social media texts obtained via web crawling, the Python-based jieba segmentation tool was employed for preprocessing. This included sentence splitting, word segmentation, removal of stop words, and synonym merging. Subsequently, BERTopic topic modeling and term frequency analysis were applied to analyze the textual data, extract residents' perceived dimensions of CES, and identify the corresponding driving factors and their indicators. Spatial data for the EBE and ILC were processed in ArcGIS 10.8 to derive, for each park, metrics such as surrounding building density (SBD), transportation accessibility (TA), population density (PD), facility richness (FR), park area (PA), and water feature (WF). Subsequently, the continuous variables for EBE and ILC were transformed into five-point ordinal variables (1 = very low to 5 = very high). This transformation was based on two primary considerations. First, it ensured consistency in measurement scales, aligning the environmental variables with CES satisfaction, which was measured using a five-point Likert scale, thereby facilitating straightforward interpretation of the path coefficients. Second, it supported the analysis of moderating effects, as the categorical nature of RSB allowed for clearer logic when constructing interaction terms with the ordinal environmental variables, thereby facilitating the translation of findings into planning practice. Questionnaire data were analyzed using SPSS 27.0.1. Descriptive statistics summarized the respondents' socio-demographic profiles. Reliability and validity analyses were performed to evaluate the measurement quality of both the intra-park environment management (IEM), visitor experience management (VEM), cultural promotion management (CPM), and residents' perceived satisfaction with CES. Relationships among different CES dimensions were examined using Pearson correlation analysis to identify trade-offs and synergies. Friedman tests were conducted to determine if significant differences existed in satisfaction levels across the five CES categories. Finally, we employed PLS-SEM to assess the measurement and structural models, evaluating reliability, validity, path coefficients, and model fit. The moderating effect of RSB was tested through regression analysis.

Table 2. Data sources.

Data Classification	Data Name	Data Source
Social media data	Weibo textual comments	Web crawler collection
Residents' socioeconomic backgrounds	Demographic characteristics	Questionnaire survey
Residents' perceptions of CES	Perceived satisfaction with CES	Questionnaire survey
External built environment	Surrounding building density Transportation accessibility Population density	2024 Open Street Map Data https://www.openstreetmap.org/ National Earth System Science Data Center http://www.geodata.cn
Internal landscape composition	Facility richness Park area Water feature	Peking University Open Research Data Platform 2024 Open Street Map Data https://www.openstreetmap.org/
Quality of service management	Intra-park environment management Visitor experience management Cultural promotion management	Questionnaire survey

https://doi.org/10.3390/su18052578

2.3. Methods

2.3.1. Research Design

This study was conducted following the three methodological steps outlined below, with the corresponding technical workflow illustrated in Figure 2.

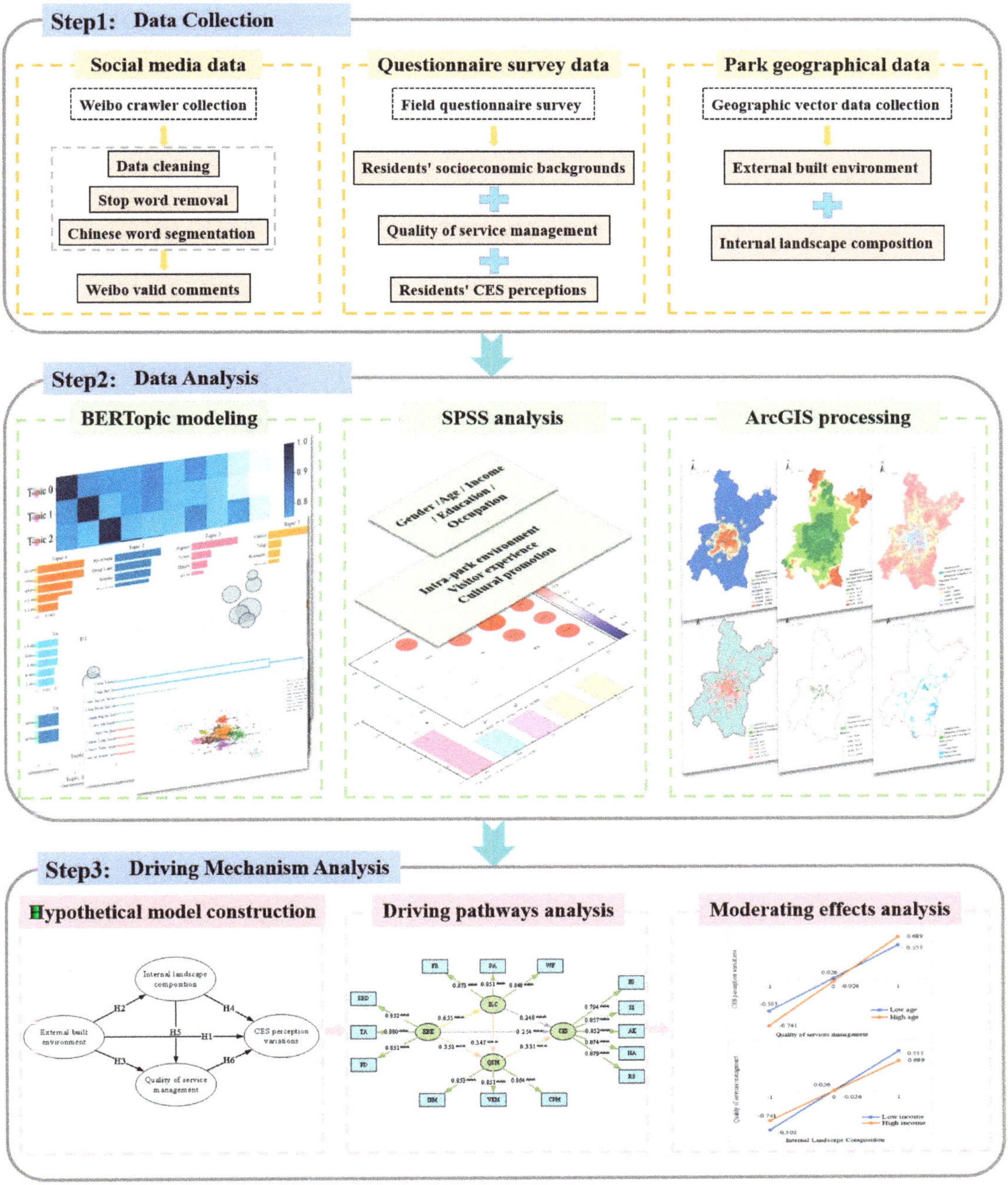

Figure 2. Research workflow.

https://doi.org/10.3390/su18052578

Step 1: Data collection. Social media data were collected from user-generated reviews on the Weibo platform. Questionnaire survey data encompassed RSB, QSM, and their evaluations of CES. Geospatial data included metrics related to the EBE and ILC.

Step 2: Data processing. The social media comments were analyzed using the BERTopic model to extract key CES perception dimensions and to identify underlying driving factors along with their observable indicators. Questionnaire data were processed using SPSS 27.0.1 software, involving correlation and difference analyses concerning CES perceptions. Geospatial data were processed and analyzed in ArcGIS 10.8, yielding specific metrics such as SBD, TA, PD, FR, PA, and WF.

Step 3: Analysis of driving mechanisms. First, a hypothesized model was constructed to explain the differences in residents' CES perceptions. Second, PLS-SEM was applied to examine the driving mechanisms behind these perceptual differences. Finally, hierarchical regression analysis was employed to assess the moderating effects of residents' socioeconomic backgrounds on the identified driving pathways.

2.3.2. Social Media Data Collection and Preprocessing

This study used Weibo as the data source for social media content related to UPGS. Textual reviews were collected from 20 parks in central Wuhan between July 2024 and July 2025 via web crawling of the platform's API, resulting in an initial dataset of 17,836 comments.

During data cleaning, blank, duplicate, and clearly irrelevant entries were initially removed. A subsequent manual review was conducted to further exclude non-original texts, such as advertisements and automatically generated content, resulting in 12,083 valid original user comments retained for analysis. The inclusion criteria required that comments be original, contain specific park names, and exceed five characters in length. Entries identified as advertisements, completely irrelevant content, or duplicate posts were filtered out accordingly.

In the text preprocessing stage, the Jieba segmentation tool in Python 3.13.6 was employed for sentence segmentation and word tokenization. Multiple Chinese stop word lists and customized dictionaries were then integrated to remove semantically insignificant terms. Text normalization procedures, including traditional-simplified Chinese conversion and spelling correction, were also performed. To further reduce noise, emoticons and URLs embedded in the text were systematically removed.

Based on this processed corpus, word frequency analysis was conducted to calculate the occurrence frequency of each term, thereby assessing its relative importance. To avoid ambiguity, only multi-character words with substantive semantic meaning were retained in the high-frequency keyword statistics, while single-character terms were excluded. Word frequency analysis was subsequently applied to identify keywords from high-frequency terms, enabling the summarization of representative driving factors and their corresponding indicators. Topic extraction and analysis were performed using the BERTopic model. In alignment with the conceptual framework of CES, five core CES perception dimensions were ultimately derived.

The BERTopic modeling procedure was systematically calibrated to suit the characteristics of the dataset, which consisted of 12,083 Weibo texts. Specifically, a pre-trained BERT model was initially employed to generate high-dimensional semantic embeddings. When applying UMAP for dimensionality reduction, the parameter configuration was deliberately selected to balance dataset size with the need to preserve global topic structure. The n_neighbors parameter was set to 50, which, given the dataset size, captures broad semantic associations by considering sufficiently large local neighborhoods, thereby helping to prevent overly fragmented clustering. The dimensionality reduction target, n_components, was set to 5 to achieve a balance between computational efficiency and

 https://doi.org/10.3390/su18052578

the retention of meaningful semantic information. Additionally, min_dist was set to 0.0, allowing the embedded points to be densely arranged and creating favorable conditions for the subsequent density-based clustering performed by HDBSCAN.

For clustering analysis, the HDBSCAN algorithm was selected due to its capacity to identify clusters of varying densities without requiring the number of topics to be predetermined. The parameters were determined through an iterative process aimed at optimizing both the coherence and granularity of the resulting topics. The min_cluster_size parameter was set to 50 to ensure that each topic is supported by a sufficient number of documents, aligning to identify meaningful, non-spurious topics within the corpus of 12,083 texts. The min_samples parameter was set to 25 to provide a conservative estimate of cluster stability, thereby reducing the likelihood of outliers being incorrectly assigned to clusters. HDBSCAN was applied using the Euclidean distance metric, as it is the standard metric for the UMAP-reduced space.

To assess the robustness of the parameter configuration, a sensitivity analysis was conducted by systematically adjusting key parameters. The final selected configuration (n_neighbors = 50, min_cluster_size = 50) produced the most interpretable and stable topic clusters, characterized by lower topic overlap and higher average topic coherence scores. In contrast, other parameter configurations either resulted in excessively fragmented topics (with smaller min_cluster_size values) or led to the merging of conceptually distinct topics (with larger n_neighbors values). Ultimately, the c-TF-IDF method was applied to extract salient keywords and characterize each topic. The complete workflow is illustrated in Figure 3.

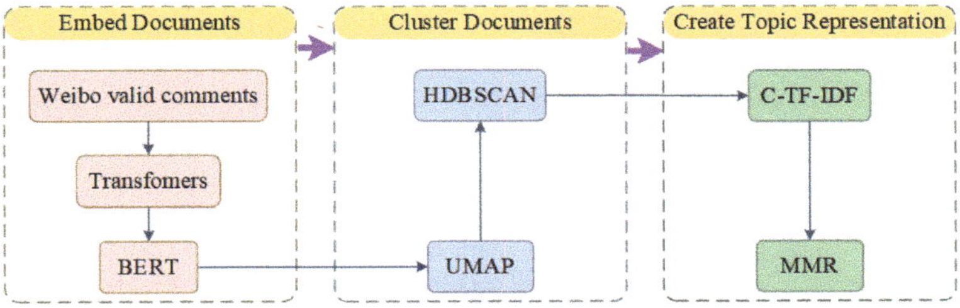

Figure 3. BERTopic topic modelling framework.

To ensure the ecological validity of the automated topic modeling results, a manual coding approach was employed to validate the thematic classification. First, a random sample of 500 comments was drawn from the full dataset as the coding sample. Two independent coders were trained to familiarize themselves with the definitions, characteristics, and boundaries of the ten initial topics. Prior to formal coding, a preliminary coding exercise was conducted to establish a shared understanding of the thematic content. During the formal coding process, the two coders independently classified the 500 sampled comments according to a predefined coding manual. A forced-choice approach was adopted, whereby each comment was assigned to the most appropriate topic based on its core semantic content. Comments that could not be clearly assigned to any existing topic were categorized as "other", with the reason for this designation documented. Throughout the coding process, the coders made independent judgments without discussion. Upon completion of the coding, inter-coder agreement was assessed using Cohen's Kappa coefficient to account for chance-corrected consistency. If an acceptable Kappa value was achieved, this indicated satisfactory inter-coder reliability. If the reliability did not meet the threshold, the coders collaboratively discussed the discrepant items to reach consensus, and a second round of coding was conducted on the disputed samples until overall reliability was deemed

https://doi.org/10.3390/su18052578

acceptable. Subsequently, the results of the manual coding were compared with the output of the automated topic modeling to validate the model's effectiveness.

After obtaining the ten initial topics through the BERTopic model, the thematic results were systematically interpreted and refined by drawing upon internationally recognized classification frameworks, such as the Millennium Ecosystem Assessment (MA) and the Common International Classification of Ecosystem Services (CICES). Taking into account the specific characteristics of UPGS and the actual perceptions of residents, the initial topics were transformed into theoretically grounded CES dimensions. The specific process and criteria for merging the topics are outlined as follows.

As a first step, a semantic similarity matrix was calculated among all initial topics. For each topic, all representative keywords were input into a pre-trained Chinese BERT model to obtain an average vector representation. The pairwise cosine similarity between topics was then computed based on these vector representations. Topic pairs with similarity scores exceeding a preset threshold (>0.70) were identified as candidate pairs for merging, indicating a high degree of proximity in the latent semantic space. Subsequently, an in-depth qualitative content review was conducted on these candidate topic pairs to ensure that any merger was supported by substantive textual grounding. The review process included the following steps. First, the overlap and semantic association between the top 10 representative keywords of each candidate topic pair were examined. Second, the 20 most representative original comments under each topic were manually reviewed to determine whether the core meanings or experiential descriptions expressed were consistent. Only when a high degree of conceptual alignment was observed in both keyword composition and representative textual content were the topics considered for subsequent merging. Finally, it was ensured that the final merged results remained compatible with authoritative theoretical frameworks of CES. The merged CES dimensions were systematically aligned with service categories defined within these frameworks, thereby ensuring that each dimension had a clear theoretical basis and corresponding nomenclature.

The Weibo data used in this study consist of publicly available user-generated content. In accordance with platform regulations and privacy ethics, personal identity information of the commenters could not be obtained. To assess the stability of CES dimension identification across different user groups, a proxy variable was employed for user segmentation and sensitivity analysis. User groups were delineated based on the time of comment posting (daytime versus nighttime). The BERTopic model was then reapplied to each of these sub-datasets, and the resulting topic structures were compared with those derived from the global model to evaluate their similarity.

As a comparative and complementary measure, the principles of sampling representativeness were strictly adhered to during the subsequent collection of questionnaire data. In both the pilot and formal survey stages, stratified sampling was employed to ensure a balanced distribution of respondents across key demographic characteristics, including age, gender, occupation, and income. This approach was intended to complement the social media data by providing a more demographically representative perspective on CES perceptions.

2.3.3. Questionnaire Design and Implementation

A pilot study preceded formal data collection to ensure methodological robustness. The preliminary questionnaire was designed with three structured sections. First, comprehensive socioeconomic and demographic data collection encompassing gender, age, income, educational attainment, and occupation. Second, evaluation of QSM, operationalized through 5-point Likert scales (1 = lowest to 5 = highest appraisal) across three domains: intra-park environment, visitor experience, and cultural promotion. Third, assessment of

 https://doi.org/10.3390/su18052578

CES satisfaction, quantifying perceived CES quality through parallel 5-point Likert scales (1 = completely dissatisfied to 5 = extremely satisfied). The research team administered 40 questionnaires at Jiefang Park on 12 October 2025, yielding 35 valid responses. The respondents were also asked for feedback on the clarity and user-friendliness of the preliminary questionnaire, including the comprehensibility of its content and the appropriateness of its format, to improve the validity of the survey results. Reliability and validity analyses were subsequently performed on the pilot data using SPSS 27.0.1 software. The overall Cronbach's alpha coefficient was 0.896, indicating strong internal consistency reliability. The KMO measure was 0.818, reflecting adequate structural validity. These results confirm the suitability of the questionnaire for the subsequent formal investigation and analysis.

Based on the pilot study findings, the preliminary questionnaire was optimized through iterative refinements, including linguistic clarification of items and response options, as well as enhanced layout design, culminating in the finalized survey instrument. The formal questionnaire retains the tripartite structure, with detailed content specifications provided in Appendix A. According to Robert, the sample size for the questionnaire survey should meet a minimum threshold of ten participants per predictor variable to ensure robust statistical power [48]. The conceptual model incorporated nine factors that influence residents' CES perception variations. Formal surveys were conducted across 13 selected sample parks in Wuhan from October 14 to 28 October 2025. A total of 325 questionnaires were distributed, with 313 valid responses collected, meeting the predetermined sample size requirements for robust statistical analysis. During the survey process, all respondents were actively engaged in park activities and demonstrated sufficient familiarity with CES-related concepts, thereby ensuring the collection of authentic and reliable data. Reliability and validity tests were conducted on the finalized survey data using SPSS 27.0.1.

2.3.4. Driver Model Construction and Validation

(1) Hypothetical Model Construction

Structural Equation Modeling (SEM) is a multivariate statistical technique used to analyze complex relationships among observed and latent variables. It consists of a measurement model, which links indicators to their underlying constructs, and a structural model, which estimates the causal pathways between latent variables. PLS-SEM is particularly suitable for theory development and exploratory research.

In this study, PLS-SEM was selected for several reasons. First, the sample size (N = 313) is relatively modest, and PLS-SEM is recognized for its capacity to handle smaller samples effectively. Second, the research objectives involve predicting and exploring complex mediating and moderating relationships, an area in which PLS-SEM excels. Third, PLS-SEM does not require strict multivariate normality of the data, making it robust for real-world data distributions. By simultaneously estimating both components of the model, it effectively handles complex causal structures and is adept at uncovering theoretical relationships that can be challenging to identify with covariance-based methods. The following hypotheses are proposed based on driving factors and observed indicators derived from social media data, and a structural model of the driving mechanisms for residents' perception differences in UPGS is constructed.

(i) Hypotheses Concerning the External Built Environment of Parks.

The external built environment of UPGS functions as a key exogenous latent variable, playing a foundational role in shaping residents' perceptions of CES. Empirical evidence indicates that high transportation accessibility not only significantly increases residents' frequency of park use and improves experiential quality but also directly enhances the perceived intensity of recreational, aesthetic, and cultural identity services by strengthening the connection between parks and communities [49]. When parks are located within easily

https://doi.org/10.3390/su18052578

accessible distances for residents, as envisioned in the "15-min city" planning concept, their role as providers of cultural services becomes fully activated and utilized. Rational planning of the surrounding built environment can also provide functional and structural support for the internal landscape, thereby improving its overall scenic value and ecological coherence [50]. Furthermore, a high-quality external built environment may indirectly elevate park management responsiveness and service quality [51]. Synthesizing these findings, we posit the following hypothesis:

H1. *The EBE has a positive effect on residents' perception variations of CES in UPGS.*

H2. *The EBE has a positive effect on the ILC.*

H3. *The EBE has a positive effect on the QSM.*

(ii) Hypotheses Concerning the Internal Landscape Composition of Parks.

Within the formative mechanism of perceived disparities in CES in UPGS, the internal landscape composition acts as a key material carrier that directly shapes residents' experiential perceptions and influences the implementation of management services. Research indicates that the diversity of facilities, the scale of the area, and the configuration of natural elements such as water bodies collectively constitute the core physical dimensions affecting CES provision. Specifically, a high richness of amenities and landscape heterogeneity not only helps attract user groups with diverse preferences but also significantly enhances the perceived intensity of multidimensional services, including aesthetic appreciation, recreation, and cultural education [52]. Furthermore, well-conceived internal landscape design provides the essential physical foundation and spatial framework for management services: a clearly arranged facility system can reduce operational and maintenance costs, scientifically zoned functional areas improve service response efficiency, and ecological landscape structures can effectively mitigate maintenance pressures through their self-regulating capacity. Thus, the internal landscape composition is not only a direct source of CES perceptions but also, through its synergistic integration with the management service system, collectively shapes the overall quality and sustainability of the socio-ecological services provided by UPGS. Based on this synthesis, the following hypothesis is proposed:

H4. *The ILC has a positive effect on residents' perception variations of CES in UPGS.*

H5. *The ILC has a positive effect on the QSM.*

(iii) Hypothesis Concerning the Quality of Service Management.

Within the formative mechanism of CES perception disparities, service management quality does not function as an independent exogenous variable. Rather, it acts as a critical mediating factor that is jointly driven by the external built environment and the internal landscape composition, thereby directly influencing residents' CES experiences. Different dimensions of service management strategies exhibit notable divergence in their cultural dissemination effects: digital media promotions demonstrate stronger appeal to younger demographics, while heritage-based immersive programs are more likely to gain value recognition from groups with high cultural capital. Optimized service management enhances CES perception quality primarily through two mechanistic pathways: first, by providing targeted cultural content that strengthens emotional resonance among visitors; and second, by implementing professional service that improves on-site experiential quality. Such improvements in management efficacy not only directly boost visitor satisfaction but may also reinforce overall CES perception intensity by prolonging dwell time and

https://doi.org/10.3390/su1805257

other behavioral responses. Based on this synthesis, the following hypothesis is proposed regarding park service quality management:

H6. *The QSM has a positive effect on residents' perception variations of CES in UPGS.*

(iv) The Moderating Effect of Residents' Socioeconomic Backgrounds.

Within the mathematical modeling framework, a moderator describes a specific role characterizing the relationship between variables. When examining the influence of an independent variable on a dependent variable, the presence of a moderating variable may alter this relationship. This study collected data on residents' socioeconomic backgrounds via questionnaires, treating them as potential moderators, including gender, age, income, educational attainment, and occupation. Prior research indicates that these factors correlate with perceived differences in CES [18,27]. However, categorical variables such as gender and occupation are unsuitable for inclusion as latent variables in the model structure. Given this understanding, we utilized hierarchical regression to examine the significance of socioeconomic background as a moderator of the model relationships.

Accordingly, the conceptual model depicted in Figure 4 delineates the linkages between EBE, ILC, QSM, and residents' perception variations of CES. Within this model, RSB is incorporated as a moderating variable to examine its influence on the aforementioned pathways.

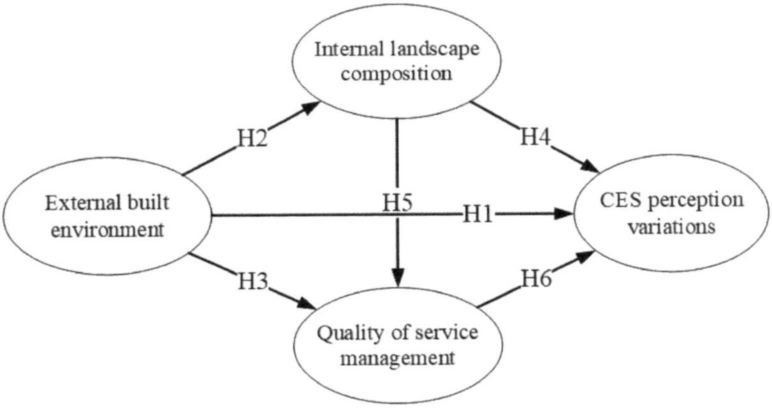

Figure 4. A hypothetical structural model.

In theoretical modeling, a moderating effect refers to the influence exerted by a variable (the moderator) on the strength or direction of the relationship between an independent variable and a dependent variable. To accurately identify such an effect, this study employs a regression-based analytical approach, the core of which lies in examining the statistical significance of the interaction term between the moderator and the independent variable. The regression equation utilized to test for this moderating effect is specified as follows:

$$Y = aX + bM + cXM + e \tag{1}$$

where X is the independent variable, Y is the dependent variable, and M represents the moderator. XM is calculated as their product.

(2) Validation of Driving Model

This study evaluated the measurement model by examining standardized factor loadings, Cronbach's alpha, Composite Reliability (CR), and Average Variance Extracted (AVE) to assess convergent validity and reliability [53–55]. Discriminant validity, which

https://doi.org/10.3390/su18052578

reflects the extent to which a latent variable is empirically distinct from other constructs in the structural model, was verified using both the Fornell-Larcker criterion and the Heterotrait-Monotrait ratio of correlations (HTMT) [55,56]. Specifically:

$$\text{AVE} = (\sum \gamma^2)/\text{n} \tag{2}$$

where γ is the factor loadings, n is the number of measurement indicators for that factor.

$$\text{CR} = (\sum \gamma)^2/((\sum \gamma)^2 + \sum \delta) \tag{3}$$

where γ is the factor loadings, δ is the residual variances.

The structural model was evaluated based on the significance and magnitude of the standardized path coefficients.

Given that both Service QSM and CES perception data were collected from the same questionnaire, a rigorous assessment of common method bias (CMB) was conducted. Procedurally, several measures were implemented to minimize potential sources of bias. These included anonymous questionnaire administration, instructions emphasizing that there were no right or wrong answers, and randomized presentation of items to reduce response patterns. These steps were designed to mitigate common method bias arising from the survey context itself. Statistically, this study employed the full collinearity variance inflation factor (VIF) test proposed by Kock to diagnose common method bias. By calculating VIF values for all latent variables in the model, this approach effectively identifies potential systematic distortion caused by the presence of a single latent factor influencing the observed relationships. Compared to the traditional Harman's single-factor test, this method offers a more rigorous and accurate diagnostic criterion [57]. The overall model fit was assessed using the coefficient of determination (R^2), and predictive relevance was examined via the Q^2 statistic, providing a comprehensive validation of the model's explanatory and predictive strength [58,59]. Additionally, the global model fit was evaluated with the Goodness-of-Fit (GoF) index [60]. Additionally, f^2 indicates the assessment of effect magnitude and is used to evaluate whether an omitted construct exerts a substantial influence on an endogenous construct [58]. Specific evaluation criteria for these indices are summarized in Table 3.

Table 3. PLS-SEM evaluation criteria.

Evaluation Metric	Interpretation	Threshold Description	
Standardized factor loadings	Strength of association between an indicator and its corresponding latent variable.	>0.7	Good
Cronbach's alpha	Internal consistency reliability within a latent variable.	>0.7	Good
CR	Internal consistency reliability indicator.	>0.7	Good
AVE	An indicator of convergent validity for a latent variable.	>0.5	Good
HTMT	Indicator for discriminant validity.	<0.85	Good
VIF	Indicator for multicollinearity among latent variables.	<3	Good
R^2	Proportion of variance explained in a latent variable.	0.20–0.33 0.34–0.67 >0.67	Weak model Moderate model Strong model
Q^2	Predictive ability of the model.	>0 0.02–0.15 0.15–0.35 >0.35	Having predictive relevance Weak predictive Moderate predictive Strong predictive

https://doi.org/10.3390/su18052578

Table 3. *Cont.*

Evaluation Metric	Interpretation	Threshold Description	
GoF	Overall model fit quality.	0.1–0.25 0.25–0.36 >0.36	Poor Moderate Good
f^2	Assess the magnitude of the effect.	0.02–0.15 0.15–0.35 >0.35	Small effect Moderate effect Large effect

3. Results

3.1. Descriptive Findings from Dual Sources

3.1.1. Identification of CES Perception Dimensions

Based on the BERTopic topic modeling of collected Weibo comments, ten thematically distinct topics with high semantic salience were identified. As illustrated in Figure 5, the topic bar chart visually displays the top five representative keywords for each topic. The analysis reveals that residents' perceptions of CES in UPGS exhibit the following structured characteristics. First, recreational activities demonstrate significant temporal dependency, correlating with specific diurnal periods and seasonal cycles (Topics 0, 6). Second, distinctive landscapes and natural features constitute the core attractions of park experiences (Topics 3, 4, 5). Third, parks function dually as physical venues for offline social interaction and as virtual platforms for online engagement (Topics 2, 9), while also promoting physical and mental well-being and facilitating collective activities (Topic 7). Furthermore, although topics directly related to "technology" account for a relatively low proportion (Topics 1, 8), their emergence indicates that science-popularization and cultural dissemination experiences have gradually entered residents' perceptual scope, reflecting a contemporary extension of the functional connotations of UPGS.

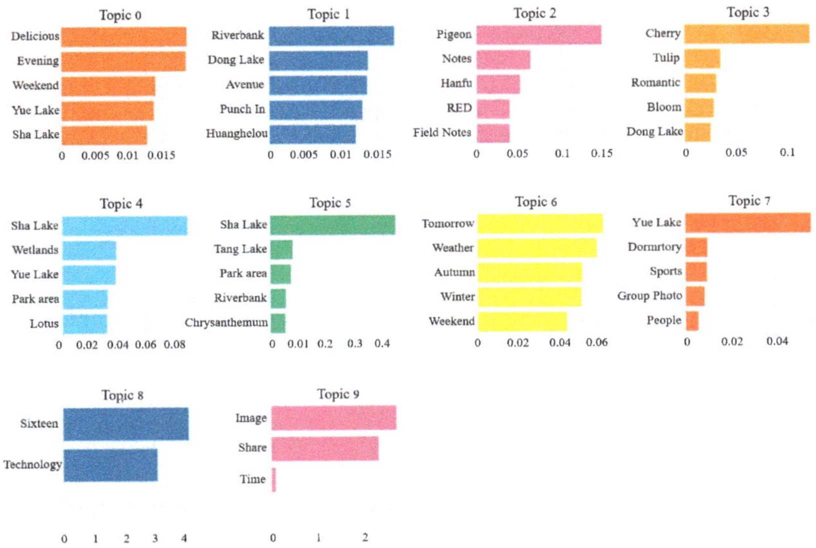

Figure 5. Visualization of topic keywords from Weibo social media data related to UPGS. The values to each keyword indicate their relative importance weights within the topic, with larger values signifying a greater semantic contribution to the topic's characterization.

https://doi.org/10.3390/su18052578

This study integrated multiple analytical methods to deepen the understanding of inter-topic relationships and to refine the interpretation of comment-based CES perceptions. Hierarchical clustering of topics, mapping of document-topic distributions, inter-topic distance analysis, and examination of keyword-weight variations were employed to structurally refine and validate the topic framework. The outcomes of these analyses are visualized in Figure 6.

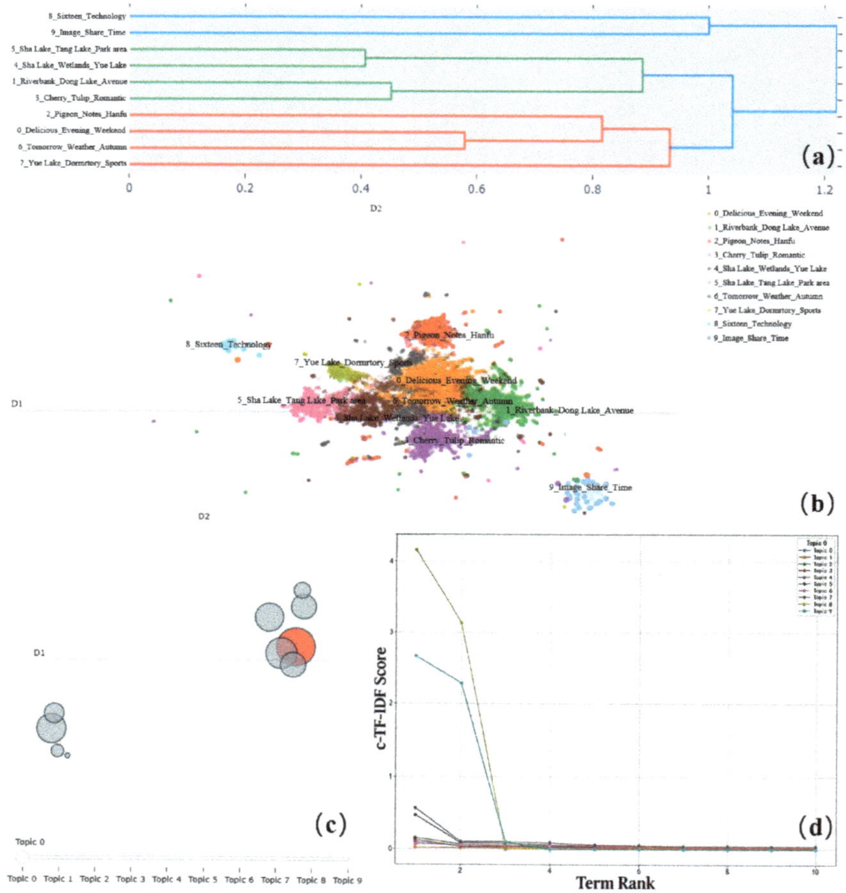

Figure 6. This figure presents the CES analysis results derived from the BERTopic model applied to Weibo text data. It consists of four standard BERTopic output subplots, each designed to convey specific analytical information. A brief explanation of each subplot follows. (**a**) Topic Hierarchical Clustering Diagram: This subplot illustrates the initial topics identified by the HDBSCAN algorithm and their hierarchical clustering relationships. (**b**) Document-Topic Distribution Map: After dimensionality reduction using UMAP, each point represents a single comment, with colors indicating its primary topic assignment. This visualization reveals how comments are clustered within the semantic space. (**c**) Inter-topic Distance Map: This plot visualizes the cosine distances between topics, calculated based on their word vector representations. The red circle denotes Topic 0. (**d**) Semantic Weight Decline of Topic Keywords: This graph displays the cumulative c-TF-IDF weight contributed by the top N keywords for each topic. The inflection point in the curve helps determine the optimal number of keywords needed to characterize a given topic.

https://doi.org/10.3390/su18052578

To elucidate the hierarchical relationships among topics, hierarchical clustering was applied, with the results presented in Figure 6a. The figure visually reveals semantic associations between topics. Specifically, Topics 0 and 6 are closely linked, indicating that temporal and meteorological factors, including time of day, weather conditions, and seasonal changes, constitute significant contextual variables shaping leisure behavior and experiential evaluations in parks. Topics 4 and 5 are directly correlated, jointly illustrating the role of parks as integrated environments for immersive natural encounters and landscape appreciation. Furthermore, Topic 3 exhibits an indirect connection to Topic 4, suggesting that activities such as flower viewing have evolved into a specialized and distinct experiential dimension within the spectrum of park services.

Figure 6b presents a two-dimensional document-topic distribution map generated through dimensionality reduction in BERTopic. The well-separated clustering of topics demonstrates robust topic distinctiveness and model stability. Based on the observed distribution patterns, the textual content was synthesized into five distinct CES perception dimensions: (1) Temporally contextualized recreation, represented by Topics 0 and 6, which reflect leisure activities strongly conditioned by diurnal, seasonal, and meteorological factors; (2) Aesthetic and nature-based engagement, captured by Topics 3, 4, and 5, focusing on aesthetic experiences associated with specific landscapes and natural features; (3) Physical activity and health promotion, corresponding to Topic 7, highlighting participatory behaviors such as sports and wellness-oriented practices; (4) Social interaction and online engagement, derived from Topics 2 and 9, illustrating the dual role of parks as physical venues for offline socialization and platforms for online content sharing; and (5) Cultural and science-informed experience, encompassing Topics 1 and 8, emphasizing the function of parks in facilitating cultural communication and public knowledge dissemination.

To further elucidate the relationships among topics, an inter-topic distance map was generated, as shown in Figure 6c. In this visualization, each circle represents a topic, with its size proportional to its document frequency and the distance between circles reflecting their semantic similarity. This analysis identifies Topic 0 ("Delicious_Evening_Weekend") and Topic 1 ("Riverbank_Dong Lake_Avenue") as the most frequently occurring themes.

Figure 6d displays the descending trend of feature word weights within each topic. The clear declining pattern observed for most topics confirms that core keywords dominate the semantic representation of each cluster. This result validates the internal coherence of the topics and supports the robustness and interpretability of the clustering outcome.

Figure 7 visualizes the semantic similarities among latent topics in the form of a heatmap. In the matrix, higher similarity values indicate a greater likelihood that topics would be grouped into the same category in subsequent merging processes. As shown in the figure, the inter-topic similarity exhibits a distinctly uneven distribution. Specifically, Topic 0 and Topic 6, Topic 3 and Topic 4, as well as Topic 4 and Topic 5, demonstrate strong semantic coherence, suggesting they may belong to a shared semantic category or functional dimension. In contrast, Topics 8 and 9 exhibit relatively low similarity with the other topics in the matrix. This pattern indicates a degree of semantic distinctiveness, suggesting they may represent distinct perceptual perspectives within the comment data.

This study applied the BERTopic model to analyze Weibo text comments, identifying ten initial topics. The analysis revealed that several topics exhibited high semantic similarity and close inter-topic relationships. Guided by the theoretical framework of CES and following criteria including proximity in hierarchical clustering, high inter-topic similarity, and spatial adjacency in the document-topic distribution, related topics were merged.

As shown in Figure 7 and Appendix B Table A4, several topic pairs exhibit high semantic similarity, including Topic 0 and Topic 6 (similarity coefficient = 0.818), Topic 3 and Topic 4 (0.808), Topic 4 and Topic 5 (0.847), and Topic 5 and Topic 7 (0.815). Additionally,

https://doi.org/10.3390/su18052578

strong associations are observed between Topic 2 and Topic 6 (0.831) and between Topic 2 and Topic 8 (0.776). Other notable correlations include Topic 1 with Topic 0 (0.820) and Topic 1 with Topic 4 (0.824). These relationships form an interconnected network centered around Topic 4. Although substantial semantic overlap exists among these topics, each retains distinct thematic emphases in terms of specific content. Based on the visual presentation of the BERTopic modeling results (Figure 6) and a semantic interpretation of the topic keywords in relation to each CES dimension, the initial topics were reclassified into five more interpretable dimensions. Dimension 1 (Topics 0 and 6): These two topics exhibit high similarity (0.818) and share a semantic focus on recreational services (RS). Dimension 2 (Topics 3, 4, and 5): Although Topic 3 and Topic 5 have a relatively low direct similarity (0.525), they are semantically bridged through Topic 4. Together, they form a thematic cluster encompassing aesthetic experiences (AE). Dimension 3 (Topic 7): Despite showing strong correlations with other topics, such as Topic 5 (0.815), its distinctive keywords justify its treatment as an independent dimension related to health-promoting activities (HA). Dimension 4 (Topics 2 and 9): Although their statistical similarity is low (0.363), both topics converge on the theme of social interaction. Topic 2 emphasizes both online and offline social contexts (e.g., Hanfu gatherings, Xiaohongshu sharing), while Topic 9 focuses on digital media sharing behaviors. Semantically, they can be consolidated into a single dimension representing social interaction (SI). Dimension 5 (Topics 1 and 8): These topics also exhibit low statistical similarity (0.380). However, Topic 1 relates to cultural landmark visitation and documentation, while Topic 8 pertains to science and technology education. Together, they fall within the broader domain of educational and cultural services (ES), encompassing both humanities and science-oriented cultural content.

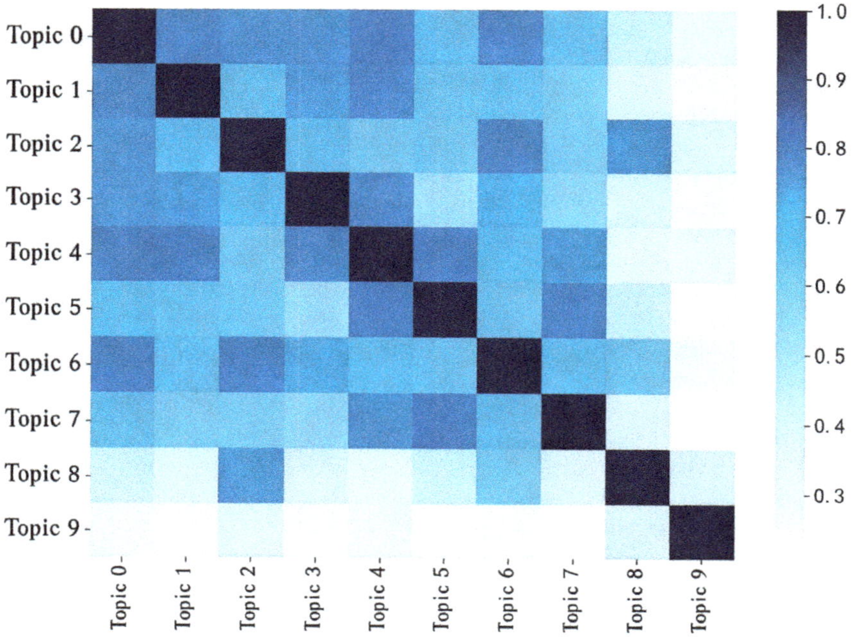

Figure 7. Topic similarity heatmap.

Following this data-driven consolidation, manual semantic validation was conducted to ensure conceptual consistency. To further ensure the reliability and ecological validity

https://doi.org/10.3390/su18052578

of the analyses, an inter-coder reliability check was performed both before and after the formal coding process, with the coding manual detailed in Appendix B, Table A5.

During the pilot coding phase, two independent coders conducted trial coding on a randomly selected sample of 50 comments, achieving an agreement rate of 76.0% and a Cohen's Kappa of 0.73. In the formal coding phase, the same two coders independently classified the full sample of 500 comments, yielding an agreement rate of 65.4% and a Cohen's Kappa of 0.61. Both Kappa values fell within the acceptable range, indicating satisfactory reliability of the coding results.

Using manual coding as the benchmark, the BERTopic model achieved an overall accuracy of 76.0% on the 500-comment test set, with a weighted average F1 score of 0.76 and a Cohen's Kappa of 0.7119. At the topic level, Topics 2 (social sharing), 5 (sense of place), 7 (physical activity), and 8 (culture and history) demonstrated excellent performance, with F1 scores exceeding 0.83. Topics 0 (food-related leisure), 1 (landmark check-ins), 6 (seasonal recreation), and 9 (family-oriented social interaction) showed good performance, with F1 scores ranging from 0.70 to 0.78. Topic 4 (wetland ecology) obtained a moderate F1 score of 0.58. Overall, the classification results of the model showed satisfactory agreement with manual coding, supporting its applicability for subsequent analysis.

Ultimately, the topics were integrated and categorized into five CES perception dimensions: Recreational Services (RS), Aesthetic Experiences (AE), Health-promoting Activities (HA), Social Interactions (SI), and Educational Services (ES). The detailed classification is presented in Table 4. As detailed in Appendix B Table A6, the CES connotations, keywords, document counts, and representative comments for both the ten initial topics and the five consolidated CES dimensions are comprehensively presented. The results confirm that the consolidation process preserved essential semantic information without any loss of critical content.

To assess the stability of CES dimension identification across different user groups, two subsets of Weibo users were delineated based on the time of comment posting: daytime and nighttime. The BERTopic model was then reapplied to each of these subsets. As shown in Appendix B, Figure A1, topics related to the CES dimensions, namely RS, AE, HA, SI, and ES, consistently emerged across both subsets. This finding indicates the robustness of the dimensional identification results to variations in user subgroups.

Table 4. Summary of Residents' Perception Dimensions of CES in UPGS.

CES	Topics	Keywords	Classification Basis
RS	Topic 0	Delicious, evening, weekend, Yue Lake, Sha Lake	Reflects comprehensive recreational behaviors, representing fundamental leisure services.
	Topic 6	Tomorrow, weather, autumn, winter, weekend	Describes conditional and seasonal leisure planning, such as sunbathing based on weather or time of year.
AE	Topic 3	Cherry, tulip, romantic, bloom, Dong Lake	Centers on spring flower appreciation, evoking a romantic emotional response and exemplifying natural aesthetic engagement.
	Topic 4	Sha Lake, wetlands, Yue Lake, park area, lotus	Involves appreciation of wetland ecosystems, floral diversity, and curated exhibitions, highlighting horticultural and ecological aesthetics.
	Topic 5	Sha Lake, Tang Lake, park area, riverbank, chrysanthemum	Focuses on viewing thematic horticultural displays like chrysanthemum exhibitions, often accompanied by social media documentation.
HA	Topic 7	Yue Lake, dormitory, sports, group photo, people	Dominated by the term "sports", which directly indicates health-promoting physical activity; other terms provide contextual support.

https://doi.org/10.3390/su18052578

Table 4. *Cont.*

CES	Topics	Keywords	Classification Basis
SI	Topic 2	Pigeon, notes, Hanfu (traditional Chinese Han clothing), RED (a social media platform in China), field notes	Activities such as community-organized Hanfu gatherings reflect offline social engagement, while terms like "RED" and "Notes" indicate online social interaction through content sharing.
	Topic 9	Image, share, time	Combines "share" and "image" to illustrate online social behavior centered on content dissemination via digital media.
ES	Topic 1	Riverbank, Dong Lake, avenue, punch In, Huanghelou	Keywords refer to well-known landmarks, natural landscapes, and flora, reflecting visitation, learning, and documentation of local cultural and natural heritage.
	Topic 8	Sixteen, technology	The term "technology" explicitly points to activities involving scientific knowledge and technological exhibitions, aligning with science education and technological culture promotion.

3.1.2. Derivation of Potential Driving Factors

From the perspective of environmental psychology, perception arises from the complex interactions among individual characteristics, social contexts, and physical environments. Based on preprocessed Weibo comment data, this study employed Python and Excel to conduct statistical and content analysis of high-frequency words. Through this process, key influencing factors and their corresponding indicators reflecting residents' differential perceptions of CES in UPGS were extracted, as summarized in Table 5.

Table 5. Driving factors of residents' CES perception heterogeneity in UPGS and their corresponding high-frequency keywords.

Factors	Indicators	High-Frequency Words
Residents' socioeconomic backgrounds	Gender	Auntie, uncle, young lady, handsome guy
	Age	Children, teenagers, university students, middle-aged, elderly, retired
	Income	High-end restaurant, buffet, convenience store, food stall, free
	Educational attainment	University student, postgraduate
	Occupation	Photographer, security guard, barista, tour guide, cleaner, retired
External built environment	Surrounding building density	Pedestrian street, business district, high-rise building, residential compound, shop, residence, storefront, office building
	Transportation accessibility	Parking lot, main road, subway station, geographical location, parking space, bus, ferry, shared bicycle
	Population density	Tourists, crowd, residents, crowded, pedestrian flow, gathering, high pedestrian flow, peak hours
Internal landscape composition	Facility richness	Street light, fitness equipment, coffee shop, public toilet, children's playground, vending machine, bench, trash bin
	Park area	Site area, garden path, trail, lawn, square, park area, expansive green space, open, view
	Water feature	Lake, river beach, lotus, boating, water cypress, lotus pond, reed, fountain, small bridge over flowing water

https://doi.org/10.3390/su1805257

Table 5. *Cont.*

Factors	Indicators	High-Frequency Words
Quality of services management	Intra-park environment management	Security, cleaning, greening maintenance, repair, inspection, upkeep, weeding, waste sorting
	Visitor experience management	Photography, check-in, guided tour, volunteer, selfie, wedding photography, explanation, visitor center, interactive experience
	Cultural promotion management	Intangible cultural heritage, exhibition, poetry, Hanfu, lantern, folk custom, historical stone inscription, pagoda forest, information board

3.1.3. Characteristics of Driving Factors

Statistical analysis of the socioeconomic backgrounds of 313 respondents is presented in Table 6. The sample exhibited a balanced gender distribution, with females comprising 48.9% and males 51.1% of respondents. Age distribution skewed toward adults, with significant proportions in the 18–30 cohort (29.1%) and seniors over 60 (23.6%). Respondents demonstrated relatively high educational levels: 25.6% held associate degrees, and 45.6% possessed bachelor's degrees or higher qualifications. Using monthly income as the metric, the majority (49.8%) reported earnings between CNY2000 and CNY5000, while 28.4% earned below CNY2000. Only 4.1% fell into the high-income bracket (≥CNY9000). The sample comprised retirees (33.5%), employed professionals (28.4%), freelancers (18.5%), and students (12.7%). Reliability analysis yielded a Cronbach's alpha coefficient of 0.859, indicating strong internal consistency. The KMO measure verified sampling adequacy (KMO = 0.897), supporting the scale's factorial validity.

Table 6. Socioeconomic Backgrounds of Respondents.

Variables	Category	Sample	
		N	%
Gender	Female	153	48.9
	Male	160	51.1
Age	<18	3	1
	18–30	91	29.1
	31–45	73	23.3
	46–60	72	23
	>60	74	23.6
Income	<CNY2000	89	28.4
	CNY2000–CNY5000	156	49.8
	CNY5000–CNY9000	55	17.5
	>CNY9000	13	4.1
Educational attainment	High school or below	90	28.8
	College	80	25.6
	Bachelor's degree	104	33.2
	Master's degree or above	39	12.4
Occupation	Students	40	12.7
	Corporate employees	47	15
	Public sector employees	42	13.4
	Freelancers	58	18.5
	Retirees	105	33.5
	Others	21	6.7

https://doi.org/10.3390/su18052578

Table 7 presents the evaluation results of 313 respondents on the service management quality of the sampled parks. A majority of the residents gave favorable ratings to the park's service management. Specifically, 47.2% of respondents expressed satisfaction with the IEM, 44% reported a positive experience with the VEM, and 45.3% held a positive attitude toward the CPM. Furthermore, among these three service dimensions, CPM received the highest mean score (M = 3.30), which was close to the score for IEM (M = 3.27). The VEM had the lowest mean score (M = 3.24). Notably, the average ratings for all three services exceeded the neutral midpoint of 3.

Table 7. Respondents' comments on the services management quality of sampled parks.

Variables	Category	Sample	
		N	%
Intra-park environment management	Very bad (1)	16	5.1
	Bad (2)	93	29.7
	Neutral (3)	56	17.8
	Good (4)	85	27.1
	Very good (5)	63	20.1
Visitor experience management	Very bad (1)	16	5.1
	Bad (2)	82	26.1
	Neutral (3)	77	24.6
	Good (4)	86	27.4
	Very good (5)	52	16.6
Cultural promotion management	Very bad (1)	12	3.8
	Bad (2)	86	27.4
	Neutral (3)	73	23.3
	Good (4)	79	25.2
	Very good (5)	63	20.1

The external built environment and internal landscape composition of the sampled parks were analyzed using geospatial techniques in ArcGIS 10.8. First, a set of spatial indicators for Wuhan was generated, encompassing surrounding building density, transportation accessibility, population density, facility richness, park area, and water features. Subsequently, these indicators were spatially matched with their corresponding sampled parks. The results are presented in Figure 8.

The EBE and ILC of the 13 sampled parks were analyzed and classified according to specific metrics. For the indicator grading, SBD, PD, FR, PA, and WF were graded on a scale from 1 (lowest) to 5 (highest), while TA was graded inversely from 1 (highest) to 5 (lowest). The resulting environmental classification for all parks is summarized in Table 8.

As evidenced by Figure 8 and Table 8, significant spatial inequity exists in the EBE. Parks in central urban areas exhibit higher SBD, greater TA, and elevated PD, with these metrics displaying a clear declining gradient from the urban core to peripheral zones. Among the sampled parks, Jiefang Park, Zhongshan Park, and Qiaokou Park exhibited the highest levels of SBD, TA, and PD, respectively. These were followed by Guanshan Park, Chuwangtai Historical Park, and the Botanical Science Park. Due to their distinct attributes and functional orientations, certain parks exhibit significant variations in their EBE. For instance, Houguanhu Park prioritizes ecological conservation, while Wuhan Zoo features expansive habitats. Both of them are renowned destinations with high TA but low SBD. In contrast, Qianchuan Park and Chaibohu Leisure Square, located in peri-urban areas, demonstrate uniformly low levels of SBD, TA, and PD.

As shown in Table 8, parks such as Jiefang Park, Zhongshan Park, and Jiangxia Central Park show comparative advantages in both PA and FR. Caidian Riverside Park

https://doi.org/10.3390/su18052578

and Houguanhu Park, located adjacent to major water bodies, feature abundant aquatic landscapes but offer relatively limited facilities. Parks, including Qiaokou Park, Guanshan Park, Chaibohu Leisure Square, and Wuhuan Square, generally have smaller PA, with their WF and FR levels largely shaped by surrounding urban conditions. In contrast, certain parks exhibit notable spatial and hydrological heterogeneity aligned with their specialized functions, while consistently maintaining higher FR ratings across different park types.

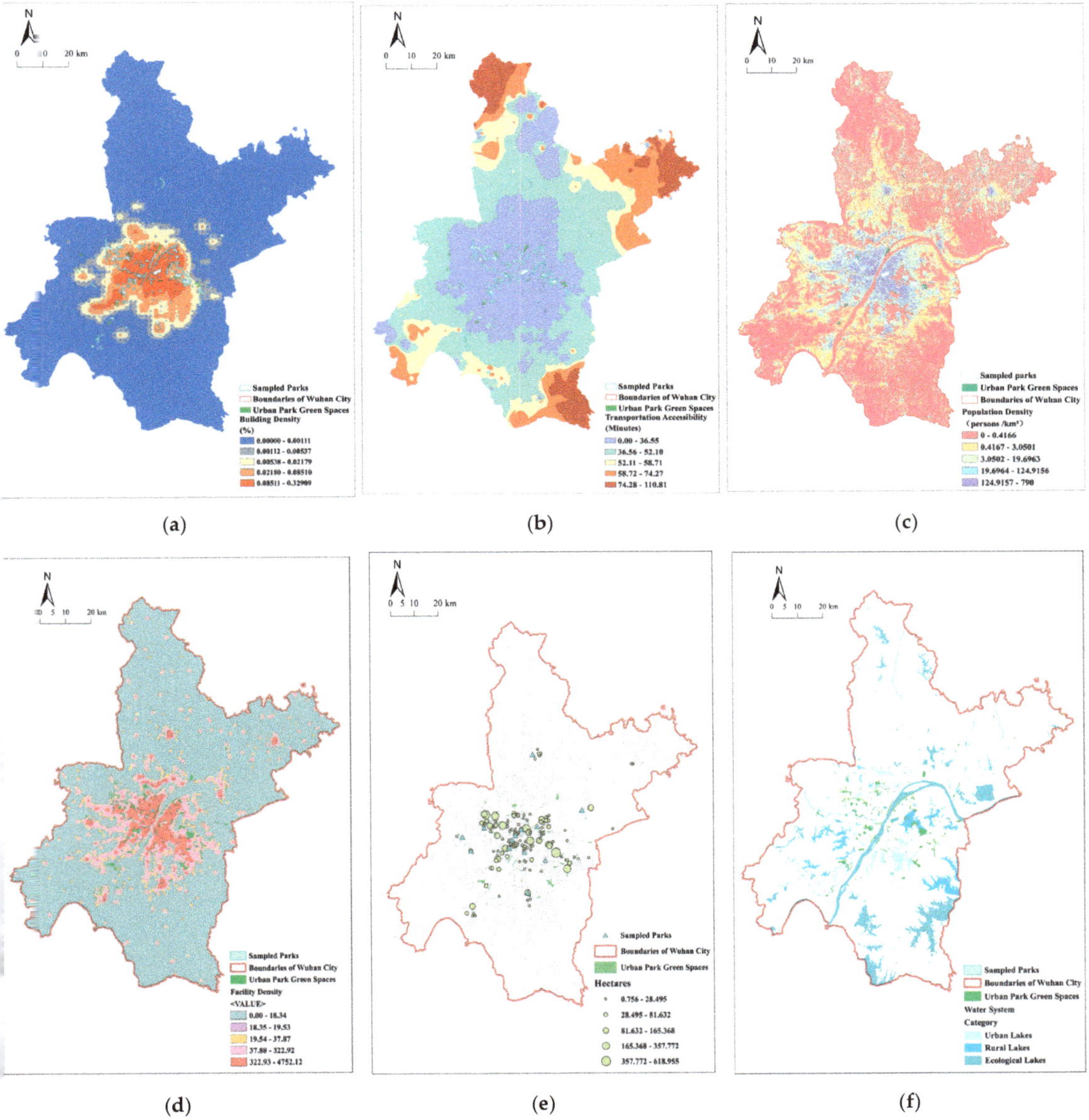

Figure 8. Geospatial analysis of the parks: (**a**) Surrounding building density; (**b**) Transportation accessibility; (**c**) Population density; (**d**) Facility richness; (**e**) Park area; (**f**) Water feature.

https://doi.org/10.3390/su18052578

Table 8. Classification of environment levels for sampled parks.

Parks	SBD	TA	PD	FR	PA	WF
Jiefang Park	5	5	5	5	4	2
Zhongshan Park	5	5	5	5	3	2
Jiangxia Central Park	3	4	3	4	5	3
Caidian Riverside Park	2	4	2	2	3	5
Houguanhu Park	1	4	1	3	3	5
Qianchuan Park	1	3	2	1	3	1
Qiaokou Park	5	5	5	3	1	3
Guanshan Park	4	5	4	4	2	1
Chaibohu Leisure Square	1	3	3	1	2	5
Wuhuan Square	2	5	5	3	1	1
Wuhan Zoo	3	5	1	5	5	4
Chuwangtai Historical Park	5	5	5	4	1	3
Botanical Science Park	4	5	5	4	2	3

3.1.4. Patterns in CES Perception Satisfaction

Pearson correlation analysis revealed significant synergistic associations among all resident-perceived CES dimensions ($p < 0.001$). As shown in Figure 9, strong positive correlations were observed between AE and SI (r = 0.93), as well as between HA and RS (r = 0.91). This pattern suggests a close relationship between residents' appreciation of scenic beauty and their engagement in social behaviors within parks: aesthetically appealing landscapes create favorable settings for social activities, while positive social interactions can, in turn, enrich the perceived aesthetic value. Furthermore, HA and RS show a high degree of functional integration in actual park use, suggesting that residents often experience these two dimensions concurrently. Park environments conducive to positive exercise experiences are strongly associated with enhanced recreational atmosphere and overall satisfaction.

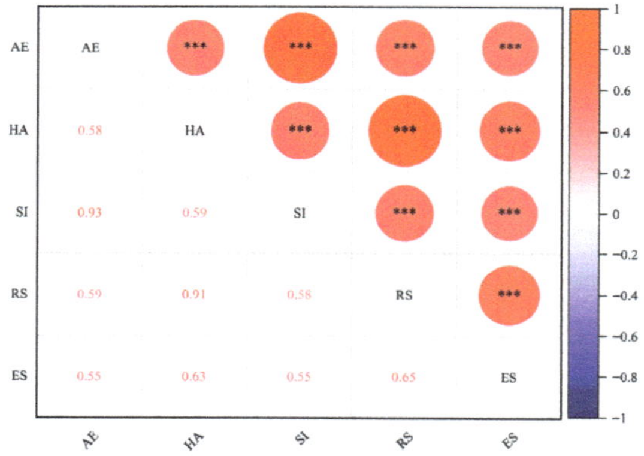

Figure 9. Correlation analysis of CES perceptions. *** $p \leq 0.001$.

Additionally, ES maintained consistently moderate to strong positive correlations with all other CES indicators (r > 0.50). These findings underscore that residents perceive the various CES functions in a highly integrated and synergistic manner. The different dimensions reinforce one another, collectively contributing to a high-quality experience in UPGS.

The Friedman rank-sum test was conducted to assess differences in residents' perceived satisfaction across the five CES categories. The overall test results revealed statisti-

https://doi.org/10.3390/su1805257

cally significant differences ($\chi^2 = 48.086$, df = 4, $p < 0.001$). Post pairwise comparisons were performed, with results displayed in Figure 10.

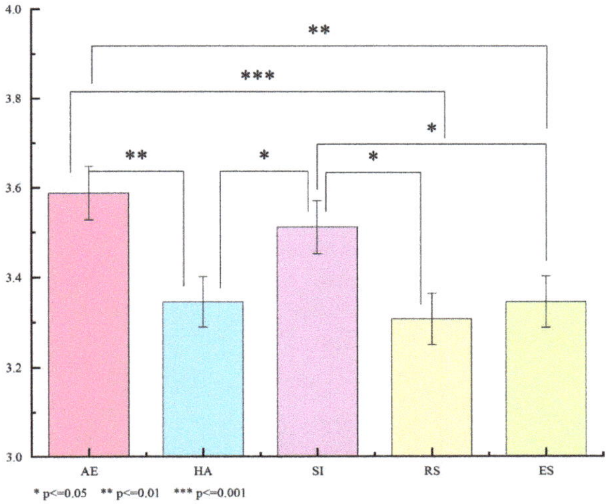

Figure 10. Variability assessment of residents' perceptions regarding CES.

As shown in Figure 10, significant disparities in satisfaction levels were observed among the CES dimensions. AE garnered the highest mean satisfaction score (M = 3.59, SD = 1.059), followed by SI (M = 3.51, SD = 1.044). In contrast, HA (M = 3.35, SD = 0.995), ES (M = 3.35, SD = 1.017), and RS (M = 3.31, SD = 1.010) yielded the lowest mean scores, which were statistically comparable to each other. These results suggest that while parks are perceived to deliver strong aesthetic and social value, their performance in supporting health, recreation, and educational functions is rated relatively lower. This pattern may reflect limitations in facility diversity or a resident preference for unstructured, experiential engagement over formally organized programs.

3.2. Validation of the Driver-Based PLS-SEM Model

3.2.1. Evaluation of the Measurement Model

In alignment with the PLS-SEM evaluation criteria outlined in Table 3, the findings detailed in Table 9 confirm full compliance with all specified criteria. This substantiates a measurement model characterized by robust internal consistency and adequate convergent validity.

Table 9. Reliability and validity indicators.

Construct	Variable	Factor Loading	Cronbach's Alpha	CR	AVE
EBE	SBD	0.852	0.826	0.828	0.742
	TA	0.880			
	PD	0.851			
ILC	FR	0.873	0.820	0.820	0.735
	PA	0.851			
	WF	0.848			
QSM	IEM	0.853	0.818	0.819	0.733
	VEM	0.851			
	CPM	0.864			

https://doi.org/10.3390/su18052578

Table 9. *Cont.*

Construct	Variable	Factor Loading	Cronbach's Alpha	CR	AVE
CES	ES	0.794	0.905	0.907	0.725
	SI	0.857			
	AE	0.852			
	HA	0.874			
	RS	0.879			

In this study, discriminant validity was cross-validated using the Fornell-Larcker criterion and HTMT. As summarized in Table 10, all HTMT values were below the threshold of 0.85, and the square root of the AVE for each latent variable exceeded its correlations with all other constructs. Both criteria were satisfactorily met, thereby confirming that the measurement model exhibits robust discriminant validity and that the constructs are statistically distinct and empirically separable.

Table 10. Discriminant validity.

Constructs	CES	QSM	ILC	EBE
CES	**0.852**	0.621	0.606	0.608
QSM	0.719	**0.856**	0.578	0.579
ILC	0.701	0.704	**0.857**	0.655
EBE	0.702	0.701	0.795	**0.861**

Note: Diagonal values (in bold) represent $\sqrt{AVE}$, values above the diagonal represent inter-structural correlations, and values below the diagonal represent the HTMT statistics.

3.2.2. Assessment of the Structural Model

All VIF values between latent constructs ranged from 1.000 to 1.960, falling below the critical threshold (Table 3), which indicates the absence of severe collinearity issues and supports the robustness of subsequent analyses. The model fit results show that the R^2 values for CES, QSM, and ILC were 0.510, 0.404, and 0.429, respectively, demonstrating satisfactory explanatory power. Further predictive relevance testing revealed Q^2 values of 0.364 for CES, 0.330 for QSM, and 0.424 for ILC, indicating adequate predictive capability of the model. The GoF index reached 0.573, further confirming the high overall fit quality of the model. Additionally, the effect sizes (f^2) of the paths in the model ranged from 0.065 to 0.752, revealing a clear gradient in the strength of influence across different relationships. Specifically, the paths EBE $\rightarrow$ QSM (f^2 = 0.119), ILC $\rightarrow$ QSM (f^2 = 0.115), and QSM $\rightarrow$ CES (f^2 = 0.133) all exhibit effects approaching the medium threshold. In contrast, EBE $\rightarrow$ CES (f^2 = 0.067) and ILC $\rightarrow$ CES (f^2 = 0.065) were categorized as small effects. Notably, the EBE $\rightarrow$ ILC pathway exhibited a substantial effect, with an f^2 value of 0.752, indicating that the external built environment exerts a dominant influence on internal landscape composition. In planning practice, internal landscape design should be adapted to the characteristics of the external built environment. For instance, in high-density urban areas, priority may be given to buffering landscapes that mitigate environmental stressors, while suburban areas may benefit more from inviting landscape designs that enhance accessibility and engagement. The pathways EBE $\rightarrow$ CES (f^2 = 0.067) and ILC $\rightarrow$ CES (f^2 = 0.065) exhibited relatively small effect sizes, suggesting that improving the external environment alone or making localized adjustments to internal landscape composition has limited effectiveness in directly enhancing cultural service experiences. In contrast, the pathways EBE $\rightarrow$ QSM (f^2 = 0.119), ILC $\rightarrow$ QSM (f^2 = 0.115), and QSM $\rightarrow$ CES (f^2 = 0.133) exhibited moderate effect sizes, thereby revealing the core transmission mechanism: these elements influence cultural service experiences primarily through service management quality as a

https://doi.org/10.3390/su18052578

mediating factor. This finding highlights that practical efforts in park management should prioritize the optimization of service quality.

To assess the statistical significance of the path coefficients (β), a bias-corrected and accelerated bootstrap procedure with 5000 resamples was employed to derive parameter estimates and their 95% confidence intervals (CIs). The 95% CIs for all hypothesized paths excluded zero, confirming their statistical significance.

For the total effect of the EBE on CES perceptions, the bootstrap analysis yielded a coefficient of β = 0.608, with a 95% CI of [0.523, 0.685]. The relatively narrow confidence interval indicates that improvements in park surrounding conditions can exert a substantial positive influence on residents' CES perceptions. In contrast, the direct effect of QSM on CES perceptions was β = 0.331, with a 95% CI of [0.204, 0.464]. The wider confidence interval suggests that, while improving service management quality is effective, the precise magnitude of its effect may vary across different park types or resident groups.

The final path coefficients, confidence intervals, significance levels, and other relevant indicators are presented in detail in Table 11.

Table 11. Standardized path coefficients of the hypothesized model.

Path Effects	Relationship	β	Standard Deviation	95% Bootstrap CI	T Statistics	*p* Values	VIF	f^2	H	Conclusion
Direct effects	EBE → QSM	0.352	0.061	[0.226, 0.466]	5.740	0.000	1.752	0.119	H3	Support hypothesis
	EBE → ILC	0.655	0.040	[0.573, 0.730]	16.319	0.000	1.000	0.752	H2	Support hypothesis
	EBE → CES	0.254	0.065	[0.126, 0.380]	3.916	0.000	1.960	0.067	H1	Support hypothesis
	ILC → QSM	0.347	0.060	[0.233, 0.466]	5.828	0.000	1.752	0.115	H5	Support hypothesis
	ILC → CES	0.248	0.066	[0.118, 0.374]	3.761	0.000	1.954	0.065	H4	Support hypothesis
	QSM → CES	0.331	0.066	[0.204, 0.464]	4.980	0.000	1.679	0.133	H6	Support hypothesis
Indirect effects	EBE → ILC → CES	0.163	0.045	[0.075, 0.251]	3.601	0.000				
	EBE → ILC → QSM	0.227	0.043	[0.149, 0.317]	5.275	0.000				
	EBE → ILC → QSM → CES	0.075	0.022	[0.039, 0.124]	3.436	0.001				
	EBE → QSM → CES	0.116	0.032	[0.061, 0.186]	3.672	0.000				
	ILC → QSM → CES	0.115	0.032	[0.062, 0.184]	3.588	0.000				
Total effects	EBE → CES	0.608	0.042	[0.523, 0.685]	14.587	0.000				
	ILC → CES	0.363	0.060	[0.248, 0.481]	6.08	0.000				
	QSM → CES	0.331	0.066	[0.204, 0.464]	4.980	0.000				

3.3. *Analysis of the Driving Mechanisms*

3.3.1. Pathways of Key Drivers and Their Interplay

Based on the validated PLS-SEM model (Figure 11), this study reveals a coherent, sequentially transmitted mechanism underlying residents' perception differences in CES within UPGS. All path coefficients reached statistical significance ($p < 0.001$), fully supporting research hypotheses H1–H6.

Analysis of the direct effects reveals a clear hierarchy of proximal drivers. QSM emerged as the most potent direct driver (β = 0.331, $p < 0.001$; $f^2 = 0.133$), indicating that service management experience proximately shapes final evaluations. The ILC also showed a significant positive direct effect (β = 0.248, $p < 0.001$), functioning as the essential physical medium for CES experiences. While the EBE exhibited a direct effect of comparable magnitude (β = 0.254, $p < 0.001$), its significantly greater total effect (β = 0.608) indicated a more complex, mediation-dominant mechanism of influence.

The perceptual differences are primarily explained by a robust multi-mediation transmission system. The EBE's influence was largely indirect, with 58.2% of its total effect

https://doi.org/10.3390/su18052578

(β = 0.354) mediated through other constructs. Two significant indirect pathways were identified: EBE → ILC → CES (β = 0.163) and EBE → QSM → CES (β = 0.116). Critically, the chained mediation pathway EBE → ILC → QSM → CES (β = 0.075) delineates the complete logical sequence from environmental foundation to perceptual outcome. This path illustrates how favorable external conditions foster superior internal landscapes, which in turn create both the necessity and the capacity for high-standard management services, thereby collectively driving elevated CES perceptions.

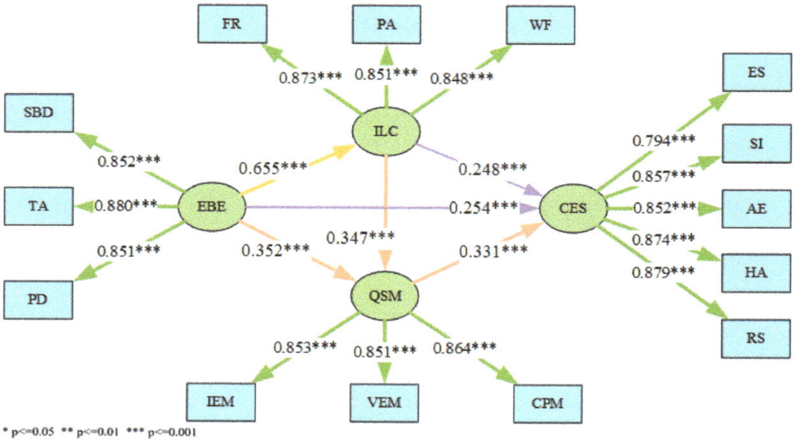

Figure 11. Driving pathways of factors influencing residents' CES perception differences.

Furthermore, the model elucidates a synergistic interaction between core physical and managerial attributes. ILC significantly amplified its total impact (Total effect = 0.363) through QSM (ILC → QSM → CES, β = 0.115). This interaction highlights that even with comparable physical foundations, divergent management performance is a key determinant of varied perceptual outcomes. The strength of this landscape-management synergy fundamentally explains the heterogeneity in perception levels observed across different parks.

It is noteworthy that the total effect of EBE on CES perceptions (β = 0.608) substantially exceeds its direct effect (β = 0.254). This marked discrepancy underscores the critical mediating role of indirect pathways in translating the advantages of the external built environment into perceptible benefits for residents. From the perspective of urban resource allocation, this strong mediation effect suggests that investment strategies should not be limited to the unidimensional optimization of the external built environment. Instead, a coordinated approach along the EBE–ILC–QSM pathway is warranted, wherein improvements to a park's external surroundings are complemented by concurrent upgrades to its internal landscape composition. Such alignment helps prevent efficiency losses that may arise from decoupled interventions. Regarding locational inequality, the mediating effect significantly amplifies the perceptual gap between parks in core urban areas and those on the periphery. Parks in core areas benefit from favorable external built environments, which, combined with high-quality internal landscape composition and service management, continuously reinforce their advantages. In contrast, parks in peripheral areas face a dual constraint: they are disadvantaged not only by inferior external conditions but also by weak mediating pathways. This structural disparity suggests that policy resources should be redirected toward peripheral areas. Increased investment in internal landscape composition and service management in these areas may help mitigate the entrenched perceptual gap arising from locational differences. From the perspective of fiscal and

https://doi.org/10.3390/su18052578

institutional support, it is essential to establish a coordinated mechanism that links improvements in the external built environment with enhancements in internal landscape composition and service management quality. Incorporating the synergistic upgrading of ILC and QSM into the performance criteria for urban park development, while allocating dedicated funding specifically for strengthening mediating pathways, would enable more effective transmission of EBE advantages through robust mediating processes. Such a strategy would ultimately contribute to more equitable and high-quality provision of CES in UPGS.

In summary, the findings indicate that CES perception differences do not stem from isolated factors but emerge from an integrated, sequentially transmitted driving system. Within this system, the EBE acts as the foundational contextual precondition, the ILC serves as the core physical medium that manifests these conditions, and QSM functions as the pivotal transformational agent that converts latent environmental and landscape potentials into tangible experiential outcomes.

3.3.2. Moderating Effects on the Driving Pathways

To systematically investigate the moderating role of RSB within the theoretical model, five key variables were examined, including gender, age, monthly income, education level, and occupation. Moderating effects for binary and ordinal variables were assessed using linear regression models with interaction terms, while multi-categorical variables were analyzed via group comparison approaches. The influence of RSB on the hypothesized driving pathways is visually summarized in Figure 12. The analysis revealed distinct moderating patterns. Specifically, for hypothesis H6, the interaction between QSM and age reached statistical significance ($\beta = 0.093$, $p = 0.037$), indicating that age moderates the relationship between QSM and residents' CES perception variations. Older residents exhibit a stronger positive response to improvements in service quality. Regarding hypothesis H5, the interaction between ILC and monthly income approached marginal significance ($\beta = -0.079$, $p = 0.069$), suggesting a potential, though not fully conclusive, moderating role of income in the pathway from ILC to QSM. As income increases, the positive influence of landscape optimization on perceived service quality appears to diminish, indicating that higher-income groups may be less sensitive to enhancements in management services resulting from landscape improvements. No statistically significant moderating effects were detected for the remaining socioeconomic variables across any pathways in the model, based on both *p*-value and z-score criteria.

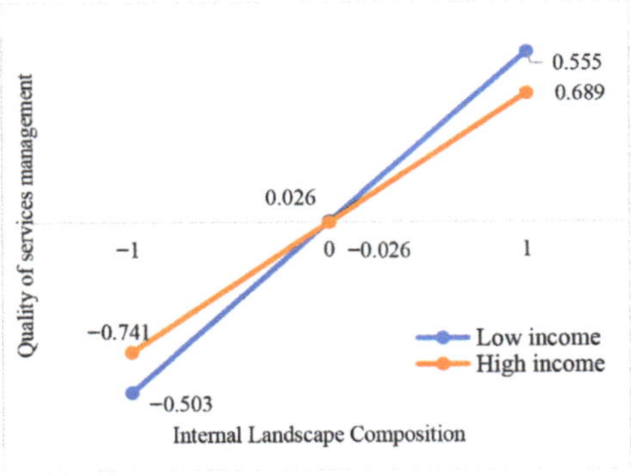

Figure 12. Moderating effects of residents' socioeconomic background on the driving pathways: (**a**) Moderating Effect of Age; (**b**) Moderating Effect of Income.

https://doi.org/10.3390/su18052578

4. Discussion

4.1. Integrated Insights into CES Perceptions from Dual-Source Data

Environmental psychology theory indicates that human perception of the environment involves a dual process, consisting of both spontaneous, context-dependent, immediate responses and reflective, integrated, systematic judgments [61]. Traditional questionnaire surveys can gather reflective data with demographic representativeness. However, they are often influenced by recall bias and social desirability effects. Consequently, they struggle to capture real-time psychological experiences in dynamic environmental settings [62]. In contrast, large-scale social media data can record authentic instant behaviors and emotional expressions from the public [63]. Nonetheless, such data are subject to challenges like sample bias (e.g., an overrepresentation of younger user groups). This may limit the generalizability of findings, and these data often do not directly uncover underlying causal relationships [64]. Therefore, this study employs a sequential mixed-methods design that integrates online social media data, offline questionnaire surveys, and spatial data to systematically investigate the differences in residents' perceptions of CES in UPGS and their underlying driving mechanisms. Social media captures spontaneous emotional expressions, while questionnaires generate reflective evaluations, making the two approaches highly complementary. Social media data provide fine-grained, real-time emotional responses within specific contexts, whereas questionnaire data offer structured interpretations from sociocultural and demographic perspectives [65,66].

Although the physical environments of the studied parks are relatively homogeneous, the mixed-methods approach effectively differentiates between "immediate reactions" and "systematic evaluations". This integration enhances both the breadth and depth of insight, leading to a more comprehensive understanding of residents' perceptions and experiences of CES in parks [67].

Analysis based on spontaneous expressions from social media reveals that residents' perceptions of CES exhibit notable dynamism and non-linear characteristics [68]. Such a process of value formation is characterized by its real-time generation and high context dependency. These qualities are often difficult to capture through traditional survey methods, which rely on retrospective accounts or preset response options [69]. This finding suggests that many critical cultural service experiences do not exist as static attributes of parks, but rather emerge spontaneously during real-time, context-specific interactions between individuals and their environment, fluctuating dynamically with situational factors [70]. Consequently, this insight prompts a conceptual shift in understanding CES from a static classification of services toward a theoretical perspective that emphasizes dynamic generation and process-based construction, thereby highlighting the immediacy, contextual embeddedness, and interactive nature of perception.

This study proposed and validated a sequential mixed-methods analytical framework that integrates BERTopic topic modeling with PLS-SEM. The framework first employs BERTopic, a deep learning-based technique. It inductively identifies core dimensions of residents' perceptions of CES and their associated driving factors from unstructured social media texts, thereby generating quantitative indicators grounded in authentic contextual expressions. While these dimensions conceptually align with traditional typologies such as those of the MA and CICES, their underlying logic of generation differs fundamentally [2,71]. Traditional classification systems are derived from top-down expert synthesis and present static knowledge structures [72]. In contrast, the dimensions identified in this study emerge from spontaneous, context-specific expressions by the public, reflecting a bottom-up and dynamically evolving perceptual construct. Thus, the theoretical contribution of this study is twofold. First, it is not merely about proposing new classification labels. Second, it lies in employing a data-driven approach to reveal how CES dimensions

https://doi.org/10.3390/su18052578

are dynamically perceived and constructed in real-time within authentic experiences. In doing so, it advances the evolution of related research from static classification toward a dynamic, process-oriented understanding.

Building on this foundation, the data-driven dimensions derived from BERTopic were used to construct a structural model, which was systematically tested using PLS-SEM with integrated multi-source data to examine causal pathways and underlying mechanisms among the variables. This framework constitutes a complete methodological cycle. BERTopic extracts empirically grounded constructs and hypotheses from large-scale natural language data. Meanwhile, PLS-SEM rigorously validates the structural relationships among these constructs. This integration not only enhances the ecological validity and theoretical explanatory power of the measurement model but also strengthens the robustness of causal inferences. Consequently, it offers a systematic approach that combines exploratory and confirmatory methods for investigating the core question of how residents dynamically perceive CES in real-world settings.

This study acknowledges the limitation of anonymity inherent in social media data and, through sensitivity analyses based on proxy user segmentation, demonstrates that the core CES dimensions remain robust across different behavioral groups. However, it is crucial to further clarify that the Weibo data are characterized by a user base skewed toward younger and more digitally engaged populations [73]. This bias is not merely a matter of sample composition but fundamentally shapes the semantic boundaries and conceptual connotations of the CES themes identified by BERTopic [62].

The linguistic style of younger user groups tends to favor concise, emotionally expressive language, and they are more inclined to share experiences with strong social appeal, such as social interactions and leisure check-ins [74]. This tendency leads the BERTopic clustering process to assign greater weight to CES dimensions that are easily disseminated through short, rapidly shared texts. These dimensions include aesthetic experiences and recreational services. As a result, these dimensions become more prominent within the resulting topic structure [75,76]. In contrast, cultural services that require nuanced description or deeper reflection, such as emotional restoration or intergenerational heritage transmission, may be underrepresented in the corpus [77]. Consequently, they may be dimensionally reduced or merged. This suggests that the CES thematic dimensions identified by the model represent a structured representation of explicitly articulated CES experiences within the digital public sphere, shaped jointly by the expressive preferences of specific user groups and the discursive practices of the platform. This precisely underscores the necessity of integrating structured questionnaire surveys with social media data. Questionnaires, through their predefined demographic frameworks, ensure that potentially silent groups and the potentially implicit CES dimensions they may hold are systematically incorporated into the analysis. While Weibo data reveal what is publicly articulated and how thematic structures are shaped by user characteristics, questionnaires respond to how different groups evaluate CES [78]. The complementarity between the two provides a critical entry point for understanding the selective construction of CES within social contexts.

At a theoretical level, by connecting data-driven discovery with theory-driven validation, this framework advances CES research beyond static classification frameworks toward a more dynamic and experiential theoretical understanding. From a practical standpoint, this evidence-based analytical tool enables the systematic identification of key factors shaping public perception and satisfaction. It thereby provides a scientific basis and decision-making support for the precise planning, adaptive management, and service enhancement of UPGS.

https://doi.org/10.3390/su18052578

4.2. Heterogeneity and Synergy in CES Perceptions

Elucidating the heterogeneity in residents' perceptions of CES in UPGS holds critical theoretical and practical significance for accurately identifying diverse public needs and optimizing park service provision. This study confirms that while overall perception intensity is high, significant differences exist across CES dimensions, revealing a structured perceptual hierarchy. Aesthetic experience emerged as the most strongly perceived dimension, consistent with its established role as the core visual attractor of green spaces [20]. In contrast, perceptions of social interactions and health-promoting activities demonstrated greater volatility, being more susceptible to contextual constraints such as environmental disturbances, facility conditions, and crowding [79,80]. The lower satisfaction associated with recreational services and educational service points to deficits in their current provision, particularly regarding content depth, format innovation, and alignment with heterogeneous user expectations [27,81,82].

Notably, the analysis revealed a significant synergistic correlation between recreational services and educational services. This finding suggests a complementary relationship wherein recreational settings foster a relaxed atmosphere conducive to cultural enrichment, which in turn adds intellectual or spiritual value to the leisure experience. Such complementarity helps mitigate perceptional disparities between these two dimensions. This nuance challenges perspectives that treat CES dimensions as largely independent, indicating instead that the strength of inter-CES synergies is likely context-dependent. Divergent preferences for CES bundles among residents may be explained by variations in cultural backgrounds, leisure habits, and attitudes toward "edutainment".

These patterns collectively underscore that residents' evaluations are not uniform but are structured by both the inherent properties of service types and their interrelationships. The relational analytical framework employed here thus provides a sharper lens for capturing the intrinsic connections within the CES construct, moving beyond siloed assessments toward a more integrated understanding of park experiences.

The specific CES combinations identified through correlation analysis raise a critical theoretical question: do the observed synergies, such as those between aesthetic experience and social interaction or between health activity and recreational service, reflect patterns unique to UPGS, or do they represent generalizable phenomena with broader explanatory value? UPGS characterized by high accessibility, social vibrancy, and semi-natural environments may amplify certain service synergies [83]. Dense visitation and visually appealing landscapes foster the integration of social interaction into aesthetic experiences [84], while recreational infrastructure directly supports the coupling of health activity with leisure. Such configurations may be less salient in wilderness or productive landscapes [85].

Conversely, some synergies may originate from universal psychological or social foundations. The link between aesthetic experience and social interaction may reflect a fundamental human inclination toward shared emotional engagement [86], while the alignment of health activity with recreation may embody a holistic conception of well-being integrating physical and mental dimensions. These underlying associations may persist across contexts, manifesting through diverse service expressions. Thus, a more nuanced interpretation suggests that intrinsic relationships among CES may follow certain universal logics, yet their dominant configurations and synergistic strength are significantly moderated by situational factors, including ecosystem type, user demographics, cultural norms, and management practices [23]. The patterns observed in this study offer an instructive case for understanding the complexity of CES dynamics in human-nature interactions. Future comparative research across ecosystems and cultural settings is needed to delineate which synergies are universal and which are context-dependent, thereby supporting the development of more refined CES theory and context-sensitive management strategies.

https://doi.org/10.3390/su18052578

4.3. Driving Mechanisms Through Sequential Pathways

While existing studies have confirmed that perceptions of CES are influenced by multiple factors [37,87], the intrinsic mechanisms underlying the formation of residents' CES perception variations in UPGS remain insufficiently explored. Based on PLS-SEM, this study systematically quantified the structural relationships and influence pathways among various driving factors. It thereby elucidated the intrinsic mechanisms underlying the formation of residents' perception differences in CES within UPGS. The research identified the EBE, ILC, and QSM as key factors influencing residents' CES perception variations. Results from PLS-SEM and mediation path analysis indicate that the EBE exerts both a direct effect on CES perception variations and indirect effects through a serial mediation pathway involving ILC and QSM. This finding aligns with existing research on the spatial heterogeneity of CES perceptions. It further substantiates the systemic and transmissive characteristics of parks as carriers of ecological well-being [88]. As noted in related studies, parks located in areas of high development intensity often possess more complete landscape amenities and managerial resources [89]. This may exacerbate CES perception disparities among residents from different neighborhoods.

This study further identifies the mediating role of the ILC between the EBE and QSM, subsequently influencing residents' CES perception variations. This result aligns with the views that landscape attributes are key variables affecting visitor experiences [90]. Specifically, high-quality internal landscapes often reflect external resource investment and locational advantages. They directly shape residents' usage experiences. Consequently, they serve as crucial mediating variables influencing service management quality and CES perception differences [52]. Furthermore, the EBE also independently influences CES perception variations through QSM. This suggests that even with comparable landscape conditions, differences in the EBE can indirectly widen experience gaps among residents through uneven allocation of management resources [91,92]. This reflects how urban parks in practice often form service gradients due to unequal external resource input. Such structural differences may further affect the equitable use of UPGS by different social groups, thereby shaping CES perception disparities across multiple dimensions.

Notably, within the overall model, QSM emerges as the central node influencing residents' CES perception differences, with its effect strength surpassing other pathways. This highlights the pivotal role of management factors in explaining perception disparities. High-quality service management can directly enhance residents' identification with and perceived intensity of cultural services by improving activity experiences [93,94]. Even under similar landscape conditions, differences in management levels can lead to significant distinctions in resident perceptions [95]. The study also identified a synergistic effect between the EBE and QSM. Differences in external conditions can affect QSM through resource allocation mechanisms. This thereby amplifies disparities in residents' cultural service perceptions. This finding aligns with the conclusion that physical conditions form the foundation of CES cognition, while management optimization is an effective pathway to enhance perceptions [29]. Urban expansion and socioeconomic disparities systematically affect the accessibility and service quality of green spaces. This exacerbates inequalities in ecological well-being among different groups [96]. Consequently, park planning must emphasize spatial balance in resource investment and service provision to alleviate the structural contradictions underlying CES perception differences.

Based on the PLS-SEM analysis results, this study systematically tested the six proposed hypotheses. Regarding direct effects, the external built environment exhibited significant positive influences on residents' CES perception differences (H1: $\beta = 0.254$, $p < 0.001$), internal landscape composition (H2: $\beta = 0.655$, $p < 0.001$), and perceived service quality (H3: $\beta = 0.352$, $p < 0.001$). These results provide support for H1, H2, and H3. This

https://doi.org/10.3390/su18052578

finding confirms that the external built environment not only directly shapes residents' perceptions of cultural services. It also operates through multiple pathways by influencing the intrinsic qualities of parks. Among these, H2 demonstrated an exceptionally strong effect size ($f^2 = 0.752$). This indicates that the external built environment possesses substantial explanatory power regarding internal landscape composition. It thereby serves as a strategic leverage point yielding the greatest multiplier effect.

Internal landscape composition exhibited significant direct effects on both CES perception differences (H4: $\beta = 0.248$, $p < 0.001$) and perceived service quality (H5: $\beta = 0.347$, $p < 0.001$). These results provide support for H4 and H5. These findings indicate that the internal landscape elements of parks serve a dual role. They are not only direct sources of residents' cultural experiences but also a fundamental foundation for the formation of perceived service quality. Perceived service quality exerted the most pronounced influence on CES perception differences (H6: $\beta = 0.331$, $p < 0.001$). This provides the strongest support among all hypotheses. The 95% confidence interval for this effect ranged from 0.204 to 0.464. This underscores the central role of managerial factors in explaining variations in resident perceptions. Specifically, with 95% confidence, an improvement of one standard deviation in perceived service quality is expected to result in an increase of between 0.204 and 0.464 standard deviations in residents' CES perceptions.

Mediation analysis further elucidated the chain-mediated relationships among the hypothesized pathways. The total effect of the external built environment on CES perception differences was $\beta = 0.608$ ($p < 0.001$). Specifically, the external built environment influenced CES perception differences through three indirect pathways. First, a single mediation pathway through internal landscape composition (EBE $\rightarrow$ ILC $\rightarrow$ CES, $\beta = 0.163$, $p < 0.001$). Second, a single mediation pathway through perceived service quality (EBE $\rightarrow$ QSM $\rightarrow$ CES, $\beta = 0.116$, $p < 0.001$). Third, a sequential mediation pathway through both internal landscape composition and perceived service quality (EBE $\rightarrow$ ILC $\rightarrow$ QSM $\rightarrow$ CES, $\beta = 0.075$, $p < 0.01$). The statistical significance of this sequential mediation pathway ($p < 0.01$) provides a mechanistic explanation. It shows how the direct effect of the external built environment (H1) is amplified through the indirect pathways H2–H5–H6. These findings reveal that the external built environment carries substantial spillover value and chain-based multiplier effects. This suggests that its comprehensive benefits should be systematically accounted for in planning evaluations.

In summary, all six hypotheses were empirically supported. Notably, H2 (EBE $\rightarrow$ ILC) emerged as the most influential pathway within the model ($\beta = 0.655$, $f^2 = 0.752$). This indicates that optimizing the external built environment represents a strategic leverage point capable of generating the greatest multiplier effect. Meanwhile, H6 (QSM $\rightarrow$ CES) functioned as a central node within the model ($\beta = 0.331$, $f^2 = 0.133$). Its effect size and mediating role underscore the pivotal position of managerial factors in explaining variations in CES perception. This hypothesis-testing framework serves two purposes. It not only validates the theoretical model but also quantifies the differential contributions of each factor to perceptual divergence. It thereby provides a quantitative benchmark for translating statistical relationships into concrete planning practices.

4.4. Socioeconomic Contingencies in Perception Shaping

The analysis identified a significant positive moderating effect of age on the relationship between QSM and CES perception variations, indicating that older residents demonstrate heightened responsiveness to improvements in service quality. This demographic group exhibits a more pronounced increase in CES perception levels in response to enhanced QSM, which aligns with existing research on the distinctive environmental perception patterns and affective attachments formed by older adults [97]. This cohort

https://doi.org/10.3390/su18052578

tends to engage with natural settings through physical activity and social interaction, rendering them particularly sensitive to service enhancements that facilitate such engagements, including maintenance standards, safety provisions, and organized activities. Theoretically, older individuals often possess greater discretionary leisure time and develop stronger emotional bonds with urban parks. Consequently, optimizations in QSM directly address their core needs for comfort, safety, and socialization, thereby amplifying non-material CES benefits such as sense of place and subjective well-being [67].

In contrast, monthly income exhibited a marginally significant negative moderating effect along the ILC → QSM pathway. Specifically, as income rises, the positive influence of landscape optimization on perceived QSM appears attenuated. This pattern may be explained by the broader leisure options available to higher-income groups, including access to private green spaces and travel opportunities, which can dilute the marginal perceived utility derived from landscape improvements within any single public park [98]. Elevated expectation thresholds among higher-income residents may further reduce their sensitivity to incremental enhancements in park services management. This finding, although not reaching the conventional threshold for statistical significance ($p = 0.069$), provides preliminary evidence suggesting that socioeconomic status may exert multidimensional effects. While higher income may reinforce recognition of the intrinsic economic value of landscapes, it can simultaneously weaken the translation of landscape improvements into perceived service quality. Such differential moderation highlights the complex interplay between objective environmental attributes, subjective service evaluations, and residents' socioeconomic positioning, suggesting that uniform management strategies may not equally address the expectations of diverse demographic segments.

Given that this moderating effect only reached marginal significance, it should be regarded as a preliminary finding requiring further validation in future research with larger sample sizes. Therefore, future studies should employ expanded samples to examine the robustness of this effect and further investigate the specific mechanisms through which income influences residents' perceptions of CES in UPGS.

4.5. Limitations and Future Research

This study has several limitations that point to meaningful avenues for future inquiry.

First, while social media text data were leveraged to capture public discourse, the analysis did not incorporate visual content (e.g., photographs or videos). This omission may constrain a fully dimensional understanding of residents' visual engagement with CES. Future research could employ computer vision techniques, such as convolutional neural networks (CNNs) or visual sentiment analysis, to systematically examine image data from platforms. Integrating multimodal data would offer a more holistic perspective on how aesthetic and experiential qualities of landscapes are perceived, shared, and valued, thereby enriching the ecological validity and interpretive power of CES assessments.

Second, while this study investigated environmental and social drivers of CES perception differences, the role of temporal dynamics, such as diurnal, seasonal, or event-based fluctuations, remains underexplored. Perceptions of CES are likely sensitive to temporal contexts, yet the current cross-sectional design limits insights into these dynamics. Longitudinal or repeated-measures designs, coupled with time-series analysis, could elucidate how temporal factors moderate the relationships between drivers and CES perceptions, offering a more process-oriented understanding of ecosystem service experience.

Third, this study examined CES perception differences and their driving mechanisms at the park scale, focusing primarily on the internal and external environmental attributes of individual parks. However, it did not account for the connectivity, synergistic effects, or service coverage of urban green spaces as an integrated network. Within the context of the

 https://doi.org/10.3390/su18052578

"park city" and "15-min city" planning concepts, residents' access to CES in UPGS is often shaped more by the spatial configuration of green space networks and the accessibility of these spaces within residential catchments than by the attributes of individual parks alone. Residents' cumulative CES benefits may depend more on the density and connectivity of green spaces within their daily activity radius than on the performance of individual parks. Future research should extend the spatial scale to the urban green space network or the 15-min living catchment, exploring how the overall configuration, connectivity, and equitable distribution of green spaces influence residents' CES perceptions. Such an extension would provide more targeted insights for optimizing urban green space system planning and promoting equity in the provision of ecosystem services.

Fourth, this study employed a cross-sectional questionnaire design, and both QSM and CES data were obtained from the same respondents, which may have introduced a slight overestimation of path coefficients due to common method bias. However, the full collinearity test (with all VIF values below 1.96) and the independent replication of key findings using social media data collectively indicate that this bias did not substantially compromise the robustness of the core conclusions. Future research could further strengthen the validity of causal inferences by adopting three methodological strategies: separating the measurement time points for predictor and outcome variables, incorporating objective or multi-source measurements, and pre-specifying and including marker variables in the survey design.

Fifth, several methodological limitations warrant further reflection. Although the merging and labeling of BERTopic topics were cross-validated by multiple coders, the process may still involve some degree of subjective interpretation. Future research could incorporate multimodal semantic analysis to enhance the objectivity of thematic classification. In addition, discretizing continuous variables into ordinal categories may affect the precision of parameter estimates. Future studies could re-estimate the model using the original continuous forms of these variables and compare the resulting path coefficients and significance levels across the two data preprocessing approaches to assess the robustness of the findings. Finally, this study did not conduct an in-depth analysis of heterogeneity across different park types. Patterns of CES perception may vary among distinct categories of green spaces, and future research could employ multi-group analysis to examine the moderating role of park typology.

5. Conclusions

This study employs a sequential two-phase analytical framework to systematically investigate the heterogeneity in residents' perceptions of CES within UPGS and their underlying driving mechanisms, addressing a critical gap in urban sustainability research. In the initial phase, BERTopic modeling and keyword frequency analysis were applied to Weibo review texts. This process identified five core CES perception dimensions: recreational services, aesthetic experiences, health-promoting activities, social interactions, and educational services. Concurrently, it extracted four primary categories of key drivers along with their measurable indicators: residents' socioeconomic background, the external built environment, internal landscape composition, and service quality management. In the subsequent phase, questionnaire data from 313 respondents across 13 representative urban parks in Wuhan were integrated with park-related geospatial datasets. A PLS-SEM was constructed and validated to systematically examine the pathways through which the identified drivers influence residents' CES perception differences. This sequential mixed-methods design effectively bridges exploratory data-driven pattern discovery with confirmatory model testing, providing a comprehensive understanding of the perceptual drivers and their complex interrelationships.

https://doi.org/10.3390/su18052578

By integrating social media data with questionnaire survey data, this study reveals their synergistic and complementary value in capturing residents' perceptions of CES, offering a methodological advancement for socially inclusive sustainability research. The two data sources capture distinct yet interrelated dimensions of perceived experiences, namely spontaneous public expressions versus structured reflective evaluations, thereby offering a more integrated and comprehensive understanding of CES perception. Social media text data, derived from large-scale user-generated content, exhibit characteristics of real-time expression, contextual specificity, and emotional valence. These data effectively capture residents' behavioral tendencies and affective responses in authentic usage contexts, particularly reflecting the perspectives of younger populations and frequent park users. In contrast, questionnaire survey data are valued for their structured format, reflective nature, and demographic representativeness, providing systematic coverage across diverse socioeconomic backgrounds. This makes them particularly well-suited for testing theoretical hypotheses and examining causal mechanisms while ensuring that no social group is systematically excluded from the analysis. The main findings are as follows:

First, significant synergistic correlations exist among different CES types, with the strongest interrelationships observed between aesthetic experiences and social interactions ($r = 0.93$), as well as between health-promoting activities and recreational services ($r = 0.91$). These synergies suggest that well-designed urban green spaces can deliver multiple co-benefits simultaneously, maximizing the socio-ecological returns from limited land resources, which is a key principle of sustainable urban development. Satisfaction analysis revealed statistically significant differences across CES dimensions: aesthetic experiences received the highest satisfaction score ($M = 3.59$), followed by social interactions ($M = 3.51$), while recreational services scored the lowest ($M = 3.31$); health-promoting activities and educational services attained intermediate and comparable levels of satisfaction ($M = 3.35$). This satisfaction hierarchy provides empirical benchmarks for prioritizing resource allocation in sustainability-oriented park management.

Second, PLS-SEM analysis elucidated a multi-level transmission mechanism underlying the perception differences, with all path coefficients statistically significant ($p < 0.001$). The external built environment functions as a foundational contextual factor, exerting indirect effects through three mediation pathways: EBE $\rightarrow$ ILC $\rightarrow$ CES ($\beta = 0.163$), EBE $\rightarrow$ QSM $\rightarrow$ CES ($\beta = 0.116$), and the chained mediation path EBE $\rightarrow$ ILC $\rightarrow$ QSM $\rightarrow$ CES ($\beta = 0.075$). The total indirect effect of EBE on CES ($\beta = 0.354$) accounted for 58.2% of its total effect ($\beta = 0.608$), indicating that the influence of the built environment is predominantly transmitted through internal landscape and service management. Internal landscape composition serves as the physical carrier of CES perceptions, contributing both directly (ILC $\rightarrow$ CES: $\beta = 0.248$, $p < 0.001$) and indirectly via service management (ILC $\rightarrow$ QSM $\rightarrow$ CES: $\beta = 0.115$, $p < 0.001$). Notably, quality of service management emerged as the most critical direct driver of perception differences (QSM $\rightarrow$ CES: $\beta = 0.331$, $p < 0.001$), with the largest effect size among direct predictors ($f^2 = 0.133$), and played a central mediating role throughout the transmission chain, participating in two of the three significant indirect pathways from EBE to CES. This finding identifies QSM as a high-leverage intervention point for urban planners seeking to enhance the sustainability outcomes of park investments, as improvements in service management can amplify the benefits of both built environment and landscape features.

Third, the moderating effects of socioeconomic background variables displayed differentiated patterns. A significant positive moderating effect of age was found on the relationship between service quality management and CES perception differences ($\beta = 0.093$, $p = 0.037$), indicating that older residents tend to respond more positively to improvements in service quality. This finding suggests that age-friendly service enhancements, such as

https://doi.org/10.3390/su18052578

accessible facilities, seating, and restrooms, can yield particularly significant sustainability benefits for an aging urban population, thereby addressing a demographic trend with profound implications for SDG 11's commitment to "inclusive" cities. Monthly income showed a marginally significant negative moderating trend along the pathway from internal landscape composition to service quality management ($\beta = -0.079$, $p = 0.069$), implying that the influence of landscape optimization on perceived service quality weakens with higher income levels. This income-based disparity raises concerns about environmental justice: if higher-income residents are less sensitive to landscape improvements, while lower-income groups rely more heavily on park quality for their CES benefits, then uniform investment strategies may inadvertently exacerbate existing inequities in access to urban green space amenities. Other socioeconomic variables, including gender, education level, and occupation, did not demonstrate statistically significant moderating effects.

In summary, by integrating social media big data with survey data, this study develops and validates a theoretical model that explains the formation mechanisms of residents' CES perception differences in UPGS, contributing to the evidence base for sustainable urban planning. These findings provide a scientific basis for evidence-based policy implementation in park city construction. In high-density old urban areas, priority should be given to enhancing landscape quality through small-scale renewal interventions to compensate for limited ecological space. Compact, mixed-use functional zones should be developed to strengthen the synergy between recreation and services, while vertical greening may be increased to enrich aesthetic experiences. Furthermore, optimizing the deployment of cleaning and mobile services during peak hours can leverage the mediating role of QSM in improving overall perceived quality. In newly developed urban areas, ecological spaces and infrastructure should be integrated at the planning stage to foster long-term CES provision capacity. Large-scale ecological landscapes can be strategically aligned with public transport networks to amplify the functions of aesthetic experiences and health-promoting activities. For different population groups, differentiated strategies should be implemented to advance social sustainability. For older adults, priority should be given to optimizing age-friendly facilities and supporting services. For low-income groups, equitable access to free public amenities and inclusive services should be ensured, alongside improvements to the external built environment of surrounding parks to mitigate spatial injustice. These targeted interventions align with the core principle of "leaving no one behind" embedded in the 2030 Agenda for Sustainable Development. Collectively, these findings provide scientific support for urban planning and land management strategies that balance spatial justice with the diverse needs of population groups, ultimately contributing to the transition toward more resilient, equitable, and sustainable cities in line with SDG 11.

Author Contributions: X.L.: Conceptualization, Writing—original draft, Writing—review & editing, Funding acquisition, Methodology, Supervision. Z.Y.: Writing—original draft, Writing—review & editing, Data curation, Investigation, Visualization, Formal analysis. L.L.: Conceptualization, Methodology. Y.W.: Data curation, Investigation. J.H.: Data curation, Investigation. All authors have read and agreed to the published version of the manuscript.

Funding: This work was supported by the Department of Education of Hubei Province [grant number 24Q176]; and the National Natural Science Foundation of China [grant number 72404082].

Institutional Review Board Statement: This study was conducted in accordance with the Declaration of Helsinki and approved by the Scientific Research Ethics and Technology Safety Committee of Hubei University of Technology (protocol code HBUT20250073, 9 October 2025).

Informed Consent Statement: Informed consent was obtained from all participants involved in the study. Prior to participation, each individual was presented with a consent statement and explicitly confirmed their agreement via a checkbox mechanism.

https://doi.org/10.3390/su18052578

Data Availability Statement: Due to the privacy and confidentiality of the respondents, the questionnaire data used in this study cannot be made publicly available. However, the data can be accessed upon reasonable request and under the condition that participant privacy is ensured, by contacting the corresponding author.

Acknowledgments: We thank all participants of the questionnaire survey for their time and valuable input, which greatly contributed to this study.

Conflicts of Interest: Author Lin Lei was employed by the company Changjiang Ecology (Hubei) Technology Development Co., Ltd. The remaining authors declare that the research was conducted in the ab-sence of any commercial or financial relationships that could be construed as a potential conflict of interest.

Abbreviations

The following abbreviations are used in this manuscript:

RS	Recreational services
AE	Aesthetic experiences
HA	Health-promoting activities
SI	Social interactions
ES	Educational services
EBE	External built environment of parks
ILC	Internal landscape composition of parks
QSM	Quality of services management
SBD	Surrounding building density
TA	Transportation accessibility
PD	Population density
FR	Facility richness
PA	Park area
WF	Water features
IEM	Intra-park environment management
VEM	Visitor experience management
CPM	Cultural promotion management

Appendix A

Survey Questionnaire on the Driving Mechanisms of Residents' Perception Variations in Cultural Ecosystem Services of Urban Park Green Spaces.

Dear Sir/Madam,

Hello! Thank you very much for helping us complete this questionnaire survey. This questionnaire is solely for academic research purposes and will not be used for any other purpose.

Cultural ecosystem services (CES) in urban parks refer to the direct or indirect benefits that humans derive from the various ecosystems within urban parks. These services represent humans' subjective perceptions of objective ecosystems and primarily include types such as aesthetic experiences, health-promoting activities, social interactions, recreational services, and educational services. This questionnaire has been designed to explore the driving mechanisms behind the differences in residents' CES perceptions provided by urban park ecosystems. During the survey process, your privacy will not be compromised. There are no right or wrong answers to the questionnaire; please simply mark the appropriate box based on your personal experiences in the park. Once again, thank you for your patience and assistance.

https://doi.org/10.3390/su18052578

Table A1. Basic information.

Variable	Categories					
Gender	Male	Female				
Age	<18	18–30	31–45	46–60	>60	
Income	<CNY2000	CNY2000–CNY5000	CNY5000–CNY9000	>CNY9000		
Education attainment	High school and below	College	Bachelor's degree	Master's degree or above		
Occupation	Students	Corporate employees	Public sector employees	Freelancers	Retirees	Others

Table A2. Assessment scale for quality of park service management.

Types of Service Management	Questions	Quality Evaluation (from Low to High)				
Intra-park environment management	How would you rate the quality of environmental management services in the park, such as plant maintenance, hygiene, and facility maintenance?	1	2	3	4	5
Visitor experience management	How would you rate the quality of tour experience management services, such as road guidance, facility distribution, and safety protection in the park?	1	2	3	4	5
Cultural promotion management	How would you rate the quality of cultural promotion management services, such as the establishment of cultural landscapes, the organisation of related activities, and park promotion?	1	2	3	4	5

Table A3. Perceptions of Cultural Ecosystem Services Provided by Park Ecosystems.

CES Types	Questions	Perceived Satisfaction (from Low to High)				
Aesthetic experiences	Your perceived satisfaction with the park's aesthetic aspects, including landscape design, vegetation, and water features	1	2	3	4	5
Health-promoting activities	Your perceived satisfaction with the park's fitness and walking facilities	1	2	3	4	5
Social interactions	Your perceived satisfaction with the park's interactive aspects, including making friends and strengthening relationships with family and friends	1	2	3	4	5
Recreational services	Your overall satisfaction with the park's recreational experiences, including leisure, entertainment, and relaxation, is:	1	2	3	4	5
Educational services	Your perceived satisfaction with the park's educational aspects, including animal and plant science and landscape knowledge	1	2	3	4	5

https://doi.org/10.3390/su18052578

Appendix B

Table A4. Topic similarity matrix.

	Topic 0	Topic 1	Topic 2	Topic 3	Topic 4	Topic 5	Topic 6	Topic 7	Topic 8	Topic 9
Topic 0	1.00	0.82	0.79	0.78	0.84	0.67	0.82	0.66	0.44	0.30
Topic 1	0.82	1.00	0.67	0.77	0.82	0.63	0.66	0.59	0.38	0.26
Topic 2	0.79	0.67	1.00	0.68	0.61	0.62	0.83	0.59	0.78	0.36
Topic 3	0.78	0.77	0.68	1.00	0.81	0.53	0.72	0.54	0.38	0.25
Topic 4	0.84	0.82	0.61	0.81	1.00	0.85	0.68	0.76	0.33	0.30
Topic 5	0.67	0.63	0.62	0.53	0.85	1.00	0.65	0.82	0.44	0.24
Topic 6	0.82	0.66	0.83	0.72	0.68	0.65	1.00	0.68	0.62	0.26
Topic 7	0.66	0.59	0.59	0.54	0.76	0.82	0.68	1.00	0.40	0.20
Topic 8	0.44	0.38	0.78	0.38	0.33	0.44	0.62	0.40	1.00	0.39
Topic 9	0.30	0.26	0.36	0.25	0.30	0.24	0.26	0.20	0.39	1.00

Table A5. Coding manual.

Topic	Topic Name	Description	Keywords	Distinguishing Features
Topic 0	Food-related leisure	Engaging in leisure activities such as dining, culinary experiences, and nightlife within or in the vicinity of parks.	Delicious, evening, weekend, Yue Lake, Sha Lake, cuisine, restaurant, snacks, barbecue, picnic	Distinction from Topic 3: Topic 0 focuses on food-related leisure, whereas T3 centers on recreational activities.
Topic 1	Culture landmark check-ins	Visiting and documenting experiences at urban cultural landmarks or science-education sites, often accompanied by social media check-ins.	Riverbank, Dong Lake, avenue, punch In, Huanghelou, Yangtze River Bridge, landmark, scenic spot, tourism	Distinction from Topic 5: Topic 1 focuses on cultural and science landmark check-ins, whereas Topic 5 emphasizes emotional attachment and sense of belonging.
Topic 2	Social sharing	Documenting and sharing park experiences through social media platforms such as RED and Weibo.	Pigeons, notes, Hanfu, RED, personal journals, photography, sharing, documenting, vlog, video	Distinction from Topic 8: Topic 2 emphasizes the act of sharing itself, whereas Topic 8 focuses on the aesthetic experience of nature.
Topic 3	Natural aesthetics	Appreciation of natural landscape aesthetics in parks, such as flowers and seasonal changes.	Cherry blossoms, tulips, romance, blooming, East Lake, lotus, plum blossoms, ginkgo, red maple, osmanthus.	Distinction from Topic 4: Topic 3 focuses on the aesthetic appreciation of flowers, whereas Topic 4 emphasizes wetland and aquatic ecology.
Topic 4	Wetland ecology	Experiencing the wetland, lake, and other aquatic ecological environments within parks.	Sha Lake, wetland, Yue Lake, park area, lotus, lake water, water body, ecology, environment, green	Distinction from Topic 3: Topic 4 focuses on the overall aquatic ecosystem, not merely visual appreciation.
Topic 5	Sense of place and belonging	Developing a sense of place identity, emotional attachment, and belonging toward a park.	Sha Lake, Tang Lake, park area, riverbank, chrysanthemums, memories, recollection, nostalgia, childhood, hometown	Distinction from Topic 1: Topic 5 emphasizes emotional depth and long-term connection to place, rather than one-time check-in behaviors.
Topic 6	Seasonal recreation	Recreational activities undertaken in parks during specific seasons, holidays, or festive periods.	Tomorrow, weather, autumn, early winter, weekend, start of autumn, Mid-Autumn Festival, spring, summer, winter	Distinction from Topic 0: Topic 6 emphasizes seasonal recreational activities, whereas Topic 0 focuses on food-related leisure experiences.

https://doi.org/10.3390/su18052578

Table A5. *Cont.*

Topic	Topic Name	Description	Keywords	Distinguishing Features
Topic 7	Physical activity and fitness	Engaging in physical activities such as sports, fitness, and running within parks.	Yue Lake, dormitory, sports, group photos, people, running, cycling, fitness, walking, strolling	Distinction from Topic 0: Topic 7 emphasizes physical activities and exercise, whereas Topic 0 focuses on food-related leisure.
Topic 8	Culture and history	Experiencing the historical and cultural significance, humanistic depth, and artistic atmosphere of parks.	Sixteen, technology, museum, history, culture, heritage site, ancient architecture, memorial, martyr, revolution	Distinction from Topic 1: Topic 8 emphasizes historical and cultural connotations, whereas Topic 1 focuses on landmark appearances and check-in behaviors.
Topic 9	Family-oriented social interaction	Social interactions including parent–child activities, family outings, and gatherings with friends.	Photos, sharing, time, child-accompanying, children, parent–child, family, friends	Distinction from Topic 2: Topic 9 emphasizes interpersonal interaction, whereas Topic 2 focuses on the act of sharing through digital media.

Table A6. Comparison of 10 initial topics and 5 merged CES dimensions.

CES	Connotation	Topic	Keywords	Document Count	Representative Comments
RS	Provide spaces for outdoor leisure and recreational activities.	Topic 0	delicious, evening, weekend, Yue Lake, Sha Lake	2848	"On a lovely weekend, Beibei and I took a taxi to the park. We ended up getting in the wrong cab, but luckily, we were with good friends, and everywhere is fun with them, haha. In the evening, we had fried chicken and pan-fried buns at the park. They were so, so tasty."
		Topic 6	tomorrow, weather, autumn, winter, weekend	736	"Embracing the fullness of autumn with all its warmth and sweetness, I went to Guishan Park on a sunny weekend to gather the essence of the season."
AE	Offer scenic natural and cultural landscapes that deliver spiritual enjoyment and aesthetic pleasure.	Topic 3	cherry, tulip, romantic, bloom, Dong Lake	1554	"A casual shot at Qingshan Park, with plum blossoms in full bloom as spring approaches."
		Topic 4	Sha Lake, wetlands, Yue Lake, park area, lotus	1225	"I happened to pass by Tanghu Park and stopped for a quick visit, only to be completely captivated by its beauty. The park features wetland scenery reminiscent of Jiangnan water towns. With numerous small peninsulas along the lake, every step offers a new and captivating view, making it hard for visitors to leave."
		Topic 5	Sha Lake, Tang Lake, park area, riverbank, chrysanthemum	1184	"It had been a long time since I last visited Jiefang Park. Wandering among the vibrant blossoms, strolling and stopping along the way, it felt as though I had returned to my childhood."

https://doi.org/10.3390/su18052578

Table A6. *Cont.*

CES	Connotation	Topic	Keywords	Document Count	Representative Comments
HA	Serve as venues for physical exercise and mental relaxation, contributing to both physical and psychological well-being.	Topic 7	Yue Lake, dormitory, sports, group photo, people	525	"A snippet of daily life: While others have been at work for an hour already, I've already run 5 kilometers around Moon Lake Park. I consoled myself by thinking, at least I really made the most of this beautiful weather and lovely spring day."
SI	Function as settings that facilitate social connections and enable the sharing of experiences through online platforms.	Topic 2	pigeon, notes, Hanfu (traditional Chinese Han clothing), RED (a social media platform in China), field notes	1610	"My park check-in is seriously belated, and my procrastination is getting out of hand. To celebrate reaching 10,000 followers on my notes, I went to Zhongshan Park to take some nice photos and share them. The schedule was super tight. I managed to check out four spots in just one day."
		Topic 9	image, share, time	48	"I had planned to take some parent-child photos in traditional Hanfu costume, but the heavy makeup and the big headpiece completely freaked my kid out. They burst into tears, thinking they could no longer find their mom. We had to give up on the idea, take off the headpiece, and, as compensation, ended up going to Heping Park's amusement park while still wearing that heavy makeup. So much for that plan!"
ES	Provide access to places of cultural and historical significance and support environmental education activities.	Topic 1	riverbank, Dong Lake, avenue, punch In, Huanghelou	1874	"After leaving Jianghan Road, we headed to the Guishan Scenic Area. By the riverbank, we spent about an hour enjoying the breeze, watching trains cross the Yangtze River Bridge, and quietly taking in the beauty of the Yangtze. Later, we climbed to the highest point of Guishan Park. From this vantage point, we enjoyed the sweeping views, a truly refreshing and delightful experience."
		Topic 8	sixteen, technology	306	"On Saturday, we took the two little ones to a tech exhibition at Moon Lake Park. They had so much fun. The older brother was busy playing on his own the whole time and even made a new friend. It was the younger one's first visit, and he was so excited he could only squeal with delight."

https://doi.org/10.3390/su18052578

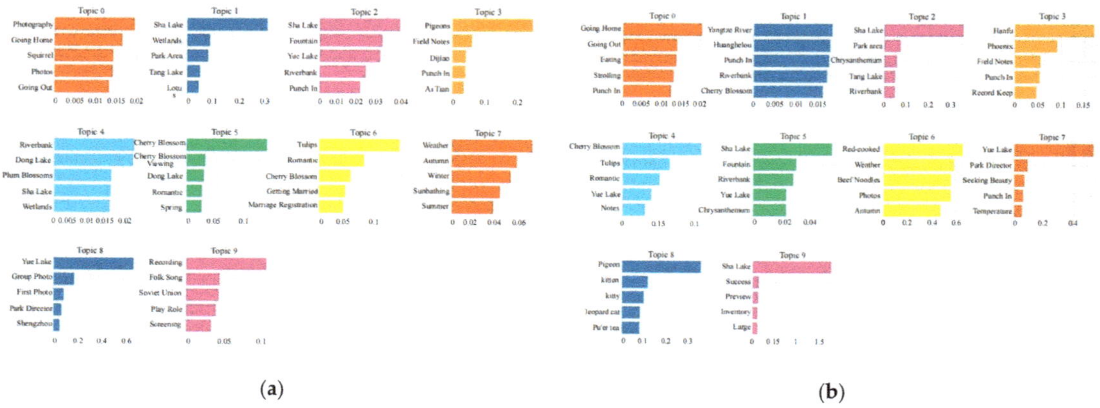

Figure A1. Visualization of BERTopic-Derived Topic Keywords by Daytime and Nighttime Weibo User Groups: (**a**) Daytime Period; (**b**) Nighttime Period.

References

1. Olaiya, A. Transforming our world: The 2030 agenda for sustainable development: International. *Civ. Eng. Siviele Ingenieurswes* **2015**, *24*, 26–30.
2. MEA. *Millennium Ecosystem Assessment: Ecosystems and Human Well-Being: Synthesis*; Island Press: Washington, DC, USA, 2005
3. Luo, J.; Xing, X.; Wu, Y.; Zhang, W.; Chen, R.S. Spatio-temporal analysis on built-up land expansion and population growth in th Yangtze River Delta Region, China: From a coordination perspective. *Appl. Geogr.* **2018**, *96*, 98–108. [CrossRef]
4. Wei, H.; Qi, W.; Liu, S.; Li, Y.; Yin, Y. Uncovering the edge urban villages using social media big data: A case study in Beijin China. *Habitat Int.* **2025**, *159*, 103357. [CrossRef]
5. Zhou, W.; Yu, W.; Qian, Y.; Han, L.; Pickett, S.T.A.; Wang, J.; Li, W.; Ouyang, Z. Beyond city expansion: Multi-scale environmenta impacts of urban megaregion formation in China. *Natl. Sci. Rev.* **2022**, *9*, 79–91. [CrossRef]
6. Wang, S.; Zhang, J.; Wang, F.; Dong, Y. How to achieve a balance between functional improvement and heritage conservation A case study on the renewal of old Beijing city. *Sustain. Cities Soc.* **2023**, *98*, 104790. [CrossRef]
7. Li, K.; Hou, Y.; Randall, M.T.; Skov-Petersen, H.; Li, X. The spatio-temporal trade-off between ecosystem and basic public service and the urbanization driving force in the rapidly urbanizing region. *Sustain. Cities Soc.* **2024**, *111*, 105554. [CrossRef]
8. Ferreira, C.; Kalantari, Z.; Pereira, P. Liveable cities: Current environmental challenges and paths to urban sustainability. *J. Enviror Manag.* **2021**, *277*, 111458. [CrossRef]
9. *CJJ/T85-2002*; Standard for Classification of Urban Green Space. China Government Online: Beijing, China, 2018.
10. Wang, Z.; Miao, Y.; Xu, M.; Zhu, Z.; Qureshi, S.; Chang, Q. Revealing the differences of urban parks' services to human wellbein based upon social media data. *Urban For. Urban Green.* **2021**, *63*, 127233. [CrossRef]
11. Paudel, S.; States, S.L. Urban green spaces and sustainability: Exploring the ecosystem services and disservices of grassy lawn versus floral meadows. *Urban For. Urban Green.* **2023**, *84*, 127932. [CrossRef]
12. Wang, L.; Xu, Y.; Zhai, Y.; Xu, D.; Fang, J.; Yao, Y.; Ren, Z.; Ye, Z. The ideal characteristics of landscape pattern and morphologica spatial pattern for seasonal cool island regulation in urban park green space. *Urban For. Urban Green.* **2025**, *107*, 128813. [CrossRef
13. Haque, M.N.; Sharifi, A. Barriers and solutions for just and equitable access to urban ecosystem services: A critical analysis. *Int. Sustain. Dev. World Ecol.* **2025**, *32*, 370–386. [CrossRef]
14. Yang, Y.; Jiang, L.; Ma, X.; Liu, S.; Wang, L. A comprehensive approach to identifying the supply and demand of urban park cultural ecosystem services in the Megalopolis area of Shanghai, China. *Land* **2025**, *14*, 455. [CrossRef]
15. De Luca, C.; Calcagni, F.; Tondelli, S. Assessing distributional justice around Cultural Ecosystem Services (CES) provided by urban green areas: The case of Bologna. *Urban For. Urban Green.* **2024**, *101*, 128556. [CrossRef]
16. Wei, J.; Chen, M.; Chu, C.; Zhao, C.; Xia, X.; Li, Y. Assessing cultural ecosystem services supply–demand balance of urban parks in the context of old and new urban districts. *Ecol. Indic.* **2024**, *159*, 111688. [CrossRef]
17. Xu, S.; Li, J.; Gao, X.; Zhao, H.; Hu, J.; Yuan, S. Framework for analyzing the relationship between supply, demand, and flow o recreational services in urban park green spaces. *Ecol. Indic.* **2024**, *166*, 112403. [CrossRef]
18. Blake, D.; Carver, S.; Ziv, G. Demographic, natural and anthropogenic drivers for coastal Cultural ecosystem services in the Falkland Islands. *Ecol. Indic.* **2021**, *130*, 108087. [CrossRef]

https://doi.org/10.3390/su1805257

Li, J.; Huang, Z.; Zhu, Z.; Ding, G. Coexistence Perspectives: Exploring the impact of landscape features on aesthetic and recreational values in urban parks. *Ecol. Indic.* **2024**, *162*, 112043. [CrossRef]

Arslan, E.S.; Orucu, O.K. MaxEnt modelling of the potential distribution areas of cultural ecosystem services using social media data and GIS. *Environ. Dev. Sustain.* **2021**, *23*, 2655–2667. [CrossRef]

Inácio, M.; Gomes, E.; Bogdzevič, K.; Kalinauskas, M.; Zhao, W.; Pereira, P. Mapping and assessing coastal recreation cultural ecosystem services supply, flow, and demand in Lithuania. *J. Environ. Manag.* **2022**, *323*, 116175. [CrossRef]

Wan, C.; Shen, G.Q.; Choi, S. Effects of physical and psychological factors on users' attitudes, use patterns, and perceived benefits toward urban parks. *Urban For. Urban Green.* **2020**, *51*, 126691. [CrossRef]

Tian, T.; Huang, S.; Wu, Y.; Zeng, P.; Liu, Y.; Che, Y. Non-ecological factors affect human interaction with urban nature and perception of cultural ecosystem services. *Sustain. Cities Soc.* **2024**, *112*, 105643. [CrossRef]

Johansson, M.; Pedersen, E.; Weisner, S. Assessing cultural ecosystem services as individuals' place-based appraisals. *Urban For. Urban Green.* **2019**, *39*, 79–88. [CrossRef]

Sandarić, N.; Ives, C.D.; Watkins, C. Can we plan for urban cultural ecosystem services? *J. Urban Ecol.* **2020**, *6*, juaa016. [CrossRef]

Zhang, K.; Tang, X.; Zhao, Y.; Huang, B.; Liu, M.; Luo, E.; Li, Y.; Jiang, T.; Zhang, L.; Wang, Y.; et al. Differing perceptions of the youth and the elderly regarding cultural ecosystem services in urban parks: An exploration of the tour experience. *Sci. Total Environ.* **2022**, *821*, 153388. [CrossRef]

Gai, S.; Fu, J.; Rong, X.; Dai, L. Users' views on cultural ecosystem services of urban parks: An importance-performance analysis of a case in Beijing, China. *Anthropocene* **2022**, *37*, 100323. [CrossRef]

Zhou, L.; Guan, D.; Huang, X.; Yuan, X.; Zhang, M. Evaluation of the cultural ecosystem services of wetland park. *Ecol. Indic.* **2020**, *114*, 106286. [CrossRef]

Xie, X.; Wang, Q.; Wu, J.; Wang, H.; Chen, Z.; He, J. Effectively enhancing perceptions of cultural ecosystem services: A case study of a karst cultural ecosystem. *J. Environ. Manag.* **2022**, *315*, 115189. [CrossRef]

Cabana, D.; Ryfield, F.; Crowe, T.P.; Brannigan, J. Evaluating and communicating cultural ecosystem services. *Ecosyst. Serv.* **2020**, *42*, 101085. [CrossRef]

Liu, Z.; Hanley, N.; Campbell, D. Linking urban air pollution with residents' willingness to pay for greenspace: A choice experiment study in Beijing. *J. Environ. Econ. Manag.* **2020**, *104*, 102383. [CrossRef]

Meng, S.; Huang, Q.; Zhang, L.; He, C.; Inostroza, L.; Bai, Y.; Yin, D. Matches and mismatches between the supply of and demand for cultural ecosystem services in rapidly urbanizing watersheds: A case study in the Guanting Reservoir basin, China. *Ecosyst. Serv.* **2020**, *45*, 101156. [CrossRef]

Baumeister, C.F.; Gerstenberg, T.; Plieninger, T.; Schraml, U. Exploring cultural ecosystem service hotspots: Linking multiple urban forest features with public participation mapping data. *Urban For. Urban Green.* **2020**, *48*, 126561. [CrossRef]

Dang, H.; Li, J. Supply-demand relationship and spatial flow of urban cultural ecosystem services: The case of Shenzhen, China. *J. Clean. Prod.* **2023**, *423*, 138765. [CrossRef]

Ghermandi, A.; Camacho, V.; Trejo-Espinosa, H. Social media-based analysis of cultural ecosystem services and heritage tourism in a coastal region of Mexico. *Tour. Manag.* **2019**, *77*, 104002. [CrossRef]

Riechers, M.; Barkmann, J.; Tscharntke, T. Diverging perceptions by social groups on cultural ecosystem services provided by urban green. *Landsc. Urban Plan.* **2018**, *175*, 161–168. [CrossRef]

Dade, M.C.; Mitchell, M.G.E.; Brown, G.; Rhodes, J.R. The effects of urban greenspace characteristics and socio-demographics vary among cultural ecosystem services. *Urban For. Urban Green.* **2020**, *49*, 126641. [CrossRef]

Li, X.; Gan, X.; Jiang, Z.; Li, Z.; Li, J. How do landscape patterns in urban parks affect multiple cultural ecosystem services perceived by residents? *Sci. Total Environ.* **2024**, *946*, 174255. [CrossRef]

Valánszki, I.; Kristensen, L.S.; Jombach, S.; Ladányi, M.; Filepné Kovács, K.; Fekete, A. Assessing relations between cultural ecosystem services, physical landscape features and accessibility in central-eastern Europe: A PPGIS empirical study from Hungary. *Sustainability* **2022**, *14*, 754. [CrossRef]

Chen, C.; Luo, W.; Li, H.; Zhang, D.; Kang, N.; Yang, X.; Xia, Y. Impact of perception of green space for health promotion on willingness to use parks and actual use among young urban residents. *Int. J. Environ. Res. Public Health* **2020**, *17*, 5560. [CrossRef]

Cheng, X.; Van Damme, S.; Li, L.; Uyttenhove, P. Cultural ecosystem services in an urban park: Understanding bundles, trade-offs, and synergies. *Landsc. Ecol.* **2022**, *37*, 1693–1705. [CrossRef]

Aranguren, M. Adding experiments to the classical survey-interview mix: A three-method design for supporting micro-to-macro explanations in sociology. *Res. Soc. Stratif. Mobil.* **2024**, *91*, 100933. [CrossRef]

Bradley, V.C.; Kuriwaki, S.; Isakov, M.; Sejdinovic, D.; Meng, X.-L.; Flaxman, S. Unrepresentative big surveys significantly overestimated US vaccine uptake. *Nature* **2021**, *600*, 695–700. [CrossRef] [PubMed]

Fan, Y.; Hashim, N.H.; Goh, H.C. A Systematic Review of Place-Based Cultural Ecosystem Service Assessments: Categories, Methods, and Research Trends. *Sustainability* **2026**, *18*, 644. [CrossRef]

 https://doi.org/10.3390/su18052578

45. Su, R.; Ye, S.; Yu, L.; Wu, J.; Kang, Y.; Chen, R. Social and ecological factors on the perception of cultural ecosystem services an disservices: Insights from Shanghai, China. *Landsc. Urban Plan.* **2025**, *259*, 105373. [CrossRef]
46. Li, J.; Li, K.; Wang, Y.; Jiao, R. Comparative Study on the Perception of Cultural Ecosystem Services in Taibai Mountain Nationa Forest Park from Different Stakeholder Perspectives. *Land* **2024**, *13*, 2207. [CrossRef]
47. WMSB. *Statistical Communiqué on the National Economic and Social Development of Wuhan City*; Wuhan Government Online: Wuha China, 2024.
48. DeVellis, R.F. *Scale Development: Theory and Applications*; Sage Publications: Thousand Oaks, CA, USA, 2016.
49. Phillips, A.; Plastara, D.; Khan, A.Z.; Canters, F. Integrating public perceptions of proximity and quality in the modelling of urba green space access. *Landsc. Urban Plan.* **2023**, *240*, 104875. [CrossRef]
50. Leng, S.; Sun, R.; Yang, X.; Chen, L. Global inequities in population exposure to urban greenspaces increased amidst tree an nontree vegetation cover expansion. *Commun. Earth Environ.* **2023**, *4*, 464. [CrossRef]
51. Buonocore, C.; Carlucci, F.; Ciciarelli, L.; Papola, A.; Tinessa, F.; Tocchi, D.; Trincone, B. Accessibility analysis in spatial plannin A case of special economic zones (SEZs) in Campania, Southern Italy. *Land Use Policy* **2023**, *132*, 106763. [CrossRef]
52. Wang, Y.; Shi, X.; Cheng, K.; Zhang, J.; Chang, Q. How do urban park features affect cultural ecosystem services: Quantifie evidence for design practices. *Urban For. Urban Green.* **2022**, *76*, 127713. [CrossRef]
53. Edeh, E.; Lo, W.-J.; Khojasteh, J. Review of Partial Least Squares Structural Equation Modeling (PLS-SEM) Using R: A Workboo *Struct. Equ. Model.-A Multidiscip. J.* **2023**, *30*, 165–167. [CrossRef]
54. Monecke, A.; Leisch, F. semPLS: Structural Equation Modeling Using Partial Least Squares. *J. Stat. Softw.* **2012**, *48*, 1–3 [CrossRef]
55. Fornell, C.; Larcker, D.F. Evaluating structural equation models with unobservable variables and measurement error. *J. Mark. Re* **1981**, *18*, 39–50. [CrossRef]
56. Henseler, J.; Dijkstra, T.; Sarstedt, M.; Ringle, C.; Diamantopoulos, A.; Straub, D.; Ketchen, D.J., Jr.; Hair, J.; Hult, G.T.M Calantone, R. Common Beliefs and Reality About PLS: Comments on Rönkkö and Evermann (2013). *Organ. Res. Methods* **2014**, *1* 182–209. [CrossRef]
57. Kock, N. Common Method Bias in PLS-SEM. *Int. J. E-Collab.* **2015**, *11*, 1–10. [CrossRef]
58. Hair, J.; Hult, G.T.M.; Ringle, C.; Sarstedt, M. *A Primer on Partial Least Squares Structural Equation Modeling (PLS-SEM)*, 2nd ed Springer International Publishing: Berlin/Heidelberg, Germany, 2016.
59. Sarstedt, M.; Ringle, C.M.; Hair, J.F. Partial Least Squares Structural Equation Modeling. In *Handbook of Market Research*; Homburg C., Klarmann, M., Vomberg, A., Eds.; Springer International Publishing: Cham, Switzerland, 2017; pp. 1–40.
60. Hair, J.; Ringle, C.; Sarstedt, M. Partial Least Squares Structural Equation Modeling: Rigorous Applications, Better Results an Higher Acceptance. *Long Range Plan.* **2013**, *46*, 1–12. [CrossRef]
61. Cavalcante, S.; Maciel, R.H. Métodos de avaliação da percepção ambiental. In *Métodos de Pesquisa nos Estudos Pessoa-Ambiente*, 1s ed.; Pinheiro, J.Q., Günther, H., Eds.; Casa do Psicólogo: São Paulo, Brazil, 2008.
62. Huai, S.; Liu, S.; Zheng, T.; Van de Voorde, T. Are social media data and survey data consistent in measuring park visitation, par satisfaction, and their influencing factors? A case study in Shanghai. *Urban For. Urban Green.* **2023**, *81*, 127869. [CrossRef]
63. Zhao, X.; Lu, Y.; Huang, W.; Lin, G. Assessing and interpreting perceived park accessibility, usability and attractiveness throug texts and images from social media. *Sustain. Cities Soc.* **2024**, *112*, 105619. [CrossRef]
64. Ghermandi, A.; Depietri, Y.; Orenstein, D.E. Human-nature interactions through a digital prism: Heterogeneity in user socio demographics and content across social media platforms. *Landsc. Urban Plan.* **2026**, *268*, 105568. [CrossRef]
65. Liang, H.; Yan, Q.; Yan, Y.; Zhang, L.; Zhang, Q. Spatiotemporal Study of Park Sentiments at Metropolitan Scale Using Multipl Social Media Data. *Land* **2022**, *11*, 1497. [CrossRef]
66. Kong, L.; Liu, Z.; Pan, X.; Wang, Y.; Guo, X.; Wu, J. How do different types and landscape attributes of urban parks affect visitors positive emotions? *Landsc. Urban Plan.* **2022**, *226*, 104482. [CrossRef]
67. Wu, Y.; Tang, L.; Huang, C.B.; Shao, G.; Hou, J.; Sabel, C.E. Enhancing human well-being through cognitive and affective pathways linking landscape sensation to cultural ecosystem services. *Landsc. Ecol.* **2024**, *39*, 175. [CrossRef]
68. Keleş Özgenç, E.; Dönmez, A.H.; Özgenç, E. Evaluating Cultural Ecosystem Services Through Geospatial Social Media Data: A Study of Edirne City. *J. Geovis. Spat. Anal.* **2024**, *8*, 30. [CrossRef]
69. Wykowski, J.H.; Starks, H. What Type of Self-Assessment Is Best for Your Educational Activity? A Review of Pre-Post, Now-Then and Post-Only Designs. *J. Gen. Intern. Med.* **2025**, *40*, 1010–1015. [CrossRef] [PubMed]
70. Yee, T.B.L.; Carrasco, L.R. Applying deep learning on social media to investigate cultural ecosystem services in protected areas worldwide. *Sci. Rep.* **2024**, *14*, 13700. [CrossRef] [PubMed]
71. Haines-Young, R.; Potschin, M. *Common International Classification of Ecosystem Services (CICES): 2011 Update*; European Environment Agency: Copenhagen, Denmark, 2011.
72. Hubatova, M.; McGinlay, J.; Parsons, D.J.; Morris, J.; Graves, A.R. Assessing Preferences for Cultural Ecosystem Services in the English Countryside Using Q Methodology. *Land* **2023**, *12*, 331. [CrossRef]

https://doi.org/10.3390/su18052578

Sultan, M.; Tump, A.N.; Ehmann, N.; Lorenz-Spreen, P.; Hertwig, R.; Gollwitzer, A.; Kurvers, R. Susceptibility to online misinformation: A systematic meta-analysis of demographic and psychological factors. *Proc. Natl. Acad. Sci. USA* **2024**, *121*, e2409329121. [CrossRef]

Pang, D.; Guntuku, S.C.; Sherman, G.; Liu, T.; Rai, S.; Cho, Y.-M.; You, R.; Bakker, A.B.; Ungar, L.H. Understanding gender and age differences in language use: Cross-cultural insights from Weibo and Facebook. *Humanit. Soc. Sci. Commun.* **2025**, *12*, 1667. [CrossRef]

Lu, J.; Zhang, H.; Zhang, X. Cultural ecosystem services in China's national parks and their impact on public online engagement—Analysis of Douyin short videos data based on BERTopic modeling. *J. Nat. Conserv.* **2025**, *87*, 126969. [CrossRef]

Souza, C.N.; Martínez-Arribas, J.; Correia, R.A.; Almeida, J.A.G.R.; Ladle, R.; Vaz, A.S.; Malhado, A.C. Using social media and machine learning to understand sentiments towards Brazilian National Parks. *Biol. Conserv.* **2024**, *293*, 110557. [CrossRef]

Oguro, M.; Shibata, R. Quantifying effects of demographic biases on estimation of cultural ecosystem services using social media in Japan. *Ecosyst. Serv.* **2025**, *76*, 101777. [CrossRef]

Sohail, S.S.; Khan, M.M.; Madsen, D.Ø.; Alam, M.A.; Irshad, R.R. Geospatial and Linguistic Analysis of Twitter Behavioral Trends: Examining the Impact of Socioeconomic Development on Social Media Use. *Hum. Behav. Emerg. Technol.* **2025**, *2025*, 1376983. [CrossRef]

Rivera, E.; Timperio, A.; Loh, V.H.; Deforche, B.; Veitch, J. Important park features for encouraging park visitation, physical activity and social interaction among adolescents: A conjoint analysis. *Health Place* **2021**, *70*, 102617. [CrossRef]

Veitch, J.; Ball, K.; Rivera, E.; Loh, V.; Deforche, B.; Best, K.; Timperio, A. What entices older adults to parks? Identification of park features that encourage park visitation, physical activity, and social interaction. *Landsc. Urban Plan.* **2022**, *217*, 104254. [CrossRef]

Wang, Y.; Hayashi, K. Methodological development of cultural ecosystem services evaluation using location data. *J. Clean. Prod.* **2023**, *396*, 136523. [CrossRef]

Koh, Y.F.; Loc, H.H.; Park, E. Towards a "city in nature": Evaluating the cultural ecosystem services approach using online public participation GIS to support urban green space management. *Sustainability* **2022**, *14*, 1499. [CrossRef]

Dushkova, D.; Taherkhani, M.; Konstantinova, A.; Vasenev, V.I.; Dovletyarova, E.A. Understanding Factors Affecting the Use of Urban Parks Through the Lens of Ecosystem Services and Blue–Green Infrastructure: The Case of Gorky Park, Moscow, Russia. *Land* **2025**, *14*, 237. [CrossRef]

Wan, C.; Shen, G.Q.; Choi, S. Eliciting users' preferences and values in urban parks: Evidence from analyzing social media data from Hong Kong. *Urban For. Urban Green.* **2021**, *62*, 127172. [CrossRef]

Sikorski, P.; Gawryszewska, B.; Sikorska, D.; Chormański, J.; Schwerk, A.; Jojczyk, A.; Ciężkowski, W.; Archiciński, P.; Łepkowski, M.; Dymitryszyn, I.; et al. The value of doing nothing—How informal green spaces can provide comparable ecosystem services to cultivated urban parks. *Ecosyst. Serv.* **2021**, *50*, 101339. [CrossRef]

Cross, E.S.; Darda, K.M.; Moffat, R.; Muñoz, L.; Humphries, S.; Kirsch, L.P. Mutual gaze and movement synchrony boost observers' enjoyment and perception of togetherness when watching dance duets. *Sci. Rep.* **2024**, *14*, 24004. [CrossRef]

Lyu, B.; Gao, Z.; Wang, Y.; Liu, J.; Zhang, L.; Song, J.; Pan, Y.; Cheng, M.; Liu, S.; Chen, Q.; et al. Seasonal Dynamics and Trade-Offs/Synergies of Cultural Ecosystem Services in Urban Parks: A Case Study of Chengdu, China. *Land* **2025**, *14*, 2126. [CrossRef]

You, W.; Xu, H.; Duan, J.; Zhou, Y.; Li, J.; He, D.; Kizos, T. Assessing spatial differences of perceptions of cultural ecosystem services for coastal cultural landscape management: A case study from rural and urban areas in Quanzhou, China. *J. Environ. Manag.* **2025**, *395*, 127674. [CrossRef]

Zhang, Y.; Zhao, L.; Hu, S. Assessing perception and equity of cultural ecosystem services in urban parks using social media data. *Sci. Rep.* **2025**, *15*, 33127. [CrossRef] [PubMed]

Li, J.; Luo, J.; Deng, T.; Tian, J.; Wang, H. Exploring perceived restoration, landscape perception, and place attachment in historical districts: Insights from diverse visitors. *Front. Psychol.* **2023**, *14*, 1156207. [CrossRef] [PubMed]

Liu, T.; Yu, L.; Chen, X.; Chen, Y.; Li, X.; Liu, X.; Cao, Y.; Zhang, F.; Zhang, C.; Gong, P. Identifying Potential Urban Greenways by Considering Green Space Exposure Levels and Maximizing Recreational Flows: A Case Study in Beijing's Built-Up Areas. *Land* **2024**, *13*, 1793. [CrossRef]

Vierikko, K.; Gonçalves, P.; Haase, D.; Elands, B.; Ioja, C.; Jaatsi, M.; Pieniniemi, M.; Lindgren, J.; Grilo, F.; Santos-Reis, M.; et al. Biocultural diversity (BCD) in European cities—Interactions between motivations, experiences and environment in public parks. *Urban For. Urban Green.* **2020**, *48*, 126501. [CrossRef]

Cheng, Y.; Browning, M.H.E.M.; Zhao, B.; Qiu, B.; Wang, H.; Zhang, J. How can urban green space be planned for a 'happy city'? Evidence from overhead- to eye-level green exposure metrics. *Landsc. Urban Plan.* **2024**, *249*, 105131. [CrossRef]

Guo, R.; Ann Diehl, J.; Zhang, R.; Wang, H. Spatial equity of urban parks from the perspective of recreational opportunities and recreational environment quality: A case study in Singapore. *Landsc. Urban Plan.* **2024**, *247*, 105065. [CrossRef]

Zhong, J.; Lin, Z.; Jia, F.; Schoenherr, T. Smooth or sticky? An analysis of service variability. *Humanit. Soc. Sci. Commun.* **2025**, *12*, 113. [CrossRef]

96. Yang, H.; Jin, C.; Li, T. A paradox of economic benefit and social equity of green space in megacity: Evidence from Tianjin i China. *Sustain. Cities Soc.* **2024**, *109*, 105530. [CrossRef]
97. Wang, C.; Luo, F. Differing perceptions on cultural ecosystem service values of urban forest landscapes for children and youth A case study of Changsha, Central China. *Urban For. Urban Green.* **2025**, *112*, 129005. [CrossRef]
98. Sun, X.; Liu, H.; Liao, C.; Nong, H.; Yang, P. Understanding recreational ecosystem service supply-demand mismatch and soci groups' preferences: Implications for urban–rural planning. *Landsc. Urban Plan.* **2024**, *241*, 104903. [CrossRef]

 https://doi.org/10.3390/su1805257

sustainability

MDPI

rticle

How Do External Environments Shape the Cultural Ecosystem Services of Urban Parks to Promote Sustainable Urban Development? An Empirical Study of Multi-Travel Scenes in 15-Min Living Circles in Chengdu, China

idi Dong [1], Binzhu Wang [1], Mingming Chen [1], Jiaxi He [1] and Yingyin Yang [2,*]

[1] School of Art and Design, Xihua University, Chengdu 610039, China; dqd@mail.xhu.edu.cn (Q.D.); wangbinzhu@stu.xhu.edu.cn (B.W.); chenmingming@stu.xhu.edu.cn (M.C.); hejiaxi@stu.xhu.edu.cn (J.H.)

[2] Deyang City Territorial Spatial Planning Compilation and Research Center, Deyang 618099, China

* Correspondence: 2019110003@stu.sicau.edu.cn

Abstract

In light of the accelerating process of global urbanization, the quality of cultural ecosystem services (CES) in urban parks has become a core metric for efforts to promote urban livability and sustainable cities. However, previous research has failed to consider the differential impacts of the external environment across various travel scenes. In this study, 32 parks in Chengdu serve as the empirical data, and public CES perception data are extracted from social media comments via text mining. Based on a unified 15 min time threshold, we delineate the service scope for four travel scenes and employ geographically weighted regression and piecewise regression models to analyze the spatial heterogeneity, driving mechanisms and threshold effects associated with the relationship between external environmental factors and park CES. The findings indicate that the external environment's influence on CES exhibits a "scene-factor-scale" adaptation pattern. Walking scenes are influenced primarily by land-use and population factors; in contrast, cycling scenes rely on the availability of shared bicycle facilities, and public transport and driving scenes are driven by economic vitality and traffic-support factors, respectively. Five critical thresholds are identified, including a 40% impervious surface area. This research proposes scene-based optimization strategies and helps enhance the "external environment–travel behavior–spatial characteristics" coupling framework, thereby serving as a scientific reference for efforts to improve 15 min living circles.

Keywords: sustainable urban development; urban park; cultural ecosystem services; 15 min city; travel behavior; external environment

1. Introduction

As global urbanization continues to accelerate, the global urbanization rate is expected to reach 68% by 2050 [1]. The rapid concentration of the urban population has made the contradiction between cities' service capacity and residents' continually increasing living needs increasingly prominent [2]. Against this backdrop, the question of how to maximize the ecological recreational service supply capacity of urban parks through scientific planning, thereby enhancing the sustainable well-being of urban residents, has become an important scientific and practical issue in the field of sustainable urban development. The concept of the "15 min city", which defines the scope of services based on a specific time threshold and ensures equal access to daily public services for residents, has become the core framework for global livable city development [3].

Urban parks, as a core component of urban green infrastructure, provide a range of cultural ecosystem services (CES), including landscape aesthetics, leisure and recreation, and spiritual healing. These services represent the intangible benefits that humans can obtain through spiritual perception, leisure and entertainment, and cultural inheritance [4]. They are directly related to public health, social cohesion, and urban resilience [5] and are the core dimensions used to assess the construction of livable cities worldwide and the extent to which United Nations Sustainable Development Goal (SDG) 11.7 has been achieved [1,6]. In the context of developing a 15 min living circle, residents' perceptions and experiences of the park's CES do not focus merely on the park's landscape. Instead, such perceptions and experiences encompass the entire process by which individuals travel from their residences to the park, using various modes of transportation. However, urban park construction currently faces a widespread problem, namely, "emphasizing the creation of internal landscape but neglecting the adaptation of the external travel environment" [7]. The park's external environment is not adequately designed to meet the needs of different travel scenarios, thereby directly weakening residents' perceptions of the park's CES and restricting the full use of the livability value offered by the 15 min living circle.

On this basis, this study explores whether the overall influence of the external environment on CES within a 15 min living circle varies across different travel scenarios in terms of the relevant driving factors, directions and intensities How do the spatial heterogeneity and scale-dependent patterns that characterize the external environment support the precise optimization of park CES in different travel scenarios within a 15 min living circle? Two scientific questions guide this research to provide practical references to support the sustainable development of urban parks and the construction of 15 min living circles.

We selected the Chengdu Ring City Ecological Zone as the empirical area of study for this research. This area is a benchmark of China's ecological protection green belt; it extends 500 m on both sides of the Ring Expressway (G4202) and features a total area of 187.15 km^2. It forms a green necklace across the core urban area. The geographical location and distribution of park spaces in this region are shown in Figure 1. Our choice of this area for this research offers three major advantages. First, after parks have been uniformly planned, the level of homogeneity in their internal functions and landscape bases is relatively high (i.e., the Normalized Difference Vegetation Index (NDVI) is between 0.47 and 0.56), which can reduce the interference from internal environmental differences to some extent. Second, each park is located 8–15 km away from the city center. Given that this region relies on a dense road network and greenways, it can accommodate four types of travel scenarios and mitigate interference from differences in the distance between the parks and the city center. Third, under the new urban development strategy of Chengdu, the external environment of the area has come to exhibit a rich spatial environmental gradient, thus providing good spatial support for efforts to interpret differences in the external environment.

We considered the following conditions in the process of selecting the parks included in the sample: (1) the park must be free of charge to avoid interference from special types of parks, such as theme parks and special parks; (2) the spatial attribute data for the park must be complete; and (3) the number of valid social media comments associated with the park must be $\geq$30, thus ensuring the representativeness of the CES perception data. Finally, we selected 32 for the sample (see the green section in Figure 1). After verification through a spatial autocorrelation analysis, Moran's I for the selected samples was -0.061 ($p = 0.871 > 0.05$), indicating that the sample distribution was relatively uniform and lacked significant clustering or dispersion; furthermore, the representativeness was good.

https://doi.org/10.3390/su18094177

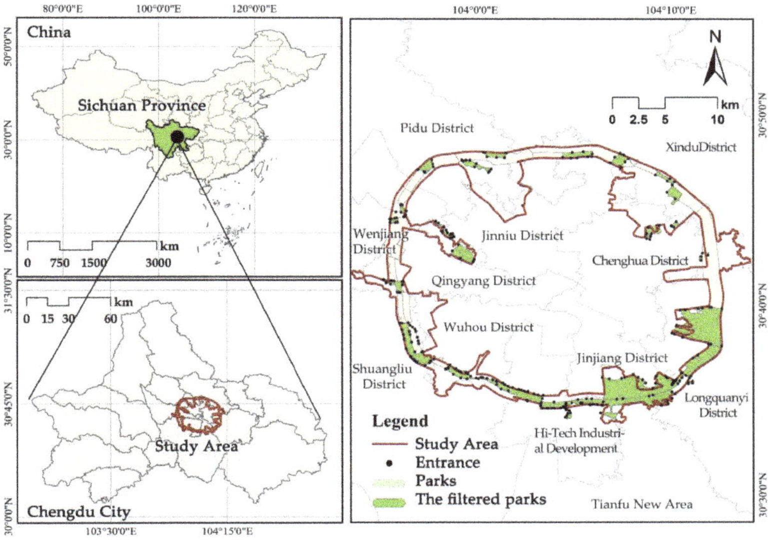

Figure 1. Overview of the study area.

2. Theoretical Foundations and Literature Review

2.1. CES Evaluation

CES is derived from ecosystem services theory, which posits a chain of "service supply–human use–well-being realization," thereby clarifying the transmission pathway associated with CES [8]. The UK National Ecosystem Assessment (NEA) refined the classification by distinguishing between supply and demand, thereby reinforcing its direct linkage to human well-being [9,10]. These contributions have provided a theoretical foundation for constructing the CES evaluation framework in this study and for analyzing the mechanism underlying the external environment's influence on CES perception.

Regarding the connotations and classification of CES, the academic community has reached a multidimensional consensus. Plieninger, Chan, and colleagues divided CES into dimensions such as aesthetic value, social connection, freedom, responsibility, and memory [11,12]. Based on a systematic review, Andra et al. identified three core dimensions of CES: spiritual experience, leisure experience, and cultural identity [13]. These studies have provided the classic classification system of CES, which serves as the basis for the classification employed in this research. In terms of assessment methods, current mainstream approaches can be divided into three categories: field investigations, questionnaires, and the emerging category of text mining. Field surveys and questionnaires result in highly accurate data but entail high costs and limited sample sizes.

In contrast, text mining of social media user-generated content offers advantages such as large samples, broad coverage, and strong timeliness [14–16], and the measures of public sentiment associated with this approach are closely in line with the results of social surveys [17], thereby providing more efficient and extensive data support for CES research. At present, the evaluation of comments pertaining to CES relies mainly on natural language processing techniques and perceptual dictionary recognition [6,14]. However, Chinese perceptual dictionaries remain relatively scarce, and most related studies have used general dictionaries to support their analyses, resulting in insufficient accuracy in capturing the CES perception characteristics of residents in local urban park scenarios (e.g., dialects and colloquial expressions).

https://doi.org/10.3390/su18094177

2.2. *Influence of the External Environment*

The supply–demand matching theory of ecosystem services holds that the core of value realization for ecosystem services lies in the production of accurate matches between spatial supply at the supply end and residents' use at the demand end. The external environment surrounding the park serves as the medium that connects the park's ecosystem service supply with the residents' demand [7,18]. The environment thus serves as a crucial link that determines whether residents can reach the park conveniently and comfortably and complete the process of using ecosystem services. Its characteristic elements directly determine the efficiency and effectiveness of the match between supply and demand for ecosystem services and provide a theoretical basis for constructing an external environment index system and analyzing the collaborative influence mechanism, which involves multiple factors in this study. The "15 min living circle" planning concept [3] further highlights the importance of researching this connection.

Previous studies on the external environmental impacts of parks initially divided the external environment into factors such as land, population, facilities, transportation, and economic vitality. A series of basic consensuses emerged in the literature on this topic: notably, land-use is a core category of park vitality and satisfaction. A balanced land-use mix improves the convenience of park visits by enhancing the quality of supporting facilities [19,20]. Among the relevant population and facility factors, population density and the diversity and mix of facilities directly shape CES perceptions [21]. Transportation links users to parks, and the density of road networks and bus stops affects visitors' experiences in this context [7]. Indicators of economic vitality, such as commercial density (e.g., catering and retail), reflect regional economic development and the quality of urban construction [2,22], and these factors indirectly increase the convenience that users experience in parks [23].

However, such previous studies have not yet explored the relationship between CES and the external environment. Moreover, most studies on the external environment have analyzed individual factors in isolation, and no integrated analysis of the collaborative effects of multiple elements, such as "spatial development–population agglomeration–facility support", has yet been performed [14,24]. Furthermore, such studies have relied largely on macroscale analyses, which can obscure microscale spatial heterogeneity and thus fail to explain the "differing impacts of the same external environment on CES at different spatial scales." Third, the extant research has given insufficient attention to the nonlinear characteristics of external environmental influences, failed to identify the effective thresholds of environmental factors, and has been unable to provide practical, quantifiable standards for planning practices. The theory of spatial heterogeneity is an important theoretical basis for explaining this phenomenon. This theory can effectively explain the unevenness and complexity of factors across different geographical spaces [24], thus providing an important way in which these differences in influence can be understood in this study.

2.3. *Travel Scenarios with Scale Effects*

According to the 15 min living circle model, accessibility and spatial scale are key to analyzing the impact of the external environment on CES. However, a gap remains in applying this concept in the field of CES. Travel behavior theory indicates that human travel decisions follow the logic of "demand-facility matching" (slow travel) or "cost-efficiency" (motorized travel) [25]. This rule determines the natural coupling of different travel scenarios and spatial scales. This point constitutes the theoretical premise for the analysis of travel scenarios presented in this paper.

https://doi.org/10.3390/su18094177

This situation leaves us unable to separate the independent influence of the spatial scale effect of the travel scene from the interference of the temporal scale. Travel–behavior theory indicates that travel decisions follow either a demand–facility matching logic for slow modes or a cost-efficiency logic for motorized modes [26]. This rule defines the intrinsic link between travel scenes and spatial scales, thereby underpinning the separate analysis of travel scenes presented in this paper.

Previous research has confirmed the existence of significant differences in spatial scales that correspond to different travel scenarios and has revealed that the influence of external environmental factors also changes significantly with spatial scale. Specifically, short-distance modes such as walking and cycling are highly sensitive to local roads and community support facilities [26]. In contrast, motorized travel, including public transit and driving, operates across meso- to transregional scales and thus emphasizes regional transportation links and infrastructure [27]. The variation in focus across different scenes reflects how external environmental factors change substantially alongside the spatial scale [26,28,29]. However, previous studies have not fully linked travel scenarios to spatial scale and thus have failed to accurately analyze the scale-dependent patterns of external environmental influences on CES. Such studies have also failed to link the planning characteristics of the 15 min life circle to the CES influence mechanism in depth, leading to a disconnect between this research's conclusions and the planning practices associated with the 15 min life circle.

2.4. Research Gaps, Research Objectives and Points of Innovation

In summary, previous research has exhibited three major deficiencies in terms of its analysis of park CES, thus highlighting the entry point of this study. First, in terms of evaluation dimensions, previous research has lacked a localized CES perception dictionary that is tailored to Chinese urban park scenarios. Second, regarding the influencing mechanism, the existing CES research has given insufficient attention to the external environment of parks and lacked an integrated analysis of multifactor synergy effects, as well as a clarification of the nonlinear threshold effect of environmental factors on CES. Third, in terms of scene adaptability, previous research has not linked travel scenarios with the spatial scale based on a unified 15 min living circle time threshold and thus failed to reveal the spatial heterogeneity and scale-dependent laws that characterize the external environment's influence on CES in different travel scenarios; the adaptability of the conclusions of such research to planning practices is thus insufficient.

Based on these research gaps, we take the external environment of parks, multiple travel scenarios, and CES as our main line of focus to investigate a sample consisting of 32 parks in the Chengdu Ring City Ecological Zone. We use text mining to extract public CES perception data and construct a multidimensional CES assessment system that is tailored to Chinese urban park scenarios, which can offer a precise overview of the spatial differentiation characteristics of park CES in the Chengdu Ring City Ecological Zone. We integrate multisource spatial data, combine the four types of travel scenario analysis units defined by the unified 15 min time threshold, analyze the differentiated mechanisms underlying the influence of external environmental factors on CES, clarify the driving factors associated with different scenarios, identify the effective threshold for relevant influencing factors, reveal the "scene-factor-scale" adaptation laws by which the external environment affects CES, and propose park external environment optimization strategies based on travel scenarios, thereby providing a scientific reference for efforts to improve the livability and quality of the 15 min living circle.

Based on this analysis, it is possible to conduct a quantitative analysis of the spatial heterogeneity and threshold effects of the external environment on CES, thereby reveal-

https://doi.org/10.3390/su18094177

ing how the public perceives and utilizes these cultural services. The main innovative contributions of this study are as follows:

(1) Contribution in terms of the research perspective: This research expands the boundaries of the literature on this topic to encompass travel scenarios pertaining to parks within the 15 min living circle and use it as a related condition for the scale of CES perception., thus filling the gap according to which insufficient attention has been given to travel scenarios and scale effects in previous research.

(2) Methodological contribution: Constructing an analytical framework that involves "text mining–accessibility spatial division–spatial regression–segmented regression", thereby not only facilitating the precise capture of public CES perceptions through text mining but also revealing the spatial heterogeneity of the corresponding influencing effect based on the geographically weighted regression (GWR) model and identifying the effective threshold for influencing factors through segmented regression; on this basis, a complete logical loop of "global test–local analysis–threshold verification" is established.

(3) Practical value innovation: The conclusions of this research can be adapted directly to the planning practices associated with the 15 min living circle, and the proposed optimization strategies that are differentiated by scenario and region have clear quantitative thresholds and operability and can thus serve as replicable reference models to facilitate the planning and construction of similar parks.

3. Materials and Methods

3.1. Research Design

On the basis of ecosystem service supply–demand matching theory (which supports analyses of the supply–demand matching mechanism associated with), travel behavior theory (which supports efforts to interpret the correspondence between travel scenarios and spatial scales), and spatial heterogeneity theory (which supports the study of differences at multiple scales), we constructed the following conceptual analytic framework: "external environment–travel scenario–CES perception". We included the CES perceptions of urban parks as the dependent variable, the external environment of the parks as the independent variable, and the four travel scenarios within a 15 min time threshold (i.e., walking, cycling, public transportation, and self-driving) as the boundary adjustment conditions for this effect. We believe that the influence of external environmental factors on the CES perceptions associated with parks significantly varies alongside changes in the spatial scale corresponding to travel scenarios. Different driving factors are relevant in different travel scenarios, and the relevant influencing factors have nonlinear threshold effects.

The specific process involved in this operation is as follows (Figure 2). ① Data preparation: Multisource data concerning CES perception, the external environment, and accessibility are collected, and standardized preprocessing is performed. ② Determine the analytical unit: Based on a unified 15 min time threshold, the accessibility ranges for four travel scenes are identified, and the units for the analysis are established. ③ Construction of the index system: A tailored index framework for the external environment and CES is developed, and core variables are selected. ④ Analysis of the influence law: Spatial regression models and threshold identification are used to assess how the external environment affects CES and how these effects are exhibited in spatial heterogeneity across different travel scenes (i.e., different spatial scales). ⑤ Strategy proposal: Targeted optimization strategies for park external environments are derived from the research law to inform the construction of 15 min living circles.

https://doi.org/10.3390/su18094177

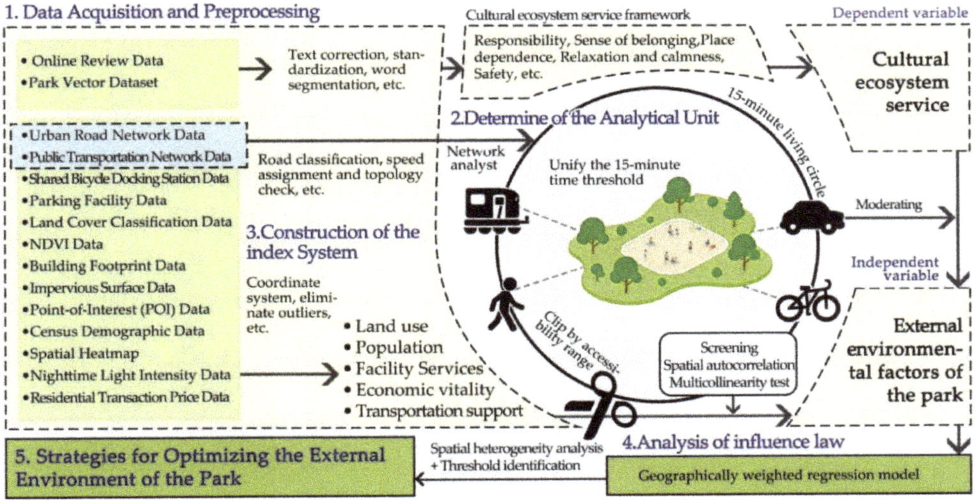

Figure 2. Method and process diagrams.

3.2. Dependent Variable: Construction and Processing of the CES System

Social media comments reflect tourists' immediate perceptions of and feedback regarding their travel to and experiences in parks [6]. This study uses text-mining methods and retrieves comment data—including comment text, score, and posting time—via the open API of the most widely used public review website in China (https://www.dianping.com/ (accessed on 8 August 2025)). This website is the core channel through which Chinese users share their consumption experiences. The age of users on this platform is higher than that of users on other platforms (a network analysis revealed that more than 45% of Dianping's users are over 40 years old) [30]. These data cover tourists of different ages who travel for different purposes, thereby avoiding sample bias.

Most parks in the region did not open until 2022, and some were not completed until 2023. To ensure data integrity and alignment with external spatial datasets, we selected review data from January 2023 to September 2025, a period spanning four seasons, thereby reducing seasonal effects on CES perceptions.

3.2.1. CES Data Preprocessing

The text cleaning process involved three levels of filtering: ① typos were corrected using the text correction tool on the Baidu AI intelligent platform (https://cloud.baidu.com/product/nlp_apply/text_corrector (accessed on 15 October 2025)); ② missing entries, advertisements, comments under 10 words, and duplicate comments were manually removed; ③ a merged thesaurus containing existing standardized vocabulary was used (https://github.com/liuhuanyong/ChineseSemanticKB (accessed on 27 October 2025)), dialectal and online terms were normalized (e.g., "yyds" was adjusted to "very great"), and synonymous terms were merged (e.g., "green meadows" and "lawn" were combined as "plants"). After filtering, 23,843 valid comments remained for analysis.

Keyword extraction was performed based on Jieba word segmentation (from the open source Python 3.12.4 (Python Software Foundation, Wilmington, NC, USA) Chinese word segmentation library). We observed that starting with the 500th word, the frequency of words decreased sharply and entered the long-tail range, which featured many low-frequency noise words. Given the size and lexical complexity of the corpus, we selected the top 500 highest-frequency words as the core corpus for the CES perception dictionary to ensure its representativeness and effectiveness.

3.2.2. CES Framework Construction

We referred to ecosystem service theory and previous studies to define CES. A CES framework consisting of 11 indicators was constructed; it included multiple dimensions, such as responsibility, sense of belonging, and place dependence (Table 1). In light of the scarcity of Chinese perception dictionaries for extracting CES-related Chinese words from Chinese review texts, we invited 11 researchers working in related fields to participate in multiple rounds of dictionary classification evaluation. We classified words that appeared frequently in the review texts in accordance with their semantics into the CES framework. In cases involving ambiguous or unclear terms, the research team carefully reviewed and reclassified such terms until a preliminary consensus was reached. Ultimately, a perception dictionary that included 179 words was formed and exhibited good classification consistency (Kappa = 0.78).

Table 1. CES framework and perceptual dictionary construction.

Indicator	Abbr.	Conceptual Interpretation	Schematic Diagram of the Perceptual Dictionary	Indicator Source
Responsibility	C1	I feel responsible for this place	Responsible, civilized, mission, etc.	Chan et al. [11]; Bryce et al. [31]; Fish et al. [32]
Sense of belonging	C2	This place evokes some kind of memory for me	Childhood, longing, memories, etc.	Dallimer et al. [33]; Bryce et al. [31]; Fish et al. [32]
Place dependence	C3	This place represents an integral part of my daily life	Daily, regular, convenient, etc.	NEA [9]; Chen et al. [34]; Fish et al. [32]
Relaxation and calmness	C4	This place relaxes me	Relaxation, happiness, contentment, etc.	
Safety	C5	This place makes me feel safe	Safety, security, patrol, etc.	Bryce et al. [31]; Li et al. [6]; Nowak Olejnik et al. [35]
Aesthetic value	C6	This place is beautiful	Beautiful, limpid, attractive, etc.	
Nature connection	C7	This place makes me closer to nature	Freshness, fragrance, vitality, etc.	
Social connection	C8	This place offers a lot of social opportunities	Family, dates, parties, etc.	Chan et al. [11]; Bryce et al. [31]
Reflection	C9	This place makes me think	Think, reflect, reflect, etc.	Dallimer et al. [33]; Bryce et al. [31]; Fish et al. [32]
Inspiration	C10	This place inspires me	Inspiration, creativity, epiphany, etc.	Chan et al. [11]; Bryce et al. [31]; Fish et al. [32]
Knowledge	C11	This place teaches me knowledge	Popular science, knowledge, education, etc.	Bryce et al. [31]; Fish et al. [32]

Subsequently, we used Python regular expressions to identify whether the relevant words contained in the CES perception dictionary existed in the review text (in which context the cumulative frequency of word occurrences was used to reflect the intensity of the perceived CES) and excluded the context window limitation to improve the accuracy of the identification (for example, to avoid classifying "no birdsong" as "birdsong"). Finally, we matched the recognition results to spatial data concerning parks obtained from the Chengdu Park City Construction and Development Research Institute.

To verify the validity of the CES perception information provided by the review texts, we randomly recruited tourists to participate in a questionnaire (a sample item was as follows: Did you feel this CES? Indicate the intensity of your perception on a scale ranging

https://doi.org/10.3390/su18094177

from 1 to 10) in Bai Luwan Wetland Park, which is located in the southern part of the area. We collected 89 valid questionnaires. We then compared the results with the review texts. The kappa test revealed that the overall kappa value of the social media data and the offline survey was 0.61 ($p < 0.001$), and the kappa values for the various dimensions ranged from 0.56 to 0.67, all of which met the relevant consistency standards.

3.2.3. Standardization and Dimensional Extraction of CES

To eliminate the influence of comment length differences on CES evaluation, we followed prior studies [14] by computing a weighted average adjustment of CES scores. This calculation was based on the lower and upper quartiles (*Q1*, *Q3*) of the number of effective comment words per park. The formula used in this context is as follows:

$$S_{\mathrm{j}} = \frac{\sum(s_{ij} \times w_i)}{\sum w_{\mathrm{i}}} \tag{1}$$

where S_j is the weighted score of a CES index for the j-th park; s_{ij} is the original score of the index for the i-th comment in the j-th park; and w_i is the weight of comment i. The weighting assignment method used in this regard is as follows: $w_i = 0.5$ when the comment length $\leq$ Q1, $w_i = 1.0$ when $Q1 <$ the comment length $< Q3$, and wi = 1.5 when the comment length $\geq Q3$ [14]; $\sum w_i$ indicates the total weight of all the valid comments for the *j*th park.

To verify the internal consistency of the index system and the feasibility of dimensionality reduction, we first conducted reliability and validity tests using SPSS 28.0 (IBM Corporation, Armonk, NY, USA). Moreover, this study included 11 initial dependent variables. To mitigate the problem of multicollinearity among the variables, which could result in deviations in the subsequent model estimation, as well as to simplify the data structure, we conducted a principal component analysis (PCA) for dimensionality reduction processing. Based on the factor loading matrix and index connotations of the principal components, the extracted principal components were named and weighted, thus allowing us to calculate the dimensional scores of each CES.

3.3. *Independent Variable: Construction and Treatment of the External Environment Index*

3.3.1. Construction of the External Environment Index System

In this study, we were guided by the review of external environmental influences in the Introduction [3,7,18,29] and the principles of "theoretical relevance–data availability–quantitative feasibility" to develop an external environment index system. The model includes land-use, population, facility services, transportation support, and economic vitality and features 16 secondary indicators. Among these factors, the "land-use" dimension represents the built environment base variable, which reflects people's willingness to engage in slow travel and the continuity of landscape perception. The "population" dimension serves as a bridge that facilitates the matching of CES supply with demand, and the clustering characteristics directly affect the structure of the park service population and the level of CES perception. The "facility services" dimension reflects the completeness of the surrounding living facilities and is associated with the convenience of residents' visits and travel experience; the "transportation support" dimension is the carrier of transportation services for multiple travel scenarios and is the key path underlying the influence of the external environment on the transmission of CES. Finally, the "economic vitality" dimension represents the regional development and commercial support level and is an important explanatory variable for differences in CES perceptions across regions.

For the definitions and calculation methods used for the index, see Table 2.

https://doi.org/10.3390/su18094177

Table 2. Construction of external environmental indicators.

Classification	Indicator Name	Abbr.	Index Definition	Data Name and Source	Computing Method
land-use	Land function mixing degree	LU1	Reflects the functional complexity of the land.	10 m land cover data in 2023 (https://esa-worldcover.org (accessed on 15 May 2025))	Entropy method: $H = -\sum_{i=1}^{5} P \cdot lnP_i$ (*PI* is the area proportion of various types of land)
	Proportion of green space	LU2	Reflects the spatial agglomeration intensity of park.	Annual NDVI grid data in 2024 (https://www.earthdata.nasa.gov (accessed on 17 May 2025))	Area of green space in the unit/unit area (percentage)
	Building density	LU3	Reflects the regional development intensity	2024 China multiattribute building dataset [36] (https://figshare.com/articles/dataset/CMABThe_World_s_First_NationalScale_MultiAttribute_Building_Dataset/27992417 (accessed on 17 May 2025))	Total area of building basement in the unit/unit area (percentage)
	Proportion of impervious surface	LU4	Reflects the degree of regional hardening	Global 30 m resolution impervious surface grid for 2024 (https://data.casearth.cn/dataset/6628a088819aec27fa7d0eea (accessed on 16 May 2025)	Total area of impervious surface of roads, building roofs, parking lots, etc., in the unit/unit area (percentage)
Population	Residential density	P1	Reflects the daily community population base	POI data in 2025 (https://lbs.amap.com (accessed on 3 July 2025)	Number of residential areas in the unit/unit area (piece/km^2)
	Resident population density	P2	Reflect stable flow distribution	Census data regarding streets in Chengdu (https://cdstats.chengdu.gov.cn/ (accessed on 12 June 2025))	Number of permanent residents in the unit/unit area (person/km^2)
	Active population density	P3	Reflects the characteristics of the floating population	Monthly average data for the thermal map in May 2024 (https://lbsyun.baidu.com/ (accessed on 1 June 2025))	The spatial agglomeration intensity of the average daily active population in the unit (excluding the interference of holidays and nighttime)
Facility services	External facility density	FS1	Reflects the convenience of daily support	POI data	Total number of facilities in the unit/unit area (piece/km^2)
	Degree to which external facilities are mixed	FS2	Reflects the richness of facility types		Entropy method calculation
Transportation support	Public transport density	TF1	Reflects the richness of facility types		Total number of bus/subway stops in the unit/unit area (piece/km^2)
	Road density	TF2	Reflects road network support capacity	Road network data in 2025 (https://www.openstreetmap.org (accessed on 8 June 2025))	Total length of roads in the unit/unit area (km/km^2)
	Shared bicycle docking density	TF3	Matches the level of shared bicycles	Shared single vehicle data in 2024 (https://mp.weixin.qq.com/s/GBQ0mKUUP_JskgYSCp_V2A (accessed on 13 May 2025))	Total number of shared bicycle docking spots in the unit/unit area (piece/km^2)
	Parking density	TF4	Reflects traffic-support capacity	POI data	Total number of parking lots in the unit/unit area (piece/km^2)

https://doi.org/10.3390/su18094177

Table 2. *Cont.*

Classification	Indicator Name	Abbr.	Index Definition	Data Name and Source	Computing Method
Economic vitality	Mean nighttime light intensity	EV1	Reflects the vitality of regional development	Night light data in 2024 (https://eogdata.mines.edu/products/vnl/ (accessed on 7 June 2025))	Mean value of the night light grid in the cell
	Regional business vitality index	EV2	Reflects the quality of supporting services	POI data (The quantity of POIs offsets the impact of area and can represent the commercial vitality of a region.)	Total number of commercial POI facilities/unit area (piece/km^2).
	Spatial interpolation of house price	EV3	Reflects regional value	2025 housing price data (https://www.anjuke.com (accessed on 15 June 2025))	Housing price points in a unit

Note: the phrase "within unit" mentioned above refers to the unit that is composed of the accessibility results regarding different travel scenes.

3.3.2. External Environment Data Processing and Standardization

We first unified all the data using the WGS84 coordinate system. Raster datasets were unified to a 100 m resolution with the assistance of "Resample" for ArcGIS 10.7 (Environmental Systems Research Institute, Inc., Redlands, CA, USA). We then used the raster calculator to remove anomalous values (for example, night-light pixels with radiation values > 100) to prevent interference from traffic hubs and other special areas. For the road network, we corrected and supplemented branch and internal park roads, completed topology checks, and classified the various roads into five levels following the relevant urban road engineering design specifications. Point of interest (POI) data were consolidated: starting from the original 14-category scheme (including sports and fitness, leisure and entertainment, companies and enterprises) and taking into account the index requirements for this study, we simplified the taxonomy to five categories—public services, transportation facilities, tourist attractions, commercial finance, and commercial services [7]. Finally, the park entrance and exit locations were crosschecked against POI records and urban street-view imagery to ensure consistency with on-the-ground conditions.

To eliminate dimensional differences among indicators (for example, the unit of "road density" is km/km^2, whereas the unit of "facility density" is pieces/km^2) and to ensure that the model coefficients for subsequent trip scenes were comparable, we standardized all the indicators based on the extreme-value method. We rescaled each index to the interval [0, 1]. The formula used in this context is as follows:

$$K_{\mathrm{i}} = \frac{(X_{\mathrm{i}} - X_{\mathrm{min}})}{(X_{\mathrm{max}} - X_{\mathrm{min}})} \quad (2)$$

where X_i is the measured value of index i, X_{max} is the maximum value of index i, X_{min} is the minimum value of index i, and K_i is the standardized value of the indicator.

3.4. Research Unit: Definition of the 15-Min Scope of Service in Multitrip Scenes

Based on the preprocessed road data, we constructed three types of network datasets that corresponded to various travel scenes: walking/cycling roads (excluding expressways), roads for motor vehicles (excluding cycling greenways), and public transport roads (vehicular lanes plus subway lines).

We determined the speed parameters for these various modes of transportation based on the common parameters that have been used in previous studies on 15 min living circles and the accessibility of urban parks [20,24,29], and we sought to match the general characteristics of leisure travel for urban residents in China. The specific parameters that

https://doi.org/10.3390/su18094177

we used are as follows: (1) a walking speed of 80 m/min, which is the common standard value used in studies on slow traffic and suitable for the average travel characteristics of residents' leisure walks; (2) a cycling speed of 200 m/min, which is the mainstream setting used for urban cycling scenarios and avoids the numerical deviation associated with extreme competitive cycling; (3) a public transportation speed of 1300 m/min for metropolitan railways and 800 m/min for buses, which are the common values used for the average travel speed (including stops) of urban public transportation in similar domestic studies; and (4) a driving speed of 600 m/min for main roads, 400 m/min for secondary roads, and 250 m/min for branch roads, which indicate the peak and off-peak travel speed ranges of these different types of roads, as stipulated in urban road design specifications.

Moreover, given that the waiting time at different levels of road intersections can vary (for example, from 40–90 s at small intersections to 60–120 s at large intersections [22,37]), we established a 60 s average waiting time at the intersections to reflect the time losses faced during the actual travel process and to simulate the real environment in further detail.

This study employed network analysis, in which park entrances and exits were used as origins, to generate 15 min accessibility ranges for four travel scenes in ArcGIS. The results reveal distinct spatial configurations (Figure 3): walking accessibility exhibits a roughly circular pattern with the smallest average radius (approximately 1.61 km). Cycling extends this reach beyond the level achievable through walking. Public-transport accessibility follows linear and irregular corridors along the road network. Driving covers the largest area, featuring an average radius of approximately 5.27 km and a cross-regional distribution. These notable spatial differences among the four travel scenes highlight the need to conduct impact analyses for each.

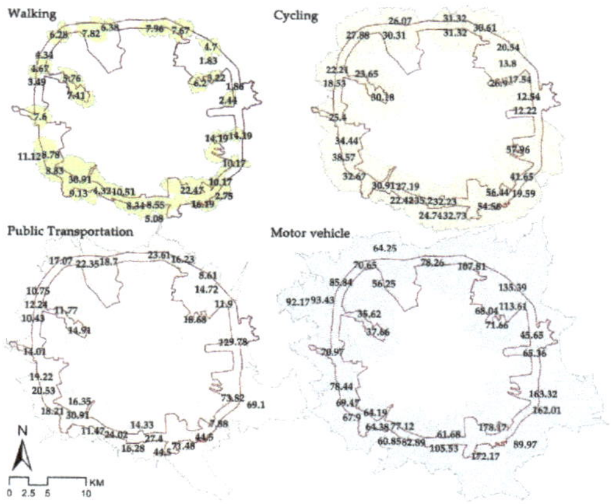

Figure 3. The 15 min accessibility of four types of travel scenes (represented as the accessibility area of each park; unit: km^2).

Finally, we used the accessibility ranges of the four travel scenes in each park to extract 16 external environmental factors aiming to generate an external environment dataset for each scene in each park.

3.5. Analytical Methods

3.5.1. Selection of Independent Variables

To improve the explanatory power and relevance of the subsequent models, we used the CES dimensions extracted through PCA as dependent variables. We conducted correlation analyses based on the independent variables of the park's external environment

https://doi.org/10.3390/su18094177

across different travel scenarios. Following the screening process, we established an exclusive indicator set.

The procedure that we used in this context was as follows. (1) Spatial correlation analysis: We used GeoDa 1.22 (by Dr. Luc Anselin) to perform bivariate spatial autocorrelation and retain statistically significant (999 permutations, $p < 0.05$) independent variables, as well as to eliminate the variables that lacked substantial correlations to avoid increasing modeling redundancy. (2) Multicollinearity test: Ordinary least squares (OLS) regression models were constructed by retaining the independent variables and the corresponding dependent variables retained in the previous step, retaining the independent variables without collinearity problems (VIF < 7.5 [38]), and evaluating the Koenker significance ($p < 0.05$) to confirm the existence of spatial heterogeneity. Moreover, the AICc value was used as an output to facilitate subsequent comparison with the spatial regression model [39], thereby ensuring the reliability of the conclusion.

3.5.2. Construction of the Spatial Regression Model

Given the complex external spatial environment and scale variations, accurate identification of external environmental factors is a prerequisite for efforts to enhance CES perceptions. The GWR model, as a spatial regression analysis method, differs from traditional global regression models. It can embed regression coefficients in space and effectively reveal regional differences in the impact of external environmental factors on CES. This model has been widely applied in fields of spatial analysis, such as urban planning and geography [39]. It is well-suited to the complex external environment of parks.

We used ArcGIS to fit a GWR model to capture the local spatial heterogeneity of external factors. In this model, we took the scores of each CES dimension as the dependent variable and the external environmental indicators pertaining to different travel scenes as the independent variables. The basic form of the model is as follows:

$$Y_i = \beta_0(u_i, v_i) + \sum_{k=1}^{p} \beta_k(u_i, v_i)x_{ik} + \varepsilon_i \tag{3}$$

where Y_i is the observed value of the sample point and is used as the dependent variable in the GWR model. (u_i, v_i) are the coordinates of the i-th sample point, and $\beta_k(u_i, v_i)$ is the i-th regression coefficient on each sample point. x_{ik} is the k-th explanatory variable of the i-th observation point, p is the total number of explanatory variables, ε_i is the regression residual, and $\beta_0(u_i, v_i)$ is the regression intercept term of the model at the i-th sample point.

The model bandwidth was selected using a fixed kernel and AICc minimization. This approach establishes a balance between the fit and complexity of the model and is the standard method used for bandwidth selection in GWR models in spatial metrology research [40]. Model validity was evaluated by reference to goodness of fit and spatial autocorrelation, and the corrected R^2 was used to measure the performance of the model. The residual Moran's I was used to assess spatial autocorrelation; Moran's I ≈ 0 combined with $p > 0.05$ indicated that spatial autocorrelation was effectively removed and that no "pseudoregression" problem remained.

3.5.3. Threshold Effect Recognition Based on Piecewise Regression

To determine how changes in the external environment affect the CES across different travel scenes and spatial scales, we retained the significant factors (with $p < 0.05$) obtained from the GWR model results, which were used as independent variables; in turn, we used the corresponding CES values as dependent variables to verify their basic linear relationships. We used the segmented package in the R language to perform piecewise regression and detect potential threshold points. This method automatically determines the

https://doi.org/10.3390/su18094177

optimal threshold position by optimizing the piecewise goodness of fit. The basic formula for the model is as follows:

$$Y = \begin{cases} \alpha_1 + \gamma_1 X + \mu_1, X \leq T \\ \alpha_2 + \gamma_2 (X - T) + \mu_2, X > T \end{cases} \tag{4}$$

where Y is the dependent variable, X is the independent variable, α_1 and α_2 are the local intercept terms before and after the threshold, respectively, γ_1 and γ_2 are local regression coefficients before and after the threshold, respectively, μ_1 and μ_2 are random errors, and T is the optimal threshold.

The traditional threshold identification model used only significance (as evaluated on the basis of the Davies test performed to compare the differences between the segmented model and the linear model) for discrimination. By comparing the differences in fit between the segmented regression model and the single linear regression model, the statistical significance of the threshold was verified [41]. This process might also reveal weakly effective factors. To address these issues, this study used multiple comprehensive criteria to judge the validity of the threshold. We drew on the research practices used in the fields of environmental psychology and landscape ecology [42,43], specifically requiring the adjusted R^2 of the segmented model to increase by $\geq$20% before and after the threshold was set in comparison with the original linear model and requiring the relative difference in the regression coefficients of the two segments before and after the threshold to be $\geq$30%. Only when the threshold met both the Davies significance test ($p < 0.05$) and the dual criteria mentioned above could it be recognized as an effective threshold and included in the subsequent analysis.

4. Results

4.1. Spatial Differentiation Characteristics of CES

The reliability and validity tests performed for the initial CES indicators yielded Cronbach's $\alpha = 0.745$, KMO = 0.621 (>0.6), and Bartlett's sphericity test $p < 0.001$, thus indicating satisfactory internal consistency. Figure 4 on the left shows the minimum/average and maximum values of each indicator., the CES most frequently mentioned by tourists include location dependence (C3), relaxation and happiness (C4), appreciation and connection with nature (C7), and social connection (C8) (the means of which range from 0.43 to 1.98). In terms of other aspects, reflection (C9) and inspiration (C10) have the lowest means (both <0.05). These results suggest that tourists report strong perceptions of natural experiences with respect to CES but relatively weak cognitive perceptions in this context.

The spatial distributions of the initial CES indicators are illustrated in Figure 4a. High-value areas for most indicators (for example, C6, C9, and C11) are concentrated in parks located in the east and south of the ecological zone surrounding Chengdu, whereas low-value indicators are clustered in the north. However, there are exceptions to these general trends: for example, a park in the southeast appears as a high-value area for most indicators but registers a low value for sense of belonging (C2). This pattern may reflect the weak cultural attributes of some natural parks, which do not readily elicit a sense of belonging among visitors. In addition, the global spatial autocorrelation test reveals that only C10 exhibits a significant positive spatial correlation ($p < 0.05$); the spatial distributions of the other indicators do not exhibit significant clustering.

We extract four principal components with eigenvalues greater than 1 through PCA, thus yielding a cumulative variance explanation of 78.791%, which effectively represents the original index information (Table 3). Based on each component's characteristics, these factors are labeled as follows: comfort and nature-derived experience (PC1), environmental

https://doi.org/10.3390/su18094177

adaptation and landscape appreciation experience (PC2), emotional belonging and higher-order cognitive experience (PC3), and social interaction experience (PC4).

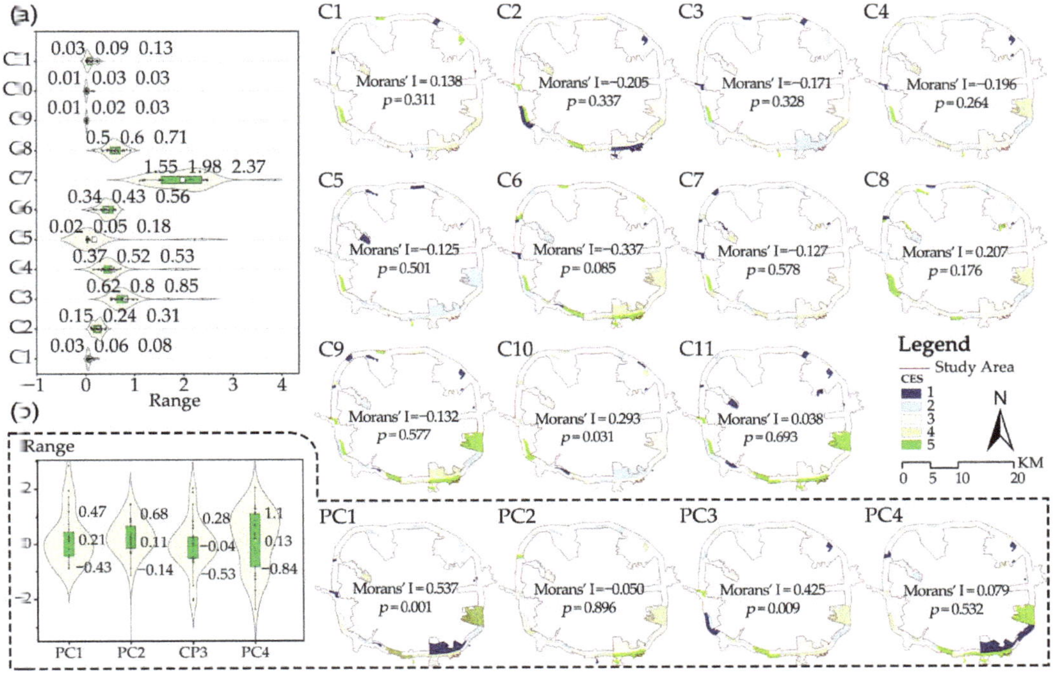

Figure 4. (**a**) Statistics and spatial distribution of the CES indicators; (**b**) statistics and spatial distribution of the CES dimensions.

Table 3. Principal component analysis of CES.

Index	Number	Characteristic Root	Variance Interpretation Rate%	Factor Loading	Subject of Dimension	Dimension Number
Responsibility	C1	3.863	35.117	0.515	Comfort and nature-derived experience	PC1
Relaxation and calmness	C4			0.908		
Safety	C5			0.851		
Nature connection	C7			0.705		
Inspiration	C10			0.548		
Place dependence	C3	2.059	18.716	0.693	Environmental adaptation and landscape appreciation experience	PC2
Aesthetic value	C6			0.573		
Sense of belonging	C2	1.621	14.739	0.616	Emotional belonging and higher-order cognitive experience	PC3
Reflection	C9			0.556		
Knowledge	C11			0.705		
Social connection	C8	1.124	10.219	0.794	Social interaction experience	PC4

https://doi.org/10.3390/su18094177

The spatial distribution of the four thematic dimensions indicates a pronounced "high in the south and low in the north" gradient, which is characterized by notable spatial heterogeneity (Figure 4b). PC1 exhibits the largest high-value area and a strong positive spatial autocorrelation (Moran's I = 0.537, p = 0.001). PC3 features a relatively low overall score but a fairly uniform spatial distribution. In contrast, PC2 and PC4 are more fragmented and unevenly distributed, including nonsignificant Moran's I values (−0.050 and 0.079, respectively; $p > 0.05$). These contrasts confirm the spatial heterogeneity of CES.

4.2. Independent Variable Screening Results and Impact

The indicator screening produced 16 "travel scene–CES dimension" combination models. The spatial autocorrelation and OLS results concerning these models were significant ($p < 0.05$), and the variance inflation factors (VIFs) were less than 7.5. These results indicate significant spatial correlation between the dependent and independent variables throughout the study area, thus revealing no multicollinearity among the independent variables. Table 4 provides a detailed report concerning the selected independent variables in each model.

Table 4. OLS analysis results of the "travel scene–CES dimension" combined model.

Travel Scene	CES Dimension			
	PC1	PC2	PC3	PC4
Walking	LU4 (−1.56, 0.036 *), P1 (1.96, 0.017 *), P3 (−1.06, 0.184), TF3 (2.11, 0.001 *)	LU1 (0.61, 0.309), LU2 (0.16, 0.717), LU4 (−1.63, 0.085), P2 (3.97, 0.066), FS1 (−0.84, 0.026 *), TF3 (−4.13, 0.115), EV2 (1.91, 0.027 *)	LU3 (−2.07, 0.001 *), P1 (0.55, 0.434), P2 (1.52, 0.000 *), FS1 (1.94, 0.003 *), FS2 (2.04,0.001 *), EV3 (−1.24, 0.014 *)	LU1 (1.42, 0.015 *), FS2 (0.79, 0.126), EV2 (1.44, 0.002 *)
Cycling	LU4 (−0.96, 0.093), P3 (0.49, 0.335), TF3 (3.79, 0.000 *), EV2 (−0.93, 0.307), EV3 (−1.21, 0.001 *)	LU1 (2.10, 0.021 *), LU4 (−1.64, 0.003 *), P3 (−5.88, 0.007 *), TF1 (6.91, 0.009 *), TF3 (−5.79, 0.011 *), EV2 (2.82, 0.001 *), EV3 (3.57, 0.002 *)	LU3 (−1.13, 0.033 *), P3 (−0.14, 0.784), FS2 (1.36, 0.039 *), TF3 (2.07, 0.001 *), EV3 (−0.95, 0.161)	LU2 (0.19, 0.824), LU4 (1.26, 0.151), P3 (−1.80, 0.021 *), TF3 (1.87, 0.003 *)
Public transportation	P1 (2.67, 0.005 *), P2 (1.38, 0.021 *), EV1 (−1.42, 0.014 *), EV2 (−1.04, 0.185)	P2 (3.36, 0.043 *), TF3 (−4.29, 0.033 *)	LU3 (−0.64, 0.048 *), P2 (0.33, 0.813), FS2 (0.69, 0.448), TF3 (−0.50, 0.266), EV2 (2.44,0.774), EV3 (−0.64, 0.348)	LU1 (2.97, 0.000 *) FS1 (3.81, 0.012 *), TF4 (−5.14, 0.012 *), EV1 (−3.53, 0.005 *), EV3 (3.82, 0.0006 *)
Motor vehicle	LU2 (−1.64, 0.007 *), TF4 (2.38, 0.001 *)	LU3 (2.67, 0.082), P1 (0.97, 0.272), FS2 (1.43, 0.097), TF3 (−5.44, 0.038 *), EV1 (−4.99, 0.029 *), EV3 (3.48, 0.067)	EV1 (−3.11, 0.000 *), EV3 (3.23, 0.000 *)	LU4 (−1.65, 0.024 *), FS1 (2.21, 0.005 *)

Note: the coefficients and p values (* $p < 0.05$) of each index of the OLS model following the screening process are briefly reported in parentheses.

The impact factors for each travel scene are characterized as "economic vitality dominant, facility services weak, and land-use balanced with population and transportation support factors". In every scene, the economic vitality dimension appears more than three times, thus identifying it as the dominant influence on the CES.

From the perspective of travel scenes, perceptions of CES in the context of walking are driven by both land-use-related environmental factors, such as building density (LU3) and the proportion of impervious surfaces (LU4), as well as economic factors, such as the regional business vitality index (EV2) and the spatial interpolation of housing prices (EV3). These factors have significant effects in different directions on various dimensions of CES (the absolute coefficients are all >1.2). Shared bicycle docking density (TF3) has

https://doi.org/10.3390/su18094177

the strongest and most significant influence across CES dimensions (an absolute coefficient magnitude > 1.8), indicating a pronounced dependence on transport infrastructure. For public transport, variables associated with economic vitality are observed most frequently and consistently have significant effects (absolute coefficient magnitudes > 1), thus indicating a pattern of clear dominance. Driving is primarily shaped by land-use and transport-support factors and is notably affected by traffic-related variables, such as parking lot density (TF4).

Within the same CES dimension, the impact factors associated with different travel scenes are clearly correlated. For instance, the types of factors that influence walking and cycling are largely similar. However, differences are also evident; cycling exhibits increases in two economic vitality indicators relative to walking, a pattern that may reflect spatial heterogeneity resulting from the expansion of the spatial scale.

4.3. Spatial Rules Regarding the Intensity of the Impact of the External Environment on CES for Different Travel Scenes

Based on the four core dimensions of CES extracted via principal component analysis, we use the GWR model to assess the strength of external environmental factors' influence on each CES dimension across four travel scenarios, and we analyze the spatial differentiation characteristics and scene differentiation rules in greater depth.

Firstly, the results of the Koenker test indicate that for all 16 "travel scene–CES dimension" combination models, $p < 0.05$, suggesting that these models are suitable for GWR model construction. A comparison of the fits of the OLS and GWR models (see the details in Appendix A, Table A1) reveals that 81.25% of the GWR models exhibit superior performance—except for PC3 in public transport scenes and PC3 and PC4 in driving scenes. Specifically, AICc decreases by more than 3, R^2 increases by 4.5% to 22.9%, and the residual sum of squares decreases to varying degrees; model residuals also lack significant spatial clustering. These results indicate that the GWR model captures external environmental effects more accurately.

We integrate the results from all effective GWR models to reveal the direction, magnitude, and spatial distribution of the effects of external environmental indicators on each CES dimension across the four travel scenes. The unstandardized regression coefficients are displayed at the center of each map, whereas the coefficient mean and standard deviation appear in the lower-left corner. Different colors are used to explain the impacts of the external environment on CES.

4.3.1. Walking Scene

The contributions of the external environmental indicators in the walking scene are illustrated in Figure 5. Overall, the degree of land-use mixing (LU1), residential density (P1), and degree of external facility mixing (FS2) significantly positively impact all dimensions of CES. All three factors exhibit a consistent spatial pattern of "high in the south and low in the north," which may reflect Chengdu's emphasis on development in the southern region (see Section 5.2).

Among the 4 CES dimensions, only PC1 shows significant spatial heterogeneity for each indicator. Particularly for LU4, the north–south heterogeneity is particularly pronounced. In the southwestern parks, it strongly negatively affects PC1 (with a mean coefficient of −1.67), whereas the negative impact in the northern parks is significantly weaker. This finding is highly consistent with the planning for the north in the study area, which is focused mainly on ecological conservation and agricultural landscapes. The indicators of the other CES dimensions do exhibit spatial heterogeneity, but the manifestation of such heterogeneity is not obvious, as their regression coefficients change only in the thousandth place.

https://doi.org/10.3390/su18094177

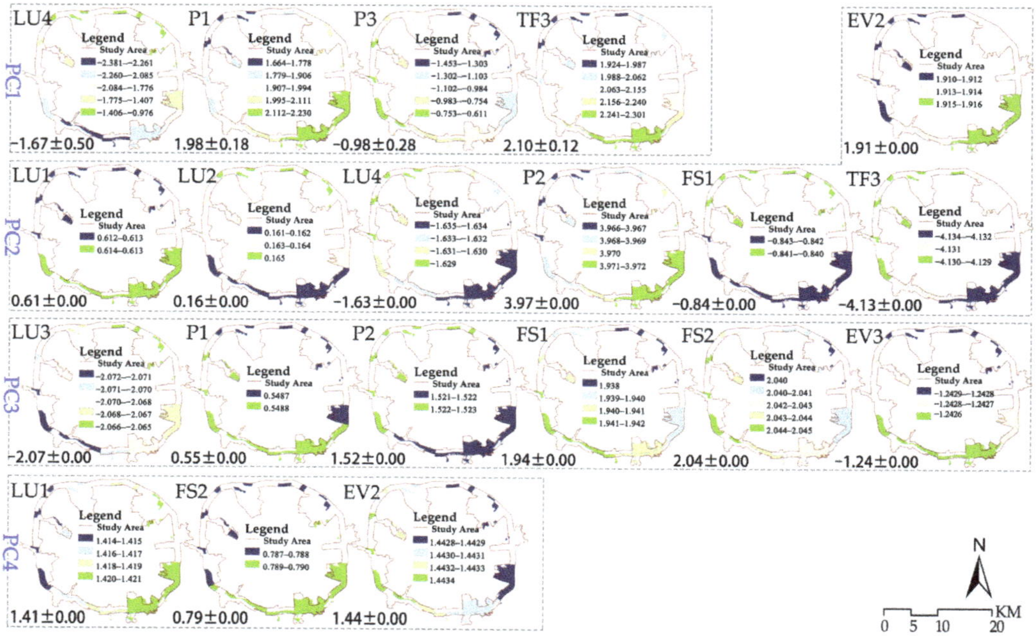

Figure 5. Spatial heterogeneity of relevant influencing factors in walking scenes.

Some indicators exhibit opposing spatial patterns across CES dimensions; for example, shared bicycle docking density (TF3) has a positive effect on PC1 but a negative effect on PC2, which may result from the visual clutter resulting from high bicycle parking density and its adverse impact on landscape experience.

4.3.2. Cycling Scene

In the cycling scene (Figure 6), TF3, a key source of cycling tools, positively influences the CES in terms of every principal component other than PC2. This finding is directly related to the transportation planning of the "all-region greenway" in Chengdu's ring city ecological zone. The southern greenway connects business districts with residential areas, thus giving rise to a high level of demand for daily cycling. The improvement in the completeness of slow-moving facilities has a more notable positive effect on perceptions of CES, whereas the northern greenway focuses on ecological conservation and is located far from residential and commercial clusters; thus, the intensity of influence of cycling facilities is significantly weaker in this context.

Active population density (P3) is significantly negatively correlated with the CES in terms of all dimensions other than PC1 and is characterized by a north–high, south–low pattern. LU4 exhibits heterogeneous effects: it is correlated negatively with PC1 and PC2, including east–west differences, but it is correlated positively with PC4.

4.3.3. Public Transportation Scene

The public-transport scene highlights three influential CES dimensions (Figure 7). Specifically, resident population density (P2) is significantly positively associated with both PC1 and PC2 and has a stronger influence on the Southern Park. The mean nighttime light intensity (EV1) generally has a negative effect on CES in PC1 and PC3, which is characterized by a spatial pattern of "low in the southeast–high in the northwest." Furthermore, TF3

 https://doi.org/10.3390/su18094177

and parking-lot density (TF4) are negatively correlated with PC2 and PC3, respectively, and their strongest negative effects are concentrated in the southeast.

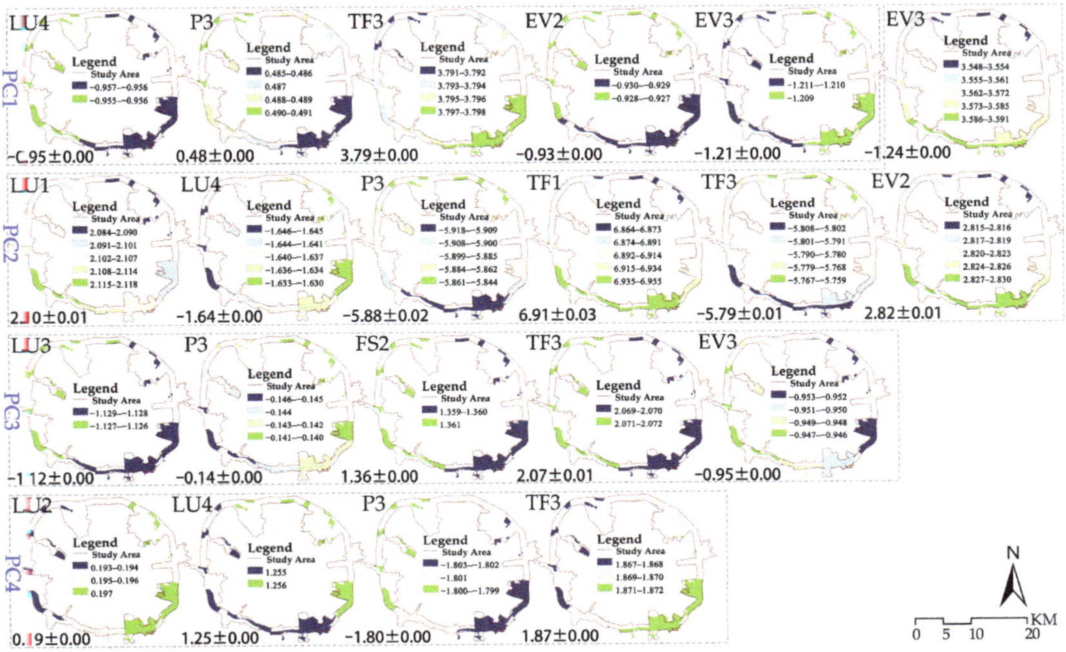

Figure 6. Spatial heterogeneity of relevant influencing factors in cycling scenes.

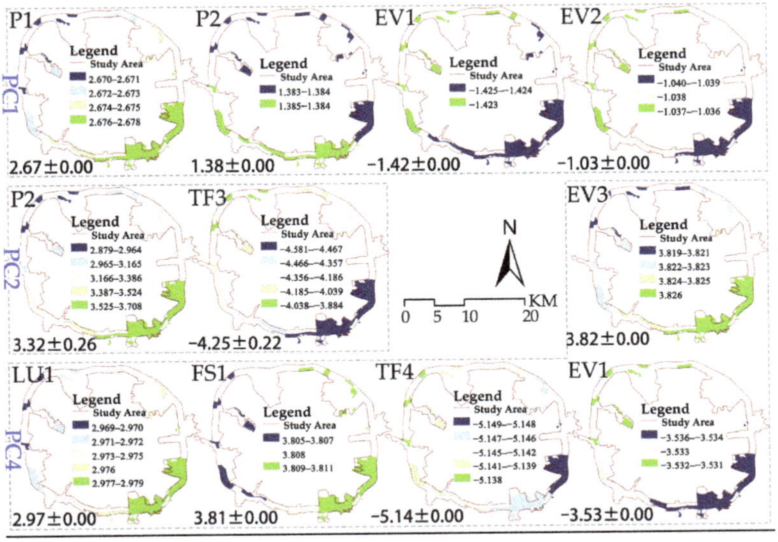

Figure 7. Spatial heterogeneity of relevant influencing factors in public transport scenes.

https://doi.org/10.3390/su18094177

These results indicate that public transportation, as a cross-regional means of mobility, enables tourists to travel across a wider area. Their perceptions of CES are thus more dependent on overall transportation accessibility and the level of economic development in the region, and spatial stability significantly improves in this context.

4.3.4. Motor Vehicle Scene

Two valid CES dimensions are retained for the driving scene (Figure 8). The impact factor of the PC1 dimension is TF4, which exhibits a significant positive correlation; its influence is strong in the south and weak in the north. In PC2, TF3 and EV1, the negative driving factors are clustered in the southeast. Other factors, including FS2, P1, and the spatial interpolation of housing prices (EV3), likely indicate the positive supporting effect of facility quality and environmental conditions in the region's high-price southern areas on landscape experience.

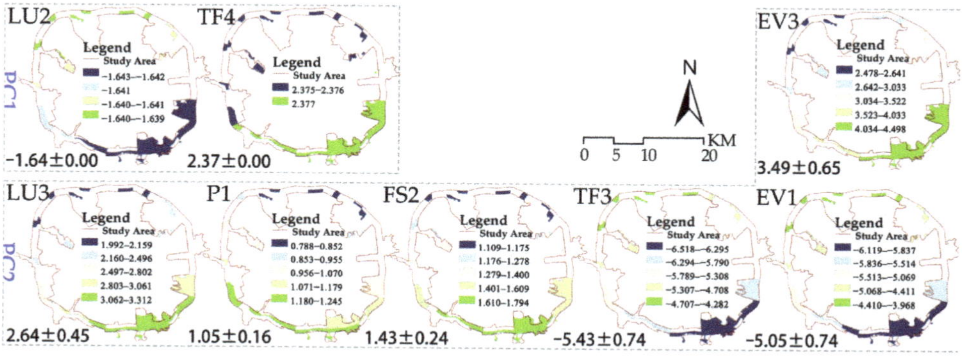

Figure 8. Spatial heterogeneity of relevant influencing factors in driving scenes.

This pattern is highly consistent with the large-scale characteristics of self-driving travel. The travel radius of self-driving tourists is larger, and they are more dependent on the connectivity of high-level road networks for park accessibility. This finding validates the influence pattern associated with long-distance motorized travel.

4.4. Spatial Correlation Rules for Cross-Trip Scenes

Further analysis of the spatial correlation rule of external environmental indicators across various travel scenes reveals the following.

(1) The spatial extent of environmental influence on CES varies notably, and this factor is strongly correlated with the spatial scale of the corresponding travel scenes.

The coefficient for short-distance slow travel exhibits greater volatility. The coefficient of variation (standard deviation ÷ average mean) for the indicators in the walking scene ranges from 0.01% to 29.99%, and LU4 reaches the maximum value (29.99%). Negative correlations are concentrated in parks in the southwest, whereas the negative effect of parks in the north is notably weaker. These findings indicate that small-scale slow-scene settings are more strongly influenced by heterogeneity in the local environment.

The factor coefficients for cross-regional motorized scenes exhibit only minor fluctuations. The coefficient of variation for the indicators in the public transport scene ranges from 0.03% to 7.82%. In particular, EV1 equals 0.06%, and the north–south region occurs only at the thousandth level, thus indicating a balanced, contiguous spatial distribution. These results indicate that as the spatial scale expands, the external environmental effects on CES exhibit strong coefficient stability.

https://doi.org/10.3390/su18094177

(2) The external environmental factors that impact CES exhibit clear travel scene differentiation and are characterized by a "factor scale" adaptation pattern.

Short-distance travel is influenced primarily by land-use and by facilities and services. In walking scenes, indicators such as land-use (LU1, LU4) and facilities and services (FS2) occur more frequently and have greater effects (average coefficient magnitude > ±1.0). In contrast, these indicators do not appear in public transportation and driving scenes, reflecting the fact that slow-mode travelers are highly sensitive to local land-use conditions and nearby service convenience.

Cross-regional travel relies on factors such as traffic support and economic vitality. In the public-transport scene, traffic-support (TF3, TF4) and economic vitality (EV1, EV2) are notable, thus emphasizing the demand for traffic accessibility and a high level of regional development among motorized travelers.

The intensity of the factor also differs notably across scenes. For instance, along PC1, the P1 impact on walking (mean coefficient 1.98) is lower than the corresponding impacts on public transport (2.67), and the impact associated with TF3 for walking (2.10) is much lower than that associated with TF3 for cycling (3.79). The TF3 indicator in PC2 exhibits notable negative fluctuations in correlation coefficients across the four travel scenes. In PC3, the contrast is even more pronounced between the FS2 and EV3 indicators, where the impact intensity for walking substantially exceeds the corresponding impact intensity for cycling. These observations highlight the heterogeneous effects resulting from scale differences among travel scenes and suggest the existence of a threshold effect that warrants further verification.

4.5. Threshold Effect of External Environmental Impact

Based on the external environmental factors identified by the GWR model in the previous text, which have significant linear effects on CES, this study employs a segmented regression model to conduct a threshold effect test, thereby clearly identifying the critical turning point at which the external environmental factors cause significant structural change in CES and providing a hierarchical quantitative reference for actors involved in the planning and control of the park's external environment.

We observed that the threshold effects of the 5 factors were significant (Figure 9). These factors featured 3 stable planning thresholds and 2 exploratory thresholds (which were verified by the Davies local significance test, although the fitting situation was weaker than before). The regression characteristics of each factor under the threshold exhibited significant heterogeneity and a nonlinear trend. A total of 2 invalid factors (TF3 in the PC1 dimension and LU1 in the PC4 dimension) were identified, whose model explanatory power and the intensity of factor influence did not change substantially; these factors were thus not included in the analysis.

4.5.1. Stable Planning Thresholds

LU4 in the PC1 dimension, EV2 in the PC2 dimension, and FS2 in the PC3 dimension all achieved significant regression results before and after the threshold ($p < 0.05$).

Among these factors, LU4 exhibited a segmented linear characteristic. When the proportion of impervious surface area was <40%, its impact on PC1 was negative, and the coefficient decreased from the original 2.37 to -2.814, thus indicating that a lower impervious surface area can help increase the degree of comfort and natural-derived experience; in contrast, when this value was above this threshold, the factor exhibited a significant positive correlation with PC1, and the threshold effect gradually stabilized. This finding may highlight the heterogeneous roles played by impervious surfaces in different regions.

https://doi.org/10.3390/su18094177

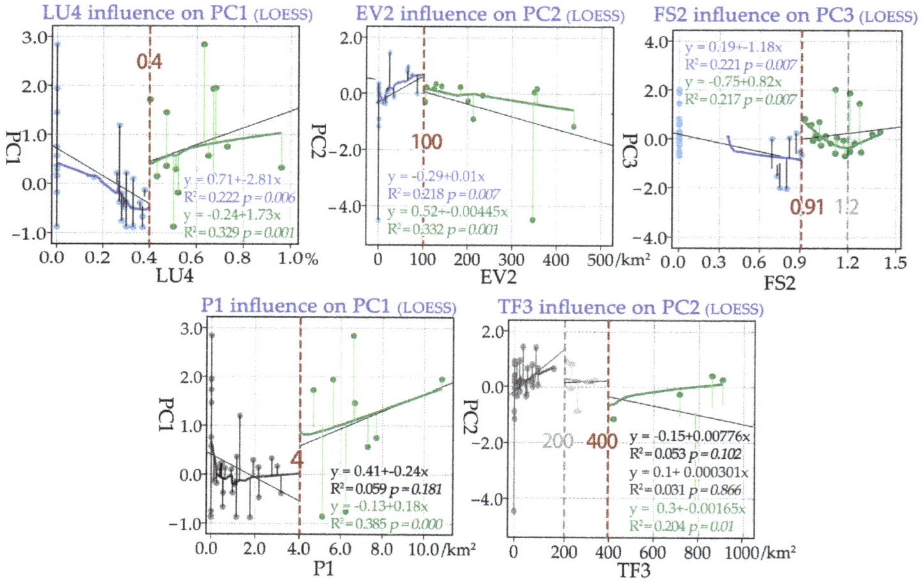

Figure 9. Effective threshold scatter and segment fitting of effective external environmental factors. The optimal threshold points are clearly marked with red vertical dotted lines, and the fitting equations, significance levels and goodness of fit both before (blue) and after (green) the segmentation are marked. Factors without significant results are marked in gray.

EV2 exhibits a single-peak threshold characteristic. The R^2 change was not obvious before the threshold, but it increased by 23.02% after the threshold. When the commercial facility density was $<100/km^2$, this factor had a continuous positive effect on PC2, and an improvement in commercial facilities could significantly increase the travel convenience and leisure experience of tourists in public transportation scenarios; in contrast, when the commercial facility density was $>100/km^2$, its impact on PC2 gradually decreased, thus indicating that an excessively high commercial facility density may give rise to environmental disturbances such as noise and overcrowding, which actually reduce the environmental adaptation and landscape appreciation experience.

FS2 exhibited a fluctuating upward threshold characteristic. We identified two thresholds of 0.91 and 1.2, but after 1.2 was added, the regression results in the middle and later stages of the threshold were not significant; thus, this threshold was deemed invalid. When the degree of functional mixing was <0.91, R^2 increased by 56.74%, and this factor had a negative effect on PC2 (coefficient of -1.179); when this value was above this threshold, the regression coefficient significantly increased. These results indicate that only when the facility types surrounding the park are sufficiently diverse and when the degree of functional mixing is above 0.91 does this factor have significant positive effects on tourists' emotional attachment and higher-level cognitive experience.

4.5.2. Exploratory Thresholds

The two exploratory thresholds, i.e., P1 on PC1 and TF3 on PC2, exhibited nonsignificant regression coefficients before the threshold ($p > 0.05$) and significant coefficients after the threshold. For P1, when the number of residential areas exceeds $4/km^2$, the postthreshold coefficient increases by 14.91%, thus suggesting that a higher residential area density may increase the levels of comfort and experience that can be derived from

https://doi.org/10.3390/su18094177

natural elements. For TF3, two thresholds were identified (200/km and 400/km). When the shared bicycle parking density is less than 400/km^2, TF3 has no significant effect on PC2; in contrast, when this value exceeds 400/km^2, the postthreshold coefficient increases by 19.39%, and both the explanatory power and effect size of the models improve slightly, thus indicating that an appropriate level of shared bicycle parking density can facilitate environmental adaptation.

We observed that these threshold nodes are not merely statistical inflection points but rather align with the spatial balance critical point of the "ecological protection–urban development" in the Chengdu Ring City Ecological Zone. The difference between the effective and exploratory thresholds essentially implies the existence of a "demand scale threshold" (see Section 5.2) with respect to the impact of the external environment on CES. This rule both corroborates the spatial heterogeneity results and provides core data support for efforts to analyze the mechanisms that generate heterogeneity and to design targeted optimization strategies.

5. Discussion

5.1. Marginal Contributions and Academic Dialog Concerning CES Assessment in Urban Parks

A precise assessment of urban parks CES is crucial for supporting the construction of livable cities. This study, which is based on an analytical framework that focuses on 15 h living circle multitravel scenarios, fills the core gap in previous research, which has ignored the heterogeneity of travel scales and failed to distinguish between environmental effects inside and outside parks. The specific contributions made by this research are as follows.

In terms of boundary expansion, previous studies on the mechanism underlying the influence of CES have focused mostly on a single spatial scale or mode of travel and failed to separate the scale effects of different travel scenarios, resulting in a disconnect between the results of such research and the actual park usage behaviors exhibited by residents. We did not rely on traditional scale division methods such as administrative boundaries and geographical distances. From the perspective of the 15 min living circle, we innovatively explored the influence of the external environment on perceptions of CES in different travel scenarios. We identified the "factor-scale" adaptation rule that "short-distance slow travel is dominated by land-use and public facilities, whereas long-distance motorized travel is dominated by transportation support and economic vitality," and expanded the research on the mechanism underlying the influence of CES in parks from a single scale to a multiscale analytical framework that links multiple scenarios. This design specifies spatial scales for each travel scene, thus facilitating a precise alignment of "time-threshold-travel scenes-spatial analysis unit", in line with the idea—which is embedded in spatial heterogeneity theory—that local microfactors govern small-scale travel dependence, whereas macrofactors govern large-scale regions [22,27]. Based on a spatial econometric paradigm of "global test–local analysis–threshold verification" [40], this study demonstrates that models that incorporate the broader urban environment outside parks outperform OLS, which is consistent with the findings reported by Tang et al. [39]. The theoretical approach highlighted above significantly enhances the ability of the model to explain the factors influencing CES and further verifies the applicability of spatial heterogeneity theory in "ecological-built" integrated regions. This research has thus provided a scientific and rigorous methodological reference for similar studies.

In terms of cognitive supplementation, previous studies have mostly identified demographic and sociological characteristics as the core driving factors of parks [7]. However, after the interference of the internal environment of the park was eliminated, the cross-scenario impact intensity of economic vitality observed in this study was significantly greater than the corresponding intensity of the impact of population density, and the effect

https://doi.org/10.3390/su18094177

of the facility service factor was much weaker than has been reported in previous studies. This difference is the result of the particular context of the "ecological barrier + livelihood space" of the Chengdu Ring City Ecological Belt. The external environmental effect observed in this context is more strongly regulated by the urban development pattern and planning orientation. These findings supplement our understanding of the mechanism underlying CES in the scenario of the ring city ecological belt, thus providing a local theoretical reference to support the planning of ecological belts in China's megacities.

In terms of verifying the results, from the perspective of external influencing factors, it is interesting to note that the findings of our study on Chengdu are consistent with some of the results that have been reported in studies conducted in other parts of the world. For example, we reveal that environmental factors pertaining to land use—such as the degree of functional mixing and the proportion of impervious surfaces—are stable planning thresholds; these findings align with the core driving role of land-use in residents' well-being described by Chen et al. [34]. Similarly, the patterns underlying the governance of walking scenes, which focus mainly on factors related to land use, whereas driving scenes are governed mainly by traffic factors, are in agreement with the research on the adaptability of travel scenes in parks and support the claim made in studies on travel scene decision-making that short-distance travel emphasizes immediate accessibility, whereas long-distance travel emphasizes regional convenience [25]. However, by analyzing multiple travel scenes simultaneously, we overcome the scale limitations of traditional research. At broader scales, we reveal that CES is more strongly affected by economic vitality, a result that stands in contrast to the demographic emphasis that was reported in the Australian study [44]. Moreover, with respect to every dimension of CES, the influence of functional-facility features appears weaker than has previously been assumed [45]. Taken together, these findings emphasize the distinctiveness of this study.

5.2. Heterogeneity Attribution of External Environmental Factors Associated with CES

5.2.1. The Mechanism Underlying the Emergence of Heterogeneity in Multiple Travel Scenarios and the Attribution of North–South Differentiation

The influence of the external environment of the Chengdu Ring City Ecological Zone Park on perceptions of CES highlights a composite feature of heterogeneity across multiple travel scenarios and spatial differentiation. This pattern is not the result of the random effects of external environmental factors. Rather, it is largely influenced by urban structure, population characteristics, and development strategies and fundamentally reflects the connections among "ecological protection–urban development–residents' needs" in the study area.

The differences in factor-driven effects observed in various travel scenarios are fundamentally the result of the matching requirements between the spatial scale of travel behavior and the urban spatial structure. We observed that walking, as a frequent core scenario within the 15 min living circle, depends strongly on land-use and population density in terms of CES perceptions, which is consistent with the conclusion proposed by Fang et al. regarding the high sensitivity of walking scenarios to local land-use environments [27]. These phenomena are highly consistent with the characteristics of population concentration in the southern area of Chengdu (such as the region's population density of approximately 5500 people/km^2, which is approximately 1.7 times that of the northern area), as well as the highly intertwined characteristics of built-up space and ecological space. The northern area represents the junction of the city core and the suburbs, and it exhibits a core orientation toward ecological protection, low levels of land development intensity, and sparse population. The effects of relevant factors are significantly limited in this context.

https://doi.org/10.3390/su18094177

Furthermore, we revealed that the cycling scenario is strongly dependent on the density of shared bikes, not merely on simple demand for facility provision; in addition, this scenario is also deeply connected to the planning layout of urban ring greenways throughout the entire region. The southern greenways connect business districts and residential areas, thus giving rise to strong cycling demand, and the effect of supporting facilities is stronger in this context; in turn, the northern greenways focus on ecological conservation and are located away from areas with high residential and commercial concentration, thus leading to lower cycling adaptability and weaker effects of relevant factors. This pattern echoes the conclusions reported by Hou et al. regarding the impact of urban greenway layouts on the experience of slow-moving parks [2], and this study links this context with population distribution and regional differences in travel demand, thereby supplementing the analysis of travel behavior heterogeneity in the ring-shaped greenway.

With respect to other aspects, public transportation and motorized travel scenarios, such as those involving driving, in which residents' perceptions of the local environment are weaker, focus more on the convenience and commercial completeness of regional transportation. Therefore, economic vitality and transportation support factors become the core drivers in this context. These findings are in line with Dong et al.'s claim that motorized travel is driven by regional development levels and transportation support [22]. This situation might be due to the fact that the southern area, as a new urban center, contains two to three times more subway stations than the northern area does, resulting in more active economic activities in the former region. These spatial differentiation patterns are in line with the results of studies on the ring-shaped green belts of cities such as Beijing and London. For example, the external space of the park ring in Beijing is characterized by north–south differentiation because of differences in urban development strategies [22]. The differences in the ring-shaped green belts of London and Frankfurt are directly related to population density and development intensity [46]. However, the uniqueness of this study lies in the fact that it combines multiple travel scenarios, threshold effects, and spatial differentiation, thereby revealing the "scale–threshold–space" composite law that governs the mechanism underlying the ecological service impact of ring-shaped cities and supplementing the micromechanism analysis of travel behavior in the ring-shaped ecological area.

5.2.2. The Essence and Applicable Boundaries of Threshold Effects

The threshold effects of external environmental factors further increase the influences described above. Their essence lies in a balance between urban development and ecological protection rather than an absolute criterion for judgment. The core stable planning threshold (such as 40% coverage of impervious surfaces) corresponds with the requirements for the construction area for land-use stipulated in the planning of the Chengdu ring ecological zone, thus indicating that the threshold is not a statistical "turning point" but rather a "management boundary" in urban planning. The threshold effects of residential point density (4 per km^2) and shared bike parking density (400 per km^2) are directly related to the north–south development gradient of the study area. The northern area has not reached the threshold triggering conditions for population and cycling demand, and ecological control constraints have weakened the regulatory effect of the facility mix, thus supplementing the conclusion of existing research that "population size affects the effectiveness of park factor thresholds" [28]; in turn, this finding clarifies the intrinsic relationship between population characteristics and threshold effects.

In summary, the mechanism through which the external environment of the Chengdu ring ecological zone affects CES is the result of the combined effects of urban development patterns, population distribution characteristics, and ecological protection planning.

https://doi.org/10.3390/su18094177

The "south high, north low" pattern observed in the CES driving effect is in line with the "multicenter + ring city-ring belt" urban structure and the north–south differentiated population-industry layout of the area. These factors have shaped the factor-driven heterogeneity of multiple travel scenarios and given rise to threshold effects that are in line with local requirements for development control. This situation not only reflects the general rule that governs urban parks but also highlights the unique regional characteristics of Chengdu's "park city" development policy background. This research thus provides core logical support for the subsequent development of strategies aimed at the improvement and optimization of CES.

5.3. Strategies for Enhancing and Optimizing the CES of the External Environment of Urban Parks

On the basis of the previously highlighted influence mechanism and threshold characteristics, we considered the regional characteristics of the Chengdu Ring City Ecological Zone and the universal needs of urban parks, focusing in particular on two core dimensions of precise policy implementation pertaining to different scenarios and the optimization of threshold-based spatial zoning; on this foundation, we proposed targeted improvement strategies that can serve as a reference for efforts to optimize the CES of similar urban parks.

5.3.1. In Accordance with the Factor-Driven Differences Among the Four Types of Travel Scenarios, Scenario Optimization Should Be Implemented on the Basis of the Threshold Intensity

Based on three strong and effective robust thresholds of no more than 40% impervious surface coverage, a $\geq$0.91 mix of land-use, and a $\leq$100/km^2 commercial density, these factors can directly serve as rigid control requirements for spatial planning around parks in the Chengdu Ring City Ecological Belt. At the pedestrian scenario level, we suggest that the impervious surface coverage within the walking scale around the park should be strictly controlled and that the connection between facilities and the park should be strengthened to lay a solid foundation for the people-oriented services associated with the 15 min life circle. At the public transportation scenario level, the commercial density should be regulated to prevent excessive commercial development, which can erode the ecological base of the park, and the layout of public transportation stations and parking lots should be optimized to reduce the negative impact of excessive nighttime lighting on CES. At the driving scenario level, the need to retain the original landscape while improving traffic connections should be emphasized.

On the basis of the two weakly effective exploratory thresholds of $\geq$4/km^2 residential density and $\geq$400 per km^2 shared bicycle parking density, these factors serve only to guide the optimization of the surrounding environment of the parks. At the pedestrian scenario level, in the southern population-concentrated area, we suggest ensuring that the residential density around the park remains above the reference threshold to take full advantage of the positive effect of population concentration. At the cycling scenario level, in combination with the increase in the ring greenway network, the shared bicycle parking density should be managed within the cycling reach of the park to approach the reference threshold, the gaps in the slow travel facilities in the northern area should be addressed, and the connection of cycling paths between the greenway and residential areas and small commercial districts should be optimized.

https://doi.org/10.3390/su18094177

5.3.2. Based on the Core Thresholds and the Spatial Heterogeneity of the North and South, Three Functional Zones Are Divided for Optimization in Different Forms

The southern optimization area focuses on making core improvements to facilities, promoting diverse development around the park, and enhancing the degree of mixing to create a multifunctional and composite CES service scenario; in turn, the northern conservation area adheres to the principle of ecological priority, strictly controlling impervious surface coverage through planning restrictions, relying on ecological restoration measures to implement control requirements, and providing basic slow travel service facilities only to a moderate degree; and the north–south transition zone optimizes the allocation of shared bicycle parking density and commercial facilities based on the benchmark of the shared bicycle parking density and regional commercial vitality index, thereby connecting the development intensity and service gradient between the north and south. This "threshold control + spatial differentiation adaptation" zoning optimization approach can provide a universal reference for the "dynamic balance between ecological protection and facility support" in the context of CES improvement practices in global cities.

5.4. Research Limitations and Prospects

Although we have systematically revealed the mechanism through which the external environment of urban parks affects CES in various travel scenarios, room for further improvement and expansion remains.

This study focused on external environmental effects and selected urban ecological zone parks in Chengdu, which exhibit relatively homogeneous internal functions and landscape bases, as an empirical case. Although this design helps separate the impacts of different internal park characteristics, it also prevents this research from fully reflecting the differences in the gradient of the external environment across different types of urban parks in the results, and it was difficult for this research to cover the scene characteristics associated with other types of parks. In the future, on the basis of comparative studies of multiple types and different cities, the coverage range of park samples can be expanded. With the goal of controlling the consistency of the internal environment of the parks, samples of different types of parks, such as strip parks, linear parks, community parks, and rural parks, can be included, and different development levels and planning models for cities can be selected to facilitate cross-city comparisons. This approach can help clarify the common patterns and differentiated characteristics associated with external environmental effects across different park types and different urban contexts, thus enhancing the generalizability of the conclusions of this research.

Moreover, the threshold effect analysis presented in this study remained at the static cross-sectional level, thereby capturing the threshold characteristics only at a single time point; accordingly, it could not reflect the dynamic change patterns associated with these thresholds across different periods and different development stages, nor could it clarify the long-term impacts of external environmental changes such as urban renewal and park renovation on CES. In the future, the dynamic evolution of the influence threshold of park CES should be explored from a temporal perspective, with a particular focus on analyzing the changes in such thresholds during peak and off-peak traffic periods, weekdays and holidays, and the annual urban renewal process. The core method suggested here can combine intelligent sensing devices, long-term sequence multisource monitoring data, and panel threshold models to construct a dynamic simulation analytical framework. This approach can overcome the limitations of static cross-sectional analysis and facilitate the development of a dynamic and full-cycle threshold control system, thereby providing support for the long-term operation and optimization of parks.

https://doi.org/10.3390/su18094177

In addition, the network accessibility analysis presented in this study cannot fully verify the degree of match between the accessibility analysis construct and the actual travel behavior exhibited by residents, and it cannot fully reflect the entire causal chain transmission of "external environment–travel behavior–CES perception". Future research can integrate first-hand data such as GPS tracking of residents' travel, sensor data from smart wristbands, and in-depth interviews to measure actual travel behaviors directly. It can also couple multi-scale mediation effect models or incorporate travel behavior as a variable in regression models. This approach can help verify the degree of match between the accessibility construct and actual travel behavior, clarify the causal logic underlying the relationships among the variables, and enhance the practical guidance offered by the conclusions of this research.

6. Conclusions

In this study, the urban park in the Chengdu Ring City Ecological Belt is taken as an empirical object, and a multiscale coupling analytical framework of "external environment–multiple travel scenarios–CES" is constructed. This approach systematically responds to the core research question, which focuses on how the external environment shapes the CES of urban parks across different travel scenarios in the context of a 15 min living circle. The findings confirm that the external driving effect of the CES of urban parks is significantly scale dependent and that the core driving logic exhibits essential differences between short-distance walking scenarios and long-distance motorized scenarios.

This study incorporates travel scenarios into the CES analytical framework as specific scale-related conditions, thus filling the core research gap exhibited by previous studies that have ignored scale heterogeneity and failed to distinguish between the effects of the park's internal and external environments. At the methodological level, this research establishes an integrated paradigm for CES assessment and effect identification that is adaptable to multiple travel scenarios, thereby correcting the methodological defects of previous threshold studies that have merely identified turning points while failing to verify their validity. Finally, from an empirical perspective, this study contributes to the local understanding of the driving mechanism underlying CES in the context of the Chengdu Ring City Ecological Belt of megacities in China.

Based on the core conclusions presented throughout this paper, this study focuses on two dimensions of core inspirations. At the level of park planning and traffic design in practice, the core of this research focuses on the two main lines of differentiated optimization by scenario and hierarchical control, which entails rigid constraints with respect to the robust effective boundary that has been verified in terms of robustness. This research also identifies a central balance between ecological protection and people's livelihood services with regard to spatial development around parks in the Chengdu Ring City Ecological Belt, while simultaneously addressing the core needs of different travel scenarios and the differences in urban development between the north and the south, thus identifying appropriate optimization strategies without establishing uniform, rigid compliance requirements. At the academic level of CES assessment methods, future research on this topic should abandon the traditional analytical paradigm of a single scale, fully incorporate the scale heterogeneity of residents' travel scenarios, construct a multiscale assessment framework that is in line with residents' actual park usage behavior, and implement a complete quality control process to facilitate the cross-validation of multiple data sources to improve the robustness of the assessment results.

This study can serve as a scientific reference for efforts to improve CES in parks within a 15 min living circle in megacities, as well as for the planning and construction of parks.

https://doi.org/10.3390/su18094177

Author Contributions: Conceptualization, Q.D.; methodology, Q.D. and M.C.; software, Q.D. and M.C.; validation, M.C.; formal analysis, Y.Y. and M.C.; investigation, Q.D. and J.H.; data curation, Y.Y.; writing—original draft preparation, Q.D., Y.Y., J.H. and B.W.; writing—review and editing, Q.D., Y.Y. and B.W.; visualization, Q.D., J.H. and B.W.; project administration, Q.D.; funding acquisition, Q.D. All authors have read and agreed to the published version of the manuscript.

Funding: This work was supported by the Sichuan Province Social Science Key Research Base—Sichuan Center for Rural Development Research (NO. CR2426) and the Research Center for Beautiful City Construction with Harmonious Coexistence between Humans and Nature (NO. MLCS2025ZC13).

Institutional Review Board Statement: The requirement for ethical review was waived for this study as this project does not fall within the scope of ethical review as stipulated in the "Ethical Review Measures for Biomedical Research Involving Humans" and the "Regulations on the Management of Laboratory Animals", and meets the conditions for ethical exemption set by the Institutional Committee.

Informed Consent Statement: Informed consent was obtained from all subjects involved in the study.

Data Availability Statement: The data presented in this study are available on request from the corresponding author.

Acknowledgments: We would like to express our gratitude to Chengdu Park City Construction and Development Research Institute for providing data support.

Conflicts of Interest: The authors declare no conflicts of interest.

Abbreviations

The following abbreviations are used in this manuscript:

CES	Cultural Ecosystem Services
SDG	Sustainable Development Goal
NEA	National Ecosystem Assessment
GWR	Geographically Weighted Regression
PCA	Principal Components Analysis
POI	Point of Interest
OLS	Ordinary Least Squares

Appendix A

Table A1. Comparison between the OLS model and the GWR model.

Travel Scenes	Dimension	Model	Regression Model Parameters				Global Spatial Autocorrelation Parameters of Residuals		
			R^2	AIC	Residual Sum of Squares	*p* Value	Moran's I	Z Value	*p* Value
Walking	PC1	OLS	0.638	69.172	9.73	0.000 *	0.013	0.26	0.795
		GWR	0.71	64.833	8.041	-	0.003	0.206	0.837
	PC2	OLS	0.449	98.197	12.139	0.046 *	−0.336	−1.774	0.116
		GWR	0.49	94.228	6.018	-	−0.24	−1.229	0.167
	PC3	OLS	0.625	73.735	5.64	0.000 *	−0.091	−0.329	0.742
		GWR	0.666	69.127	2.024	-	−0.03	−0.247	0.884
	PC4	OLS	0.333	94.185	24.173	0.001 *	−0.176	−0.797	0.425
		GWR	0.354	89.866	18.187	-	−0.15	−0.563	0.546

https://doi.org/10.3390/su18094177

Table A1. *Cont.*

Travel Scenes	Dimension	Model	Regression Model Parameters				Global Spatial Autocorrelation Parameters of Residuals		
			R^2	AIC	Residual Sum of Squares	p Value	Moran's I	Z Value	p Value
Cycling	PC1	OLS	0.646	71.756	12.216	0.000 *	0.010	0.182	0.756
		GWR	0.676	66.467	9.836	-	0.007	0.101	0.825
	PC2	OLS	0.572	90.144	17.226	0.000 *	−0.358	−1.886	0.119
		GWR	0.604	85.124	8.096	-	−0.255	−1.429	0.154
	PC3	OLS	0.406	84.846	4.362	0.002 *	0.102	0.761	0.447
		GWR	0.462	80.649	3.023	-	−0.032	−0.677	0.698
	PC4	OLS	0.297	98.939	9.788	0.000 *	−0.327	−1.641	0.101
		GWR	0.316	95.365	5.051	-	−0.316	−1.577	0.115
Public transportation	PC1	OLS	0.610	71.526	9.518	0.000 *	0.066	0.551	0.582
		GWR	0.631	66.537	5.019	-	0.032	0.343	0.620
	PC2	OLS	0.411	83.673	22.826	0.041 *	−0.232	−1.124	0.261
		GWR	0.430	78.592	18.405	-	−0.102	−0.973	0.316
	PC3	OLS	0.367	86.904	9.449	0.000 *	−0.038	−0.031	0.975
		GWR	0.378	85.918	9.791	-	−0.014	0.106	0.916
	PC4	OLS	0.452	94.245	10.673	0.006 *	−0.371	−1.895	0.098
		GWR	0.483	90.269	6.026	-	−0.265	−1.463	0.123
Driving	PC1	OLS	0.405	79.148	7.328	0.000 *	0.198	1.308	0.191
		GWR	0.431	75.158	3.013	-	0.137	1.122	0.233
	PC2	OLS	0.485	92.155	17.422	0.050 *	−0.228	−1.128	0.259
		GWR	0.561	87.552	9.087	-	−0.154	−1.108	0.291
	PC3	OLS	0.383	76.888	3.548	0.000 *	−0.057	−0.144	0.886
		GWR	0.383	76.389	3.013	-	−0.061	−0.167	0.867
	PC4	OLS	0.133	97.122	2.252	0.034 *	−0.178	−0.808	0.419
		GWR	0.138	97.119	2.009	-	−0.176	−0.796	0.426

Note: Since GWR does not have a unified global model, it is impossible to calculate the significance for the whole model in the same manner as OLS, so this item is left blank; * $p < 0.05$.

References

1. United Nations. The 17 Goals: Sustainable Cities and Human Settlements. Available online: https://sdgs.un.org/goals/goal11 (accessed on 12 January 2026).
2. Hou, K.; Zhu, L. Spatial correlations between cultural ecosystem services and human activities along the shenyang's hun river corridor. *Sci. Rep.* **2025**, *15*, 27781. [CrossRef]
3. Zhao, J.; Chen, Y.; Liu, J.; Salvadeo, P. Towards well-being in old residential areas: How health-promoting environments influence resident sentiment within the 15-min living circle. *Land* **2025**, *14*, 2035. [CrossRef]
4. Duraiappah, A.; Naeem, S.; Agardy, T.; Ash, N.; Cooper, H.; Díaz, S.; Faith, D.; Mace, G.; Mcneely, J.; Mooney, H. Millennium ecosystem assessment. In *Ecosystems and Human Well-Being: Synthesis*; Island Press: Washington, DC, USA, 2025; pp. 26–101.
5. Fornaciari, M.; Rossi, F.; Filipponi, M.; Castellani, B.; Giuseppe, A.D.; Proietti, C.; Ruga, L.; Orlandi, F. Co2 emission compensation by tree species in some urban green areas. *Sustainability* **2024**, *16*, 3515. [CrossRef]
6. Li, J.; Gao, J.; Zhang, Z.; Fu, J.; Shao, G.; Zhao, Z.; Yang, P. Insights into citizens' experiences of cultural ecosystem services in urban green spaces based on social media analytics. *Landsc. Urban Plan.* **2024**, *244*, 104999. [CrossRef]
7. Wu, H.; Gong, C.; Wang, R.; Niu, X.; Cao, Y.; Cao, C.; Hu, C. Moderating effects of park accessibility and external environment on park satisfaction in a mountainous city. *Land* **2025**, *14*, 77. [CrossRef]

https://doi.org/10.3390/su18094177

Costanza, R.; Darge, R.; Groot, R.; Farber, S.; Grasso, M.; Hannon, B.; Limburg, K.; Naeem, S.; O'Neill, R.V.; Paruelo, J.; et al. The value of the world's ecosystem services and natural capital. *Nature* **1997**, *387*, 253–260. [CrossRef]

Fish, R.; Church, A.; Willis, C.; Winter, M.; Tratalos, J.A.; Haines-Young, R.; Potschin, M. Making space for cultural ecosystem services: Insights from a study of the uk nature improvement initiative. *Ecosyst. Serv.* **2016**, *21*, 39–343. [CrossRef]

Rose, D.C. UK national ecosystem assessment follow-on: Synthesis of the key findings. *Int. J. Environ. Stud.* **2015**, *2*, 364–365. [CrossRef]

Chan, K.M.A.; Satterfield, T.; Goldstein, J. Rethinking ecosystem services to better address and navigate cultural values. *Ecol. Econ.* **2012**, *74*, 8–18. [CrossRef]

Plieninger, T.; Dijks, S.; Oteros-Rozas, E.; Bieling, C. Assessing, mapping, and quantifying cultural ecosystem services at community level. *Land-Use Policy* **2013**, *33*, 118–129. [CrossRef]

Pan, Y.; Hashim, N.H.N.; Goh, H.C. A systematic review of place-based cultural ecosystem service assessments: Categories, methods, and research trends. *Sustainability* **2026**, *18*, 644. [CrossRef]

Chen, Y.; Hong, C.; Yang, Y.; Li, J.; Wang, Y.; Zheng, T.; Zhang, Y.; Shao, F.; De Meo, I. Mining social media data to capture urban park visitors' perception of cultural ecosystem services and landscape factors. *Forests* **2024**, *15*, 213. [CrossRef]

Li, X.; Ye, Z.; Lei, L.; Wen, Y.; Huang, J. What drives residents' divergent perceptions of cultural ecosystem services in urban park green spaces? A dual-source analysis synergizing social media and survey data. *Sustainability* **2026**, *18*, 2578. [CrossRef]

Li, J.; Yang, Z.; Guo, J.; Zhou, J.; Wang, J.; He, K. Seasonal variation of cultural ecosystem services in urban blue-green spaces based on the solves model and social media data. *Ecol. Indic.* **2025**, *181*, 114393. [CrossRef]

UN Sustainable Development Solutions Network. World Happiness Report 2024. Available online: https://worldhappiness.report/ (accessed on 1 May 2025).

Rahman, B.; Kautsary, J. Internal or external? Factors influence acceptance of knowledge about river functions in jelawat park visitors. *Munic. Econ. Cities* **2022**, *1*, 45–52. [CrossRef]

Tian, F.; Li, X.; Wu, X.; Gong, L. Identifying supply–demand mismatches of ecosystem services and social-ecological drivers at different scales to support land-use planning. *Ecol. Indic.* **2025**, *174*, 113462. [CrossRef]

Veisi Nabikandi, B.; Shahbazi, F.; Biswas, A. Evaluating ecosystem services under various trajectories and land-use/land cover changes in a densely populated area, iran. *Earth Sci. Inform.* **2025**, *18*, 372. [CrossRef]

Ai, M.; Chen, X.; Yu, Q. Spatial correlation analysis between human disturbance intensity (hdi) and ecosystem services value (esv) in the chengdu-chongqing urban agglomeration. *Ecol. Indic.* **2024**, *158*, 18. [CrossRef]

Tian, D.; Wang, J.; Xia, C.; Zhang, J.; Zhou, J.; Tian, Z.; Zhao, J.; Li, B.; Zhou, C. The relationship between green space accessibility by multiple travel modes and housing prices: A case study of beijing. *Cities* **2024**, *145*, 104694. [CrossRef]

Dong, Q.; Cai, J.; Chen, S.; He, P.; Chen, X. Spatiotemporal analysis of urban green spatial vitality and the corresponding influencing factors: A case study of chengdu, china. *Land* **2022**, *11*, 1820. [CrossRef]

Pickett, S.T.A.; Cadenasso, M.L. Landscape ecology: Spatial heterogeneity in ecological systems. *Science* **1995**, *269*, 331–334. [CrossRef]

Malekzadeh, H.; Morency, C. How will individuals travel post-COVID? A statistical framework to identify the determinants of different travel behaviors. *Transp. Res. Procedia* **2025**, *82*, 2075–2088. [CrossRef]

Siba, A.; Agyei, V.; Iddrisu, S.; Adjei, S.A.; Ibrahim, H. Distance and perception of safety of the built environment are predominant factors influencing walking and bicycling to school: A systematic review. *Transp. Res. Part F Psychol. Behav.* **2025**, *109*, 1137–1149. [CrossRef]

Fang, D.; Liu, D.; Kwan, M.P. Evaluating spatial variation of accessibility to urban green spaces and its inequity in chicago: Perspectives from multi-types of travel modes and travel time. *Urban For. Urban Green.* **2025**, *104*, 128593. [CrossRef]

Lei, K.; Wenzhong, Z.; Yunxiao, D.; Zhaoping, Y.; Caicai, L. Research on influencing factors of residents' perception of social equity in low-income communities in beijing: Based on spatial equity perspective. *World Reg. Stud.* **2022**, *31*, 201–213. [CrossRef]

Zoderer, B.M.; Stanghellini, P.S.L.; Tasser, E.; Walde, J.; Wieser, H.; Tappeiner, U. Exploring socio-cultural values of ecosystem service categories in the central alps: The influence of socio-demographic factors and landscape type. *Reg. Environ. Change* **2016**, *16*, 2033–2044. [CrossRef]

Wanbingbing. Who Are the Majority of Users on Dianping? Are They People of a Certain Age Group? Available online: https://m.dianping.com/ugcdetail/390407104?sceneType=0&bizType=29&msource=baiduappugc (accessed on 28 March 2026).

Bryce, R.; Irvine, K.N.; Church, A.; Fish, R.; Ranger, S.; Kenter, J. Subjective well-being indicators for large-scale assessment of cultural ecosystem services. *Ecosyst. Serv.* **2016**, *21*, 258–269. [CrossRef]

Fish, R.; Church, A.; Winter, M. Conceptualising cultural ecosystem services: A novel framework for research and critical engagement. *Ecosyst. Serv.* **2016**, *21*, 208–217. [CrossRef]

Martin, D.; Irvine, K.N.; Skinner, A.M.J.; Davies, Z.G.; Rouquette, J.R.; Maltby, L.L.; Warren, P.H.; Armsworth, P.R.; Gaston, K.J. Biodiversity and the feel-good factor: Understanding associations between self-reported human well-being and species richness. *Bioscience* **2012**, *62*, 47–55. [CrossRef]

 https://doi.org/10.3390/su18094177

34. Chen, M.; Lin, G. How perceived sensory dimensions of urban green spaces affect cultural ecosystem benefits: A study on haiz wetland park, china. *Urban For. Urban Green.* **2023**, *86*, 127983. [CrossRef]
35. Nowak-Olejnik, A.; Dzialek, J.; Hibner, J.; Liro, J.; Madej, R.; Sudmanns, M.; Haase, D. The benefits and disbenefits associat with cultural ecosystem services of urban green spaces. *Sci. Total Environ.* **2024**, *926*, 17092. [CrossRef]
36. Zhang, Y.; Zhao, H.; Long, Y. Cmab: A multi-attribute building dataset of china. *Sci. Data* **2025**, *12*, 430. [CrossRef]
37. Dong, Q.; Lu, H.; Luo, X.; He, P.; Li, D.; Wu, L.; Wei, Y.; Chen, X. Evaluation and optimization of green space fairness in urb built-up areas based on an improved supply and demand model: A case study of chengdu, china. *Sustainability* **2023**, *15*, 150 [CrossRef]
38. Esri. Exploratory Regression Help Archive. Available online: https://desktop.arcgis.com/en/arcmap/latest/tools/spati statistics-toolbox/exploratory-regression.htm (accessed on 28 March 2026).
39. Zhang, Y.; Ge, J.; Bai, X.; Wang, S. Blue-green space seasonal influence on land surface temperatures across different urb functional zones: Integrating random forest and geographically weighted regression. *J. Environ. Manag.* **2025**, *374*, 1239 [CrossRef]
40. Shi, T.; Xu, H. Study on ecosystem service values of urban green space systems in suzhou city based on the extreme gradie boosting geographically weighted regression method: Spatiotemporal changes, driving factors, and influencing mechanism *Land* **2025**, *14*, 564. [CrossRef]
41. Davies, R.B. Hypothesis testing when a nuisance parameter is present only under the alternative. *Biometrika* **2002**, *89*, 484–48 [CrossRef]
42. Yang, B.; Ma, D.; Wang, X.; Dong, W.; He, S.; Zhou, Y.; Dong, D.; Shi, Y.; Wang, Y.; Zeng, S. From benefit to burden: Assessing th full range of health impacts in urban green spaces using a threshold model. *J. Environ. Manag.* **2025**, *375*, 124408. [CrossRef]
43. Zhang, Y.; Van den Berg, A.E.; Van Dijk, T.; Weitkamp, G. Quality over quantity: Contribution of urban green space neighborhood satisfaction. *Int. J. Environ. Res. Public Health* **2017**, *14*, 535. [CrossRef]
44. Lichter, D.T.; Parisi, D.; Ambinakudige, S. The spatial integration of immigrants in europe: A cross-national study. *Popul. R Policy Rev.* **2019**, *39*, 465–491. [CrossRef]
45. Rivera, E.; Timperio, A.; Loh, V.H.Y.; Deforche, B.; Veitch, J. Critical factors influencing adolescents' active and social park use: qualitative study using walk-along interviews. *Urban For. Urban Green.* **2020**, *58*, 126948. [CrossRef]
46. Xie, X.; Kang, H.; Behnisch, M.; Baildon, M.; Krüger, T. To what extent can the green belts prevent urban sprawl?—A comparati study of frankfurt am main, london and seoul. *Sustainability* **2020**, *12*, 679. [CrossRef]

https://doi.org/10.3390/su18094177

sustainability

MDPI

Article

Does the Ecological Conservation Redline Policy Enhance Multidimensional Ecosystem Services? A Causal Assessment of Mechanisms and Governance Pathways

Hao Liu [1], Guangcheng Ma [2,*], Mahamane Famanta [2] and Yiru Chen [3]

1 School of Law, Shanghai University of Finance and Economics, Shanghai 200437, China; liuhao_sufe@163.com
2 School of Business, Nanjing Normal University, Nanjing 210023, China; 31213003@njnu.edu.cn
3 School of Economics, University of Leeds, Leeds LS2 9JT, UK; yc56n23@soton.ac.uk
* Correspondence: mgc@njnu.edu.cn

Abstract

This paper develops a dynamic multidimensional ecosystem service value index for 280 prefecture-level cities in China from 2000 to 2023. The index is constructed by integrating remote sensing, GIS, and ecological–economic indicators, with machine learning used as a data-driven tool to aggregate multidimensional ecological information. Building on this measurement framework, the paper applies a staggered Difference-in-Differences (DID) model to evaluate the impact of the ecological conservation redline policy on regional ecosystem service value. The results show that the policy significantly increases urban ecosystem service value and that the effect is cumulative over time. Mechanism analysis suggests that the policy mainly works through three channels: ecological benefit improvement, ecological spatial reconstruction, and community public participation. Heterogeneity analysis further shows that the effect is stronger in early pilot cities and in high-ecological-function zones. In addition, policy coordination and local governance capacity significantly strengthen policy effectiveness. By combining multidimensional ecosystem service measurement with causal policy evaluation, this study extends existing research on ecological conservation redline and provides empirical evidence for improving land spatial governance and ecological protection policy design in China.

Keywords: ecosystem service governance; ecological benefit feedback; ecological spatial reconstruction; community public participation; staggered DID; machine learning

1. Introduction

Ecosystem services, as the natural foundation upon which humanity depends for survival and development, encompass multiple dimensions, including environmental regulation, biodiversity support, material production, and cultural experiences, and are of great significance to regional sustainable development [1–3]. However, China, as a typical example of an area where ecologically fragile regions overlap with densely populated areas, has experienced a sharp decline in its natural capital stock due to urbanization and industrialization over the past four decades. Imbalanced development of national land space has led to a continuous weakening of the capacity to supply ecosystem services, urgently requiring stringent institutional constraints to reverse the path of resource depletion. Against this backdrop, the ecological conservation redline (ECR) policy has emerged as a core tool of national spatial governance, aiming to establish a legally binding ecological space management system by delineating areas with important ecological functions and service values [4,5]. This policy not only reflects the national strategic orientation of prioritizing ecology and green development but also embodies the inherent requirement of ecological protection, promoting high-quality development [6]. At the same time, the policy also speaks to broader international debates on ecosystem service governance, spatial ecological planning, and sustainability-oriented land-use regulation. In many countries, conservation policy is no longer limited to setting aside protected land; it increasingly involves how ecological space is identified, how multiple ecosystem functions are balanced, and how ecological goals are integrated with regional development planning [7]. Therefore, revealing the profound impact pathways of the ECR on enhancing the value of regional ecosystem services and on achieving scientific optimization of ecological space allocation is crucial for modernizing China and, by extension, the global ecological governance system.

Ecosystem services encompass multiple dimensions, including regulatory, provisioning, cultural, and support services. These services are directly related to the quality and sustainability of regional socio-economic development. This policy not only reflects the national strategic orientation of prioritizing ecology and green development but also embodies the inherent requirement of ecological protection driving high-quality development [6]. However, to date, relevant research has mostly focused on macro-level aspects of policy design, implementation effects, and improvements in ecological environment quality, and has lacked a systematic quantitative assessment of ecosystem service value based on multi-source data [8,9]. Existing research is limited by data acquisition, single indicators, and analytical methods, making it difficult to reveal the profound changes in the structure and quality of ecosystem service provision caused by the Ecological Conservation Redline, and even more difficult to understand its inherent policy effect mechanisms and impact pathways [10,11]. In ecosystem service valuation methods, traditional market value methods, preference disclosure methods, and stated preference methods each have their strengths and weaknesses, making it difficult to address the spatiotemporal heterogeneity and multidimensional attributes of ecosystem service value [12,13]. With the integration of remote sensing technology, geographic information system (GIS), and ecological economic indicators, constructing a multi-level, multidimensional evaluation indicator system has become crucial for accurately identifying ecological value [14,15]. Chen et al. [16] introduced the environmental restoration cost method and the PSER model into the evaluation of urban forest ecosystem services, effectively quantifying the eco-economic correlation of ecological compensation. Sun et al. [17] achieved a dynamic comprehensive assessment of land resource assets and liabilities by integrating ecological footprint and ecosystem service value. This measurement challenge is not unique to China. Internationally, one strand of the literature focuses on monetary valuation and accounting frameworks, while another stresses biophysical indicators, landscape connectivity, or bundles of ecosystem functions. These approaches often lead to different conclusions because they capture different dimensions of ecological value and operate at different spatial scales [7,13]. As a result, a key issue in the literature is not only how much ecological value exists, but also how it should be measured in a way that is sensitive to policy intervention, spatial heterogeneity, and dynamic change.

Regarding the regional ecosystem service value and ECR policy practice, current research and practice face two major challenges. First, ecosystem service value assessment and ecological compensation performance evaluation are often separated, making it difficult for policies to accurately reflect the true ecological value [16]. Second, the spatial–temporal dynamics of ecosystem services and the causal relationships between value realization are complex; traditional PSR models are insufficient for their exploration, and the understanding of dissipative dynamics in evaluation indicator systems lags [18,19]. Furthermore, how to incorporate non-traditional variables such as policy coordination and

 https://doi.org/10.3390/su18125905

governance capacity into the evaluation framework to form a more comprehensive policy performance disclosure system remains a research gap [20,21]. Regarding the mechanism of action of ECR policies, existing research suggests that processes such as ecological benefit feedback, ecological spatial reconstruction, and public participation play key roles in policy implementation and effectiveness [19,22,23]. Ecological benefits are reflected in the mutual reinforcement between ecological protection investments and the enhancement of ecological functions, forming a virtuous cycle [24]. Ecological spatial reconstruction, through spatial pattern adjustment and land use optimization, enhances ecological connectivity and service functions [25]. Community public participation, as an important aspect of social capital, contributes to co-governance and the sharing of ecological governance, thereby enhancing policy acceptance and enforcement [26,27]. These themes also appear prominently in the international literature. Studies on protected areas, spatial planning, and land-use regulation have shown that policy effects depend not only on formal protection status, but also on governance quality, landscape context, implementation incentives, and public support [28,29]. Importantly, the findings are not uniformly positive. Some studies find clear conservation gains, while others report limited, uneven, or highly conditional effects, especially when protection is weakly enforced, when leakage shifts ecological pressure elsewhere, or when policy success is judged using only a narrow set of environmental indicators [7,29,30]. Although these mechanisms have significant theoretical implications, quantitative testing and multi-mechanism integration analysis are still lacking, particularly in how to incorporate policy synergy and local governance capacity into multi-dimensional action mechanism models to reveal their moderating effect on policy outcomes [31,32]. Some scholars have also focused on the heterogeneous impact of ECR policies. Early pilot cities, leveraging their accumulated policy experience and governance resources, have seen more pronounced positive policy effects [8,33]. High ecological function zones, due to their high ecological value base, are more likely to exhibit significant value-added trends [21,34]. Meanwhile, significant differences exist across regions in policy response speed and pathways, ranging from land use and economic development levels to public participation enthusiasm, indicating that heterogeneity must be incorporated into the core of mechanism analysis [35,36]. This mixed international evidence suggests that the effect of ECR should not be taken as self-evident. It needs to be tested with a design that can better separate policy effects from background trends and that can consider the institutional conditions under which ecological gains are more likely to occur.

Existing research largely remains at the macro-descriptive stage, lacking quantitative assessments based on geographic information and eco-economic data. China's regional ecosystem types are diverse, and the spatial delineation of ecological red lines and policy implementation involves complex, multi-factor influences [32,33]. Methodologically, the literature also remains divided. Some studies rely on descriptive comparisons or before–after changes, while others use spatial models, panel methods, or quasi-experimental designs to strengthen causal interpretation. These choices matter, because policy effects can easily be overstated when pre-existing trends, policy selection, or spatial spillovers are not adequately addressed [30]. Based on data accumulated over recent years on ECR implementation, this study uses a combination of panel data models and spatial econometric analysis, which is particularly significant for accurately assessing the impact pathways and the intensity of policies on ecosystem service value. This not only enriches research on the evaluation of ecological and environmental policies but also provides empirical support for the coordinated development of ecological protection and economic growth [23,27]. More specifically, by combining a multidimensional ecosystem service value index with the staggered rollout of ECR, this paper seeks to contribute to international debates on how large-scale conservation zoning policies can be evaluated when ecological outcomes

https://doi.org/10.3390/su18125905

are multidimensional, spatially uneven, and slow-moving. In summary, systematically constructing theoretical models and empirically analyzing the impact of China's ECR on ecosystem service value, while revealing the mechanisms of policy operation and regional differences, represents a fusion of theoretical innovation and policy practice. This study aims to use key ecological red line areas in China as samples, using ecosystem service value estimation and spatial panel data methods to explore the effects of policy implementation and their impact mechanisms, and then propose optimization suggestions to provide scientific reference for promoting ecological civilization construction and regional sustainable development.

The potential contributions of this study are mainly reflected in the following aspects. First, this paper constructs a more systematic research framework for ECR and regional ecosystem service value, enriching the existing theoretical system. By integrating theories of ecological economics, regional environmental governance, and policy evaluation, it reveals the role of ECR in enhancing the value of multidimensional ecosystem services. Second, this paper develops an integrated framework that moves beyond the traditional static two-dimensional assessment framework and uses machine learning to construct a dynamic, multidimensional, and process-oriented composite index system: the dynamic multidimensional ecosystem service value index (It should also be noted that machine learning in this study is used as an empirical tool rather than as a substitute for research design or theoretical reasoning. The main contribution of the paper does not lie in the use of algorithms alone, but in the combination of a conceptually grounded multidimensional indicator system, transparent variable construction, and a causal evaluation framework linked to staggered policy rollout. In other words, machine learning is used here to help aggregate complex ecological information and capture nonlinear structure, while the substantive contribution of the study remains the design of the measurement framework, the identification of policy effects, and the interpretation of the mechanisms through which ecological zoning affects ecosystem service value). Unlike existing studies that mainly rely on simple weighting of ecological land area and a single remote sensing ecological index, this study systematically integrates four dimensions: supply quantity, ecological quality, connectivity and service flow, and multifunctionality and resilience, realizing the transformation from static stock assessment to dynamic process simulation. Combining remote sensing data, GIS, and ecological economic indicators, it effectively captures the dynamic changes and spatial differentiation of ecosystem services [15,17], while also responding to the wider international call for policy-relevant ecosystem service indicators that can reflect multiple ecological functions rather than a single proxy outcome [13,37]. Third, the paper links this measurement framework to a quasi-natural experimental setting and provides evidence on the effect of ECR using a staggered Difference-in-Differences (DID) design, which helps move beyond purely descriptive assessments of ecological policy. Fourth, it examines policy-relevant channels—ecological benefit improvement, ecological spatial reconstruction, and community public participation—and further tests whether policy coordination and local governance capacity condition the strength of these effects. In this sense, the study does not assume that ecological zoning automatically works; instead, it asks under what conditions and through which pathways a large-scale conservation policy is more likely to improve ecosystem service value.

2. Theoretical Analysis and Hypothesis

2.1. China's ECR and Regional Ecosystem Service Value

China's ECR, as an important institutional arrangement for national ecological protection, aims to delineate ecological space baselines and ensure the stability and integrity of key ecological functional zones. Ecosystem service value reflects the multi-dimensional

 https://doi.org/10.3390/su18125905

ecological benefits provided by natural ecosystems to socio-economic development and human well-being. Since the ECR can limit the intensity of human activities and maintain the integrity of ecological spatial structure and function, it has the potential to enhance regional ecosystem service values [22,38]. With the continued implementation of ecological red lines, improvements in ecological environment quality in relevant areas are expected to translate into stronger ecosystem functions and, in turn, higher ecosystem service value, thereby supporting regional ecological security and sustainable development [4,5].

Ecosystem services are not a single ecological output. They include provisioning, regulating, supporting, and cultural functions, and their realization depends on both ecological quality and the spatial integrity of ecosystems [39,40]. Based on existing policy documents and regional ecological practice, the ECR emphasizes strict protection of ecological functional zones and tighter constraints on destructive land-use intervention. In theory, such protection should reduce ecological disturbance, preserve key habitats, and improve the conditions needed for the stable supply of ecosystem services. This is especially relevant in areas where ecological functions are strong but development pressure is also high [16]. Accordingly, the baseline empirical test in this paper focuses on whether cities affected by ECR experience a relative increase in the multidimensional ESV index after policy implementation. Based on the above analysis, this paper proposes the following hypothesis:

Hypothesis 1. *China's ECR has a significant positive effect on regional ecosystem service value.*

2.2. Ecological Benefit Feedback

The ecological benefit feedback effect plays an important role in the process through which ECR influences ecosystem service value. ECR protects key ecological space, restricts excessive development, and supports ecological restoration, thereby improving ecosystem functions and generating positive ecological returns. Once ecological conditions improve, these gains can further support the supply capacity and quality of ecosystem services, creating a cumulative feedback process [17].

The formation of this feedback effect mainly depends on the restrictive protection and functional restoration of core ecological space under the ECR. Strictly delineating ecological red line areas can reduce human pressure and resource overuse, which helps improve water conservation, biodiversity protection, and environmental quality. Better ecological conditions also strengthen the self-regulation and adaptive capacity of ecosystems, which is closely related to the stability of regulating and supporting services [20,39]. In the empirical part, this mechanism is not treated as a fully identified mediation chain. Instead, it is examined through a set of channel variables—such as good air quality, industrial pollution emission intensity, and green coverage—that reflect whether ECR changes ecological conditions in the expected direction.

Secondly, the ecological benefit feedback effect has a dynamic feature. Ecological restoration usually takes time, and some policy effects emerge gradually rather than immediately. As ECR continues to be implemented, the ecological gains it induces may accumulate and become more visible over time [38]. This time pattern also matches the dynamic DID strategy used later in the paper. Based on the above analysis, this paper proposes the following hypothesis:

Hypothesis 2. *China's ECR can significantly enhance regional ecosystem service value through the ecological benefit feedback channel.*

https://doi.org/10.3390/su18125905

2.3. Ecological Spatial Restructuring

The ecological spatial restructuring effect is a key channel through which ECR may shape regional ecosystem service value. Through stricter ecological zoning and limits on incompatible land use, ECR can alter the spatial pattern of ecological land and human activity, thus affecting the long-term basis of ecosystem service provision. This matters because ecosystem services depend not only on ecological quality, but also on the amount, layout, and continuity of ecological space [37,41].

Ecological spatial restructuring is first reflected in the protection and restoration of ecological land. ECR imposes stronger restrictions on development in ecologically sensitive areas and helps maintain the continuity and integrity of key elements such as mountains, rivers, forests, farmland, lakes, and grasslands. A more coherent ecological pattern can improve connectivity, reduce ecological isolation, and support the stable operation of ecological processes such as material cycling, habitat maintenance, and energy flow [42]. This spatial adjustment is especially important for regulating and supporting ecosystem services.

Secondly, the spatial restructuring effect is also reflected in the optimization of land-use structure and the reduction of fragmentation. By slowing the outward expansion of construction land and reducing patch fragmentation, ECR can improve the carrying capacity and resilience of ecological space. Spatial optimization also helps reduce edge effects and improve the balance and stability of ecosystem service supply across space and time [43]. In the empirical analysis, this channel is examined through changes in the proportion of ecological land, construction land expansion intensity, and landscape fragmentation. Finally, ECR may encourage local governments to improve ecological corridors and the broader ecological security pattern, thereby creating better spatial conditions for ecosystem services and cross-regional ecological protection [20]. Based on the above analysis, this paper proposes the following hypothesis:

Hypothesis 3. *China's ECR significantly promotes regional ecosystem service value through ecological spatial restructuring.*

2.4. Community Public Participation Interaction

The implementation of ECR involves not only administrative ecological protection and land management, but also broader social participation. The interaction effect of community public participation is therefore another possible channel through which ECR may improve regional ecosystem service value. Active public involvement can strengthen the social basis of ecological governance, improve information flow, and help correct problems in implementation [44,45].

Public participation may enhance the effectiveness of ECR in at least two ways. First, better environmental information disclosure and stronger public supervision can improve policy transparency and accountability. Residents are often direct observers of environmental change, so their feedback may help local governments identify implementation gaps and respond more quickly. Second, when environmental social organizations become more active, ecological governance may move from one-way administration toward broader collaboration among government, communities, and civil society [46,47]. Such interaction does not replace formal regulation, but it can strengthen enforcement, reduce information asymmetry, and improve the local acceptance of ecological protection policies.

In the empirical part, this mechanism is examined through changes in environmental complaints, environmental information disclosure, and the activity of environmental social organizations. These variables are not interpreted as strict mediators in a narrow causal sense; rather, they reflect whether ECR is associated with stronger public participation and

https://doi.org/10.3390/su18125905

governance interaction. Based on the above theoretical analysis, this paper proposes the following hypothesis:

Hypothesis 4. *China's ECR, through stronger community public participation and governance interaction, significantly promotes regional ecosystem service value.*

2.5. The Moderating Effect of Policy Coordination and Local Governance Capacity

As an important strategic tool for ecological civilization construction, China's ECR affects regional ecosystem service value under different institutional and governance conditions. Its actual effect may therefore vary across cities depending on whether related policies are well coordinated and whether local governments have sufficient implementation capacity. Among these factors, policy coordination is especially important. Policy coordination reflects the consistency and alignment between ECR and other policies related to ecological protection, land management, and economic development. When these policies are more coherent in goals, instruments, and implementation, the effectiveness of the ECR is more likely to be strengthened [5].

The level of policy coordination affects not only policy design but also resource integration, information sharing, and the joint operation of different departments. High coordination can help form a cross-sectoral linkage mechanism, reduce administrative friction, and improve the match between ecological protection requirements and local development planning [48]. In this sense, policy coordination can be understood as a contextual condition that shapes whether ECR produces stronger or weaker ecological returns. Based on this, this paper proposes the following hypothesis:

Hypothesis 5. *The positive effect of China's ECR on regional ecosystem service value is stronger in cities with a higher degree of policy coordination.*

Local governance capacity is another key condition that may shape the policy effect of ECR. The environmental governance resources, implementation ability, and administrative quality of local governments directly affect how ecological red line rules are translated into practice. Local governments with stronger governance capacity usually perform better in monitoring, law enforcement, ecological management, and social mobilization, which helps improve the actual effectiveness of ecological protection [49]. By contrast, regions with weaker governance capacity may face problems such as incomplete enforcement, weak supervision, or insufficient ecological investment, which can reduce the policy effect.

Local governance capacity is also reflected in the ability to adapt ECR implementation to local conditions. Governments with stronger capacity can design more suitable protection measures, compensation arrangements, and enforcement tools according to local ecological conditions and development stages [50]. They may also be better able to mobilize public participation, improve information disclosure, and sustain long-term ecological monitoring. Therefore, local governance capacity is likely to strengthen the relationship between ECR and ecosystem service value. Based on the above theoretical analysis, this paper proposes the following hypothesis:

Hypothesis 6. *The positive effect of China's ECR on regional ecosystem service value is stronger in cities with higher local governance capacity.*

Figure 1 illustrates the theoretical framework of ECR and ecosystem service value.

https://doi.org/10.3390/su18125905

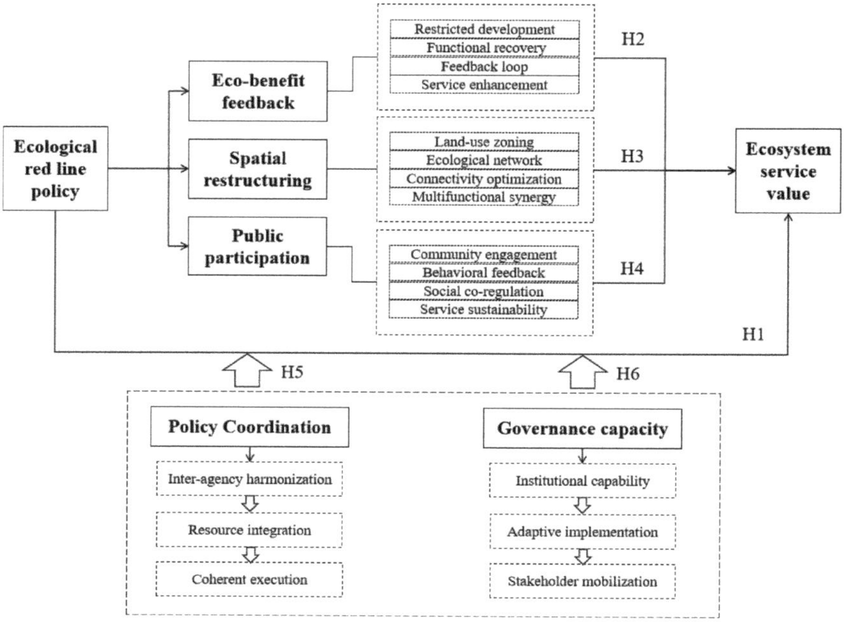

Figure 1. Theoretical framework: ecological red line policy & ecosystem service value.

3. Materials and Methods

Although the study relies on publicly available remote sensing, GIS, and statistical data, its contribution does not rest on any single dataset or software platform. Instead, it lies in integrating these heterogeneous sources into a dynamic multidimensional ecosystem service value measure and embedding that measure in a causal policy-evaluation framework. In this sense, the research design is not a simple application of existing tools, but a structured attempt to connect ecological measurement with policy identification and mechanism analysis.

3.1. Data Source

This study selected 280 prefecture-level cities in China from 2000 to 2023 as the sample, collecting data on ECR implementation and the value of regional ecosystem services to explore the impact mechanism of ECR on regional ecosystem service value. To ensure the representativeness of the sample and the accuracy of the data, we integrated and processed the data based on official documents issued by the relevant national environmental protection and ecological function zoning authorities. The main data used included the list of cities designated by the national ecological red line, ecological environment quality monitoring data, Landsat/Sentinel remote sensing imagery, land use and cover data from GIS, and environmental protection statistics published by local governments. Land use and cover data were obtained from the China Multi-Period Land Use Remote Sensing Monitoring Dataset from the Resource and Environmental Science Data Center of the Chinese Academy of Sciences, with a spatial resolution of 30 m. The multispectral data required for calculating the Remote Sensing Ecological Index were obtained from the USGS Landsat TM/ETM+/OLI. They were acquired in batches, filtered for cloud cover, radiometrically calibrated, atmospherically corrected, and synthesized using the Google Earth Engine cloud platform. The threat factor data required for the InVEST model are constructed from land-use and vector data from the National Geographic Information

Public Service Platform. The resistance surface of the minimum cumulative resistance model integrates land use type and slope data derived from the DEM. The DEM data comes from the ASTER GDEM (30 m resolution) dataset of the Geospatial Data Cloud Platform. Other data sources include the China Regional Statistical Yearbook, the China Urban Statistical Yearbook, and local economic and social development statistical bulletins. These data are not used as isolated inputs. They are harmonized across time, space, and scale to support the construction of a unified panel dataset for ecological measurement and policy evaluation. A key task of this study is precisely to connect remote sensing indicators, ecological process variables, and socioeconomic information in a form that is suitable for multidimensional ESV estimation and staggered DID analysis.

3.2. Model

Considering that the ecological conservation redline (ECR) policy was not implemented simultaneously across China, but was instead gradually promoted through pilot areas, and that the timing of policy adoption varies substantially across regions, the traditional DID model may suffer from estimation bias due to staggered treatment timing. Therefore, this paper employs the staggered DID model (CS-DID) proposed by Callaway and Sant'Anna [51] to identify the impact of ECR on regional ecosystem service value.

Let G_i denote the year in which city i is first included in the ECR pilot program. If a city is never exposed to the policy, it is assigned to the untreated group. Define $D_{it} = \mathbf{I}(t \geq G_i)$ as an indicator of whether city i has entered the treatment state in year t. The core parameter of interest in this paper is the group-time average treatment effect:

$$ATT(g, t) = E[Y_{it}(1) - Y_{it}(0) \mid G_i = g] \quad (1)$$

where $Y_{it}(1)$ and $Y_{it}(0)$ denote the ecosystem service value of city i in year t under treatment and non-treatment, respectively. By using untreated or not-yet-treated cities as the control group, this model effectively avoids the bias that may arise in the traditional TWFE-DID framework under staggered adoption settings, while also capturing the dynamic effects of the policy across different groups and periods.

On this basis, this paper further uses an event study specification to examine the dynamic changes before and after policy implementation:

$$ESV_{it} = \eta_i + \delta_t + \sum_{k \neq -1} \beta_k \mathbf{I}(t - G_i = k) + \gamma X_{it} + \varepsilon_{it} \quad (2)$$

where ESV_{it} represents the ecosystem service value level of city i in year t , and β_k represents the dynamic treatment effect relative to the year before policy implementation. If the coefficients are not significant in each period before policy implementation, the parallel trend hypothesis is supported; if the coefficients gradually increase significantly after policy implementation, it indicates that ECR has a sustained promoting effect on regional ecosystem service value. X_{it} is a series of city-level control variables, η_i and δ_t represent city and year fixed effects, respectively. ε_{it} is a random disturbance term.

3.3. Variable

Ecological conservation redline (Treatment). First, based on policy documents such as the National Ecological Conservation Redline Delineation Technical Guidelines, the National Ecological Protection Red Line—Ecological Function Baseline Delineation Technical Guidelines (Trial), and the Notice on Carrying Out Pilot Work on Ecological Protection Red Line Management, prefecture-level cities included in the national ecological protection red line were identified and defined as treatment group cities. Inner Mongolia, Jiangxi, Hubei, and Guangxi were the first batch of pilot cities in 2012; Jiangsu, Hainan, Hubei, Chongqing,

https://doi.org/10.3390/su18125905

and Shenyang were the second batch in 2015; Tianjin was the third batch in 2018; and Sichuan, Yunnan, Guizhou, Hunan, Anhui, and Shanghai were the fourth batch in 2019. Cities not included in the national ecological protection red line were used as the control group. Therefore, the year (G_i) when a city was first included in the red line was divided into the groups 2012, 2015, 2018, and 2019. The treatment status is a city-level dummy variable, taking the value of 1 if the city has entered any phase of the ecological red line pilot program, and 0 otherwise.

Ecosystem service value (ESV). This paper defines ESV as the comprehensive value of multidimensional services generated by the interaction of ecosystem supply, flow, and demand in a study area over a certain period. Instead of treating ecosystem service value as a static land-based accounting result, this study measures it as a dynamic composite index that reflects ecological stock, condition, spatial delivery, and recovery capacity [12,39]. This index overcomes the limitations of traditional area coefficient methods and single-quality indices by constructing a model across four dimensions: quantity, quality, connectivity and flow, and multifunctionality and resilience. These four dimensions are complementary: quantity reflects the ecological base, quality captures ecosystem condition, connectivity and flow describe whether services can reach benefit areas, and resilience reflects the stability of service provision under disturbance.

In terms of quantity, this paper extracts the area of ecological land such as forests, grasslands, wetlands, and water bodies using Chinese land use/cover data, and calculates their proportion in the total urban area to reflect the quantitative basis of urban ecological space. The proportion of ecological land area is calculated as follows:

$$\text{EcoArea\_share}_{it} = \frac{A_{\text{forest},it} + A_{\text{grass},it} + A_{\text{wetland},it} + A_{\text{water},it}}{A_{\text{total},it}} \tag{3}$$

Then, the Shannon diversity index for ecological land-use types is calculated as:

$$\text{Eco\_land\_diversity}_{it} = -\sum_{k} p_{k,it} \ln p_{k,it} \tag{4}$$

where $p_{k,it}$ represents the proportion of forests, grasslands, wetlands, and water bodies within the city relative to the total ecological land area.

Next, the value of ecosystem services per unit area is monetized by summing the area of each type within the city:

$$\text{ESV\_per\_area}_{it} = \sum_{k} A_{k,it} \times v_k \tag{5}$$

This monetary indicator is used as a benchmark reference rather than as the final dependent variable, which allows the proposed multidimensional index to remain comparable with the conventional equivalent-factor approach.

In terms of quality, an improved remote sensing ecological index (RSEI) is constructed based on Landsat TM/ETM+/OLI or Sentinel-2 multispectral data. This index integrates vegetation cover (NDVI/EVI), surface moisture (NDMI), aridity index (NDBSI), and surface temperature (LST). The first principal component is extracted using machine learning to reduce dimensionality, thereby obtaining the ecological quality value at the city level:

$$\text{EcoQuality}_{it} = \frac{1}{N_{it}} \sum_{p_i} \text{RSEI}_{p,t} \tag{6}$$

All remote-sensing features are standardized before estimation, and annual cloud-free composites are generated on Google Earth Engine. PCA is used as the baseline data-driven

https://doi.org/10.3390/su18125905

fusion method to derive the improved RSEI, with the sign of the first principal component adjusted so that higher values always indicate better ecological quality [52]. To further clarify the machine-learning part of the study, we conduct an external validation exercise using Random Forest (RF), XGBoost, and Support Vector Regression (SVR). These models are trained on the same remote-sensing feature set, together with auxiliary variables such as elevation, slope, impervious surface share, and nighttime lights, to test whether the selected indicators can reliably explain benchmark ecological conditions.

The sample is split into a training set (70%) and a test set (30%), and five-fold cross-validation is applied within the training set. Hyperparameters are selected by grid search. For RF, we tune the number of trees and maximum depth; for XGBoost, the learning rate, tree depth, and subsample ratio; and for SVR, the kernel, penalty parameter C, and γ. Model performance is evaluated by R^2, root mean squared error (RMSE), and mean absolute error (MAE). The test-set results of Table 1 indicate that RF, XGBoost, and SVR all provide reasonably good predictive performance, with XGBoost yielding the highest explanatory power and the lowest prediction error. These findings suggest that the selected feature set contains stable ecological information and that the machine-learning models are suitable for validation and robustness checking of the multidimensional ESV framework.

Table 1. Predictive performance of machine-learning models for validation and robustness checks.

Model	Train R^2	Test R^2	RMSE	MAE
RF	0.874	0.791	0.058	0.043
XGBoost	0.912	0.826	0.051	0.039
SVR	0.801	0.748	0.064	0.047

This indicator embodies qualitative innovation, demonstrating that ecological land of the same area can contribute differently to ESV depending on its quality. Furthermore, it utilizes the Google Earth Engine cloud platform for automated batch processing of multi-temporal imagery, enabling dynamic monitoring of long-term quality. In the connectivity and flow dimensions, this paper uses the InVEST model to simulate service flow paths such as hydrological connectivity, carbon sink flow, and habitat migration. The ecological connectivity index is calculated as: $Connectivity_{it} = f(\text{source patches, resistance surface, distance to demand sinks})$. Combining land-use data and population density, the actual flow value of supply to the benefit area is quantified.

In practice, this dimension is derived from ecological source patches, resistance surfaces, and the accessibility of service supply to populated demand areas. It therefore captures not only spatial adjacency, but also the effective transmission of ecosystem functions across the landscape [37].

In the multifunctionality and resilience dimensions, an ecological resilience index is constructed by integrating habitat quality (HQ, as a core representation of disturbance resistance), landscape diversity (reusing the quantitative Shannon diversity index), and restoration potential (based on the normalized inverse of the minimum cumulative resistance, MCR, model) calculated by the spatial InVEST model. The calculation method is as follows:

$$\text{ERI}_{it} = \frac{1}{3}[\text{normalized}(HQ_{it}) + \text{normalized}(\text{Eco\_land\_diversity}_{it}) + \text{normalized}(\text{Recovery}_{it})] \quad (7)$$

where HQ is obtained using the InVEST Habitat Quality module based on Chinese land-use data and major threat factors (e.g., construction land and roads). Recovery is calculated as the normalized reciprocal of MCR after constructing a resistance surface based on land-use and DEM data.

 https://doi.org/10.3390/su18125905

This dimension is included because ecosystem service value depends not only on current ecological status, but also on the system's ability to resist disturbance and recover after shocks. Finally, after Min–Max standardization of the four dimensions of quantity, quality, connectivity and flow, and multifunctionality and resilience, the weights are determined using the entropy weight method and linearly aggregated to obtain the dynamic multidimensional ESV:

$$\mathrm{ESV}_{it} = \alpha_1 \overline{\mathrm{EcoArea\_share}}_{it} + \alpha_2 \overline{\mathrm{EcoQuality}}_{it} + \alpha_3 \overline{\mathrm{Connectivity}}_{it} + \alpha_4 \overline{\mathrm{ERI}}_{it} \quad (8)$$

The entropy weight method is used as the baseline weighting scheme because it assigns weights according to the amount of information contained in each dimension and reduces subjective bias. The purpose of this design is not to replace ecological judgment with an automated model, but to organize different types of ecological information within a coherent and policy-relevant measurement framework. Machine learning is used here to support data fusion and validation, while the substantive contribution comes from the theoretical selection of dimensions, the construction of comparable indicators, and the linking of these indicators to policy timing and governance mechanisms.

Figure 2 illustrates the spatial distribution trend of ESV in China. The ESV distribution map shows that the southeastern coastal areas and the middle and lower reaches of the Yangtze River generally have higher ESV values, as indicated by the yellow-to-red areas in the figure. These regions have larger ecological land areas and relatively higher ecological quality, reflecting their advantages in ecological protection and land use. In contrast, the ESV values in the northwest and the Qinghai–Tibet Plateau are lower, mainly due to constraints imposed by natural conditions. Since the implementation of the ecological conservation redline policy in 2011, ecosystem service values in the southeastern coastal areas and the Yangtze River Basin have significantly improved, especially in some ecological reserves. This indicates that policy initiatives have had a positive impact on ecological quality and on expanding ecological space in these regions. The novel indicator developed in this paper, spanning both quantitative and qualitative dimensions, can reflect the dynamic relationships among policy implementation, land-use change, and ecological quality improvement, thereby providing effective data support for subsequent empirical analysis.

Control Variables. To control for other factors affecting the value of regional ecosystem services, this paper introduces a series of city-level control variables in the baseline regression. Economic development level (lnPGDP): Measured by the logarithm of per capita GDP [53]. Population size (lnPOP): Calculated as the logarithm of the city's resident population [54]. Industrial Structure (SecShare): Measured by the proportion of secondary industry value added to GDP, used to characterize the concentration of energy-intensive industries [55]. Fiscal expenditure level (lnFiscExp): Calculated by dividing the logarithm of general public budget expenditure by the resident population to obtain per capita fiscal expenditure. Environmental governance Intensity (EnvInv): Measured by the proportion of industrial pollution control investment to GDP. Urbanization level (Urban): Measured by the proportion of urban population to the resident population. Table 2 shows the descriptive statistics for the above variables.

https://doi.org/10.3390/su18125905

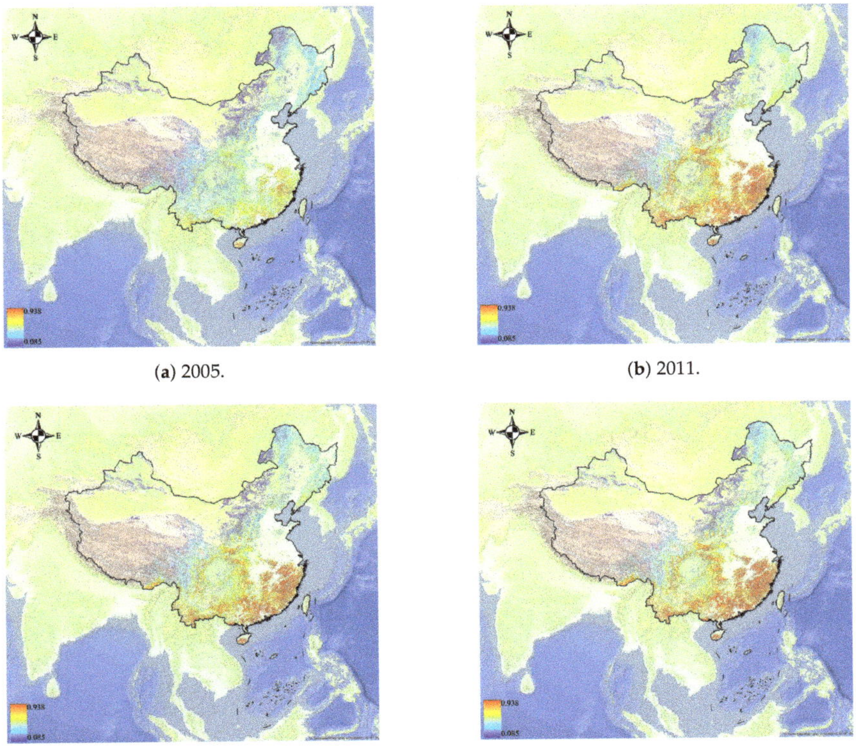

(**a**) 2005. (**b**) 2011.

(**c**) 2017. (**d**) 2023.

Figure 2. Temporal and spatial trends of ecosystem service value in China.

Table 2. Descriptive statistics of key variables.

Variable	Obs	Mean	Std. Dev.	Min	Median	Max
ESV	6440	0.486	0.162	0.085	0.472	0.938
Treatment	6440	0.138	0.345	0.000	0.000	1.000
lnPGDP	6440	10.18	0.870	8.120	10.05	12.67
lnPOP	6440	6.250	0.720	4.920	6.180	8.760
SecShare	6440	0.461	0.118	0.112	0.455	0.782
lnFiscExp	6440	8.720	1.030	6.450	8.650	11.890
EnvInv	6440	0.012	0.009	0.000	0.010	0.068
Urban	6440	0.548	0.172	0.203	0.532	0.956

4. Results

4.1. Baseline Results

Table 3 reports the simple weighted overall ATT estimates as progressively more control variables, city-level fixed effects, and year-level fixed effects are added. Column (1) shows the results without control variables, column (2) with fixed effects added, column (3) with city-level control variables added, and column (4) further controls for both city and year-level fixed effects, which is the baseline model.

The results show that the coefficient of the overall ATT of ECR is significantly positive in all four models (significant at the 1% level), indicating that after the implementation of ECR, the ESV level of the treatment group cities increased by an average of approximately 0.042–0.051 units compared to the control group, equivalent to about 8.6–10.5% of the

https://doi.org/10.3390/su18125905

sample mean. This effect remains robust after adding control variables and the two fixed effects, indicating that ECR significantly promotes improvements in regional ecosystem service value. Therefore, Hypothesis 1 is validated.

Table 3. Baseline regression results.

	(1)	(2)	(3)	(4)
ATT	0.051 ***	0.049 ***	0.047 ***	0.042 ***
	(0.009)	(0.009)	(0.008)	(0.007)
lnPGDP			0.023 **	0.018 *
			(0.011)	(0.010)
lnPOP			0.015	0.009
			(0.012)	(0.011)
SecShare			−0.112 ***	−0.098 ***
			(0.031)	(0.029)
lnFiscExp			0.008	0.006
			(0.007)	(0.006)
EnvInv			0.214 **	0.187 **
			(0.089)	(0.082)
Urban			0.037 **	0.029 *
			(0.018)	(0.016)
C-FE	No	Yes	Yes	Yes
Y-FE	No	Yes	Yes	Yes
Obs	6440	6440	6440	6440
R^2	0.214	0.253	0.287	0.356

Note: *** $p < 0.01$, ** $p < 0.05$, * $p < 0.1$. The values in parentheses are robust standard errors clustered at the city level.

4.2. Event Study Method

To verify the parallel trend hypothesis for CS-DID, this paper uses an event study to estimate the dynamic ATT in each period before and after policy implementation, with the period before the first pilot in each treatment group taken as the baseline. Figure 3 reports the coefficients and 95% confidence intervals for the four periods before and after policy implementation. The results show that the ATT coefficients in the four periods before policy implementation are not significantly different from 0 relative to the period immediately before implementation, thus satisfying the parallel trend hypothesis. In contrast, the coefficients after policy implementation are significantly positive and exhibit a gradually increasing cumulative effect, reaching 0.048 in the fourth period. This indicates that ECR has a continuous positive dynamic impact, which is consistent with the actual characteristics of the phased implementation of the policy.

4.3. Endogeneity Analysis

Although CS-DID has mitigated some endogeneity concerns within the doubly robust framework, this paper further employs an instrumental variable approach to address potential omitted-variable bias or reverse causality. The instrumental variable is the percentage of a city's area designated as a restricted or prohibited development zone in the national main functional area planning of the baseline year prior to policy implementation (*IV_eco_zone*). In terms of relevance, early ecological functional zoning directly influences the probability of subsequent red line delineation. In terms of exogeneity, the ex-ante geographical and ecological characteristics embedded in the earlier nationally unified planning are unlikely to be affected by subsequent local economic activities or changes in ESV, especially after controlling for city fixed effects.

https://doi.org/10.3390/su18125905

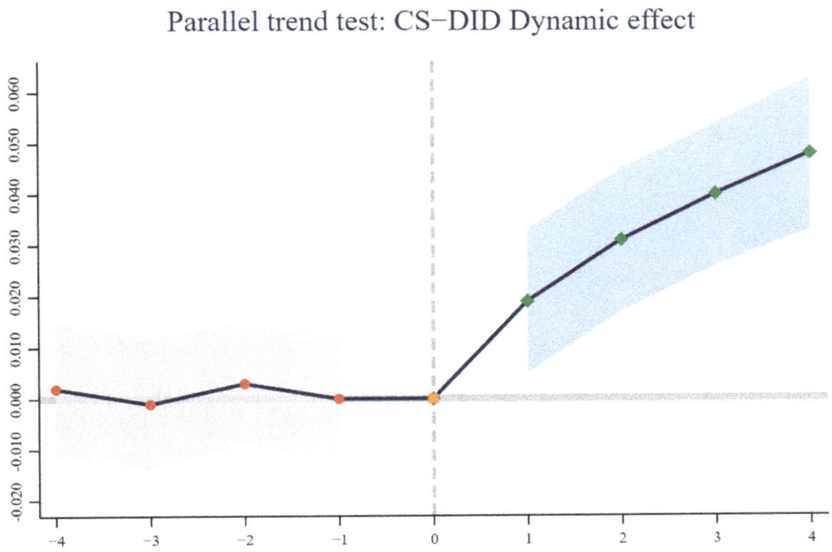

Figure 3. Pre-event effect analysis. Note: Shaded areas represent 95% confidence intervals; red dots represent the early stage of the policy; green diamonds represent the later stage of the policy; orange squares represent the year the policy was implemented.

Table 4 reports the IV-2SLS results. The first-stage F-statistic is 101.02, which is well above the weak identification threshold of 10. The underidentification and weak identification tests reject the hypotheses of underidentification and weak identification, indicating that the instrumental variable is valid. The second-stage 2SLS coefficient is 0.052, which is slightly higher than the baseline regression result, indicating that the baseline estimate is not overestimated due to endogeneity bias.

Table 4. Instrumental variable regression results.

	(1) **First-Stage**	(2) **Second-Stage**
IV_eco_zone	0.267 *** (0.032)	
ATT		0.052 *** (0.012)
Control variables	Yes	Yes
First-stage *F* value	101.02 ***	
Underidentification test		0.000
Weak identification test		249 {16.38}
C-FE	Yes	Yes
Y-FE	Yes	Yes
Obs	6440	6440
R^2	0.214	0.253

Note: The underidentification test and overidentification test report *p*-values. The weak identification test reports the Cragg–Donald Wald *F* statistic, and the values in curly braces report the Stock–Yogo weak ID test critical values at the 10% level. *** $p < 0.01$.

4.4. Placebo Test

To rule out spurious regression results caused by random factors or omitted variables, this paper conducts 500 placebo tests. Specifically, the treatment groups and policy timing are randomly shuffled, spurious treatment groups are regenerated, and the overall ATT

https://doi.org/10.3390/su18125905

is re-estimated using CS-DID. Figure 4 shows that the mean placebo ATT across the 500 simulations is close to 0, with a standard deviation of 0.007. The distribution of *p*-values further shows that only 5.2% of the simulations are significant at the 5% level. This result indicates that the positive baseline effect is not driven by random disturbances, and that the estimated policy effect is genuine.

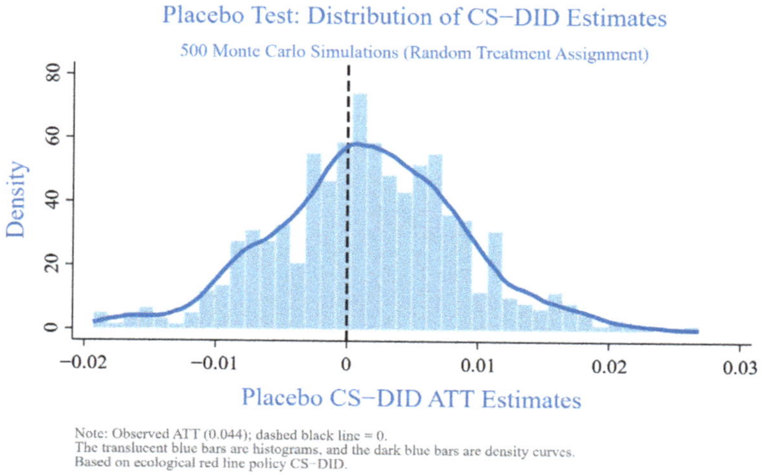

Figure 4. Placebo test.

4.5. Other Robustness Tests

This paper further conducts multi-dimensional robustness tests, and the results are summarized in Table 5. (1) Replacement of dependent variable: Column (1) is replaced with only the quantitative dimension of ESV (EcoArea_share + diversity + monetary value), and column (2) is replaced with only the qualitative dimension (RSEI). The coefficients remain significantly positive. (2) Replacement of independent variable measurement: Column (3) defines ECR as the proportion of red line coverage area (continuous variable), and the ATT coefficient remains significantly positive. (3) Change of sample scope: Column (4) excludes municipalities and provincial capitals, and column (5) retains only cities in the central and western regions. The ATT coefficients are significant in both cases. (4) Exclusion of competing policies: Column (6) simultaneously controls for the dummy variables of the carbon trading pilot and pollution rights trading pilot. The ATT coefficient is still significantly positive. (5) Change of estimation method: Column (7) uses the traditional DID method for estimation, and the results support the baseline conclusion.

Table 5. Robustness test results.

	(1)	(2)	(3)	(4)	(5)	(6)	(7)
ATT	0.038 *** (0.008)	0.029 *** (0.006)	0.039 *** (0.007)	0.045 *** (0.009)	0.051 *** (0.010)	0.040 *** (0.010)	0.050 *** (0.007)
Control variables	Yes	Yes	Yes	Yes	Yes	Yes	Yes
C-FE	Yes	Yes	Yes	Yes	Yes	Yes	Yes
Y-FE	Yes	Yes	Yes	Yes	Yes	Yes	Yes
Obs	6440	6440	6440	6440	6440	6440	6440
R^2	0.123	0.217	0.346	0.416	0.523	0.237	0.542

Note: *** $p < 0.01$.

https://doi.org/10.3390/su18125905

To assess robustness, we first recalculate the multidimensional ESV using equal weights and then construct alternative nonlinear aggregation measures with RF and XGBoost. As reported in Table 6, the descriptive statistics of the alternative indices are very similar to those of the baseline entropy-weighted ESV. In addition, the Pearson correlations with the baseline index range from 0.903 to 0.942, while the Spearman rank correlations range from 0.887 to 0.918. These results indicate that both the overall distribution and the cross-city rankings remain broadly stable across alternative specifications.

Table 6. Robustness checks for the multidimensional ESV indicator under alternative weighting and aggregation schemes.

Specification	Obs	Mean	Std. Dev.	Min	Median	Max	Pearson Corr.	Spearman Rank Corr.
Baseline ESV	6440	0.486	0.162	0.085	0.472	0.938	1.000	1.000
Equal-weighted ESV	6440	0.481	0.158	0.091	0.468	0.927	0.942	0.918
RF-based	6440	0.489	0.166	0.079	0.475	0.944	0.903	0.887
XGBoost-based	6440	0.492	0.169	0.076	0.478	0.951	0.917	0.896

We further examine validity by comparing the multidimensional ESV with benchmark ecological indicators, including conventional unit-area monetary ESV and habitat quality. As shown in Table 7, all specifications are positively correlated with these benchmark measures. In particular, the correlations with conventional unit-area monetary ESV range from 0.728 to 0.763, and those with habitat quality range from 0.798 to 0.834. Taken together, these findings suggest that the multidimensional ESV indicator is both robust and empirically valid.

Table 7. Validity checks of the multidimensional ESV indicator against benchmark ecological indicators.

Specification	Corr. with Conventional Unit-Area Monetary ESV	Corr. with Habitat Quality
Baseline entropy-weighted multidimensional ESV	0.742	0.811
Equal-weighted multidimensional ESV	0.728	0.798
RF-based nonlinear aggregation	0.756	0.826
XGBoost-based nonlinear aggregation	0.763	0.834

4.6. Heterogeneity Analysis

To further reveal the heterogeneous characteristics of the impact of ECR on regional ecosystem service value, this paper, based on the baseline results, selects two dimensions that highlight the thematic characteristics of this study to conduct heterogeneity tests according to the staggered characteristics of the phased implementation of the policy and the intrinsic matching degree with the national main functional area planning. (1) According to the first batch of pilot projects, cities are divided into the early pilot group (2012/2015 batches) and the later pilot group (2018/2019 batches). Since the policy exposure time is longer for the early pilot group, this classification can test whether there is a cumulative effect of phased pilot projects. (2) According to the proportion of restricted development zones or prohibited development zones in the 2005 national main functional area planning, cities are divided into the high ecological function zone group (the proportion is higher than the median of the sample, and the red line delineation is highly overlapping with the national ecological space management) and the low ecological function zone group (the proportion is lower than the median).

Table 8 shows that the impact of ECR exhibits significant time-cumulative heterogeneity and ecological space matching heterogeneity. The ATT of the early pilot groups

 https://doi.org/10.3390/su18125905

is significantly higher than that of the later pilot groups, indicating that cities included in the red line pilot program earlier experienced a more pronounced cumulative effect of ecological protection and restoration due to the longer policy duration. This fully demonstrates the long-term benefits of the core institutional design of phased and regional pilot programs. The ATT of the high ecological function zone group is much higher than that of the low ecological function zone group, indicating that when the red line delineation highly overlaps with the 2005 National Main Functional Zoning Plan, the policy's effect on improving ESV is stronger. This is highly consistent with the essential characteristics of the ecological red line as a bottom-line tool for national ecological space management. In areas with stricter ecological baseline constraints, the marginal protection effect of policy implementation is significantly amplified.

Table 8. Results of heterogeneity analysis.

	(1) Early	(2) Later	(3) High Function	(4) Low Function
ATT	0.051 *** (0.018)	0.037 *** (0.017)	0.056 *** (0.019)	0.029 *** (0.006)
Control variables	Yes	Yes	Yes	Yes
C-FE	Yes	Yes	Yes	Yes
Y-FE	Yes	Yes	Yes	Yes
Obs	3120	3320	3300	3140
R^2	0.372	0.349	0.385	0.336

Note: *** $p < 0.01$.

4.7. Pathway Analysis and Moderating Factor Exploration

4.7.1. Mechanism Analysis

To further identify the intrinsic mechanisms through which ECR affects regional ecosystem service value, this paper conducts mechanism testing from three dimensions: the ecological benefit feedback effect, the ecological spatial reconstruction effect, and the community public participation interaction effect. This paper uses a combination of mechanism-variable regression and dynamic effect analysis for mechanism testing. Compared with traditional mediation-effect models, this paper emphasizes whether policy shocks significantly affect the mechanism variables themselves and whether these effects exhibit evolutionary characteristics consistent with theoretical expectations over time. Specifically, it examines whether ECR significantly alters the mechanism variables across the three dimensions; subsequently, it verifies the temporal characteristics of mechanism changes by combining dynamic effect results from the CS-DID. The basic regression model for mechanism testing is specified as follows:

$$MES_{it} = \eta_i + \delta_t + \sum_{k \neq -1} \xi_k \mathbf{I}(t - G_i = k) + \gamma X_{it} + \varepsilon_{it} \quad (9)$$

where MES_{it} represents the mechanism variables corresponding to the ecological benefit feedback effect, ecological spatial reconstruction effect, and community public participation interaction effect, respectively. If ξ_k is significant and its sign is consistent with theoretical expectations, it indicates that ECR indeed operates through the corresponding mechanism variables.

The first channel concerns changes in ecological conditions and environmental pressure after ECR implementation. The underlying idea is that once high-intensity development is constrained and ecological restoration is encouraged, environmental quality may improve and the ecological system may operate under lower pressure, which is favorable to the supply of ecosystem services [39]. To capture this process, we use three indicators:

https://doi.org/10.3390/su18125905

the proportion of days with good air quality, industrial pollution emission intensity, and the green coverage rate of built-up areas. These indicators are not interpreted as ecosystem service value itself; rather, they reflect whether ECR improves environmental quality, lowers ecological stress, and expands urban green support conditions.

Table 9 reports the results. Column (1) uses the proportion of days with good air quality as the dependent variable. The coefficient on ATT is 0.037 and significant at the 1% level, suggesting that ECR is associated with a clear improvement in urban environmental quality. Column (2) uses industrial pollution emission intensity. The coefficient is −0.029 and significant at the 5% level, indicating that the policy reduces local pollution pressure. Column (3) uses the green coverage rate of built-up areas. The estimated coefficient is 0.024 and significant at the 5% level, which suggests that ECR is linked to greater urban greening and ecological restoration efforts.

Taken together, these results show that ECR changes a set of intermediate ecological conditions in the expected direction. We therefore interpret them as supportive evidence for an ecological benefit improvement channel. To avoid overstating the evidence, we no longer treat these variables as strict mediators in the classical sense. Instead, they are best viewed as observable policy-induced responses that are consistent with the idea that ECR improves environmental conditions and lowers ecological stress, thereby creating better conditions for higher ecosystem service value.

Table 9. Test of ecological benefit feedback mechanism.

	(1) Good AQ	(2) Pollution Emission Intensity	(3) Green Coverage Rate of Built-Up Areas
ATT	0.037 ***	−0.029 **	0.024 **
	(0.009)	(0.012)	(0.013)
Control variables	Yes	Yes	Yes
Bootstrap test	0.012 [0.005, 0.021]	−0.006 [−0.010, −0.002]	0.006 [0.001, 0.011]
Sobel Z-value	2.84 ***	−2.41 **	2.12 **
C-FE	Yes	Yes	Yes
Y-FE	Yes	Yes	Yes
Obs	6440	6440	6440
R^2	0.286	0.274	0.263

Note: Bootstrap sampling was repeated 5000 times, and the confidence intervals were adjusted using the bias-corrected and accelerated (BCa) method. *** $p < 0.01$, ** $p < 0.05$.

The second channel focuses on whether ECR reshapes the spatial basis of ecosystem service provision. Ecosystem services depend not only on ecological quality, but also on land-use structure, the continuity of ecological patches, and the degree to which ecological space is protected from fragmented development. ECR may therefore increase ecosystem service value by preserving ecological land, slowing the outward spread of construction land, and improving spatial continuity.

To reflect this process, we use the proportion of ecological land, construction land expansion intensity, and landscape fragmentation. These indicators are closer to the spatial conditions through which ecosystem functions are maintained and transmitted across the landscape. Compared with the previous version, we now describe them as channel variables that capture changes in ecological spatial organization, rather than as direct mediators.

Table 10 presents the results. Column (1) shows that ECR significantly increases the proportion of ecological land, with an estimated coefficient of 0.031 at the 1% significance level. Column (2) shows that construction land expansion intensity declines significantly after policy implementation, with a coefficient of −0.022 significant at the 5% level. Column (3) reports a negative coefficient for landscape fragmentation (−0.018),

https://doi.org/10.3390/su18125905

significant at the 10% level, indicating that ECR helps improve the continuity of ecological patches.

These findings suggest that ECR does not merely change environmental conditions at the margin; it also affects the spatial structure that supports ecosystem service provision. This is important because ecological land preservation, lower expansion pressure, and reduced fragmentation together create a more stable ecological network. We therefore regard these results as evidence consistent with an ecological spatial reorganization channel.

Table 10. Testing the mechanism of ecological spatial restructuring effect.

	(1) **Proportion of Ecological Land**	(2) **Construction Land Expansion Intensity**	(3) **Landscape Fragmentation**
ATT	0.031 ***	−0.022 **	−0.018 *
	(0.010)	(0.009)	(0.010)
Control variables	Yes	Yes	Yes
Bootstrap test	0.007 [0.002, 0.013]	−0.005 [−0.009, −0.001]	−0.004 [−0.008, −0.001]
Sobel *Z*-value	2.54 ***	−2.06 **	−1.97 **
C-FE	Yes	Yes	Yes
Y-FE	Yes	Yes	Yes
Obs	6440	6440	6440
R^2	0.279	0.251	0.287

Note: Bootstrap sampling was repeated 5000 times, and the confidence intervals were adjusted using the bias-corrected and accelerated (BCa) method. *** $p < 0.01$, ** $p < 0.05$, * $p < 0.1$.

The third channel concerns governance interaction. In practice, the effects of ecological regulation often depend not only on formal rules, but also on the extent to which the public receives information, responds to environmental changes, and participates in monitoring and local governance [45,47]. ECR may strengthen policy implementation by increasing environmental information disclosure, widening public supervision, and encouraging the involvement of environmental social organizations.

Based on this logic, we use the number of environmental complaints, the level of environmental information disclosure, and the activity level of environmental social organizations as channel-related variables. These indicators do not measure ecosystem service value directly. Instead, they reflect whether the governance environment becomes more participatory and whether social oversight becomes more active after ECR is introduced.

Table 11 reports the results. Column (1) shows that environmental complaints increase by 0.016 after policy implementation, significant at the 5% level. This suggests that the public becomes more responsive to environmental issues and that monitoring feedback may become more active. Column (2) shows that environmental information disclosure rises by 0.020, significant at the 5% level, indicating improved transparency and better conditions for public oversight. Column (3) shows that the activity level of environmental social organizations increases by 0.013, significant at the 10% level. This suggests that ECR is associated with broader social involvement in ecological governance.

Overall, the evidence is consistent with a community public participation channel. At the same time, we are careful not to interpret these results too strongly. Public complaints, disclosure, and organizational activity are themselves affected by policy and local governance conditions. They should therefore be read as supportive indicators of governance interaction, rather than as isolated causal mediators.

https://doi.org/10.3390/su18125905

Table 11. Testing the mechanism of community public participation and interaction effects.

	(1) Environmental Complaints	(2) Environmental Disclosure	(3) Environmental Social Organization
ATT	0.016 **	0.020 **	0.013 *
	(0.007)	(0.008)	(0.007)
Control variables	Yes	Yes	Yes
Bootstrap test	0.004 [0.001, 0.008]	0.005 [0.001, 0.009]	0.003 [0.001, 0.006]
Sobel Z-value	1.88 *	2.03 **	1.76 *
C-FE	Yes	Yes	Yes
Y-FE	Yes	Yes	Yes
Obs	6440	6440	6440
R^2	0.231	0.224	0.219

Note: Bootstrap sampling was repeated 5000 times, and the confidence intervals were adjusted using the bias-corrected and accelerated (BCa) method. ** $p < 0.05$, * $p < 0.1$.

4.7.2. Moderating Effect Analysis

Policy coordination (PS) is measured by the proportion of joint documents issued by multiple departments in city-level ecological and environmental policies. This indicator reflects the degree of coordination between ECR and related policies such as land management, territorial spatial planning, industrial development, and environmental governance at both the goal and implementation levels. A higher proportion of joint documents indicates stronger policy coordination. Relevant data were compiled annually from local government websites, government gazettes, and government information disclosure platforms.

The proportion of urban fiscal expenditure on environmental protection to general public budget expenditure measures local governance capacity (LGA). This indicator reflects the level of resource input by local governments in environmental monitoring, law enforcement supervision, pollution control, and ecological restoration. A higher proportion of expenditure indicates stronger local governance capacity. Relevant data are from the China Urban Statistical Yearbook, China Fiscal Yearbook, and local statistical bulletins.

Table 12 reports the results of the test for the moderating effects of policy coordination and local governance capacity. Overall, the interaction coefficients of both moderating variables are positive and significant at the 5% and 1% levels, respectively, indicating that both policy coordination and local governance capacity can significantly enhance the promoting effect of ECR on regional ecosystem service value. Specifically, column (1) shows that the interaction coefficient between the ECR effect and policy coordination is significantly positive, indicating that when ECR is more coordinated with policies related to land management, territorial spatial planning, and industrial development, it is easier for ECR to be transformed into an increase in ecosystem service value. This suggests that policy coordination can reduce policy fragmentation and implementation friction, improve resource allocation efficiency, and thus amplify the positive effect of ECR. Hypothesis 5 is supported. Column (2) shows that the interaction coefficient between the ECR effect and local governance capacity is significantly positive, indicating that the stronger the local governance capacity, the more significant the promoting effect of ECR on regional ecosystem service value. This suggests that fiscal resource input, law enforcement and supervision capacity, and ecological restoration capacity are important guarantees of policy effectiveness. The stronger the local governance capacity, the more conducive it is to transforming policy and institutional advantages into real ecological performance. Hypothesis 6 is supported.

https://doi.org/10.3390/su18125905

Table 12. Moderating effect test.

	(1)	(2)
ATT	0.019 **	0.017 **
	(0.008)	(0.007)
PS	−0.026 **	
	(0.010)	
LGA		−0.031 ***
		(0.008)
$ATT \times PS$	0.013 ***	
	(0.005)	
$ATT \times LGA$		0.015 ***
		(0.004)
Control variables	Yes	Yes
C-FE	Yes	Yes
Y-FE	Yes	Yes
Obs	6440	6440
R^2	0.318	0.326

Note: *** $p < 0.01$, ** $p < 0.05$.

5. Discussion

The mechanism results suggest that ECR improves regional ESV through a combined process rather than a single pathway. More specifically, the policy appears to reduce ecological pressure, stabilize ecological space, and strengthen governance interaction at the same time. This matters because conservation policies often deliver limited effects when they only restrict land conversion formally but fail to improve ecological conditions, spatial continuity, or implementation on the ground. In this sense, ECR should be understood not simply as a land-use control tool, but as a broader governance arrangement that changes how ecological space is protected and managed over time.

This interpretation is consistent with international research on area-based conservation. Recent studies have shown that conservation outcomes depend not only on the size of protected land, but also on management quality, habitat connectivity, and institutional support [56,57]. Our findings speak to this debate in a similar way. The stronger effects found in early pilot cities and in cities with higher overlap between ECR and ecological function zones suggest that conservation works better when policy duration is long enough and when regulatory boundaries better match ecological realities. This is also in line with international evidence that ecological recovery and governance learning are often cumulative and slow to emerge.

At the same time, some features of our findings are clearly shaped by the Chinese context. ECR combines spatial zoning, planning control, administrative accountability, and cross-departmental coordination within a relatively strong state-led system. This helps explain why the policy can produce measurable city-level effects. In many other countries, ecological protection is more fragmented across agencies and planning systems, which may weaken implementation or make policy effects less visible [58]. Therefore, the exact size of the estimated effect should not be directly generalized to other settings. What is more generalizable is the broader lesson that conservation is more likely to improve ecosystem-service outcomes when ecological regulation is closely linked to spatial planning and backed by sufficient governance capacity.

The study also contributes to international debates on ecological governance in a more theoretical sense. Much of the existing literature treats regulation, spatial planning, and participation as separate topics. Our evidence suggests that they are closely connected in shaping ecosystem-service outcomes. Moreover, by measuring ESV as a multidimensional outcome rather than relying only on a single monetary indicator, this paper shows that policy evaluation can better capture the composite and dynamic nature of ecosystem

https://doi.org/10.3390/su18125905

change. From this perspective, the contribution of the paper is not only to evaluate one policy case, but also to show how a spatially explicit governance model can inform broader comparative research on effective ecological governance.

6. Conclusions

This paper uses panel data from 280 prefecture-level cities in China from 2000 to 2023, employing machine learning methods to systematically integrate four dimensions: supply quantity, ecological quality, connectivity and service mobility, and multifunctionality and resilience. The CS-DID method is then used to assess the impact of ECR on the value of regional ecosystem services. The study finds that, first, ECR significantly enhances the value of regional ecosystem services, and this effect exhibits dynamic cumulative characteristics. Second, the mechanism analysis yields evidence that is broadly consistent with three plausible transmission channels, namely improvements in ecological environmental quality, adjustments in ecological spatial patterns, and stronger public participation and interaction. These intermediate results should be interpreted as suggestive rather than definitive evidence of causal mechanisms, because the observed changes capture statistically associated policy responses rather than fully verified mechanism pathways. Third, the policy effects show significant heterogeneity, with stronger effects in early pilot cities and in high-ecological-function zones. Fourth, policy coordination and local governance capacity both have significant positive moderating effects on policy effectiveness. Overall, the results suggest that ECR functions not only as a bottom-line constraint tool for ecological protection, but also as an institutional arrangement that may improve the conditions for ecosystem service provision. At the same time, given the complexity of ecological systems and the remaining uncertainty associated with observational policy evaluation, these conclusions—especially those related to policy transmission channels—should be interpreted with appropriate caution.

Based on these conclusions, this paper proposes the following policy recommendations. First, the ecological red line system should be strictly implemented, and rigid constraints and dynamic monitoring should be continuously strengthened. It is recommended that governments at all levels incorporate ecological red lines into unified national territorial spatial planning and strictly implement the hard constraints of ecological protection red lines, environmental quality baselines, and resource utilization ceilings. A sound regulatory system combining routine remote sensing monitoring and ground inspections should be established, with zero tolerance for illegal development within red line areas and accountability for such activities.

Second, policy implementation should place greater emphasis on strengthening the conditions under which ECR is more likely to generate ecological gains, rather than presuming that each policy channel has already been conclusively identified. It is recommended to systematically construct a multi-dimensional support system in policy implementation: first, increase investment in ecological restoration projects and promote integrated protection of mountains, rivers, forests, farmlands, lakes, grasslands, and deserts; second, optimize land-use structure, strictly control the disorderly expansion of construction land, and construct an ecological corridor network; third, innovate public participation models, improve environmental information disclosure platforms and community co-governance mechanisms, and stimulate social capital investment in ecological protection. These measures are supported by the empirical patterns observed in this study, but they should be understood as policy directions that are consistent with the estimated results rather than as one-to-one prescriptions derived from fully established causal mechanisms.

Third, policy coordination and local governance capacity should be enhanced to fully leverage the amplifying effect of regulatory mechanisms. It is recommended that the central

https://doi.org/10.3390/su18125905

government strengthen top-level design, promote the coordinated alignment of ECR with relevant policies on land management, industrial development, and urban–rural planning, and establish cross-departmental joint mechanisms for document issuance and enforcement to avoid policy conflicts. At the local level, the proportion of environmental protection expenditure should be increased, and environmental monitoring, law enforcement, and restoration capabilities should be strengthened. Simultaneously, differentiated protection measures and compensation pathways should be innovated based on regional realities.

Fourth, attention should be paid to regional heterogeneity, and a phased, categorized, and precisely differentiated policy optimization strategy should be implemented. Early pilot areas should focus on consolidating long-term protection achievements and exploring mechanisms for realizing the value of ecological products; later pilot areas should accelerate the improvement of infrastructure shortcomings and strengthen implementation capacities. High ecological function zones should adhere to strict protection and minimal intervention, prioritizing the construction of national parks or nature reserve systems; low ecological function zones should focus on ecological restoration and green transformation, guiding the ecological development of industries in conjunction with the rural revitalization strategy.

While this study makes progress in measuring multidimensional ecosystem service value and identifying the effect of ECR, several limitations should be noted. First, the multidimensional ESV indicator still involves uncertainty in indicator selection, weighting, and aggregation. Although we tested alternative weighting and machine learning schemes, these checks cannot fully remove measurement uncertainty. Second, the city-level framework may not fully capture spatial heterogeneity in ecological processes and policy effects. Future research could combine higher-resolution remote sensing with ecological monitoring and field surveys. Third, although the staggered DID design strengthens causal identification, some uncertainty may still remain because of omitted concurrent policies and possible measurement error in ecological and socioeconomic variables. The estimated coefficients should therefore be interpreted as credible average policy effects rather than exact magnitudes. Fourth, the machine learning framework improves flexibility in aggregating multidimensional information, but its outputs may still be sensitive to training samples, parameter choices, and model specification. Finally, the findings should not be understood only as evidence of policy success. Ecological conservation may also involve implementation challenges, uneven local capacity, and development conservation trade-offs, especially in less-developed regions. Stronger regions may benefit more quickly, while weaker regions may face higher adjustment costs. Future research should therefore pay more attention to regional inequality, implementation conditions, and the long-term social ecological trade-offs of conservation policies.

Author Contributions: Conceptualization, H.L. and G.M.; methodology, G.M. and M.F.; software, Y.C.; formal analysis, G.M. and Y.C.; data curation, G.M. and M.F.; writing—original draft preparation, H.L., G.M. and Y.C.; writing—review and editing, H.L., G.M., M.F. and Y.C.; supervision, G.M. and H.L.; funding acquisition, G.M. All authors have read and agreed to the published version of the manuscript.

Funding: This research was funded by the National Social Science Fund of China [grant number 25@ZD092 and 24ZDA097].

Institutional Review Board Statement: Not applicable.

Informed Consent Statement: Not applicable.

Data Availability Statement: The data presented in this study are available on request from the corresponding author.

https://doi.org/10.3390/su18125905

Acknowledgments: The authors sincerely thank the editorial office and reviewers for their helpful comments and suggestions about our manuscript.

Conflicts of Interest: The authors declare no conflicts of interest.

References

1. Costanza, R.; de Groot, R.; Braat, L.; Kubiszewski, I.; Fioramonti, L.; Sutton, P.; Farber, S.; Grasso, M. Twenty years of ecosystem services: How far have we come and how far do we still need to go? *Ecosyst. Serv.* **2017**, *28*, 1–16. [CrossRef]
2. Diaz, S.; Pascual, U.; Stenseke, M.; Martín-López, B.; Watson, R.; Molnár, Z.; Hill, R.; Chan, K.; Baste, I.; Brauman, K.; et al. Assessing nature's contributions to people. *Science* **2018**, *359*, 270–272. [CrossRef] [PubMed]
3. Chen, W.; Lin, R.; Ma, G. Responsibility in the shadows: How does firms' shadow banking reshape carbon emission performance? *Financ. Res. Lett.* **2026**, *103*, 110163. [CrossRef]
4. Chen, X.; He, L.; Luo, F.; He, Z.; Bai, W.; Xiao, Y.; Wang, Z. Dynamic characteristics and impacts of ecosystem service values under land use change: A case study on the Zoigê plateau, China. *Ecol. Inform.* **2023**, *78*, 102350. [CrossRef]
5. Jiang, H.; Sun, Z.; Guo, H.; Weng, Q.; Du, W.; Xing, Q.; Cai, G. An assessment of urbanization sustainability in China between 1990 and 2015 using land use efficiency indicators. *npj Urban Sustain.* **2021**, *1*, 34. [CrossRef]
6. Yan, Y.; Yang, Y.; Yang, M. Unravelling the non-linear response of ecosystem services to urban-rural transformation in the Beijing-Tianjin-Hebei region, China. *Ecol. Inform.* **2024**, *81*, 102633. [CrossRef]
7. Kremen, C. Managing ecosystem services: What do we need to know about their ecology? *Ecol. Lett.* **2005**, *8*, 468–479. [CrossRef] [PubMed]
8. Liu, J.; Yuan, Y.; Lin, C.; Chen, L. Do agricultural technical efficiency and technical progress drive agricultural carbon productivity? Based on spatial spillovers and threshold effects. *Environ. Dev. Sustain.* **2025**, *27*, 7701–7725. [CrossRef]
9. Shi, J.; Li, S.; Song, Y.; Zhou, N.; Guo, K.; Bai, J. How socioeconomic factors affect ecosystem service value: Evidence from China. *Ecol. Indic.* **2022**, *145*, 109589. [CrossRef]
10. Li, Q.; Li, D.; Wang, J.; Wang, S.; Wang, R.; Fu, G.; Yuan, Y.; Zheng, Z. Spatial heterogeneity of ecosystem service bundles and the driving factors in the Beijing-Tianjin-Hebei region. *J. Clean. Prod.* **2024**, *479*, 144006. [CrossRef]
11. Shi, C.; Yin, W.; Lv, Z.; Xiang, B.; Dou, X.; Zhang, L. Spatial-Temporal Evolution of Ecosystem Service Value in Guilin, China from 2000 to 2020: A Dual-Scale Perspective. *Remote Sens.* **2024**, *16*, 4425 . [CrossRef]
12. Costanza, R.; d'Arge, R.; De Groot, R.; Farber, S.; Grasso, M.; Hannon, B.; Limburg, K.; Naeem, S.; O'neill, R.V.; Paruelo, J.; Raskin, R.G.; et al. The Value of the World's Ecosystem Services and Natural Capital. *Nature* **1997**, *387*, 253–260. [CrossRef]
13. Bateman, I.J.; Mace, G.M.; Fezzi, C.; Atkinson, G.; Turner, R.K. Economic Analysis for Ecosystem Service Assessments. *Science* **2011**, *48*, 177–218. [CrossRef]
14. Du, H.; Zhao, L.; Zhang, P.; Li, J.; Yu, S. Ecological compensation in the Beijing-Tianjin-Hebei region based on ecosystem services flow. *J. Environ. Manag.* **2023**, *331*, 117230. [CrossRef]
15. Lin, J.; Ma, Q.; Ju, Y.; Zhang, H.; Wang, Q.; Huang, B. Relationships between urbanization, tree morphology, and carbon density: An integration of remote sensing, allometric models, and field survey. *Urban For. Urban Green.* **2022**, *76*, 127725. [CrossRef]
16. Chen, M.; Jiang, Y.; Pan, F.; Zhou, J. Urban forest ecosystem service value assessment under carbon reduction: A case study of Yanbian, China. *Environ. Res.* **2026**, *300*, 124432. [CrossRef]
17. Sun, H.; Yan, X.; Zhang, X.; Zhang, J.; Feng, T.; Qiao, R.; Yao, S.; Ren, Y.; Chen, Q. Long-term evolution of land resource assets and liabilities in Poyang Lake Basin based on ecological footprint and ecosystem service value perspectives. *Ecol. Inform.* **2026**, *95*, 103730. [CrossRef]
18. Pan, Z.; Luo, Y.; Wang, Y.; Ji, J.; Lai, B. Coupling coordination and driving mechanism between tourism development, ecological conservation and digital economy: Insights from Chinese 278 cities. *Sci. Rep.* **2025**, *15*, 37445. [CrossRef]
19. Xue, J.; Lang, J.; Guan, Y.; Lu, S. Design and measurement of small-scale regional ecological compensation model. *Econ. Anal. Policy* **2023**, *78*, 1392–1405. [CrossRef]
20. Karadeniz, E.; Sekeroglu, A.; Kaya, A.Y.; Zhang, M.; Adiguzel, F. A spatial planning approach for the sustainability of provisioning ecosystem services with economic value: The case of Türkiye. *J. Environ. Manag.* **2026**, *399*, 128652. [CrossRef] [PubMed]
21. Wang, M.; Su, M.M.; Gan, C.; Yu, Z. A coordination analysis on tourism development and resident well-being in the Yangtze River Delta Urban Agglomeration, China. *J. Clean. Prod.* **2023**, *421*, 138361. [CrossRef]
22. Hu, H.; Tian, G.; Wu, Z.; Xia, Q. Cross-regional ecological compensation under the composite index of water quality and quantity: A case study of the Yellow River Basin. *Environ. Res.* **2023**, *238*, 117152. [CrossRef]
23. Wunder, S. Revisiting the concept of payments for environmental services. *Ecol. Econ.* **2015**, *117*, 234–243. [CrossRef]
24. Barbier, E.; Koch, E.; Silliman, B.; Hacker, S.; Wolanski, E.; Primavera, J.; Granek, E.; Polasky, S.; Aswani Canela, S.; Cramer, L.; et al. Coastal Ecosystem-Based Management with Nonlinear Ecological Functions and Values. *Science* **2008**, *319*, 321–323. [CrossRef] [PubMed]

https://doi.org/10.3390/su18125905

25. Portela, A.P.; Vieira, C.; Carvalho-Santos, C.; Gonçalves, J.; Durance, I.; Honrado, J. Regional planning of river protection and restoration to promote ecosystem services and nature conservation. *Landsc. Urban Plan.* **2021**, *211*, 104101. [CrossRef]
26. Gastineau, P.; Mossay, P.; Taugourdeau, E. Ecological compensation: How much and where? *Ecol. Econ.* **2021**, *190*, 107191. [CrossRef]
27. Muradian, R.; Rival, L. Between markets and hierarchies: The challenge of governing ecosystem services. *Ecosyst. Serv.* **2012**, *1*, 93–100. [CrossRef]
28. An, X.; Kasimu, A.; Zhang, X.; Song, N.; Zhang, Y.; Shayiti, B. Integrating Revised Ecosystem Service Value, Ecological Sensitivity and Circuit Theory to Construct an Ecological Security Pattern in the UANSTM, China. *Sustainability* **2025**, *17*, 10880. [CrossRef]
29. Xu, F.; Yan, S.; Wang, X.; Wang, X. Ecological Management Zoning Through Integration of Ecosystem Service and Landscape Ecological Risk: A Case Study in Chongli, China. *Land* **2025**, *14*, 1133. [CrossRef]
30. Zhang, B.; Feng, Q.; Li, Z.; Lu, Z.; Zhang, B.; Cheng, W. Land Use/Cover-Related Ecosystem Service Value in Fragile Ecological Environments: A Case Study in Hexi Region, China. *Remote Sens.* **2024**, *16*, 563. [CrossRef]
31. Wang, P.; Wang, J.; Zhang, J.; Ma, X.; Zhou, L.; Sun, Y. Spatial-temporal changes in ecosystem services and social-ecological drivers in a typical coastal tourism city: A case study of Sanya, China. *Ecol. Indic.* **2022**, *145*, 109607. [CrossRef]
32. Zhong, J.; Li, Z.; Zhang, D.; Yang, J.; Zhu, J. An evaluation framework for urban ecological compensation priority in China based on meta-analysis and fuzzy comprehensive evaluation. *Ecol. Indic.* **2024**, *158*, 111284. [CrossRef]
33. Meng, G.; Wang, K.; Wang, F.; Dong, Y. Analysis of the tourism-economy-ecology coupling coordination and high-quality development path in karst Guizhou Province, China. *Ecol. Indic.* **2023**, *154*, 110858. [CrossRef]
34. Xie, G.; Zhang, C.; Zhen, L.; Zhang, L. Dynamic changes in the value of China's ecosystem services. *Ecosyst. Serv.* **2017**, *26*, 146–154. [CrossRef]
35. Mohammed, E.M.; Hamid, E.A.; Ndakidemi, P.A.; Treydte, A.C. The stocking density and regeneration status of Balanites aegyptiaca in Dinder Biosphere Reserve, Sudan. *Trees For. People* **2022**, *8*, 100259. [CrossRef]
36. Vatandaslar, C. Spatial assessment of ecosystem services provisioning changes in a forest-dominated protected area in NE Turkey. *Environ. Monit. Assess.* **2022**, *194*, 539. [CrossRef]
37. Mitchell, M.G.E.; Bennett, E.M.; Gonzalez, A. Linking landscape connectivity and ecosystem service provision: Current knowledge and research gaps. *Ecosystems* **2015**, *18*, 894–908. [CrossRef]
38. Li, Z.; Liu, Y.; Zeng, H. Application of the MaxEnt model in improving the accuracy of ecological red line identification: A case study of Zhanjiang, China. *Ecol. Indic.* **2022**, *137*, 108767. [CrossRef]
39. Millennium Ecosystem Assessment. *Ecosystems and Human Well-Being: Synthesis*; Island Press: Washington, DC, USA, 2005.
40. Daily, G. *Nature's Services: Societal Dependence on Natural Ecosystems*; Nature/Science; Island Press: Washington, DC, USA, 1997.
41. Zhang, Y.; Zheng, H.; Chen, X. Effects of ecological restoration projects on ecosystem services flows. *Ecosyst. Serv.* **2024**, *70*, 101681. [CrossRef]
42. Chang, Y.; Wang, Y.; Yuan, L.; Yao, X.; Zhang, R.; Liu, X.; Fan, Z.; Guo, J. Spatial scale effects on the trade-offs and synergies of ecosystem services in China's Huaihe river basin. *Sci. Rep.* **2025**, *15*, 22301. [CrossRef]
43. Xin, S.; Chen, N.; Li, Z. Enhancing ecosystem services and socio-ecological system coupling coordination through LULC optimization. *Ecol. Indic.* **2025**, *179*, 114160. [CrossRef]
44. Huber, J.M.; Newig, J.; Loos, J. Participation in protected area governance: A systematic case survey of the evidence on ecological and social outcomes. *J. Environ. Manag.* **2023**, *336*, 117593. [CrossRef] [PubMed]
45. Ostrom, E. A general framework for analyzing sustainability of social-ecological systems. *Science* **2009**, *325*, 419–422. [CrossRef]
46. Reed, M.S.; Fraser, E.D.; Dougill, A.J. An adaptive learning process for developing and applying sustainability indicators with local communities. *Ecol. Econ.* **2006**, *59*, 406–418. [CrossRef]
47. Newig, J.; Challies, E.; Jager, N.W.; Kochskämper, E.; Adzersen, A. The environmental performance of participatory and collaborative governance: A framework of causal mechanisms. *Policy Stud. J.* **2018**, *46*, 269–297. [CrossRef] [PubMed]
48. Primmer, E.; Furman, E. Operationalising ecosystem service approaches for governance: Do measuring, mapping and valuing integrate sector-specific knowledge systems? *Ecosyst. Serv.* **2012**, *1*, 85–92. [CrossRef]
49. Biermann, F.; Abbott, K.; Andresen, S.; Bäckstrand, K.; Bernstein, S.; Betsill, M.; Bulkeley, H.; Cashore, B.; Clapp, J.; Folke, C.; et al. Navigating the Anthropocene: Improving Earth System Governance. *Science* **2012**, *335*, 1306–1307. [CrossRef]
50. Chen, D.; Lu, X.; Zhang, C.; Zhang, Y.; Liao, D.; Li, J. Measuring and enhancing the effectiveness of regional integration policies for resource and environmental carrying capacity. *Humanit. Soc. Sci. Commun.* **2025**, *12*, 1102. [CrossRef]
51. Callaway, B.; Sant'Anna, P.H. Difference-in-Differences with multiple time periods. *J. Econom.* **2021**, *225*, 200–230. [CrossRef]
52. Xu, H. A remote sensing urban ecological index and its application. *Acta Ecol. Sin.* **2013**, *33*, 7853–7862. [CrossRef]
53. Ma, G.; Yang, X.; Zhong, C.; Zhang, Y. Can the digital technology implementation promote the spatial convergence of carbon emissions? Evidence from China. *Ann. Reg. Sci.* **2026**, *75*, 63. [CrossRef]
54. Ma, G.; Yang, Y. Can China's carbon market policy promote inter-city green collaborative innovation? *Energy Econ.* **2025**, *144*, 108398. [CrossRef]

 https://doi.org/10.3390/su18125905

. Chen, Y.; Ma, G. The characteristics and mechanisms of carbon finance development on green economic efficiency: An empirical analysis based on endogenous economic growth model. *Front. Environ. Sci.* **2024**, *12*, 1420708. [CrossRef]
. Maxwell, S.L.; Cazalis, V.; Dudley, N.; Hoffmann, M.; Rodrigues, A.S.L.; Stolton, S.; Visconti, P.; Woodley, S.; Kingston, N.; Lewis, E.; et al. Area-based conservation in the twenty-first century. *Nature* **2020**, *586*, 217–227. [CrossRef] [PubMed]
. Geldmann, J.; Manica, A.; Burgess, N.D.; Coad, L.; Balmford, A. A global-level assessment of the effectiveness of protected areas at resisting anthropogenic pressures. *Proc. Natl. Acad. Sci. USA* **2019**, *116*, 23209–23215. [CrossRef]
. Charles, A.; Wilson, L. Human dimensions of Marine Protected Areas. *ICES J. Mar. Sci.* **2009**, *66*, 6–15. [CrossRef]

https://doi.org/10.3390/su18125905

MDPI AG
Grosspeteranlage 5
4052 Basel
Switzerland
Tel.: +41 61 683 77 34

Sustainability Editorial Office
E-mail: sustainability@mdpi.com
www.mdpi.com/journal/sustainability